Advances in
Disease Vector Research

Advances in Disease Vector Research

Edited by

Kerry F. Harris
Virus-Vector Research Laboratory, Department of Entomology, Texas A&M University, College Station, Texas 77843, USA

Advances in Disease Vector Research

Volume 10

Edited by Kerry F. Harris

With Contributions by
El Desouky Ammar Y. Antignus
Thomas R. Burkot Erney P. Camargo
Yasuo Chinzei S. Cohen Dennis Gonsalves
Patricia M. Graves Henry H. Hagedorn
Roger Hull Edward I. Korenberg
Yurii V. Kovalevskii G.W. Otim-Nape
Michael K. Shaw Anthony J. Shelley
Walter J. Tabachnick DeMar Taylor
J.M. Thresh Franklin G. Wallace
Shyi-Dong Yeh Alan S. Young

With 48 Illustrations

Springer-Verlag
New York Berlin Heidelberg London Paris
Tokyo Hong Kong Barcelona Budapest

Kerry F. Harris
Virus-Vector Research Laboratory
Department of Entomology
Texas A&M University
College Station, TX 77843, USA

Volumes 1 and 2 of *Current Topics in Vector Research* were published by Praeger Publishers, New York, NY.

ISSN: 0934-6112

Printed on acid-free paper.

Production coordinated by Chernow Editorial Services, Inc., and managed by Terry Kornak; manufacturing supervised by Jacqui Ashri.
Typeset by Best-set Typesetter Ltd., Hong Kong.

9 8 7 6 5 4 3 2 1

ISBN-13: 978-1-4612-7596-1 e-ISBN-13: 978-1-4612-2590-4

DOI: 10.1007/978-1-4612-2590-4

Preface

Volume 10 of *Advances in Disease Vector Research* consists of seven chapters on vectors that affect human or animal health and six chapters on plant pathogens and their vectors.

In Chapter 1, Yasuo Chinzei and DeMar Taylor discuss hormonal regulation of vitellogenesis in ticks. Many blood sucking insects and ticks transmit pathogens by engorgement, which induces vitellogenesis and oviposition in adult animals. To investigate the pathogen transmission mechanism in vector animals, information on the host physiological and endocrinological conditions after engorgement is useful and important because pathogen development or proliferation occurs in the vector hosts at the same time as the host reproduction.

Chinzei and Taylor have shown that in ticks, juvenile hormone (JH) is not involved in the endocrinological processes inducing vitellogenin biosynthesis. Synganglion (tick brain) factor(s) (vitellogenesis inducing factor, VIF) is more important to initiate vitellogenesis after engorgement, and ecdysteroids are also related to induction of vitellogenin synthesis. In their chapter, based mainly on their own experimental data, the authors discuss the characterization of main yolk protein, vitellogenin (Vg), biosynthesis and processing in the fat body, and hormonal regulation of Vg synthesis in tick systems, including ixodid and argasid ticks.

Chapter 2, by Michael K. Shaw and Alan S. Young, provides a comprehensive overview of the transmission of *Theileria* species by ixodid ticks causing an economically important disease of cattle, theileriosis. The seven known species of *Theileria* affecting cattle are transmitted in the field by ticks of four genera, namely *Rhipicephalus*, *Amblyomma*, *Haemaphysalis*, and *Hyalomma*. The development cycle of the different *Theileria* species in ixodid ticks is similar and is characterized by the occurrence of transstadial transmission. Foremost in recent progress in the understanding of the development cycle is the proof of sexual recombination and that the zygote and kinete stages are diploid, whereas the sporozoite and gametocytes are haploid. The latter could have important implications in the control of theileriosis of cattle. Biologically, the development of sporozoites in the salivary glands of ticks is of particular interest since one kinete will produce thousands of sporozoites, the infective stage for the cow. The interaction between the salivary gland acinar cells and the developing parasites is important in the transmission process. The factors known to influence the transmission of *Theileria* are discussed in an effort to elucidate the dynamics of *Theileria* species in the tick. The authors also discuss where these studies may help in developing strategies for control of the disease.

Edward I. Korenberg and Yurii V. Kovalevski, in Chapter 3, present a quantitative model for tick-borne encephalitis (TBE) virus circulation that can provide the background for comprehensive computer modeling of the epizootic process. Two general assumptions, based on broad-reaching experimental data, were made during development of the model: (1) TBE virus has a certain "survival norm" (or "mortality norm") that is relatively stable and not high, and (2) the virus population has a similar basic scheme of reproduction throughout its range, and this scheme is closely associated with the developmental cycle of tick vectors. The model is based on the general mortality indices of *Ixodes persulcatus* and *I. ricinus* ticks at different stages of metamorphosis, as well as on indices of transovarial and transstadial virus transmission, virus maintenance, or death in the tick organism, and virus transmission through susceptible animal hosts.

In Chapter 4, Walter J. Tabachnick underscores the critical importance of identifying the genes, as well as environmental factors, that control vector competence for pathogens, using arboviral vectors as the model. Previous studies have shown the difficulty in determining the genetic mechanisms controlling insect per os susceptibility to arboviruses. New molecular biology methods are reviewed that will enable identification and mapping of genes controlling complex vector competence and vector capacity traits. The understanding of the genetic and environmental factors controlling vector competence and capacity will provide the means to identify and predict dangerous insect vectors at the population and individual levels. Identification of genetic and environmental factors offers the possibility of new strategies to interrupt the insect's role in the disease cycle.

In his review, Chapter 5, Henry H. Hagedorn summarizes what is known about the reproductive endocrinology of the adult female mosquito. Evidence for changes in juvenile hormone and 20-hydroxyecdysone levels is presented, and the known physiological effects of these substances are discussed in detail. The debate over the role of 20-hydroxyecdysone is reexamined in light of new data suggesting a role for juvenile hormone after a blood meal. Recent advances in methodology for the isolation and sequencing of peptide hormones have resulted in the identification of several new peptides that have been proposed to play a role in mosquito reproduction. Also discussed are peptides involved in regulating diuresis after a blood meal. Comparisons are made between the reproductive endocrinology of *Aedes aegypti*, the main mosquito studied, and several other species of mosquito. Finally, Hagedorn presents evidence showing that several parasites of mosquitoes have been shown to use host hormones as cues to regulate their own development, and suggests that given the complexity of mosquito endocrinology, a parasite has many potential cues. Thus, careful studies are needed to find evidence for effects of host hormones on parasites that cause disease.

In their treatise, Chapter 6, on "Human Malaria Transmission: Reconciling Field and Laboratory Data," Thomas R. Burkot and Patricia M. Graves examine the measurement of factors, and their interactions, that affect malaria transmission. The authors examine the recent laboratory and field-based findings relevant to malaria transmission. In their analysis, the authors identify a number of discrepancies between laboratory and field results. The basis for a number of these differences is the extrapolation from inappropriate animal models to humans and from differences in parasite densities used in the experiments compared to their densities in endemic areas. Additional problems arise from the failure of researchers to recognize that many parameters measured in the field have relative values suitable for making comparisons and are not absolute values. Despite the advances made possible by molecular biology in our understanding of malaria transmission, some basic parameters have not received the attention they deserve.

In Chapter 7, Anthony J. Shelley discusses the various factors affecting filarial transmission by simuliids. Eleven filaria species are transmitted by simuliids but most research has centered on *Onchocerca volvulus*, which infects man. The chapter principally deals with this species, correlating data on other filarial species where available. Brief descriptions are given of basic facts about the filariae and their simuliid vectors, with key references for further reading, before developments in techniques for species and strain identification are discussed. After a summary of the basic development cycle of filariae in simuliids, their transmission is reviewed. The latter part is divided into sections dealing with parasite availability in the definitive host, parasite uptake by the vector, development of the parasite in the vector and its transfer to the definitive host, host–vector contact, the effects of parasitism on vectors, and the effect of parasite and vector control on transmission. A large amount of the data used comes from research on the transmission of human and cattle onchocerciasis in West Africa, and emphasis is given to new developments on onchocerciasis research in Latin America.

In Chapter 8, J.M. Thresh and G.W. Otim-Nape consider the various possible approaches to controlling African cassava mosaic geminivirus. There are serious difficulties due to the nature and economics of the crop and the farming systems employed. Nevertheless, there are good prospects of developing an integrated set of control measures involving resistant varieties, sanitation, and cultural practices. The main problem is to ensure their adoption on a sufficiently large scale to decrease the great losses currently sustained in many African countries.

In Chapter 9, Shyi-Dong Yeh and Dennis Gonsalves provide a comprehensive overview of the development and successful large-scale application of mild strains for control of the notorious papaya ringspot virus (PRV). Although cross-protection is a general phenomenon with plant viruses, not all plant diseases caused by viruses can be controlled by a protective mild strain. The benefits and drawbacks of applying protective strains in different parts of the world are summarized. The strategy to obtain useful mild virus strains and to integrate cross-protection with other cultural practices is discussed. The major concern of the application appears to be strain-specific protection. To avoid the potential disadvantages of classical cross-protection, new approaches involving construction of attenuated strains by recombinant DNA technique and generation of coat-protein-induced resistance in transgenic papaya are discussed in detail. The development of coat-protein transgenic papayas that are resistant to PRV infection provides a promising approach. However, this innovative method shared the same problem of strain-specific protection as the classical protection.

S. Cohen and Y. Antignus critically examine and summarize, in Chapter 10, pertinent information and research data on tomato yellow leaf curl virus (TYLCV) that they have accumulated in their laboratory over the past 30 years. Their experience includes studies on the biology, epidemiology, virus–vector relationships, control, and molecular biology of this devastating viral disease. TYLCV is unique among the described whitefly-borne geminiviruses: there is no evidence to support the presence of a genomic B component; practically speaking, the virus is not mechanically transmissible, and it appears to have an active interaction with its whitefly vector, *Bemisia tabaci*, as reflected by the periodic acquisition phenomenon. TYLCV has served for many years as a model for the development of both conventional and nonconventional control measures and has stimulated the first breeding program aimed at introducing tolerance against a geminivirus into *Lycopersicun esculentum*.

In Chapter 11, El Desouky Ammar takes a fresh look at propagative transmission of plant and animal viruses by insects as well as the factors affecting vector specificity and competence. Propagative viruses, i.e., those that multiply in their arthropod vectors, include at least 49 plant viruses and 500 animal (vertebrate) viruses belonging to the following taxonomic groups: Rhabdoviridae, Reoviridae, Togaviridae, Flaviviridae, Bunyaviridae, Tenuivirus, and Marafivirus. Insect vectors of these viruses include leafhoppers, planthoppers, aphids, thrips, and mosquitoes. Comparisons between insect-borne propagative animal and plant viruses, regarding transmission characteristics, sequence of events in a propagative transmission cycle, and intrinsic factors affecting vector specificity and vector competence, are discussed. Examples are given for barriers to propagative transmission in vectors, e.g., midgut infection and escape, dissemination, and salivary gland infection and escape barriers, in addition to possible barriers to transovarial and venereal transmission. Extrinsic factors affecting vector competence, including environmental conditions and coinfection with another pathogen, are also elucidated. Finally, the questions of pathogenicity and modulation of viruses in their vectors, which have a bearing on the evolution of these viruses, on dissemination barriers in vectors, and on developing resistance to viral infection, are discussed with examples from both animal and plant viruses. Prospects for further research, to enhance our understanding of the complex problems of vector specificity and vector competence, are proposed.

Erney P. Camargo and Franklin G. Wallace discuss vectors of plant parasites of the genus *Phytomonas* in Chapter 12. Insect vectors of *Phytomonas* spp. are in the hemipteran families Lygaeidae, Coreidae, and Pentatomidae. *Phytomonas* spp. of latex plants are carried by certain lygaeids and coreids, whereas phloem *Phytomonas* of palm trees is transmitted by pentatomids of the genus *Lincus*. The only fruit parasites that have been studied (*Phytomonas* of tomatoes) have vectors in the Coreidae (*Phthia*) and Pentatomidae (*Nezara* and *Arvelius*).

The phytophagous Hemiptera are frequently infected with monoxenous trypanosomatids, most commonly of the genus *Leptomonas*. The authors present a table of 100 species of plant-feeding Hemiptera (families Lygaeidae, Coreidae, Pyrrhocoridae, Pentatomidae, Miridae, and Coromelaenidae) from which trypanosomatids have been recorded. Whereas insect parasites of the genus *Leptomonas* cannot be distinguished from *Phytomonas* microscopically, cultures may be distinguished by the following criteria: presence or absence of arginase, anti-*Phytomonas* monoclonal antibodies, and ribosomal DNA restriction analysis.

In the insect host, *Phytomonas* spp. multiply and go through cyclic changes. First the flagellates elongate in the pylorus, becoming giant forms up to 80 μm in length. Giant forms are found in the hemocoel and, in 20 days, in the salivary glands along with small aflagellate forms which are injected into the host plant.

In Chapter 13, Roger Hull suggests a unifying terminology for the interactions between plant viruses and their biological vectors, introducing the terms *externally-borne* and *internally-borne*. He reviews the different types of vectors ranging from pollen, through fungi and nematodes to arthropods, pointing out relevant features of the virus–vector interactions. The molecular interactions between viruses and insect vectors are dealt with in detail. Relatively little is known about the vector side of the interaction but this obviously involves specific sites. Depending upon the type of virus–vector interaction, the latter are either in the anterior portion of the feeding canal or in the gut wall and salivary glands.

There are accumulating data on the virus side of the interaction. The main interface with the vector involves the surface of the viral capsid which can interact either directly with the vector or involve virus-encoded helper component(s). There are relatively few examples of direct interactions, a point that is raised in the final discussion as to the selective advantage of helper components. Two groups of helper components are recognized, those that function as separate molecules from the viral coat and those that are integral with the capsid proteins. The molecular interactions of proteins in these two groups differ.

In summing up this subject, two points are raised. First, is that noted above on the evolution of these very specific virus–vector interactions. There obviously must be some important selective advantages to the evolution of these mechanisms. Second, understanding the molecular details of these interactions should lead to the development of approaches of transforming plants with molecules that interfere with the interactions. The latter would prevent or reduce virus spread and thereby protect crops against viruses.

I thank the authors for their outstanding chapter contributions as well as their patience in working with me to bring Volume 10 of *Advances in Disease Vector Research* to a most successful conclusion. I also acknowledge and appreciate the support and help of the outstanding production staff at Springer-Verlag.

Kerry F. Harris

Contents

Contributors

El Desouky Ammar
Department of Economic Entomology, Faculty of Agriculture, Cairo University, Giza, Egypt. Present address: Department of Plant Pathology, University of Kentucky, Lexington, Kentucky 40546, USA

Y. Antignus
Virus Laboratory, Agricultural Research Organization, The Volcani Center, P.O. Box 6, Bet Dagan 50250, Israel

Thomas R. Burkot
Division of Vector-Borne Infectious Diseases, National Center for Infectious Diseases, Centers for Disease Control, Foothills Campus, P.O. Box 2087, Fort Collins, Colorado 80522, USA

Erney P. Camargo
Institute of Biomedical Sciences, Department of Parasitology, University of São Paulo, São Paulo 05508, Brazil.

Yasuo Chinzei
Department of Medical Zoology, School of Medicine, Mie University, Edobashi 2-174, Tsu 514, Japan

S. Cohen
Virus Laboratory, Agricultural Research Organization, The Volcani Center, P.O. Box 6, Bet Dagan 50250, Israel

Dennis Gonsalves
Department of Plant Pathology, New York State Agricultural Experiment Station, Cornell University, Geneva, New York 14456, USA

Patricia M. Graves
Queensland Institute of Medical Research, 300 Herston Road, Brisbane Q 4029, Australia

Henry H. Hagedorn
Department of Entomology and Center for Insect Science, The University of Arizona, 430 Forbes Building #36, Tucson, Arizona 85721, USA

Roger Hull
John Innes Institute, Colney Lane, Norwich NR4 7UH, United Kingdom

Edward I. Korenberg
Gamaleya Institute for Epidemiology and Microbiology, Russian Academy of Medical Sciences, 18 Gamaleya Street, Moscow 123098, Russia

Yurii V. Kovalevskii
Gamaleya Institute for Epidemiology and Microbiology, Russian Academy of Medical Sciences, 18 Gamaleya Street, Moscow 123098, Russia

G.W. Otim-Nape
Namulonge Agricultural and Animal Production Research Institute, P.O. Box 7084, Kampala, Uganda

Michael K. Shaw
International Laboratory for Research on Animal Diseases, P.O. Box 30709, Nairobi, Kenya

Anthony J. Shelley
Department of Entomology, The Natural History Museum, Cromwell Road, London SW7 5BD, United Kingdom

Walter J. Tabachnick
Arthropod-Borne Animal Diseases Research Laboratory, USDA-ARS, University Station, P.O. Box 3965, Laramie, Wyoming 82701, USA

DeMar Taylor
Department of Medical Zoology, School of Medicine, Mie University, Edobashi 2-174, Tsu 514, Japan. Present address: Department of Biology, Texas A&M University, Koriyama, Motomachi 1-20-22, Koriyama 963, Japan

J.M. Thresh
Natural Resources Institute, Chatham Maritime, Kent ME4 4TB, United Kingdom

Franklin G. Wallace
2603 Cohansey Street, St. Paul, Minnesota 55113, USA

Shyi-Dong Yeh
Department of Plant Pathology, National Chung Hsing University, Taichung, Taiwan, Republic of China

Alan S. Young
International Laboratory for Research on Animal Diseases, P.O. Box 30709, Nairobi, Kenya

Contents for Previous Volumes

Volume 1

Volume 2

Volume 3

Volume 4

Volume 5

Volume 6

Volume 7

Volume 8

Volume 9

1
Hormonal Regulation of Vitellogenin Biosynthesis in Ticks

Yasuo Chinzei and DeMar Taylor

Introduction

Many species of ticks and mites must obtain a blood meal from a vertebrate or invertebrate host for development and reproduction to occur. This blood meal results in the transmission of various disease agents from infected animals to uninfected animals. Ticks are second only to mosquitoes as vectors of human diseases. Ticks and/or mites transmit a wide range of microorganisms: viruses, rickettsia, bacteria, and spirochetes, and also protozoa and nematoda.

After engorgement most larval and nymphal ticks molt to the next developmental stage; then they are ready to feed again. The pathogens are proliferated in the lumen of the gut, hemolymph, and tissues. Ticks that have pathogens can then transmit them transstadially by the salivary or coxal glands. In some cases, transovarial transmission of the pathogen occurs between mother and progeny. Thus engorgement and reproduction are very important in the transmission of pathogens.

Hard ticks (family Ixodidae) attach to a host and feed for long periods of time (several days to months depending on stages and species), whereas soft ticks (family Argasidae) feed in a very short time (at most a few hours). Ticks are able to intake much larger quanitities of blood than the capacity of their guts, because they concentrate the blood by excreting a serum salt solution from the salivary or coxal glands during feeding. The concentrated blood contains blood cells and plasma and is stored in the gut. The blood is incorporated into the gut epithelial cells, gradually digested in the cells, and absorbed. Argasid adult ticks feed and

Yasuo Chinzei, Department of Medical Zoology, School of Medicine, Mie University, Edobashi 2-174, Tsu 514, Japan.

DeMar Taylor, Department of Medical Zoology, School of Medicine, Mie University, Edobashi 2-174, Tsu 514, Japan. Present address: Department of Biology, Texas A&M University, Koriyama, Motomachi 1-20-22, Koriyama, 963, Japan.

lay eggs repeatedly, whereas Ixodid adult ticks feed and lay eggs only once. Larval ticks hatch from eggs after embryogenesis. In some species, larvae develop and molt to the nymphal stage without feeding, whereas in other species, larvae also require a blood meal. Nymphs molt four or five times, requiring a blood meal before each molt. The final nymphal instar molts into an adult after feeding.

Adult female ticks usually require a blood meal before they develop ovaries and lay eggs; however, a few species are autogenous, that is, the females can lay eggs before engorgement. Engorgement is normally required to supply the nutrient reserves needed for development and molting in larval and nymphal stages, and also for reproduction in the adult stage.

Ticks infected with pathogens by engorgement are infective during the subsequent feeding after ingestion of the disease agent; this will often coincide with the reproductive stage of the tick. Pathogen development or proliferation occurs concurrently with vitellogenesis and oviposition. Investigation of the physiology of reproduction, induced by engorgement, should also lay the foundation for study of the mechanisms of pathogen development. Similarly, the use of hormone antagonist and other natural products in the control of insect pests has been very successful (70, 71); however, they have not been developed in ticks. In fact, very little is known of tick endocrinology when compared to that of other arthropods. Therefore, studies on the physiology of ticks in relation to engorgement and particularly the induction mechanisms of vitellogenesis and oviposition are essential.

This chapter discusses vitellogenesis induced by engorgement and its mechanisms of hormonal regulation. Excellent reviews of earlier work in tick vitellogenesis have been published by Diehl et al. (35) and Connat et al. (28). Therefore, we will discuss more recent advances in this area, particularly showing our own data obtained from experiments with two argasid ticks, *Ornithodoros moubata* and *O. parkeri*.

Vitellogenesis in Ticks

The precursor of the main egg yolk protein, vitellogenin (Vg), is synthesized in the extraovarial tissue (liver in oviparous vertebrates and fat body in insects), secreted into the peripheral blood or hemolymph, and taken up into the developing oocytes. Vg is usually processed and deposited in the oocytes as lipovitellin (and phosvitin) or vitellin (Vn). Vn forms yolk granules together with proteins produced by follicle cells and oocytes. Matured oocytes are covered with chorion and move to the uterus just before being oviposited.

The term "vitellogenesis" is used very widely and the meaning varies depending on the author(s). In this chapter, vitellogenesis is used to describe Vg synthesis and Vg deposition in the oocytes (egg maturation) and does not include oviposition. This is important because Vg synthesis, Vg deposition, and oviposition seem to be induced by different regulatory mechanisms. Vitellogenesis is induced by estrogen in vertebrates, such as the chicken and frog, juvenile hormones in most insects and ecdysteroids in some dipteran insects. These hormones induce Vg synthesis, namely Vg gene expression in the liver, fat body, and sometimes the ovary.

Vitellogenesis in ticks is principally induced by feeding and mating in adult females. A blood meal probably stimulates the brain or nervous system and endocrine glands to secrete humoral factors, that in turn induce vitellogenesis. Soft ticks do not require mating for induction of vitellogenesis, but hard ticks require both mating and blood feeding. The factor(s) affecting vitellogenesis in the male sperm and accessory gland have not been elucidated. Most female ticks must feed before vitellogenesis, but autogeny (vitellogenesis and oviposition without feeding) has been reported in a number of Argasid species (39, 40, 58, 63). Autogeny in ticks is believed to be used for securing progeny in species which have less of a chance to feed. In *Ornithodoros parkeri*, autogeny (1) and repeated oviposition, after a normal reproductive cycle, have been observed (2). In our studies we observed 3.6% of *O. moubata* adult females to be autogenous (19).

Following a blood meal, two new proteins appear and accumulate in the hemolymph of adult females (Fig. 1.1). Similar observations have been reported for other species of ticks (3, 9, 33, 34, 55, 73). Both proteins precipitate with anti-Vn serum and are specific to reproductive female

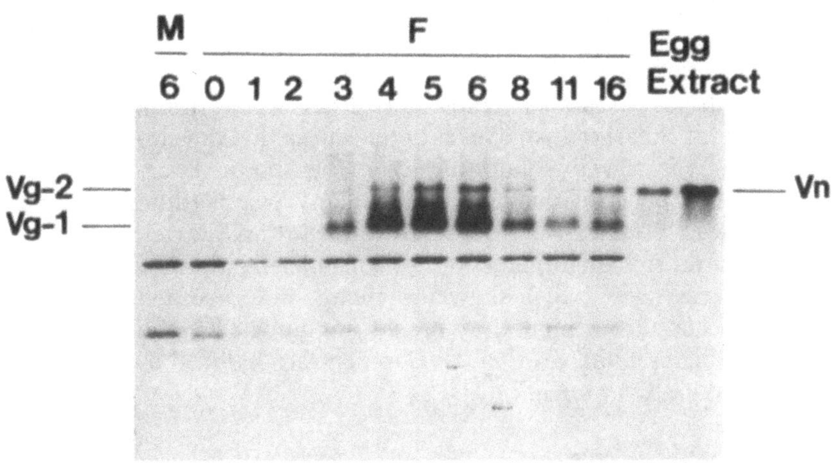

FIGURE 1.1. PAGE analysis of hemolymph proteins before and after blood meal. Hemolymph was collected from a male on day 6 and females on the days indicated. Eggs were collected within 12 hr after oviposition. Proteins were extracted from the eggs with tick Ringer. Hemolymph and crude egg extract were applied to a 2.5 to 15% acrylamide gradient PAGE gel. Gel was stained with Coomassie brilliant blue.

TABLE 1.1. New criteria for scoring ovarian development in ticks.

Balashov's stage of ovarian development stages (criteria)[a]	New modified stages and criteria of ovarian development[b]		
	Stages	Criteria	Score
I (previtellogenic)	I	No oocyte development	0
II (previtellogenic)	II	Oocyte enlargement, no yolk	1
III (vitellogenic)	III	Some oocytes with yolk	2
IV (vitellogenic-preovulate)	IV	Many immature occytes with yolk	3
V (ovulate)	V	Full grown oocytes (ovulated)	
	V-1	At least one and less than 5	4
	V-2	Over 5 full grown oocytes	5
	VI	Some eggs in oviduct or uterus	6
	VII	Some eggs oviposited	7

[a] Balashov (4).
[b] Modified from Chinzei et al. (1989).

hemolymph, indicating that they are tick Vgs. We have called the migrating bands in polyacrylamide gel electrophoresis (PAGE), Vg-1 and Vg-2, respectively. Because Vg-1 and Vg-2 react with anti-Vn serum, we can determine the changes in Vg content during reproductive development after engorgement (13). We have established a new method to determine Vg titer in the hemolymph. Using this "spotting-scanning" method, 0.25 µl of hemolymph is spotted on a filter paper and the absorbance is scanned at 400 and 450 nm. Vg titer is determined from the difference in absorbance between the two wavelenghts as compared with standard protein (purified Vn) (15). This method has made it possible to follow Vg titer in individuals over a period of time using multiple bleedings.

We have also developed a criteria to estimate ovarian development in ticks, which is presented in Table 1.1 (19). These criteria are modified from those of Balashov (4). This criteria reflects the developmental stages of "whole ovaries" more faithfully than Balashov's criteria of "oocyte" development, because ovaries contain many ova in different developmental stages. Ovaries were observed to begin rapid development within four days after tick engorgement (19). This observation is coincident with that in Chinzei (13), which showed a change in Vn content in the ovary and whole body as estimated by rocket immunoelectrophoresis. Vg titer in the hemolymph and ovarian development are induced by engorgement in adult female *O. moubata* ticks.

Vitellogenin in Ticks

Properties of Vg in many species of vertebrates and insects are reviewed by Tata (72), Hagedorn and Kunkel (44), and Wyatt and Pan (80). Several reports have been published on the egg proteins of ticks. Wigglesworth

(79) reported the presence of a brown pigment in the egg yolk of *Ornithodoros* and *Ixodes* ticks, which is derived from the host hemoglobin. The brown pigment in *Boophilus microplus* eggs has been identified as a hemoprotein, known as hemixodovin (10). Diehl (33, 34) found two egg proteins in *O. moubata*, which are equivalent to two reproductive, female-specific, hemolymph proteins as shown by electrophoresis and immunodiffusion. Tatchell (73), using electrophoresis, demonstrated two hemoglycoproteins in hemolymph and egg homogenates of female *B. microplus* but failed to demonstrate their immunological identities. Boctor and Kamel (9) purified and characterized two lipoproteins from *Dermacentor andersoni* eggs, and also demonstrated the presence of two similar lipoproteins in female hemolymph. Obenchain et al. (55) reported the presence of two lipohemoproteins in eggs of *Rhipicephalus appendiculatus* and suggested the site where these proteins many be synthesized. Araman (3) compared the egg proteins with the hemolymph proteins of *R. sanguineus* electrophoretically.

More recently, we purified Vg and Vn of a soft tick, *Ornithodoros moubata* from reproductive adult female hemolymph and from newly laid eggs (17). A rapid and efficient method for purification of Vn and Vg from the silkworm, *Philosamia cynthis*, is based on the specific precipitation of Vn or Vg at low ionic strengths (11, 12). This method can generally be applied to purify Vn and Vg from lepidopteran and orthopteran insects (46, 16). It is, however, not applicable to the purification of Vn and Vg from the tick *O. moubata*, because they are readily soluble even in water. However, as the major egg protein (85%) of the tick is Vn (13), purification can easily be achieved by column chromatography on Sepharose-4B and DEAE cellulose. The Vn preparation is homogeneous as judged by PAGE, immunoelectrophoresis, and electron microscopy. Anti-Vn serum was prepared by immunizing a rabbit with purified Vn.

Vg was purified by preparative PAGE from reproductive female *O. moubata* hemolymph. The Vg preparation displayed two components (Vg-1 and Vg-2) on PAGE, both of which reacted with anti-Vn serum. The Vn and Vg-2 preparations were homogeneous as judged by PAGE, and showed a rugby-ball shape with a cleft at the middle of the molecules, while Vg-1 appeared as half of Vn or Vg-2 in shape and size (Fig. 1.2). This together with data on molecular weights (600,000 for Vn and Vg-2 and 300,000 for Vg-1), suggests that the Vn and Vg-2 are dimers of Vg-1. Six polypeptides (P1–P6, MW 100,000–215,000) for Vgs and six polypeptides (P3–P8, MW 50,000–160,000) for Vn were demonstrated by SDS-PAGE (Fig. 1.3); P3–P6 were common to Vgs and Vn. These observations suggest proteolytic processing from larger to smaller polypeptides. Vn contains 7.6% lipids, with triacylglycerol as the predominant neutral lipid, and 12.4% carbohydrates, with mannose as the major sugar. Vn and Vgs displayed a reddish-brown coloration due to the presence of haem-moiety. The amino acid composition of Vn is high in

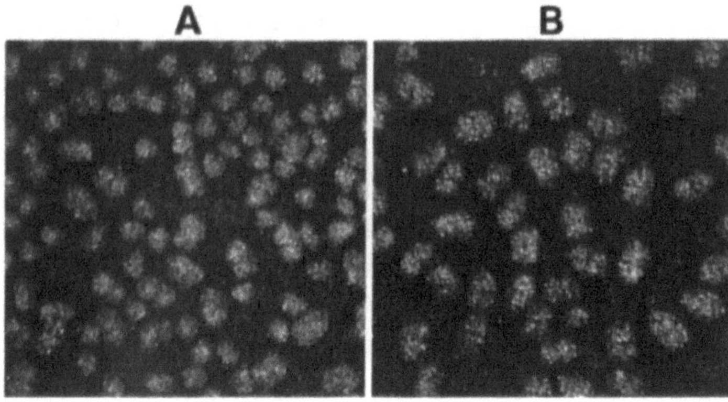

FIGURE 1.2. Electron micrograph of Vg preparation from reproductive female hemolymph (A) and Vn purified form eggs (B). Fresh samples were negatively stained with uranyl acetate.

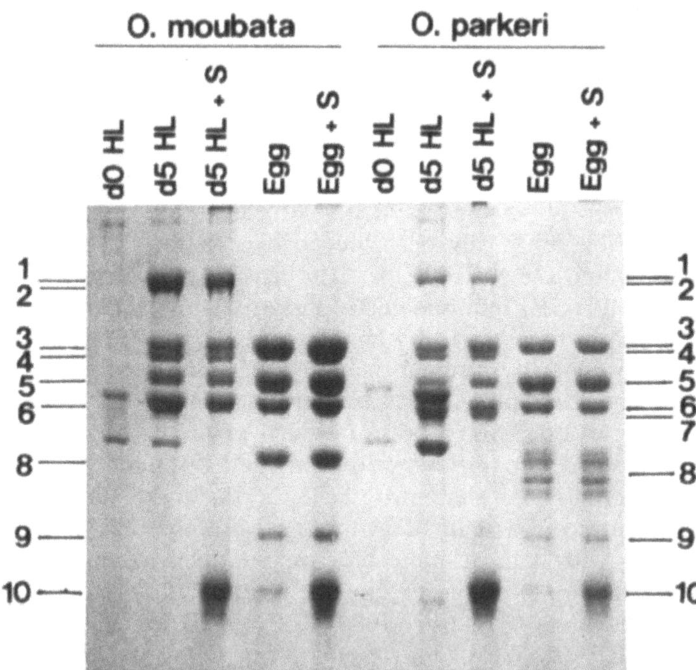

FIGURE 1.3. SDS-PAGE of hemolymph (HL) from female *Ornithodoros moubata* and *O. parkeri* ticks immediately after engorgement (d0) and day 5 (d5) after engorgement, and of crude extracts made from freshly laid eggs (Egg). Samples were also precipitated with *O. moubata* antivitellin serum (S). Numbers on both sides indicate polypeptide bands specific and common to both Vg and Vn.

glutamic acid, proline, valine, and leucine but low in methionine and isoleucine. This composition shows substantial differences from the Vn of the hard tick, *Dermacentor andersoni* (9) and that of the locust (16) and other insects (38). The isoelectric point of *O. moubata* Vn and Vgs is 6.9. Unlike Vg and Vn of insects, Vgs and Vn of ticks are water soluble.

Site of Vitellogenin Synthesis

Vgs are synthesized in the extraovarial tissue, i.e., liver in vertebrates and fat body in insects. The fat body in ticks is a very faint tissue attached to the tracheal system. Scanning electron microscopy shows that the diameter of the fat body cells expands about 10 times after engorgement (23). Obenchain and Oliver (57) also observed that the number of fat body cells does not increase during vitellogenesis but that the cell volume expands greatly after a few days of engorgement. The fine structure of fat body cells in adult female ticks changes from an inactive to very active state during protein synthesis, i.e., develops rough endoplasmic reticulum or protein granules that appear in the fat body cells, a short time after a blood meal (30, 45). Immunohistochemical staining of tissue sections using anti-Vn serum resulted in very strong fluorescence in the fat body cell cytoplasm (Fig. 1.4). Vg (Vn) was not detected immunohistochemically in

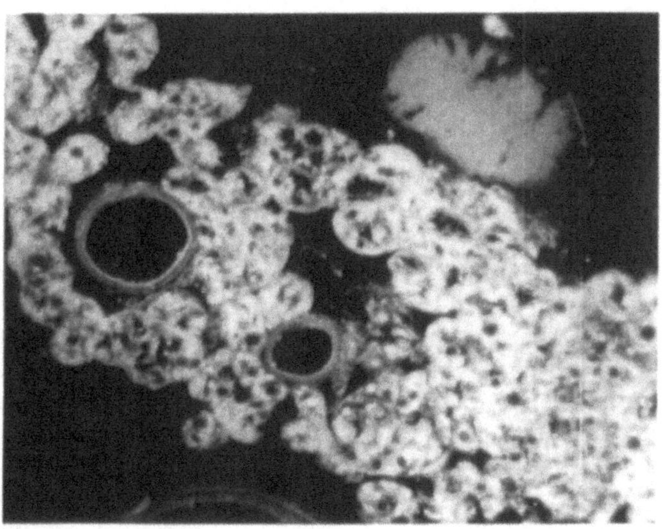

FIGURE 1.4. Immunofluorescence microscopy of fat body. Tissues dissected from females at 5 days after a blood meal were fixed and processed as described by Chinzei and Yano (23).

the cytoplasm of tissues other than fat body and oocytes (23). We did not detect any specific reaction with anti-Vn serum in the gut cells. Diehl et al. (35) also showed immunologically the presence of Vg in the fat body but not in the midgut of *R. appendiculatus*. All these observations strongly suggest that the fat body is the main site of Vg synthesis in ticks.

To confirm the site of Vg synthesis in *O. moubata*, tissue dissected from vitellogenic females was cultured in tick Ringer supplemented with $[^{35}S]$methionine. After incubation, protein was extracted from the tissue and medium, together or separately, and precipitated with anti-Vn serum. The precipitates were counted as synthetic activity of Vg and were further analyzed by SDS-PAGE and fluorography (23). Both radioactivity and fluorography of precipitates showed the fat body to be the most active tissue in synthesizing Vg. Vg synthesis occupied 40% of the total protein synthesis in the most active stage of the fat body. This value is extremely high. From these observations we conclude that the fat body is the main site of Vg synthesis in the soft tick, *O. moubata* (23). The fat body was also indicated to be the main site of Vg synthesis in *O. parkeri* (75).

A structural differentiation for protein synthesis occurs in midgut epithelium cells in adult females of *R. sanguineus* after a blood meal (30). Coons and his colleagues reported that gut epithelial cells also synthesize Vg in *R. appendiculatus*. From our observations the gut of *O. moubata* appears to have only a very minor amount of Vg synthetic activity, if any at all.

Vitellogenin Biosynthesis and Processing

We studied Vg biosynthesis and processing in the fat body by *in vitro* and *in vivo* experiments in the soft tick, *O. moubata* (14). Fat bodies from reproductive females were cultured in Ringers containing a ^{14}C-labeled amino acid mixture. Radioactivity incorporated into the fat body and medium Vg were counted separately. Synthesized Vg was accumulated linearly in the fat body for the first few hours of incubation. Vg secretion into the medium started to increase approximately 2 hr after incubation had begun. This suggests that newly synthesized Vg is processed in the cells and transfered to Golgi bodies or secretory granules in preparation for secretion and that it takes approximately 2 hr to process and begin secretion.

Immunoprecipitates of crude fat body protein extract with anti-Vn serum were analyzed by SDS-PAGE and fluorography. Six polypeptide bands (P1–P6) were observed; the radioactivities of each polypeptide increased with the incubation time and were detected in the medium approximately 2 hr after incubation (14). Pulse–chase experiments determined that polypeptide P1 and P2 are primary products of Vg mRNA and P3 to P6 are processing products in the fat body; however, Vg secreted into the hemolymph contains primary products (P1–P2) and processing products (P3–P6). Native Vg molecules are the same size as monomer Vg-1 and dimer Vg-2. This means that proteolysis does not occur in all, but only in some, Vg molecules. Therefore, polypeptides P1–P6 derived from SDS-PAGE are component polypeptides, not real subunits of native Vg molecules. Vg molecules are incorporated into the developing oocytes and processed further by proteolysis into smaller molecules, i.e., P1 and P2 disappear and P7 and P8 appear.

Similar biosynthesis and processing of Vg molecules were observed in the soft tick *O. parkeri* (75). In this species autogeny is more common in newly adult females than in *O. moubata*. Most females develop ovaries and about 40% of these females oviposit within a month after adult emergence without a blood meal. Approximately 2 months after adult ecdysis, ovaries in almost all females lack oocytes that contain yolk. This means that oocytes developed by autogeny are absorbed in the females, which do not lay eggs; however, in the hemolymph some level of Vg remains even after a long time of autogenous oviposition or oocyte absorption. The Vg remaining in the hemolymph is composed of three

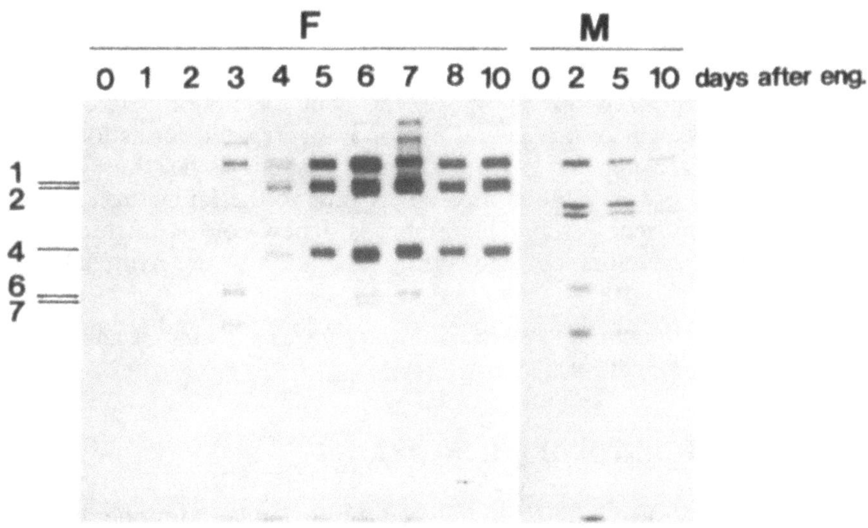

FIGURE 1.5. Fluorogram of hemolymph from ticks injected with [^{35}S]methionine. Hemolymph samples were collected 5 hr after methionine injection on the days indicated with day 0 meaning immediately after engorgement, subjected to SDS-PAGE (0.1% SDS, 7% acrylamide; run at a constant current of 30 A) and fluorography carried out as described by Taylor et al. (74). 1, 2, 4, 6, and 7 indicate the polypeptides composing Vg (see Fig. 1.3).

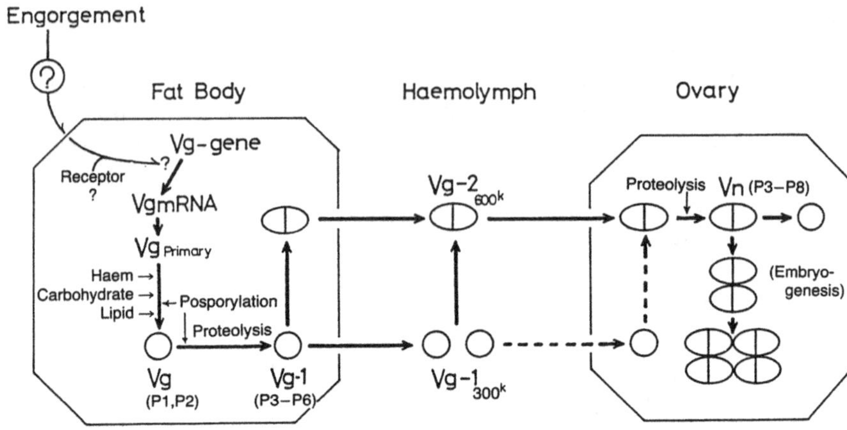

FIGURE 1.6. A model of vitellogenin (Vg) synthesis, processing, secretion, and uptake in the soft tick, *Ornithodoros moubata* (14, 17, 24).

polypeptides P3, P5, and P6, which appear to be stable forms of Vg component polypeptides. After engorgement P1, P2, P4, and P7 are newly synthesized in the fat body and processed to P3, P5, and P6 (Fig. 1.5, see Fig. 1.3). Vg incorporated in the oocytes is further processed to Vn, which contains P8, P9, and P10 and lacks P1 and P2 (75).

Vn accumulated in the yolk granules of oocytes forms polymers (dimer to octamer) during embryogenesis, creating ladder-like bands in native PAGE. Vn is utilized for embryogenesis, but the major part of Vn is conserved in the gut of larvae and nymph in aggregated forms for over 4 months after hatching. Vn is the nutrient that supports tick survival until the nymph can obtain a blood meal. This adds to the known role of yolk protein as a nutrient reserve for embryos a new role as a reserve for postembryonic development and during starvation in the nymphal stage (24).

A model for Vg synthesis, processing, translocation, and accumulation is summarized in Fig. 1.6.

Hormonal Control of Vitellogenesis

Vitellogenin (Vg) synthesis is induced by juvenile hormone in most insects that develop ovaries during the adult stage. Ecdysteroids play the same role in some dipera and holometamorphic insects that develop ovaries during the pupal stage (8, 44, 50, 64). In ticks, vitellogenesis (Vg synthesis, oocyte maturation) and oviposition in adults, like most hematophagous insects such as kissing bugs, mosquitoes, and black flies, is primarily induced by engorgement and mating with the exceptions of a few autogenous ticks (19, 28, 35, 58, 59). Both of these factors are thought to be essential in inducing hormone(s) release, which directly or indirectly stimulates Vg synthesis in the fat body. Oliver (58) and Diehl et al. (35) reported that mating is an essential factor for completion of vitellogenesis and oviposition in ticks. We have observed that virgin *O. moubata* and *O. parkeri* females have the same levels of Vg in the hemolymph and fully developed ovaries after engorgement as compared with those of mated engorged females (74). Therefore, mating may not be essential for vitellogenesis even though it is necessary for oviposition in these species.

Hormonal control of diapause and oogenesis have been observed in mites and ticks. Bassal and Roshdy (5) demonstrated that JHA terminated diapause and controlled oviposition in *Argas ardoreus*. Pound and Oliver (62) showed that ovarian development was restored by topical application of JH III to engorged adult females after treatment with precocene 2 in the soft tick, *O. parkeri*. Connat et al. (27) reported that natural JH and JHA have positive effects on ovarian development and oviposition in *O. moubata*. Oliver et al. (60) showed again that precocene 2 treated mites, *Dermanysus gallinae*, produce fewer progeny, but JH III eliminated the effect of precocene 2. These studies suggest ovarian development and oviposition are controlled by JH in the tick as in most insects.

Contrarily, in earlier studies, JHA has been reported to have no effect on diapause termination and oogenesis in *Rhipicephalus appendiculutus* (65). JHA application not only failed to stimulate oviposition but reduced oviposition and egg hatchability in engorged female, *Boophilus* ticks (52, 69). These reports did not investigate Vg synthesis or ovarian development but suggested that topical application of JHA had no effect on vitellogenesis and therefore oviposition.

We studied the effect of JH on vitellogenesis at the level of Vg synthesis and the development of the ovary. We attempted to determine (i) whether JH stimulates Vg synthesis in the fat body and ovarian development, (ii) how Vg synthetic activity and Vg titer in the hemolymph change after JH treatment, and (iii) which JH is most effective in inducing vitellogenesis. Pound and Oliver (62) used engorged precocene treated females, while Connat et al. (27) used engorged virgin females 100 days after feeding. Our preliminary studies revealed that engorged precocene-treated females and many virgin females, even 100 days after engorgement, had high Vg titer in the hemolymph and developed ovaries. Therefore, we could not use females with similar treatment and feeding schedules as Pound and Oliver or Connat to study the effect of JH on Vg induction and ovarian development.

We reported previously that a pyrethroid, cypermethrin (CyM) can induce Vg synthesis and ovarian development in unengorged adult female ticks (Table 1.2) (18, 74). These results suggest that even before

TABLE 1.2. Vitellogenin induction and ovarian development in unfed *O. moubata* mated females treated with a pyrethroid, cypermethrin (CyM).

		Vg titer (μg/μl) (M ± SE)			Ovarian development (score: M ± SE)
Treatment	*N*	Day 0	Day 5	Day 10	Day 10
Nontreatment	13	5.8 ± 2.2	8.2 ± 2.5	9.6 ± 3.5	0.30 ± 0.67
Engorgement	10		52.1 ± 26.4	32.8 ± 20.2	6.10 ± 1.37
Acetone (1 μliter)	22		7.2 ± 2.1	12.4 ± 10.1	0.55 ± 0.80
CyM (10 μg)	11		48.6 ± 18.7	94.6 ± 41.0	2.35 ± 0.94

engorgement females can produce Vg and develop ovaries if the fat body or ovary are properly stimulated. Therefore, if JH is involved in the process of stimulation of vitellogenesis, JH or JHA should be able to induce Vg in the hemolymph in "unfed" mated females as CyM was shown to do as discussed above. We investigated the effects of JH and JHA on the release of Vg into the hemolymh and ovarian development using unfed adult female ticks *O. moubata* and *O. parkeri*. We used

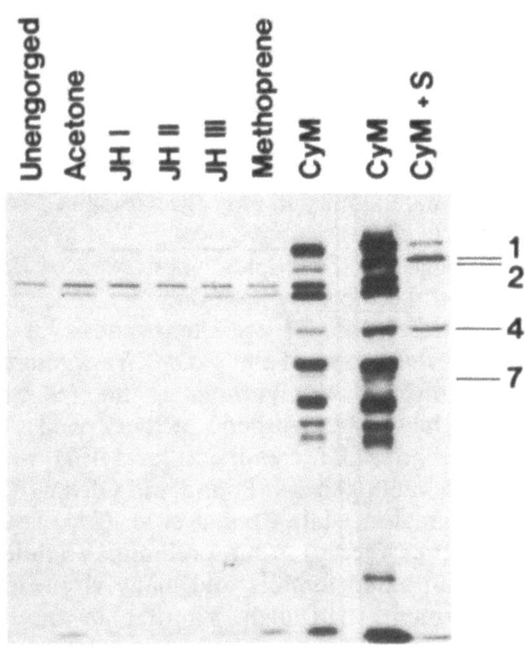

FIGURE 1.7. Fluorogram of hemolymph from ticks injected with [^{35}S]methionine on day 5 after treatment with acetone, JH I–III, methoprene, and cypermethrin. Hemolymph samples were subjected to SDS-PAGE and fluorography as described in Fig. 1.5. S, sample precipitated with *O. moubata* anti-Vn serum. 1, 2, 4, and 7 indicate the polypeptides composing Vg.

natural JH (JH I, JH II, JH III) and JHA (methoprene, S21149, S21150, and S31183 = pyriproxyfen) dissolved in acetone to test the effects of topical treatment of these compounds on vitellogenesis. None of these compounds showed positive effects on vitellogenesis in this tick (Fig. 1.7). From these experiments we concluded that JH and JHA can not stimulate Vg production (21, 75). This conclusion is contrary to that of Pound and Oliver (62) and Connat et al. (27).

JH is a terpenoid hormone specific to arthropods including insects and crustaceans. Therefore, it seems very strange that there is no report to date that JH is present in ticks and mites. Venkatesh et al. (78) showed JH esterase activity in *Dermacentor variabilis*. JH binding protein (JHBP) was found in the hemolymph of the mated female hard tick, *Dermacentor variabilis* by using photoaffinity labeling reaction (Kurcsar, Prestwitch, and Sonenshine, personal communication). These observations were believed to suggest the existence of JH, but there is no direct evidence for the presence of JH in the acari. Dr. B. Mauchamp (INRA, France) attempted to detect JH in the hemolymph of hard ticks by mass spectrometry, without success (personal communication). This information suggests that JH is not present in ticks and that vitellogenesis in ticks is regulated by a hormone completely different structurally from insect JH. Connat and Nepa (25, 29) also suggested the absence of any known JH (including JH-0 and farnesyl-methylester) in extracts of *O. moubata*, *B. microplus*, and *Amblyoma hebraeum* from their experiments that tested anti-JH agents.

Regulation Factors of Vitellogenesis and Ovarian Development

Ticks secrete coxal fluid from the coxal glands during engorgement. Initiation of coxal fluid secretion (CFS) varies in individual ticks; however, CFS initiation was found to be a very important indicator of physiological and endocrinological stages of engorgement. The ovary does not develop in females removed from the host before CFS initiation but does develop in those removed after CFS initiation. This indicates that engorgement to the time of CFS initiation is necessary and sufficient for the induction of vitellogenesis. However, ovarian development in the posterior part ligated before full engorgement was significantly lower that of unligated females. This indicates that a factor starts to be released during CFS from the anterior half of the tick, and reaches at detachment high enough levels in the hemolymph to induce vitellogenesis in the posterior half of the tick.

Ticks were removed from the host and ligated after CFS and also after detachment. Vg titer in the hemolymph of unligated, ligated posterior and ligated anterior parts of ticks was determined (Fig. 1.8) (20).

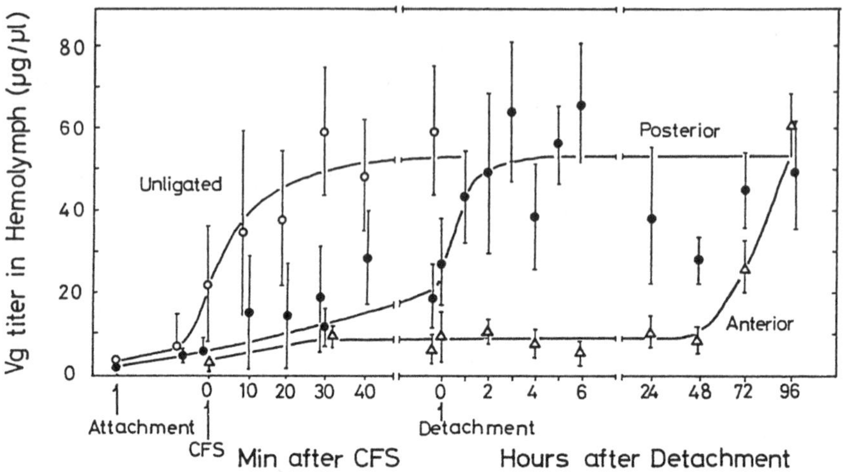

FIGURE 1.8. Vitellogenin (Vg) titer in the hemolymph of unligated and ligated adult female ticks. Ticks were removed from the host and ligated at the indicated times after coxal fluid secretion (CFS) and detachment. Vg titer in unligated (○), ligated posterior (●), and ligated anterior (△) parts were measured 5 days after removal or detachment by Spotting-scanning method (15).

Unligated adult females can produce Vg if allowed to engorge until CFS initiation. The isolated posterior part can produce the normal level of Vg if it is ligated later than 1 hr after detachment; however, the isolated anterior part does not produce Vg unless ligated 3 days after detachment.

These results suggest that the anterior part needs another factor for producing Vg. This second factor is released from a tissue in the posterior part, which is stimulated by the first factor from the anterior part (20, 77). Injection of tissue extracts into unfed females demonstrated that the synganglion (SyG) is the tissue that releases the first factor (Table 1.3). Furthermore, this factor appeared to be contained in the unfed female

TABLE 1.3. Vitellogenin induction and ovarian development in unfed *O. moubata* mated females treated with synganglion (SyG), salivary gland (SG), ovary (OV) extract, or Ringer's solution.

		Vg titer (μg/μl) (M ± SE)		Ovarian development (score: M ± SE)
Treatment	*N*	Day 5	Day 10	Day 10
Tick Ringer	28	14.9 ± 5.9	17.4 ± 8.0	0.8 ± 1.7
SyG extract	38	21.7 ± 10.3**	27.6 ± 10.5***	2.9 ± 2.3***
SG extract	24	13.7 ± 6.8	17.6 ± 8.5	1.6 ± 2.5**
OV extract	10	13.4 ± 4.7	18.2 ± 10.1	0.6 ± 0.8

, * Significantly different from the controls at $p = 0.01$ and 0.001, respectively.

SyG and activity was lost when treated with trypsin and chymotrypsin, indicating this factor is a polypeptide (22).

The presence of neurosecretory cells in the synganglion has been shown and is reviewed by Binnington and Obenchain (6, 7). Neurosecretory activity also has been shown to correlate with oogenesis and oviposition, as well as with other physiological events (32, 37, 42, 53, 54, 68). Previous papers have shown that injection of nerve ganglion extracts and ligation experiments suggest that neurosecretory activities are important in oogenesis and diapause termination (37, 41, 48, 67). Recently, several other researchers have shown that extracts from the synganglion stimulate vitellogenesis and oocyte maturation. Shanbaky et al. (68) reported that injection of synganglion extracts or hemolymph into ligated posterior portions of recently fed *Argas hermanni* stimulated oocyte maturation. Evidence that gonadotrophic factors from the synganglion stimulate egg maturation in *O. parkeri* was obtained by ligation and transplantion experiments of Oliver et al. (61). Stimulation of Vg synthesis by synganglion extracts from replete *Hyalomma dromedarii* was shown in cultured fed virgin female fat bodies (66). Similar stimulation also was observed *in vivo* in this species (66). Therefore, the presence of a vitellogenesis-inducing factor (VIF) in both soft and hard ticks as the first factor in the regulation of vitellogenesis is well supported.

Based on the known systems of regulation if vitellogenesis in the mosquito, chicken and frog, Chinzei and Taylor (20) speculated that the second factor is a steroid hormone and that it is produced in the ovary. Ecdysteroid titers have been shown to rise between mating induced engorgement and the onset of vitellogenesis in the ixodid ticks *Ammblyomma hebraeum* and *Hyalomma dromedarii* (26, 31, 47). Kaufman (47) also reports that the ovary contains approximately five times as much ecdysteroid as the hemolymph. These reports indicate that ecdysteroids could possibly function as the second factor in the regulation of tick vitellogenesis. Diehl et al. (36) reported that in *O. moubata* the injection of ecdysteroids 2–8 days after engorgement had no effect on oviposition and treated females produced a normal amount of eggs. However, injection of larger amounts of ecdysone 8 days after engorgement inhibited oviposition in 8 of 10 females. Schriefer (66) reported that 20-hydroxyecdysone stimulated Vg production *in vivo* but failed to initiate Vg production by virgin female fat bodies *in vitro* in *H. dromedarii.* Ecdysteroids were previously shown to have effects on oogenesis and molting. It is reported that they reduced oviposition and induced a supermolt in fed virgin argasid ticks (49, 51); however, combined exposure to ecdysone and high concentrations of JH III induced oviposition in 6 of 10 argasid females, with a lack of supermolting (56). We investigated the effects of ecdysteroids on the stimulation of Vg synthesis in *O. moubata* and found an increase in Vg titer in the hemolymph on day 5 of unfed mated females after injection with ecdysone or 20-hydroxecdysone (77). These results are very preliminary and further investigation as to the role of ecdysteroids in vitellogenesis is needed.

Conclusion

The hormonal regulation of vitellogenesis in ticks has been shown not to involve JH but involves a factor from the synganglion. A model of the regulation of vitellogenesis, speculated from a number of experiments, is shown in Fig. 1.9. Chinzei and Taylor (20) showed feeding to CFS initiation stimulates the SyG to release the first factor, i.e., vitellogenesis-inducing factor (VIF). The SyG continues to release VIF during feeding and at the end of engorgement VIF in the hemolymph reaches a level high enough to stimulate an unknown organ in the posterior part to produce and release the second factor, i.e., fat body-stimulating factor (FSF). FSF acts on the fat body to initiate Vg synthesis. Vg secreted into the hemolymph then is incorporated into the developing oocytes.

These processes for the induction of vitellogenesis are very similar to those of the blood sucking mosquito (43) and also oviparous vertebrates, such as the chicken and frog (72). VIF and FSF are possibly analogous to egg development neurohormone (EDNH) and ecdysone in the mosquito

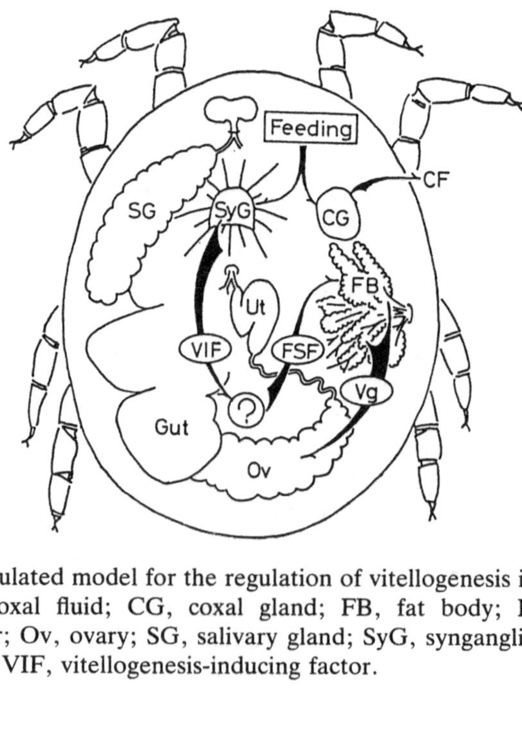

FIGURE 1.9. Speculated model for the regulation of vitellogenesis in *Ornithodoros moubata*. CF, coxal fluid; CG, coxal gland; FB, fat body; FSF, fat body-stimulating factor; Ov, ovary; SG, salivary gland; SyG, synganglion; Ut, uterus; Vg, vitellogenin; VIF, vitellogenesis-inducing factor.

and follicle-stimulating hormone (FSH) and estrogen in vertebrates. The similarities of these steps in ticks to that in the mosquito and some vertebrates causes us to further speculate that the unknown organ in the posterior part of the tick is the ovary and that FSF is a steroid hormone. We are now conducting experiments to verify these hypotheses.

Acknowledgments. The author's works were suported in part by Grant-in-Aid for Scientific Research (57560045, 58570166, 59560043, 62560042, and 014800052 to Y.C. and 01795015 to D.T.) from the Ministry of Education, Science and Culture, Japan, and in part by a Research Grant to Y.C. from "Okasan-Kato Science and Technology Foundation," Japan and Chemical Materials Research & Development Foundation, Japan. We would like to thank the research staffs in the Department of Medical Zoology, Mie University.

References

1. Aeschlimann, A. 1968. La pontechez *Ornithodoros moubata*, Murray (Ixodoidea, Argasidae). *Rev. Suisse Zool.* **75**:1033–1039.
2. Aeschlimann, A., and Grandjean, O. 1973. Observation on fecundity in *Ornithodoros moubata*. Relationships between mating and oviposition. *Acarologia* **15**:206–217.
3. Aramann S.F. 1979. Protein digestion and synthesis in ixodid females. In J.G. Rodriguez, ed., *Recent Advances in Acarology*, Vol. 1. Academic Press, New York, pp. 385–395.
4. Balashov, Yu.S. 1972. Blood sucking ticks (Ixodoidae): Vectors of diseases of man and animals. *Misc. Publ. Entomol. Soc. Am.* **8**:161.
5. Bassal, T.T.M., and Roshdy, M.A. 1974. *Argas (Persicargas) arboreus*: Juvenile hormone analog termination of diapause and oviposition control. *Exp. Parasitol.* **36**:34–39.
6. Binnington, K.C., and Obenchain F.D. 1972. Structure of the circulatory, nervous, and neuroendocrine systems of ticks. In F.D. Obenchain and R. Galun, eds., *Physiology of Ticks*. Pergamon Press, Oxford, pp. 351–398.
7. Binnington, K.C. 1986. Ultrastructure of the tick neuroendocrine system. In J.R. Sauer and J.A. Hair, eds., *Morphology, Physiology and Behavioral Biology of Ticks*. Ellis Horwood Limited, West Sussex, pp. 152–164.
8. Bownes, M. 1986. Expression of the genes coding for vitellogenin (yolk protein). *Annu. Rev. Entomol.* **31**:507–531.
9. Boctor, F.N., and Kamel, M.Y. 1976. Purification and characterization of two lipovitellins from eggs of the tick, *Dermacentor andersoni*. *Insect Biochem.* **6**:233–240.
10. Bremmer, K.C. 1959. Studies on "haemixodovin", the pigment in the eggs of the cattle tick, *Boophilus microplus* (Acarina: Ixodidae). *Aust. J. Biol. Sci.* **12**:263–273.
11. Chino, H., Yamagata, M., and Takahashi K. 1976. Isolation and characterization of insect vitellogenin. Its identity with haemolymph lipoprotein II. *Biochem. Biophys. Acta.* **441**:349–353.

12. Chino, H., Yamagata, M., and Sato, S. 1977. Further characterization of lepidopteran vitellogenin from haemolymph and mature eggs. *Insect Biochem.* **7**:125–131.
13. Chinzei, Y. 1983. Quantitative changes of vitellogenin and vitellin in adult female tick, *Ornithodoros moubata* during vitellogenesis. *Mie Med. J.* **27**:117–127.
14. Chinzei, Y. 1986. Vitellogenin biosynthesis and processing in a soft tick, *Ornithodoros moubata*. In D. Borovsky and A. Spielman, eds., *Host Regulated Development Mechanisms in Vector Arthropods*. Proceedings of the Vero Beach Symposium. University of Florida-IFAS, Florida Medical Entomology Laboratory, Vero Beach, pp. 18–25.
15. Chinzei, Y. 1988. A new method for determining vitellogenin in hemolymph of female *Ornithodoros moubata* (Acari: Argasidae). *J. Med. Entomol.* **25**:548–550.
16. Chinzei, Y., Chino, H., and Wyatt, G.R. 1981. Purification and properties of vitellogenin and vitellin from *Locusta migratoria*. *Insect Biochem.* **11**:1–7.
17. Chinzei, Y., Chino, H., and Takahashi, K. 1983. Purification and properties of vitellogenin and vitellin from a tick, *Ornithodoros moubata*. *J. Comp. Physiol. B* **152**:13–21.
18. Chinzei Y., Itoh, K., and Ando, K. 1989. Cypermethrin induction of vitellogenesis and ovarian development in unfed adult female *Ornithodoros moubata* (Acari: Argasidae). *Inv. Rep. Dev.* 15:19–26.
19. Chinzei, Y., Okuda, T., and Ando, K. 1989. Vitellogenin synthesis and ovarian development in nymphal and newly molted female *Ornithodoros moubata* (Acari: Argasidae). *J. Med. Entomol.* **26**:30–36.
20. Chinzei, Y., and Taylor, D. 1990. Regulation of vitellogenesis induction by engorgement in the soft tick, *Ornithodoros moubata*. In M. Hoshi and O. Yamashita, eds., *Advances in Invertebrate Reproduction*, Vol. 5. Elsevier, Amsterdam, pp. 565–570.
21. Chinzei, Y., Taylor, D., and Ando, K. 1991. Effects of juvenile hormone and its analogs on vitellogenin synthesis and ovarian development in the soft tick, *Ornithodoros moubata* (Acari: Argasidae). *J. Med. Entomol.* **28**: 506–513.
22. Chinzei, Y., Taylor, D., Miura, K., and Ando, K. 1992. Vitellogenesis induction by synganglion factor in adult female tick, *Ornithodoros moubata* (Acari: Argasidae). *J. Acarol. Soc. Jpn.* **1**:15–26.
23. Chinzei, Y., and Yano, I. 1985. Fat body is the site of vitellogenin synthesis in the soft tick, *Ornithodoros moubata*. *J. Comp. Physiol. B* **155**:671–678.
24. Chinzei, Y., and Yano, I. 1985. Vitellin is the nutrient reserve during starvation in the nymphal stage of a tick. *Experientia* **41**:948–950.
25. Connat, J.-L. 1988. Effects of different anti-juvenile hormone agents on the fecundity of the female cattle tick, *Boophilus microplus*. *Pestic. Biochem. Physiol.* **30**:28–34.
26. Connat, J.L., Diehl, P.A., Gfeller, H., and Morici, M. 1985. Ecdysteroids in females and eggs of the Ixodid tick *Ammblyomma hebraeum*. *Int. J. Invert. Reprod. Dev.* **8**:103–116.
27. Connat, J.L., Ducommun, J., and Diehl, P.A. 1983. Juvenile hormone-like substances can induce vitellogenesis in the tick *Ornithodoros moubata* (Acari: Argasidae). *Int. J. Invert. Reprod.* **6**:285–294.

28. Connat, J.-L., Ducommun, J., Diehl, P.A., and Aeschlimann, A. 1986. Some aspects of the control of the gonotrophic cycle in the tick *Ornithodoros moubata* (Ixodoidea, Argasidae). In J.R. Sauer and J.A. Hair, eds., *Morphology, Physiology and Behavioral Biology of Ticks*. Ellis Horwood Limited, West Sussex, pp. 194–232.
29. Connat, J.L., and Nepa, M.C. 1990. Effects of different anti-juvenile hormone agents on the fecundity of the female tick *Ornithodoros moubata*. *Pestic. Biochem. Physiol.* **37**:266–274.
30. Coons, L.B., Tarnowski, B., and Ourth, D.D. 1982. *Rhipicephalus sanguinius*: Localization of vitellogenin synthesis by immunological methods and electron microscopy. *Exp. Parasitol.* **54**:331–339.
31. Dees, W.H., Sonenshine, D.E., and Breidling, E. 1985. Ecdysteroids in the camel tick, *Hyalomma dromedarii* (Acari: Ixodidae), and comparison with sex pheromone activity. *J. Med. Entomol.* **22**:22–27.
32. Dhanda, V. 1967. Changes in neurosecretory activity at different stages in the adult *Hyalomma dromedarii* Koch 1944. *Nature (London)* **214**:508–509.
33. Diehl, P.A. 1969. Haemolymphoprotein and Vitellogenese bei *Ornithodoros moubata* Murray (Ixodoidea, Argasidae). *Bull. Soc. Entomol. (Suisse)* **42**:117–125.
34. Diehl, P.A. 1970. Zur Oogenee bei *Ornithodoros moubata* Murray (Ixodoidea, Argasidae) Unterbesonderer Berucksichtigung der Vitellogenese. *Acta. Trop.* **27**:305–355.
35. Diehl, P.A., Aeschlimann, A., and Obenchain, F.D. 1982. Tick reproduction: Oogenesis and oviposition. In F.D. Obenchain and R. Galun, eds., *Physiology of Ticks*. Pergamon Press, Oxford, pp. 277–350.
36. Diehl, P.A., Connat, J.L., and Dotson, E. 1986. Chemistry, function and metabolism of tick ecdysteroids. In J.R. Sauer and J.A. Hair, eds., *Morphology, Physiology and Behavioral Biology of Ticks*. Wiley, New York, pp. 165–193.
37. Eisen, Y., Warburg, M.R., and Galun, R. 1973. Neurosecretory activity as related to feeding and oogenesis in the fowl-tick *Argas perscius* (Oken). *Gen. Comp. Endocrinol.* **21**:331–340.
38. Engelman, F. 1979. Insect vitellogenin—identification, biosynthesis and role in vitellogenesis. *Adv. Insect Physiol.* **14**:49–108.
39. Feldman-Muhsam, B. 1973 Autogeny in soft ticks of the genus *Ornithodoros* (Acari: Argasidae). *J. Parasitol.* **59**:536–539.
40. Feldman-Muhsam, B., and Havivi, Y. 1973. Autogeny in the tick *Ornithodoros tholozozani* (Ixodoidea, Argasidae). *J. Med. Entomol.* **10**:185–189.
41. Gabbay, S., and Warburg, M.R. 1976. Neurosecretory activity as related to feeding, mating, and oogenesis in the female cave tick, *Ornithodoros tholozani*. *Insect Physiol.* **22**:1291–1301.
42. Gabbay, S., and Warburg, M.R. 1977. The diversity of neurosecretory cell types in the cave tick, *Ornithodoros tholozani*. *J. Morphol.* **153**:371–386.
43. Hagedorn, H.H. 1985. The role of ecdysteroids in reproduction. In G.A. Kerkut and L.I. Gilbert, eds., *Pysiology, Biochemistry and Pharmacology*, Vol. 8. Pergamon Press, Oxford, pp. 205–261.
44. Hagedorn, H.H., and Kunkel, J.G. 1979. Vitellogenin and vitellin in insects. *Annu. Rev. Entomol.* **24**:475–505.

45. Hecker, H., and Aeschlimann, A. 1970. Ultrastrukturelle Aspekte der Eibildung bei *Rhipicephalus bursa* (Canestrini and Fanzago) (Ixodoidea, Ixodidae). *Z. Tropenmed. Parasit.* **21**:31–45.
46. Izumi S., Tomino, S., and Chino, H. 1980. Purification and molecular properties of vitellin from the silkworm, *Bombyx mori. Insect Biochem.* **10**:199–208.
47. Kaufman, W.R. 1991. Correlation between haemolymph ecdysteroid titer, salivary gland degeneration and ovarian development in the ixodid tick, *Amblyomma hebraeum* Koch. *J. Insect Physiol.* **37**:95–99.
48. Khalil, G.M., and Shanbaky, N.M. 1976. Hormonal control of diapause in the tick *Argas arboreus. J. Insect Physiol.* **22**:1659–1663.
49. Kitaoka, S. 1972. Effects of ecdysone on ticks, especially *Ornithodoros moubata* (Acarina: Argasidae). *Proc. 4th Int. Cong. Entomol.* p. 272.
50. Koeppe J.K., Fuchs, M., Chen, T.T., Hunt, L.M., Kovalick, G., and Briers, T. 1985. The role of juvenile hormone in reproduction. In G.A. Kerkut and L.I. Gilbert, eds., *Comprehensive Insect Physiology, Biochemistry and Pharmacology*, Vol. 7. Pergamon Press, Oxford, pp. 165–203.
51. Mango, C.K.A., Odhiambo, R.T., and Galun, R. 1976. Ecdysone and the super tick. *Nature (London)* **260**:318–319.
52. Mansingh, A., and Rawlins, S.C. 1977. Antigonadotropic action of insect hormone analogs on the cattle tick *Boophilus microplus. Naturwissenschaften* **64**:41.
53. Marzouk, A.S., Moez, M.K.A., and Darwish, Z.E.A. 1987. The effect of feeding and mating on the neuroseretory activity in female *Hyalomma dromedarii* synganglion (Acari: Ixodoidea: Ixodidae). I. Changes in neurosecretory cell types in virgin and mated semifed females. *J. Egypt. Soc. Parasitol.* **17**:547–568.
54. Marzouk, A.S., Khalil, G.M., El Tunsy, M., and Darwish, Z.E.A. 1987. The effect of feeding and mating on the neuroseretory activity in female *Hyalomma dromedarii* synganglion (Acari: Ixodoidea: Ixodidae). II. Changes in neurosecretory cell types in fully engorged females. *J. Egypt. Soc. Parasitol.* **17**:657–668.
55. Obenchain, F.D., Dhadialla, T.S., Ahmad, Z., and Ojowa, R. 1978. The processing of lipovitellins in the brown ear tick *Rhipicephalus appendiculatus. Annu. Rep. ICIPE (Nairobi)* **1978**:42–44.
56. Obenchain, F.D., and Mango, C.K.A. 1980. Effects of exogenous ecdysteroids and juvenile hormones on reproductive development of female *Ornthidoros porcinus porcinus. Am. Zool.* **20**; Abstr. 1192.
57. Obenchain, F.D., and Oliver, J.H., Jr. 1973. A qualitative analysis of the form, function and interrelationships of fat-body and associated tissues in adult ticks (Acari-Ixodoidea). *J. Exp. Zool.* **186**:217–236.
58. Oliver, J.H., Jr. 1974. Symposium on reproduction of arthropods of medical and veterinary importance IV. Reproduction in ticks (Ixodoidea). *J. Med. Entomol.* **11**:26–34.
59. Oliver, J.H., Jr. 1986. Induction of oogenesis and oviposition in ticks. In J.R. Sauer and J.A. Hair, eds., *Morphology, Physiology and Behavioral Biology of Ticks*. Ellis Horwood Limited, West Sussex, pp. 233–247.
60. Oliver, J.H., Jr., Pound, J.M., and Severino, G. 1985. Evidence of a juvenile hormone-like compound in the reproduction of *Dermanyssus gallinae* (Acari: Dermanyssidae). *J. Med. Entomol.* **22**:281–286.

61. Oliver, J.H., Jr., Zhu, X.X., Vogel, G.N., and Dotson, E.M. 1992. Role of synganglion in oogenesis of the tick *Ornithodoros parkeri* (Acari: Argasidae). *J. Parasitol.* **78**:93–98.
62. Pound, J.M., and Oliver, J.H., Jr. 1979. Juvenile hormone: Evidence of its role in the reproduction of ticks. *Science* **206**:355–357.
63. Pound, J.M., Oliver, J.H., and Andrews, R.H. 1984. Effects of temperature and tick weight on expression of autogeny in the argasid tick *Ornithodoros parkeri* Cooley (Acari: Argasidae). *J. Parasitol.* **70**:279–284.
64. Raabe, M. 1986. Insect reproduction: Regulation of sucessive steps. *Adv. Insect Physiol.* **19**:29–154.
65. Sannasi, A., and Subramoniam, T. 1972. Hormonal rupture of larval diapause in the tick *Rhipicephalus sanguineus* (Lat.). *Experientia* **28**:666–667.
66. Schriefer, M.E. 1991. Vitellogenesis in *Hyalomma dromedarii* (Acari: Ixodidae): A model for analysis of endocrine regulation in Ixodid ticks. Ph.D. Dissertation. Old Dominion University and Eastern Virginia Medical School, Norfolk, Virginia.
67. Shanbaky, N.M., and Khalil, G.M. 1975. The subgenus *Persicargas* (Ixodoidea: Argasidae: Argas). 22. The effect of feeding on hormonal control of egg development in *Argas (Persicargas) arboreus. Exp. Parasitol.* **37**:361–366.
68. Shanbaky, N.M., Mansour, M.M., Main, A.J., El-Said, A., and Helmy, N. 1990. Hormonal control of vitellogenesis in *Argas (Argas) hermanni* (Acari: Argasidae). *J. Med. Entomol.* **27**:968–974.
69. Solomon, K.R., and Evans, A.A. 1977. Activity of juvenile hormone mimics in egg-laying ticks. *J. Med. Entomol.* **14**:433–436.
70. Staal, G.B. 1975. Insect growth regulators and juvenile hormone activity. *Annu. Rev. Entomol.* **20**:417–460.
71. Staal, G.B. 1986. Anti-juvenile hormone agents. *Annu. Rev. Entomol.* **31**:391–429.
72. Tata, J.R. 1978. Induction and regulation of vitellogenin synthesis by estrogen. In G. Litwack, ed., *Biochemical Actions of Hormones*, Vol. 5. Academic Press, New York, pp. 397–431.
73. Tatchell, R.J. 1971. Electrophoretic studies on the proteins of the haemolymph, saliva and eggs of the cattle tick *Boophilus microplus. Insect Biochem.* **1**:47–55.
74. Taylor, D., Chinzei, Y., Ito, K., Higuchi, N., and Ando, K. 1991. Stimulation of vitellogenesis by pyrethroids in mated and virgin female adults, male adults and fourth instar females of *Ornithodoros moubata* (Acari: Argasidae). *J. Med. Entomol.* **28**:322–329.
75. Taylor, D., Chinzei, Y., Miura, K., and Ando, K. 1991. Vitellogenin synthesis, processing and hormonal regulation in the tick, *Ornithodoros parkeri* (Acari: Argasidae). *Insect Biochem.* **21**:723–733.
76. Taylor, D., Chinzei, Y., Miura, K., and Ando, K. 1992. Effects of precocene on vitellogenesis in the adult female tick, *Ornithodoros moubata* (Acari: Argasidae). *Exp. Appl. Acar.* **14**:123–136.
77. Taylor, D., Chinzei, Y., Miura, K., and Ando, K. 1993. Timing and hormonal regulation of vitellogenesis in the soft tick, *Ornithodoros moubata* as determined of ligation experiments. *J. Insect Physiol.* (in preparation).
78. Venkatesh, K., Roe, R.M., Apperson, C.S., Sonenshine, D.E., Schriefer, M.E., and Boland, L.M. 1990. Metabolism of juvenile hormone during adult

development of *Dermacentor variabilis* (Acari: Ixodidae). *J. Med. Entomol.* **27**:36–42.
79. Wigglesworth, V.B. 1943. The fate of haemoglobin in *Rhodnius prolixus* (Hemiptera) and other blood sucking arthropods. *Proc. R. Soc. (London) Ser. B* **131**:313–339.
80. Wyatt, G.R., and Pan, M.L. 1978. Insect plasma proteins. *Annu. Rev. Biochem.* **47**:779–817.

2
The Biology of *Theileria* Species in Ixodid Ticks in Relation to Parasite Transmission

Michael K. Shaw and Alan S. Young

Introduction

In this review we consider the dynamics of transmission of *Theileria* species by ixodid ticks. *Theileria* parasites are of great economic importance since they cause theilerioses in domestic animals in many parts of the world (23, 94). These disease of cattle have been referred to as East Coast fever, Corridor disease, and January disease in the case of *T. parva* infection and Tropical theileriosis for the disease caused by *T. annulata*. Several other species cause mild disease in both cattle and other domestic livestock and some are avirulent. This review is, however, restricted mainly to *Theileria* species of cattle and in particular *T. parva*, about which most is known.

While ticks were identified as the vectors of *Theileria* during the early years of this century, a fuller understanding of the life cycle of these parasites has been obtained only in recent years. These studies have been reviewed recently by Mehlhorn and Schein (68) and Norval et al. (87). Although this review is primarily concerned with the biology of *Theileria* in its tick vectors, it must be realized that the importance of *Theileria* parasites lies in the diseases they produce in their mammalian hosts. The sporozoite stage of *Theileria* introduced into the mammalian host by the feeding tick infects lymphocytes, which become transformed by the developing schizont stage so that they become lymphoblastoid and divide rapidly, so producing infected daughter lymphoblastoid cells (43, 46). Ole MoiYoi (90) termed this transformation of the lymphocytes "reversible parasite-induced transformation" since it depends on the presence of parasite and is reversible when the intracellular parasite is eliminated

---

Michael K. Shaw, International Laboratory for Research on Animal Diseases, P.O. Box 30709, Nairobi, Kenya.
Alan S. Young, International Laboratory for Research on Animal Diseases, P.O. Box 30709, Nairobi, Kenya.

by chemotherapy (92). The transformation of lymphocytes and their destruction during merogony of the schizont and by cytolytic T lymphocytes is responsible for the pathogenic effects of several species (87) while merogony in the erythrocytes is responsible, among other effects, for the development of anemia in other *Theileria* species (87).

Several tick genera have been identified as field vectors together with additional experimental vectors for *Theileria* species of cattle and domestic animals (81, 87, 114, 115, 117). Transmission has been reported to be transstadial, that is, the larval or nymphal instar becomes infected and transmits the *Theileria* parasites in the nymphal or adult instars (3, 81). Reports of transovarian transmission have not been substantiated in *Theileria* infections (87) although transovarian transmission is usual in related *Babesia* species (36).

The economic importance of theileriosis in domestic animals caused by the different *Theileria* parasites has not been fully established. It is difficult in endemic areas to differentiate the effects of vector ticks and chronic theileriosis on the productivity of cattle, although several recent attempts have been made (22, 75, 78, 88, 91, 113). Acute theileriosis causing severe disease and high mortality is easier to cost (77, 145). The conclusion, in economic terms, is that theileriosis is a major problem in Africa, Asia, and in limited areas elsewhere, particularly in unstable epidemic areas and in areas where cattle are being upgraded for higher production.

Many of the factors affecting the transmission of *Theileiria* species by ixodid ticks are related to the biology of their development in the tick. Therefore, in this article we will, first, review the current information available regarding the development of *Theileria* parasites in their tick vectors, before discussing the major factors that control the dynamics of transmission in the field. We hope that a review of the biology of *Theileria* transmission by ixodid ticks will be of use to scientists and veterinarians attempting to develop robust control methods in many parts of the world.

Theileria Species

The taxonomy and characteristics of *Theileria* species are shown in Table 2.1 and have recently been reviewed by Norval et al. (87). At present, seven species of *Theileria* infecting cattle are recognized, namely *T. parva*, *T. annulata*, *T. mutans*, *T. velifera*, *T. sergenti*, *T. buffeli* and *T. taurotragi* (see Table 2.2). Many species have been discarded or are considered synonyms and this subject has been fully reviewed by Neitz (81), Uilenberg (117), and Norval et al. (87). However, most species of artiodactyls harbor *Theileria* parasites, many of which have not yet been identified. Thus, there is a possibility of new cattle species being

TABLE 2.1. Classification and features of *Theileria*[a].

Classification	Features
Subkingdom	Protozoa; single celled eukaryotes
Phylum	Apicomplex; apical complex present at least in some stages: reproduce sexually by syngamy
Class	Sporozoea; sporogonic stage producing sporozoites
Subclass	Piroplasmia; piroform rod-shaped or ambeoboid parasites in erythrocytes and some other cell types
Order	Piroplasmida; asexual and sexual reproduction; ticks are vectors
Family	Theileriidae; schizont stages in lymphocytes
Genus	*Theileria*; piroplasm stage in erythrocytes lacks pigment

[a] After Irvin (44).

recognized. For example, recently in Kenya a new species for that country, *T. buffeli*, was isolated during an immunization trial (145) and initially caused some confusion. In areas where large populations of wild and domestic artiodactyl coexist, transmission from wildlife to domestic animals can easily occur, particularly as both wild and domestic artiodactyls serve as hosts for the tick vectors of *Theileria*. For example,

TABLE 2.2. *Theileria* species of cattle, their ixodid vectors, and known distribution.

Theileria species	Tick vectors	Distribution
Theileria parva (Theiler, 1904)	*Rhipicephalus appendiculatus*, *Rhipicephalus zambeziensis*, *Rhipicephalus duttoni*, *Rhipicephalus nitens*	Eastern, central, and southern Africa
Theileria mutans (Theiler, 1906)	*Amblyomma variegatum*, *Amblyomma gemma*, *Amblyomma hebraeum*, *Amblyomma cohaerens*, *Amblyomma lepidum*	Western, eastern, central, and southern Africa; Caribbean Islands
Theileria taurotragi (Martin and Brocklesby, 1960)	*Rhipicephalus appendiculatus*, *Rhipicephalus pulchellus*, *Rhipicephalus zambeziensis*	Eastern, central, and southern Africa
Theileria velifera (Uilenberg, 1964)	*Amblyomma variegatum* and other *Amblyomma* species	Western, eastern, central and southern Africa, including Madagascar
Theileria annulata (Dschunkowsky and Luhs, 1904)	*Hyalomma anatolicum* and other *Hyalomma* species	Northern Africa, southern Europe, middle East, Asia including India
Theileria buffeli (Neveu-Lemaire, 1912)	*Haemaphysalis* species	Europe, eastern Africa, Asia, including Japan and Australia
Theileria sergenti (Yakimov and Dekhterev, 1930)	*Haemaphysalis* species	Japan and Korea

T. taurotragi will infect cattle, sheep, and goats as well as most wild artiodactyls tested (111) and *T. parva* will infect cattle, African buffalo (*Syncerus caffer*) and waterbuck (*Kobus* spp.) (111, 132).

The *Theileria* species of sheep, goats, and domestic buffalo have been less studied. In Africa and Asia, *T. ovis*, *T. seperata*, and *T. hirci* have been recognized as parasites of small ruminants (117). For biological and transmission reasons certain *Babesia* species such as *B. equi* have been placed in the genus *Theileria* (101) but recent studies indicate that they should occupy a seperate genus (M.T.P. Allsopp, personal communication).

The original method of identifying and differentiating *Theileria* species relied almost exclusively on the morphology of the parasite in stained blood and tissue smears from mammalian host. However, as most of the mammalian *Theileria* species have very similar morphologies, identification by such means is both highly subjective and relies on the experience of the observer. It is important to be able to recognize *Theileria* species in their hosts because, for example, East African cattle could be infected at any one time with up to five *Theileria* species, can be infested with several species of tick and at least two different *Theileria*

TABLE 2.3. Methods available for identification of *Theileria* species in their mammalian and tick hosts[a].

	Theileria species				
	T. parva	*T. annulata*	*T. mutans*	*T. velifera*	*T. taurotragi*
Morphology					
Giemsa stained blood smear	−	−	−	+	+
Giemsa stained lymph node smear	−	−	+	−	−
Giemsa stained smear of tick	+	+	+	+	+
Serological					
IFA test	±	±	±	±	±
Specific antigen ELISA Test	+	+	+	NA	+
Specific monoclonal antibody	+	+	+	NA	+
Molecular					
DNA probes	+	+	+	NA	+
Polymerase chain reaction	+	+	+	NA	+
Restriction fragment length polymorphism	+	NA	+	NA	NA

[a] NA, not available; −, negative; ±, sometimes effective; +, effective.

species may be transmitted by one species of tick vector. Table 2.3 shows the current methods available for the recognition of *Theileria* species.

The recent advances in producing specific ELISA tests for *T. mutans* using both antibodies and antigen defection (50) suggest that such reagents will be very useful tools that can possibly be applied to all *Theileria* species. This technique relies on the production of *Theileria* species-specific monoclonal antibodies that recognize antigens that are universally expressed by all stocks of the *Theileria* species. It is presumed that species-specific monoclonal antibodies, which have already been developed for mammalian stages (19, 20), could also be developed for the tick stages of the parasites. These reagents would be useful for identifying tick stages of *Theileria* from field ticks where tick vector species harboring at least two *Theileria* infections can be expected. In addition DNA and RNA technology has been developed to a level where it can be used practically to identify *Theileria* species in both ticks and mammalian hosts (10, 16, 20, 87). Hence techniques to support detailed and accurate studies of the epidemiology of theileriosis are becoming available that will enhance the accuracy of current epidemiological and particularly transmission studies.

Ixodid Vectors

The vectors of *Theileria* species of cattle are summarized in Table 2.2. The identification and distribution of the tick vectors of theileriosis have been reviewed recently by Norval et al. [87]. However, while rapid progress on the identification of *Theileria* species using newly developed technologies has been made, the taxonomy of the tick vector is still generally decided using classical taxonomy. For ixodid ticks apart from basic morphological studies, limited attempts have been made to use polymorphisms in isoenzymes and cuticular hydrocarbons as a means of differentiating tick species. DNA and RNA technologies comparable to those available for the identification and characterization of *Theileria* species have yet to be applied to the taxonomy of ixodid ticks.

In relation to *Theileria*, a major problem occurs in the taxonomy of rhipicephalid species where only a limited number of taxonomists can accurately identify them. For example, differentiating between *R. appendiculatus*, *R. zambeziensis*, *R. duttoni*, and *R. nitens*, which are all reported field vectors of *T. parva*, is difficult to the extent that even the taxonomists are not sure about the distinctions between them (125). In particular, at present, ticks specialists cannot agree on the differences between *R. appendiculatus* and *R. zambeziensis*. Furthermore all these species transmitting *Theileria* species infest a range of hosts other than cattle (131, 124). Therefore, the nontaxonomists, such as epidemiologists working in the field who have to know which species they are studying, require reliable and easy techniques to identify the tick vectors. Such easy and reliable techniques are still not available.

Development Cycle of *Theileria* in Ticks

A generalized development cycle of *Theileria* species within their tick vectors is shown in Fig. 2.1 with further details given in Table 2.4. The development cycles of *Theileria* species in ticks and cattle in Africa are similar, although they differ in detail. The most complete information available relates to *T. parva*, *T. annulata*, and *T. taurotragi* (32, 68, 138).

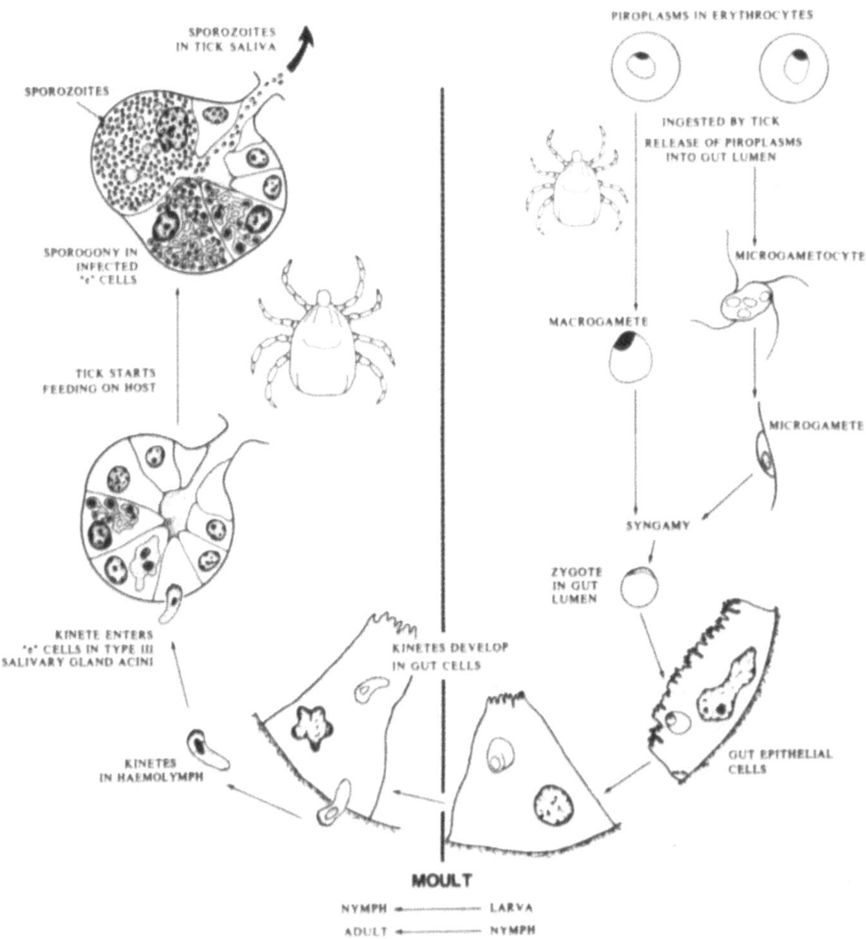

FIGURE 2.1. Diagram of the life cycle of *Theileria* parasites in their ixodid tick vectors.

TABLE 2.4. A comparison of some characteristics of the development of *Theileria* species in ticks maintained at 28°C.

	Species of *Theileria* and vector[a]				
Development characteristic	*T. taurotragi*[1] *Rhipicephalus appendiculatus* (Nymph–adult)	*T. parva*[2] *Rhipicephalus appendiculatus* (Nymph–adult)	*T. annulata*[3] *Hyalomma anatolicum* (Nymph–adult)	*T. mutans*[4] *Amblyomma variegatum* (Nymph–adult)	*T. velifera*[5] *Amblyomma variegatum* (Nymph–adult)
Length of microgamete (μm)	8	10	12	7	7
Period postrepletion to first occurrence of microgametes (days)	1–2	2	3	5	3
Size of macrogametes (diam. μm)	4	4–5	4	4	4
Mean size of mature zygote (μm)	12.2 × 9.0	9.0 diam.	10.0 diam.	6.0 diam.	12.0 diam.
Mean time postrepletion to moult of ticks (days)	14	19	21	26	26
Earliest transformation of zygote to kinete (days)	12	20	12	30	15
Mean size of kinete (μm)	22.1 × 6.5	19.0 × 5.5	17.6 × 5.5	14.0 × 4.7	18.6 × 7.0
First detection of kinete in hemolymph (days)	16	22	17	34	16
First penetration of salivary glands by kinete (days)	18	25	18	?	?

[a] Data from (1) Young et al. (139), (2) Schein et al. (104), Mehlhorn et al. (70), (3) Schein et al. (103), Schein and Fiedhoff (102), (4) Warnecke et al (125), and (5) Warnecke et al. (126).

Terminology Used in the *Theileria* Life Cycle

To discuss the life cycle of *Theileria* in the tick, it is necessary to review the terminology in current use (87). *Gametocytes* are stages within the blood of infected animals that, on ingestion by the tick, can develop into gametes in the gut lumen of larval or nymphal ticks. *Gametes*, micro- and macrogametes, are stages that develop within the gut of the tick and that subsequently undergo syngamy (fusion). *Gamonts* are the intermediate stages in the process of gametogony before the gametocytes mature into gametes. The *zygote* is the product of the fusion of the gametes (syngamy). The *kinete* (or *ookinete*) is a motile stage that develops from the zygote in the gut epithelium, enters the hemolymph, and penetrates the salivary gland of the tick. A *sporont* is a rounded form of the parasite that initiates the development of sporogony in the salivary glands. *Sporogony* is the process of multiple division of the parasite nuclei within the salivary glands that results in the production of sporozoites from the sporoblast, which is the the stage of the parasite that undergoes sporogony. The *sporozoite* is the final product of sporogony and is a single uninucleated cell, which is transmitted by the tick into the mammalian host and initiates infection in the new host.

The Piroplasm Stage

The life cycle of *Theileria* in the tick begins with the ingestion of piroplasm-infected erythrocytes with the blood meal. The ultrastructure of the piroplasm stage has been described for a number of *Theileria* species (see ref. 33 and references therein). The piroplasms lie free within the erythrocyte and not within a parasitophorous vacuole, as is the case with, for example, *Plasmodium* species. *Theileria* species do not produce residual pigments when feeding on erythrocytic cytoplasm, unlike all the malarial parasites of the genus *Plasmodium*. Several theilerial species cause changes in the characteristics of the infected erythrocyte cytoplasm with crystallized host cell material a common feature (33, 87). Intraerythrocytic multiplication has been described for several *Theileria* species, although for *T. parva*, the developmental cycle of the piroplasm within the erythrocyte is still unclear since prolonged division within the erythrocyte has not been demonstrated (18). It is also not clear whether definite *Theileria* gametocytes occur in the circulating blood as is clearly the case for *Plasmodium* (109).

Development of Sexual Stages within the Tick Gut

During feeding and at repletion many millions of infected erythrocytes are ingested by the tick even from an animal with a low piroplasm parasitemia, and infected erythrocytes are readily detectable in the gut lumen for 2–3 days after repletion.

However, recent studies at ILRAD on *T. parva* in *R. appendiculatus* nymphs have shown that, even at engorgement, the vast majority of these ingested piroplasms are rapidly destroyed within the gut lumen. The rapid destruction of infected erythrocytes and piroplasms within the gut is thought to be due to the secretion of acid phosphatases by the gut epithelial cells during feeding (120). Only in about 10% of any tick population studied does the parasite survive in large numbers and sexual stages subsequently develop in detectable numbers. This survival and continued development of the parasite could be related to differences in the rate of digestion between individual ticks. However, there appears to be no obvious correlation between the numbers of infected erythrocytes ingested and the proportion of piroplasms that survive and eventually undergo development into the sexual stages, although this is being investigated in detail at present.

No development of the intraerythrocytic piroplasm has been reported to occur within the tick gut before repletion and the process of sexual differentiation or gametocyte development, which occurs at or just after repletion, appears to be initiated by the release of the parasite from the infected erythrocyte during their lysis within the tick gut. It must be realized that there is a temperature decrease after the tick completes repletion and falls to the ground. The temperature on the mammalian skin is invariably higher than the ambient or ground temperature. Thus, the initial stages of gametogenesis in *Theileria* and *Babesia* species (136) differ from the process in other related Apicomplexan parasites (e.g., *Plasmodium* species; Dactylosomatidae), where differentiation of the sexual stages begins within the erythrocyte (5, 26, 108, 109).

In *Theileria*, the freed parasites, which have a variety of shapes and sizes, are readily seen in smears of the gut contents. A variable proportion of the freed piroplasms differentiates into the sexual stages. There are suggestions in the literature that only the larger "ring-form" piroplasms develop into gametes (68, 138), but this remains to be confirmed.

The sequence of gametogenesis is similar in *T. parva*, *T. annulata*, *T. mutans*, *T. taurotragi*, and *T. velifera*. At present, we have little indication of the factors required for the initiation of gamete development. However, the survival of the freed parasite and gametogenesis occur only at or just after engorgement, when changes in the physiology of the gut occur (e.g., the secretion of acid phosphatase into the gut lumen ceases) suggesting that physiological conditions within the gut lumen are critical to the further development of the parasite. Furthermore, parasites ingested by adult *R. appendiculatus* tick do not survive within the gut lumen for more than 2–3 days depending on the temperature (135), and there are indications that the conditions within the gut lumen of adult ticks differ from that of the larval and nymphal stages.

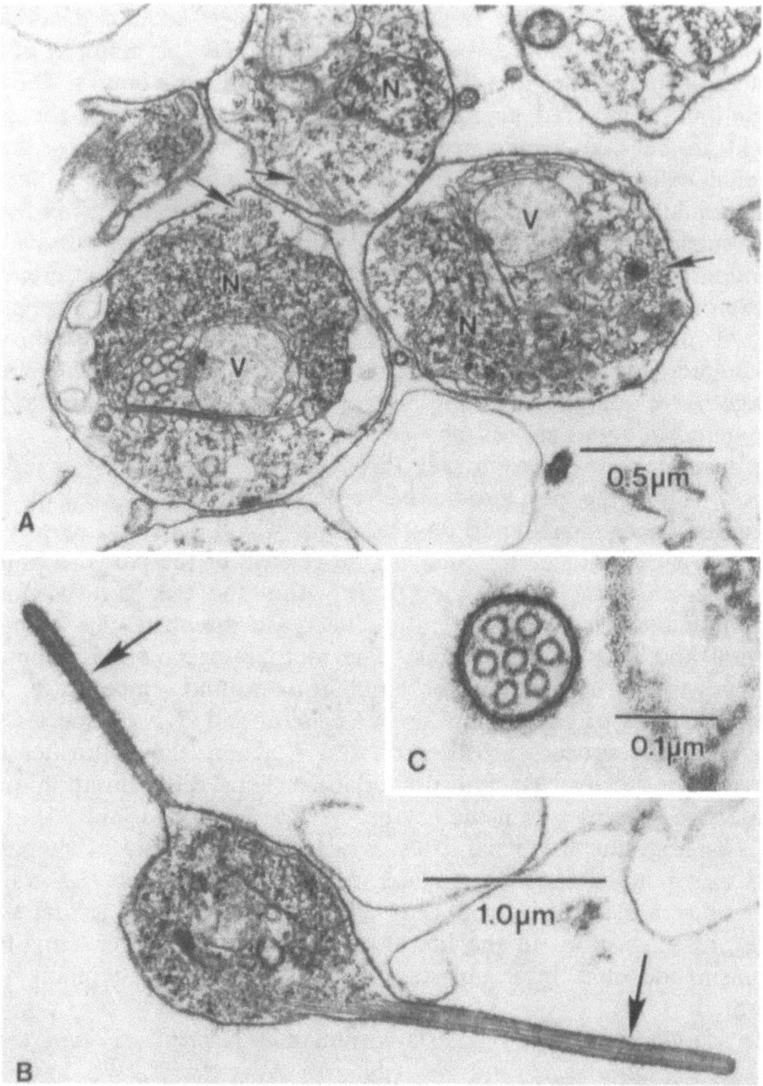

FIGURE 2.2. Electron micrographs of developing *T. parva* microgametes isolated from the gut lumen of adult female *R. appendiculatus*. (A) Section through developing microgametes. Note that the cytoplasm of each gamete contains a nucleus (N), a vacuole (V), and associated membranes structures, and numerous microtubules (arrows). (B) Longitudinal section through a uninucleate microgamete showing the microtubule-containing protrusions or axonemes (arrows) of the body. (C) Cross section through an axoneme showing the arrangment of the microtubules.

Once initiated gametogenesis occurs relatively rapidly. Beginning soon after the completion of feeding so called "strahlenkorper" or ray bodies can be seen in smear preparations. These ray bodies, which are considered to be microgametes, have been observed in gut smears of ticks harboring *T. parva*, *T. annulata*, *T. mutans*, *T. taurotragi*, and *T. velifera* infections (103, 104, 126, 127, 138). However, although ray bodies were first described by Koch (53) and confirmed by several other workers (21, 24, 39), it was not until the work of Schein and colleagues (67–69, 100) that their detailed structure and significance as the sexual stages within the *Theileria* life cycle were finally established.

Development of the ray bodies or microgametes occurs rapidly. For *T. taurotragi* in *R. appendiculatus* nymphs the microgametes form two or more fine lateral projections within 24 hr postrepletion and the nucleus rapidly divides forming two or, in many cases, four daughter nuclei. By 24–48 hr postrepletion the microgametes have completed their differentiation and appear as spindle-shaped structures with a single central nucleus, and measure between 8 and 12 μm in length and about 1.0 μm width at their mid-region. The microgametes have a short life span with no microgametes or earlier stages being detectable in the gut lumen after day 5 postrepletion at higher ambient temperature (138).

The ultrastructure of the microgametes of *T. annulata* and *T. parva* has been described from both the tick gut as well as from in vitro cultures of infected erythrocytes (68). While it is possible to isolate gametes from tick guts (Fig. 2.2), the lack of a defined culture system that will support gametogenesis is a major problem that needs to be solved before detailed investigations, comparable to those undertaken on gametogenesis in *Plasmodium* (108, 109), can be contemplated. Short-term culture systems have already been developed in which intraerythrocytic multiplication of the piroplarm stage has been observed (18). However, it has not yet been possible to adapt these systems to consistently support gametogenesis.

Between 0 and 48 hr postrepletion relatively large spherical parasites also develop from the "ring-form" piroplasms. These forms, which appear to undergo no cycle of differentiation comparable to the microgametes, are considered to be the macrogametes.

Syngamy

The fusion of the micro- and macrogametes, syngamy, occurs *in vitro* after 6 days (68) but normally occurs in a shorter time in the tick gut, although the precise timing is temperature dependent (135).

The process of syngamy and zygote formation in *T. parva* and *T. annulata in vitro* has been described briefly by Mehlhorn and Schein (68), although precise details of the culture conditions are not given. The process apparently involves two gametes (presummably a micro- and macrogamete) making initial contact followed by one gamete forming a finger-like protrusion that penetrates into the opposite gamete. At the point of contact the membranes apparently fuse, leading to an open connection between the two gametes. As this connection becomes larger the two nuclei unite and fuse to form the uninucleate zygote. The process of syngamy in *Theileria* as described by Mehlhorn and Schein (68) is clearly different from the process in *Plasmodium* (108, 109). Whether there are also significant differences in the process of syngamy between *Theileria* and other Apicomplexan parasites needs to be examined in more detail. However, such studies must again await the development of a culture system capable of supporting gametogenesis for *Theileria* species.

Some of the implications of a sexual cycle within the *Theileria* life cycle have been reviewed by Irvin and Boarer (45) and, although, syngamy has been partially described, only now is there direct evidence of genetic recombination at this stage in the *Theileria* life cycle (75). In *Plasmodium* spp. there is substantial evidence for genetic recombination being a major cause of the polymorphism exhibited by malarial parasites (119). Extensive polymorphism is also found in field isolates of *T. parva* and it is important to ascertain if this is due to genetic recombination during sexual reproduction. Recently Morzaria et al. (75) reported that sexual recombinations between different clones of *T. parva* occur. Furthermore, they have also indicated that all the parasite stages in cattle, and the gametes, and the sporozoite stage are haploid, while the zygote and the kinete stages are diploid, which is similar to *Plasmodium* (109). It is likely that meiosis in *Theileria* occurs in the salivary glands, although this still requires confirmation, particularly in view of the report of nuclear division occurring in the developing kinete while still present within the gut epithelial cells (70). The determination of the exact nature of the extensive polymorphism exhibited by *T. parva* is important in planning immunization strategies against East Coast fever and other forms of theileriosis.

Development of the Zygote and Kinete in the Gut Epithelial Cells

After syngamy, the resulting spherical zygote invades a gut epithelial cell and eventually develops into the kinete (Fig. 2.3A). However, information on the detailed structure of the zygote and on the process of entry into the gut epithelial cells is still not available. For example, it is not known whether zygotes develop an apical complex analogous to that of the sporozoite and merozoite stages or whether they have any secretory organelles, rhoptries and microspheres, that are essential in the other life cycle stages for entry and establishment in host cells. Furthermore, it is not known in what type of tick gut cell the zygotes enter and develop.

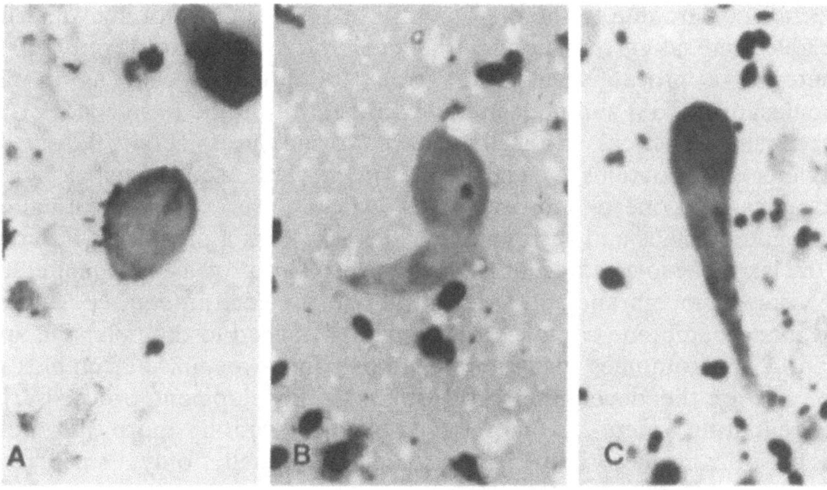

FIGURE 2.3. Light micrographs of the development stages of *T. taurotragi* in the gut epithelium of *R. appendiculatus*. (A) Zygote stage from the gut epithelium. (B) Developing zygote transforming into a kinete in the gut epithelium. (C) Mature kinete.

The tick gut is composed of a number of different cell types that undergo considerable morphological changes during feeding and thereafter (121). Within the gut lumen there are also motile phagocytic digestive cells that appear at or just after engorgement and are present throughout the period of moulting. Walker (120) suggested that these motile phagocytic cells may engulf and destroy *Theileria* zygotes. Does the *Theileria* zygote enter any gut cell that it makes contact with or does it enter and develop only in a specific type of gut cell? In the latter case it can be expected that zygotes will have specific receptors for entry into the gut cells. Whichever gut cell type(s) the zygote enters it must be one that survives tick moulting. While the actual process of zygote entry into the epithelial cell is not known, Mehlhorn et al. (70) and Mehlhorn and Schein (68) have shown that the developing parasite is not enclosed within a parasitophorous vacuole but lies free within the host cell cytoplasm. Interestingly, all the intracellular stages in the *Theileria* life cycle, namely the sporoblast within the tick salivary gland cell, the schizont stage within the bovine lymphocyte, and the piroplasm stage in the bovine erythrocyte, also lie free within their host cell cytoplasm. In the two latter cases the invading organism enters surrounded by the host cell plasma membrane from which it rapidly escapes (31, 106).

Once within the gut epithelial cells the zygotes appear to grow and often groups of zygotes can be observed in clusters within an individual gut cell. Whether these groups of zygotes are the result of multiple infections or are due to the division of a single zygote is not known. At a variable time after repletion of the tick, the zygote transforms into a kinete by a process that differs from the process occurring in the Haemosporida (68) and appears to be unique within the Protozoa. For *T. parva*, the spherical intracellular zygote measures about 9.0 µm in diameter and contains a single eccentric nucleus, some smaller dense inclusions, and one or more vacuoles. At the onset of kinete formation the zygote nucleus and surrounding cytoplasm form a protrusion into one of the large vacuoles adjacent to the nucleus. This protrusion is limited by the vacuolar membrane and at its apical pole an electron-dense polar ring and inner membrane complex are formed. Attached to the polar ring are about 40 microtubules that run beneath the inner membrane complex to the base of the developing protrusion. As development proceeds the differentiating kinete becomes larger by incorporating more and more cytoplasm from the original spherical cell. Eventually, only a small zone of cytoplasm surrounds the large vacuole containing the developing kinete, which, however, remains attached to the residual zygote by a small cytoplasmic bridge at its posterior end. At this stage the residual body ruptures releasing the motile, club-shaped kinete into the gut cell cytoplasm. Only a single motile kinete is formed form each zygote. While the majority of mature, intracellular kinetes contain a single nucleus, Mehlhorn et al. (70) observed that in several kinetes nuclear division had already started with stages containing up to four nuclei being detected.

The formation of kinetes and their appearance in the hemolymph have been related to the moulting cycle of the tick host (133, 137). Kinetes first appear in the hemolymph during or immediately after the moulting of the tick instar, although in the case of *T. annulata* and *T. velifera* kinetes have been reported in the hemolymph prior to moulting (68). Regardless of the precise timing of the appearance of kinetes in the hemolymph, entry of the kinetes into the salivary gland appears to be possible only after the glands have redeveloped following each moult (27, 28, 30). It appears that *Theileria* already present within the salivary gland of nymphal or adult ticks prior to feeding cannot survive the regression of the salivary glands to primordia that occurs after each feeding.

To date the actual process of kinete penetration of the salivary gland has not been described. Infection of the salivary glands by the kinetes may be a chance event since *in vitro* salivary glands appear to exert no chemoattractive influence on the kinetes (6). However, neither penetration nor the development of *Theileria* parasites within any tick organs other than the salivary glands has been observed, indicating that some degree of host organ recognition by the kinete may occur. Walker (120) suggested that during the replication and early differentiation of the salivary glands following the previous feeding and moulting cycle, there may be specific external markers on the various acinar cells to enable the motile kinete to selectively invade the appropriate cell type.

Furthermore, in all the *Theileria* infections described so far, the parasites appear to develop in a limited range of cells in the Type III salivary gland acini (see below). Whether the kinetes can specifically recognize and enter a selective range of salivary gland cell types or whether only selective cell types provide the appropriate intracellular environment for the survival and further development is not known.

Development of *Theileria* in the Salivary Gland Acini

Structure of the Tick Salivary Gland

Before describing the development of *Theileria* parasites within the tick salivary gland, it is necessary to give a brief description of the tick salivary gland system and outline the changes that occur during tick attachment and feeding. The general organization of the tick salivary glands, which lie in the hemolymph of the tick, has been well described in a number of light and electron microscopic studies (38, 122). The paired salivary glands of adult female *R. appendiculatus* each consists of about 1400 acini of three types (Types I–III) while in adult male ticks there are about 1350 acini of four types (Types I–IV) (122). A large central duct runs the length of each gland with several major branches and numerous branching lobular ducts arising from it. The various types of acini are differentially arranged along the length of each gland and each individual acinus is composed of a variety of different cell types.

The Type I acini are composed of nongranular cells and are thought to play a major role in water vapor uptake and in the maintenance of osmotic balance during the periods between blood meals (51, 61, 80, 122). The Type II–IV acini contain a plethora of secretory cell types usually designated by a lower case letter (e.g., "a," "b," to "h"). The identification of these various cell types at the light microscope level has been based on differences in the size, staining properties, and histochemical composition of the secretory granules (8, 27, 28). Recently, ultrastructural details of the cytoplasmic organization and additional identifying features such as the relative electron density and substructure of the secretory granules have allowed for more precise identification of the different cell types. The salivary gland cells are known to secrete numerous products including cement substances for the attachment cone, histamine blocking agents, anticoagulants, cytolysins, vasoactive mediators to increase the vascular permeability and thus facilitate blood feeding, various enzymes, and paralytic toxins whose function(s) have yet to be defined (51). However, while detailed information is available regarding the structure and to some extent histochemical composition of the various secretory granules, precise data on the chemical nature and function of the individual granules are lacking.

The salivary glands do not become fully active until the ticks start feeding, whereupon they undergo a number of dramatic structural changes. These structural and histochemical changes that occur during feeding have been described in varying amounts of detail for a number of important tick vectors (8, 27, 28, 122). Apart from changes occurring due to the discharge of the various secretory granules, the most dramatic ultrastructural alterations occur in the cells of the type III acinus and have been related to the rapid increase in fluid secretion that accompanies blood feeding (28).

Sporogony in the Salivary Glands of Adult Ticks

As noted earlier, *Theileria* parasites appear to develop in a limited range of salivary gland cells. For example, *T. parva* develops in the "e" cells of Type III acini (29, 30, 32), although Mehlhorn and Schein (68) claim that development in the "d" cells may also occur. On rare occasions infected Type II acini have also been found. Similarily, buffalo-derived *T. parva* and *T. taurotragi* appear to develop exclusively within the "e" cells of Type III acini, and in addition *T. cervi*, a parasite of north American deer, develops only in the "e" cells of Type III acini in female *Amblyomma americanum* (41). In *T. annulata* in *Hyalomma* ticks development normally occurs in the Type III acini, although development may also occur in Type II acini, particularly in cases where ticks have fed on cattle with high parasitemia and in which large numbers of kinetes develop (120). Although the process of kinete entry into the tick salivary gland cells has not been described, the established parasite is not enclosed within a parasitophorous vacuole but lies free within the salivary gland cell cytoplasm (29, 32).

A variable amount of development of the *Theileria* parasite occurs within the salivary gland cell before the ticks feed on hosts. In unfed adult ticks the parasitized host cell has often already undergone marked hypertrophy, while the parasite has become a multinucleated syncytium and occupies a significant portion of the cell cytoplasm. The amount of host cell hypertrophy and parasite development varies both from tick to tick and from acinus to acinus in the same tick, possibly reflecting varying times of kinete entry or differing degrees of dormancy of the sporoblasts. One of the factors controlling the onset of parasite development in unfed ticks is high ambient temperature (99, 137), although the numbers of sporozoites produced were much lower than when sporogony was induced by feeding ticks on rabbits (142). It is likely from these results that sporogony is under the partial control of a heat shock gene.

With the onset of tick attachment and feeding, there is a rapid increase in parasite mass and continuing hypertrophy of the host cell. Parasite development is a continuous process and is not, as suggested in the earlier literature (60, 64, 71, 76, 95, 102, 138), a step wise proces involving the production of different types of sporoblasts. Fawcett and co-workers (29, 30, 32) have now shown that the parasite develops as a ramifying, multinucleate syncytium that rapidly increases in size and complexity until it gives rise to large numbers of sporozoites during a terminal episode of cytoplasmic fission.

Initially the proliferating parasite nuclei occupy peripheral lobules that are continuous with a central labyrinth of slender branching and

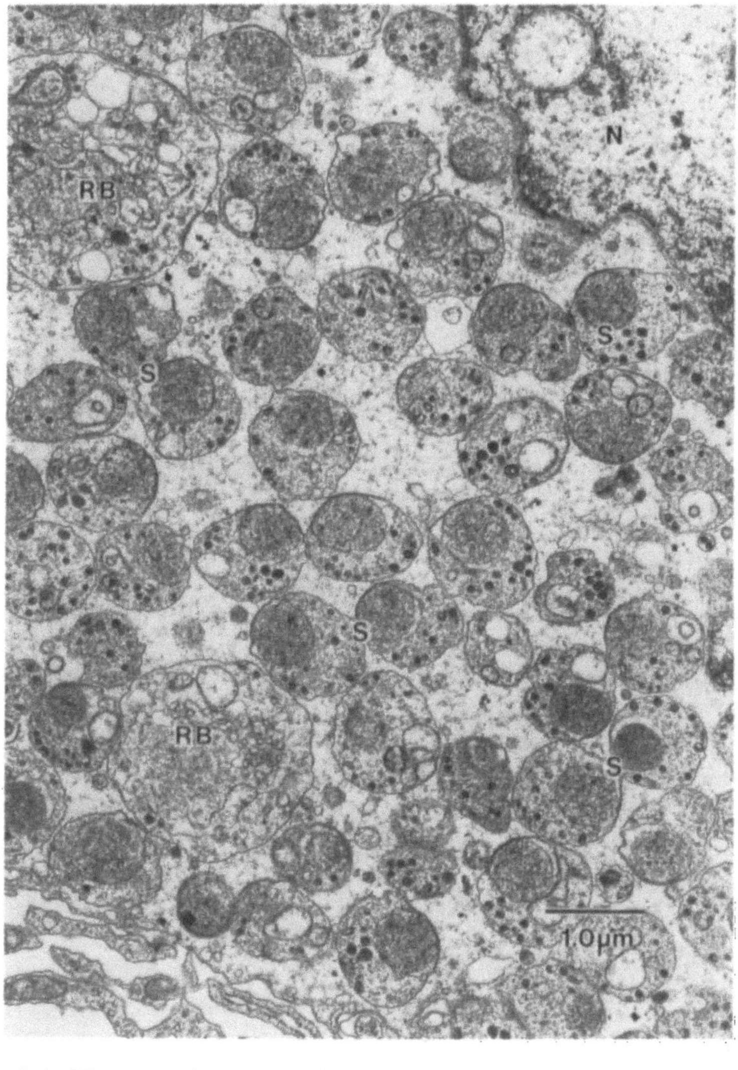

FIGURE 2.4. Electron micrograph of fully formed *T. parva* sporozoites (S) in the salivary glands of *R. appendiculatus*. Note that the host cell cytoplasm is relatively empty due to the active elimination of organelles and inclusions by the host cell itself during sporogony. RB, residual body of sporoblast; N, host cell nucleus.

anastomosing processes, which greatly increases the surface area of the developing parasite. The membrane of this central labyrinth contains numerous cytostomes or micropores and the whole structure presents a very large surface area for the interchange of metabolites with the host cell cytoplasm. Contrary to the claim that parasite growth occurred by the ingestion of host cell cytoplasm through the cytostomes or micropores (68), Fawcett et al. (30, 32) found no evidence of bulk uptake of host cell cytoplasm or organelles into food vacuoles (2).

In *T. parva*, depending on the degree of initial development in the unfed adult tick, the development of the individual uninucleate sporozoites from the syncytial sporoblast generally occurs from day 3 to 4 after the start of tick feeding (Fig. 2.4). In other species the timing of sporozoite formation differs. The number of sporozoites produced during sporogony varies from between 30,000 and 50,000 per infected acinar cell for *T. parva* in female *R. appendiculatus* (30, 32) to up to 140,000 in the case of *T. taurotragi* (32, 138). There are, however, indications that for *T. parva* the numbers of sporozoites developing within an infected "e" cell in male ticks are much lower than in female ticks. *Theileria mutans* developing in the salivary gland of *A. variegatum* produces fewer (approximately 5000 per infected cell) but much larger sporozoites (1.5–2.0 μm diameter) (Fig. 2.5A) and does not induce the same degree of

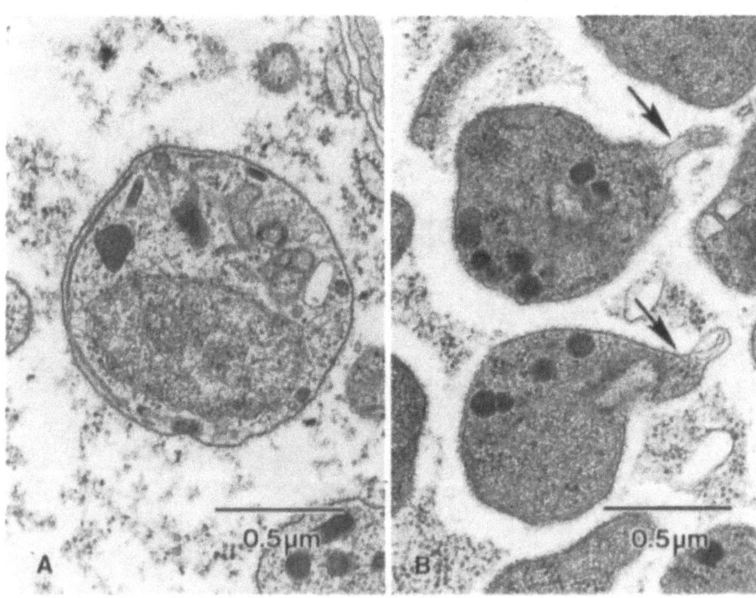

FIGURE 2.5. Electron micrographs of *Theileria* sporozoites at the same magnification to illustrate differences in size and shape. (A) *T. mutans*. (B) *T. taurotragi*. Note that *T. taurotragi* sporozoites often have a tail-like projection (arrows in B).

host cell hypertrophy as do the other *Theileria* species (12, 96, 141). The newly formed sporozoites of cattle-derived *T. parva* are uniformly ovoid or spherical and measure about 1.0 µm in diameter (Fig. 2.4). In contrast, the sporozoites of buffalo-derived *T. parva* and *T. taurotragi* are more variable in size and shape and for *T. taurotragi* many of the sporozoites have a tail-like projection (Fig. 2.5B) (32, 138).

It must be emphasized here that the majority of previous work on the development of *Theileria* parasites in the tick salivary gland has been carried out on adult ticks. However, because in the field many more larval and nymphal ticks feed to repletion as compared to adults, it is important that the biology of *Theileria* within the larval and nymphal instars should be investigated. Current studies on *T. parva* in our laboratory indicate that while the infection rates in nymphae and adult ticks are similar, the abundance and intensity of infection are highter in the adult ticks. Furthermore, many more sporozoites appear to be produced in infected "e" cells in adult ticks as compared to the nymphae. These differences in parasite development in nymphae and adult ticks appear to correlate well with differences in the size of the salivary gland between nymphae and adults, and with the different feeding behaviors of the two tick life cycle stages.

The Structure of the Mature Sporozoites

The ultrastructure of the mature *Theileria* sporozoite has been described by Fawcett and colleagues (29, 31, 32) and, with one major difference, conforms to the general description of zoites for this and other genera of the Apicomplexa (72, 105). Each sporozoite contains a single, eccentrically located nucleus that occupies one half of the zoite. Closely associated with the nuclear envelope is a single noncristate mitochondrion and a multilamellar spherical body. This spherical body, which is closely apposed to the mitochondrial membrane, is enclosed by three or more membranes and contains filamentous material resembling the nucleic acids of prokaryotic cells. Similar spherical or more elongate bodies are also closely associated with the mitochondia in the schizont, merozoite, and piroplasm stages of the life cycle. This structure resembles the spherical body described in the merozoites of malarial parasites (1, 42) and may be the source of the extrachromosomal 35 kilobase circular DNA present in both malaria parasites and other Apicomplexa (129). As yet, however, the nature and function of these structures in *Theileria* and other Apicomplexan parasites remain to be elucidated.

The sporozoite cytoplasm also contains numerous membrane-free ribosomes, a number of peripherally located, membrane-bounded secretory bodies termed microspheres (106), and a group of up to six larger, membrane-bounded rhoptries. However, unlike other genera of Apicomplexa, the sporozoites (and merozoites) of *Theileria* have no clearly defined apical complex and a conoid or similar apical structure is absent. The rhoptries, of which there may be up to six, attach to a common inwardly projected peg at the opposite pole to the nucleus, and are associated with a very much reduced inner membrane complex composed of two closely apposed membranes. No subpellicular microtubules are present and no cytostome or micropore has been seen, even though numerous cytostomes are present in the sporoblast.

Host Cell Responses to *Theileria*

The rapid development of large numbers of unicellular sporozoites within a single tick salivary gland cell invariably leads to dramatic changes in the host cell. These parasite-induced changes in the host cell occur initially with, but ultimately at the expense of the changes that normally occur as a result of tick feeding. These responses of the adult tick salivary gland host cells to parasitism by *Theileria* have been described by Fawcett et al. (30, 32).

Invasion of the host cell often results in massive hypertrophy with, in the case of *T. taurotragi*, the host cell nucleus attaining a volume calculated to be 20 times that of an uninfected "e" cell (32). Normally "e" cells of Type III acini synthesize and store large numbers of secretory granules prior to feeding. At the onset of feeding these preformed granules are released and new ones synthesised. In *T. parva*-infected cells, while there may be a preferential discharge of some of these preformed granules, as parasite development continues, host cell synthetic activity decreases as indicated by the production of smaller granules, the loss of the extensive arrays of endoplasmic reticulum, and an increase in the numbers of autophagic vacuoles. However, in the case of *T. mutans* in the salivary gland of *Amblyomma* species, recent studies in our laboratory have shown that the development of significantly fewer sporozoites clearly has a less detrimental effect on the infected salivary gland cell. Infected cells containing mature sporozoites exhibit less hypertrophy than observed in other *Theileria* infections and still contain large numbers of fully formed secretory granules (Fig. 2.6).

Accumulation of glycogen in the host cell cytoplasm is a prominent feature of sporogony in cattle-derived *T. parva* but is less frequently seen or absent in cells infected with buffalo-derived *T. parva*, and *T. taurotragi*, respectively (32).

During sporogony in *T. parva*, the gradual loss of host cell organelles is the result of the active elimination of these organelles by the host cell itself and not, as earlier light microscopy studies had suggested (see references in ref. 30), as a result of the ingestion of these organelles by the developing parasite. Throughout sporogony the host cell clearly remains viable and metabolically active, and short profiles of endoplasmic reticulum and mitochondria are found widely scattered in the extremely luscent host cell cytoplasmic matrix.

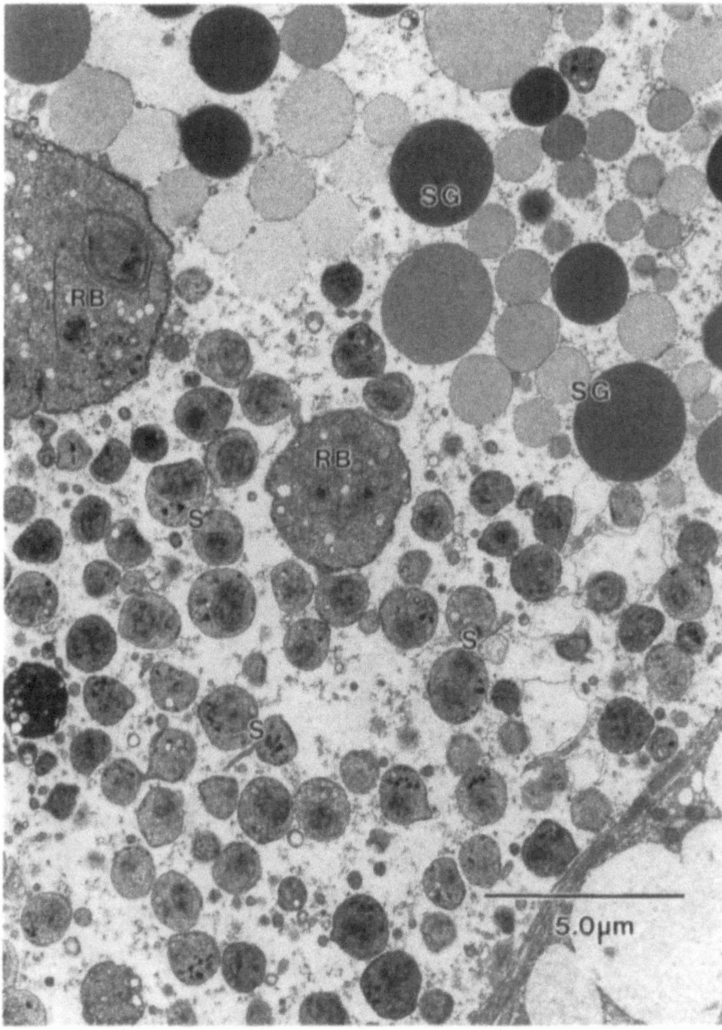

FIGURE 2.6. Electron micrograph of mature sporozoites (S) of *Theileria mutans* in the salivary gland of *Amblyomma variegatum* at the completion of sporogony. Note that the sporozoites are not tightly packed within the host cell cytoplasm and that the infected gland cell still contains numerous fully formed secretory granules (SG). RB, residual bodies of parasite syncytium that have not undergone segmental fission to form sporozoites.

Release of Sporozoites from the Salivary Gland and Infection of the Mammalian Host

The process whereby the mature, fully formed sporozoites are released from the infected salivary gland cell has not been described. However, we have some evidence to indicate that the sporozoites are released gradually

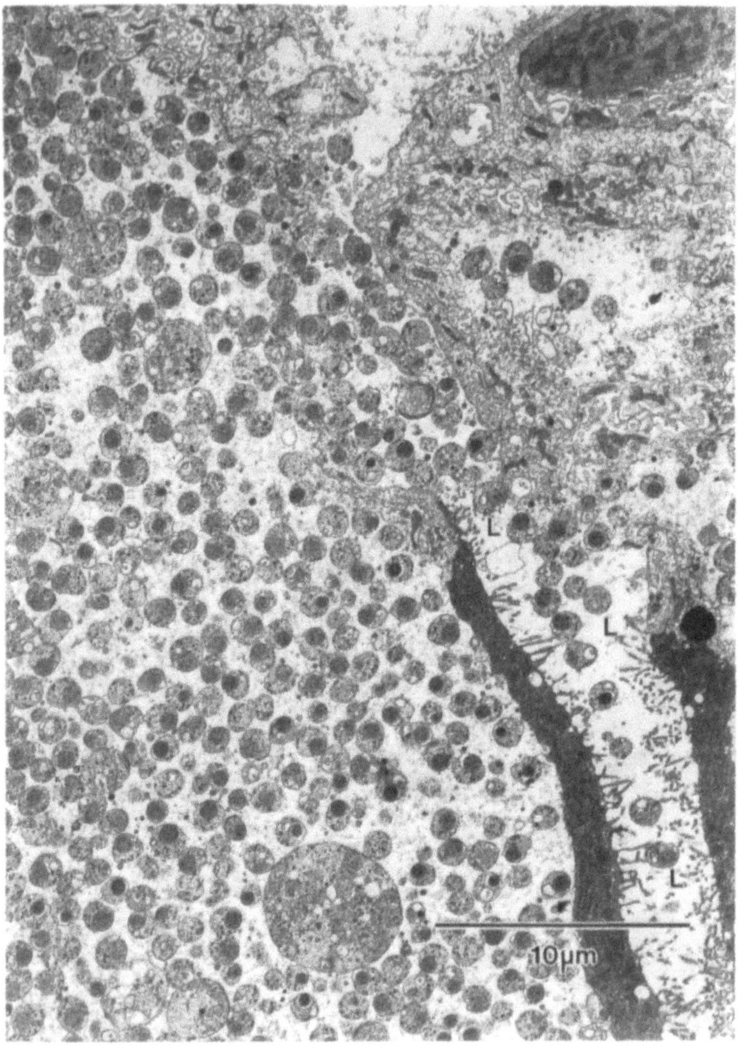

FIGURE 2.7. Electron micrograph of a *Theileria parva*-infected "e" cell in the salivary gland of *R. appendiculatus* showing the gradual release of mature sporozoites into the lumen of the acinus (L).

from an individual infected cell in a manner comparable to the release of secretory granules by apocrine secretion (Fig. 2.7).

Once the infective sporozoite has been emitted from the salivary gland during tick feeding, it enters bovine lymphoid cells. While the process of sporozoite entry into bovine lymphocytes has been described in detail (31, 106), the actual site within the mammalian host at which infection occurs remains to be determined.

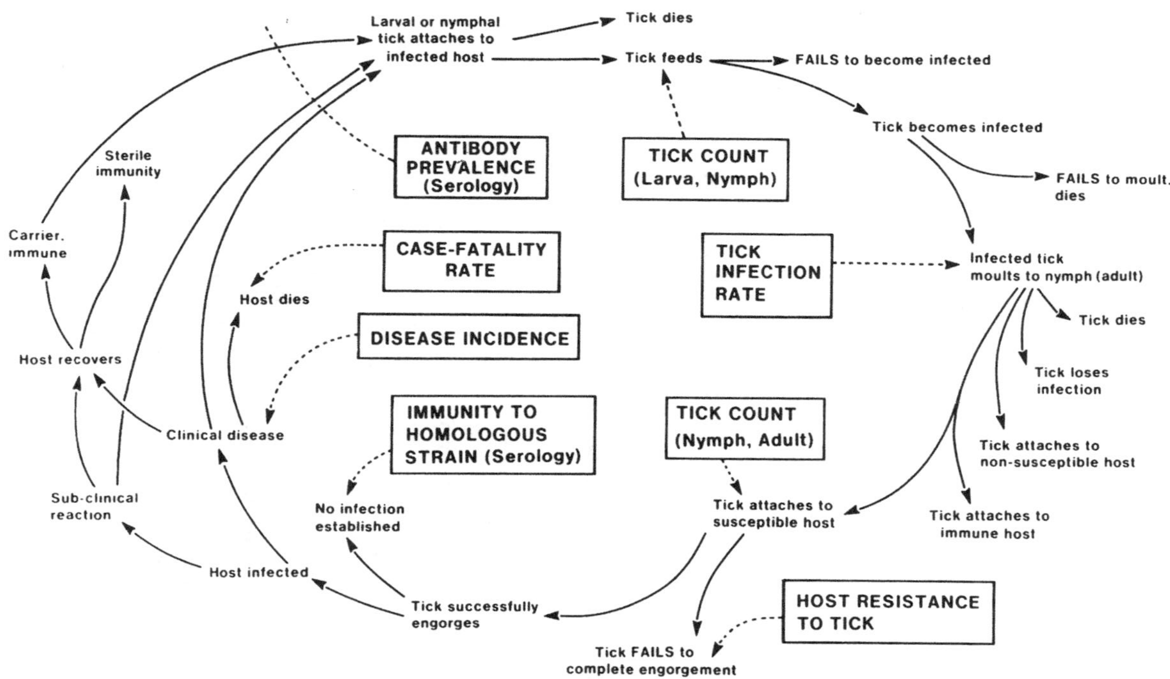

FIGURE 2.8. Dynamics of *Theileria* infections in the mammalian and tick hosts. The factors in boxes are the ones which can be readily measure. (From Norval et al. (87).)

Factors that Control the Dynamics of *Theileria* Transmission

While the factors controlling the establishment and development of *Theileria* within the individual tick are important to the survival of the parasite, the successful transmission and survival of the parasite in the field depend on a separate set of factors. Those factors that may affect the dynamics of *Theileria* infections in the mammalian and tick hosts are shown in Fig. 2.8, with, only those factors in boxes being easily measured.

There are two important interfaces between the tick and the mammalian host:

- Infection of the tick by feeding on an infected mammalian host.
- Transmission of infection by the tick feeding on a mammalian host.

However, it is the intermediate processes that control the degree of infection and transmission attainable by the tick population. The success of parasite transmission depends on several factors, including the size of the tick infestation, the tick infection rate and level of infection and the success of feeding, as well as external influences such as climate.

Dynamics of Tick Vector Populations

The dynamics of tick vector populations has been reviewed extensively by Norval et al. (87). It is unusual that more is known about the dynamics of tick vectors of *Theileria* in Africa, and in particular in eastern, central, and southern Africa, than about ticks in any other continent. The tick population dynamics drives the epidemiology of theileriosis, which in turn is driven by climate, which affects the survival, development, and behavior of the tick instars. It is the presence and abundance of the different instars infesting the mammalian hosts of *Theileria* that affect the epidemiology of theileriosis by controlling the level of transmission. To determine the population dynamics of the tick vectors appropriate techniques have to be used for their study, which may differ for both different instars and different tick species (87).

To obtain a true picture of the population dynamics of the tick vector, accurate sampling methods have to be used to quantify tick populations in the environment. This can be achieved by collecting ticks from the vegetation in a systematic manner or by attracting active, host-seeking ticks to a point where actual or simulated host stimuli are released. For example, all stages of *R. appendiculatus* and *R. zambeziensis* await hosts on the vegetation (17, 97, 107) while in *Amblyomma* species it is normally only the larvae that ascend the vegetation; the nymphs and adults of these species actively seek hosts in response to specific stimuli (87). As the unfed larvae and adults of the two-host *Hyalomma* vectors of *T. annulata* occur in farmyard walls, buildings, or in burrows it is likely that both of these stages actively seek hosts after they have been stimulated but otherwise remain concealed in favorable microhabitats (87).

Tick sampling from the host can provide information for a variety of purposes. These include comparisons of the number of ticks feeding on different host species and the age classes of hosts, estimation of host resistance to tick infestation, and the determination of patterns of seasonal occurrence. Total tick counts, however, provide only partial information and what is required is information on the number of ticks that engorge successfully. This led Wharton and Utech (128) to develop the concept of "standard" ticks, a standard tick being one that will engorge and then detach in the 24 hr following engorgement. Standard tick measurements have been developed for *R. appendiculatus* (49, 79, 88), *A. hebraeum* (88), and *A. variegatum* (49). However, there are also added difficulties in studying the levels of infestation with larvae and nymphs because of their small sizes.

The seasonal occurrence of the tick vectors in different ecological zones can be very different. Taking *R. appendiculatus* as an example, around Lake Victoria in Kenya where favorable temperature and rainfall exists all year round, trivoltine populations can exist with all instars occurring on cattle all year round (48, 93). In drier areas further away from the shore of Lake Victoria the population may become univoltine (130) due to failure of eggs to hatch during the dry period. In central and southern Africa, *R. appendiculatus* populations tend to be univoltine. This marked seasonality of each instar is controlled by behavioral diapause in the adult tick (97, 107). This diapause is stimulated by changes in photoperiod and results in the loss of host-seeking behavior so that the tick will remain on the vegetation or ground until diapause is broken. This period of diapause may last for 7 months and is thought to be a strategy for the survival of the tick during the long hot dry periods experienced in central and southern Africa. However, Berkvens (7) has suggested that in eastern Zambia, the diapausing population does not break diapause due to changes in the photoperiod but because of physiological age. Hence in southern Africa, populations of adult ticks may occur on the host for 3–4 months during the wet season ensuring that the most sensitive stages, the eggs and larvae are not killed by desiccation. Another complication is the quiescence of instars, which is induced by unfavorable environmental conditions, although the tick will became active as soon as favorable conditions return. East African tick populations do not exhibit diapause and survive in areas only where favorable conditions occur throughout most of the year.

Rhipicephalus appendiculatus is unusual among African ticks in that it is found in high numbers both on its mammalian hosts as well as on the vegetation (84, 93, 124, 131). Very heavy infestations are frequently found in small game parks in high rainfall areas (56, 59), and especially in those areas where host population densities are also high. There are several other reasons why *R. appendiculatus* becomes very abundant, in particular, it is found on a wide range of hosts, and many of these hosts do not acquire a high degree of resistance to the ticks (13, 15, 34, 82). The relationship between the abundance of *R. appendiculatus*, climate, vegetation, and host density is complex. Where ungulate populations, for example, become so high that overgrazing and environmental degradation occurs, the tick species tend to disappear. The most favorable environment for the ticks is savannah with tree cover providing a suitable microenvironment for the survival of the unfed instars.

Host Populations Dynamics

A variety of domestic and wild ungulates maintain both the *Theileria* parasites and their tick vectors. It is likely that the African buffalo (*Syncerus caffer*) was the original host for *T. parva*, *T. mutans*, and *T. velifera*, while a whole range of wild ungulates is susceptible to *T. taurotragi* infection. *Theileria annulata* and *T. buffeli* infect both cattle and asian buffalo.

Domesticated cattle (*sensu stricto*) are a relatively new introduction into Africa and cattle were unknown in southern Africa before 700 AD (25). The cattle population in Africa is heterogeneous and can be classified as humpless long horned and short horned cattle, Sanga, Zebu, and Zebu/Sanga crosses (65). In the twentieth century with European colonization of Africa, settlers from Britain, France, Germany, Belgium, and elsewhere were accompanied by the large importation and movement of cattle. Many movements were local with indigenous cattle being assembled from local sources and managed on new farm units. Indigenous cattle were used as a base for the development of commercial breeding herds, although Taurine bulls were also imported from Europe and elsewhere for breeding purposes.

The nature of cattle populations in any area has a large effect on the dynamics of tick vectors and *Theileria* population dynamics. There is clear evidence of variation in the susceptibility of different cattle types and breeds to both the tick vectors and to *T. parva* and *T. annulata* infections. Introduced cattle tend to be much more susceptible to tick infestation and to *Theileria* infection resulting in high mortality due to theileriosis. This is not seen in endemic areas where indigenous Zebu or Sanga cattle have become partially resistant to both tick infestation and theileriosis (73).

In areas endemic for *T. annulata* infection, local cattle and domestic buffalo (*Bulbulus bubulis*) tend to be highly resistant to theileriosis and the disease problem occurs when imported breeds are introduced in an attempt to increase productivity.

Dynamics of *Theileria* Populations in Ticks

Walker (120) pointed out that although hard ticks (ixodid) are often said to be efficient vectors of pathogens, this is an anthropomorphic expression that implies that ticks have become adapted for the benefit of pathogens. Ticks are undoubtedly adapted for their species survival; some of these adaptions may be to protect themselves against pathogens and some may, in turn, benefit the pathogens. Most infections of *Theileria* in ticks in the field are relatively low and do not affect the survival of the tick vector. However, Walker et al. (123) found *Hyalomma* ticks in the field in Sudan with sufficiently high infections of *T. annulata* to reduce both their feeding abilities and survival.

A fundamental difference between primary and carrier *Theileria* infections in cattle is the level of piroplasm parasitemia. This in turn affects the level of infection acquired by the feeding tick. During the primary infection high infection rates in the feeding ticks are normally produced. However, this phase of high piroplasm parasitemia in the host is usually short lived. The infections established in ticks as a result of their feeding on long-term carrier animals are likely to be more significant in determining infection rates in the tick population as a whole. In an endemically stable situation only calves exhibit primary infection while the majority of older animals are immune carriers. It is a long established feature of tick infestation that calves have lower tick numbers than older animals (4, 73). A highly tick resistant population of cattle lowers infection rates in ticks compared to a highly susceptible population (35, 55). Host resistance to tick feeding works in at least two ways: by a reduction in the size and nature of the blood meal and by reducing the success of tick feeding. African buffaloes are more efficient carriers of *T. parva* than cattle so that the infection rates of *R. appendiculatus* fed on carrier buffaloes will normally be much higher than those fed on carrier cattle (40, 62, 63, 143).

An important phase in the infection of the tick is production of gametes and syngamy. Walker (120) suggested that a series of hazards exists for the survival of *Theileria* in the tick hosts. As yet the factors that contribute to the survival of ingested parasites and that control the development of *Theileria* gametes and their subsequent syngamy are not understood. However, Young et al. (139) found that unfed nymphal *R. appendiculatus* treated with sublethal doses of irradiation developed lower infections than normal ticks when fed on *T. parva*-infected cattle. This suggested that tick reaction to *Theileria* parasites was not important for their development.

Young and Leitch (135) showed that temperatures had profound effects on the development of engorged *R. appendiculatus* nymphs as well as on the number and development of the parasites in them (Figs. 2.9, 2.10). However, in nature temperatures fluctuate diurnally which differs from

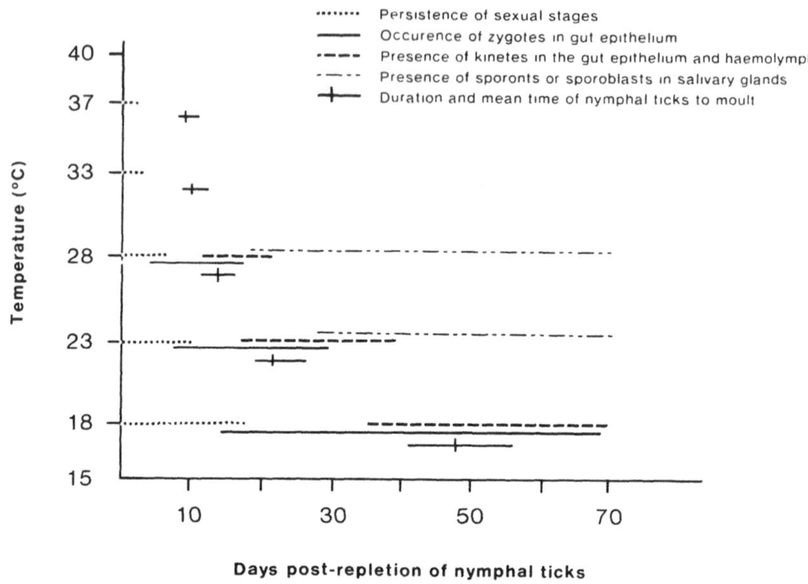

FIGURE 2.9. The rate of development of *Theileria parva* stages is dependent on temperature. (Redrawn from Young and Leitch (135).)

the constant values normally used in the laboratory. Young and Leitch (135) also obtained results that indicated that diurnal temperature rhythms produced higher *T. parva* infection rates in ticks than those produced in constant temperature studies.

An interesting feature of the parasite dynamics is the reported lack of transovarian transmission in *Theileria* infection of ticks. The life cycle of *Theileria* as it is currently understood is not adapted for transovarian transmission since, unlike *Babesia* (136), widespread merogony has not been reported in the female tick. For efficient transovarian transmission a parasite stage infective for the ovum must occur in large numbers and acquire access to the ovum of the engorged female tick. The kinete stage of *Theileria parva* has not been reported to occur in large numbers in engorged female ticks and appears to be a transient stage in the hemolymph at the time of the moult of the female tick.

There are indications that certain species and stocks of ixodid ticks are better vectors of *Theileria*. For example workers in southern Africa have reported that *R. zambeziensis* is a more efficient vector of *T. parva* in buffalo than *R. appendiculatus* (11, 54). In addition diapausing stocks of Zimbabwean *R. appendiculatus* are less efficient vectors of *T. parva* than nondiapausing stocks of *R. appendiculatus* from Kenya (47, 86). In Kenya studies have also shown that different stocks of *R. appendiculatus* from different ecological zones transmitted *T. parva* with different degrees of efficiency (52) and this could relate to differences in the biological properties of the different tick populations. This variation should allow for the selection of susceptible and refractory populations of ticks, which could be useful in determining factors that control *Theileria* infection of ticks (104).

In early studies, it had been shown by Theiler (116) in South Africa, that the infected adult tick could harbour *Theileria* infection for one year. He concluded that pasture could be cleansed of ECF by preventing animals from grazing on it for 15 months. In subsequent studies under laboratory conditions, it was shown (57, 58, 64) that *T. parva* could survive for one year in adult ticks which was slightly shorter than the survival time of the ticks themselves. In field studies in Kenya, Newson et al. (83) and Young et al. (140) showed that at an altitude of 2100 m, *R. appendiculatus* adults remained infected for virtually as long as the adult survived (up to 2 years); at 1500 m altitude, *R. appendiculatus* remained infected for 15 months (144); at sea level for nine months. These

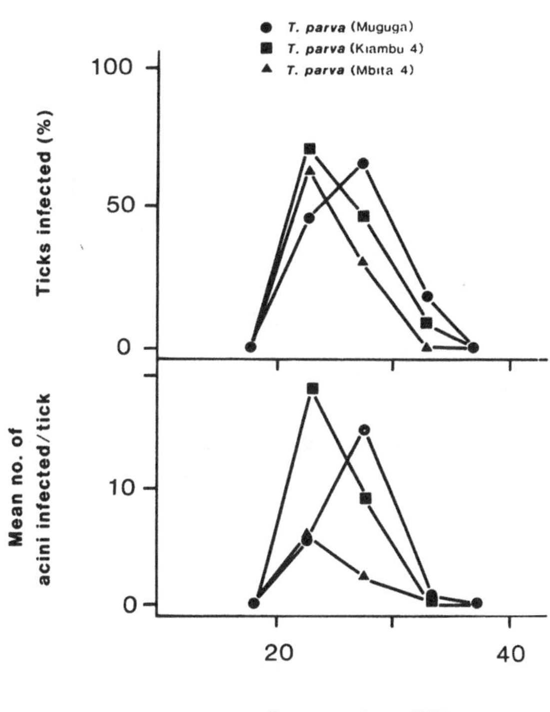

FIGURE 2.10. *Theileria* infection rates in salivary glands of ticks are dependent on incubation temperature. (Redrawn from Young and Leitch (135).)

differences in parasite survival times within their tick hosts were temperature, and possibly humidity, dependent. In contrast, nymphs and their *Theileria* infection survived for shorter periods (83).

It is well recorded in the literature that ticks transmit *Theileria* sporozoites after several days of feeding. However, Samish (49) and Young et al. (137) showed that unfed adult ticks kept for several days at 37°C developed *T. annulata* and *T. parva* sporozoites infective to cattle. Young et al. (137) demonstrated that heat stimulation was not as efficient as feeding ticks on animals and that heat stimulated sporozoites die rapidly unlike under normal ambient conditions where the sporoblasts remain dormant and survive much longer. The susceptibility of sporoblasts to heat stimulation is age dependent since older sporoblasts will not respond. Heat stimulation under natural conditions was observed by Young et al. (147) in the Trans-Mara in Kenya. Ochanda et al. (89) have demonstrated that heat stimulated ticks can transmit *T. parva* infection within 24 hr after application to cattle.

Sexual differences in the tick influence *Theileria* infection. Female ticks invariably have higher infection rates and levels of *T. parva* and *T. taurotragi* than male ticks. In *Amblyomma variegatum* males have invariably higher infection rates and levels of *T. mutans* than females. Hence the sex ratio of the adult tick population will influence transmission of *Theileria* parasites. A major difference between male and female *R. appendiculatus* is the number of "e" cells in Type III acini, which are greater in females than in males, so there are more cells available in females ticks to become infected. Sublethal irradiation of nymphal ticks reduces the number of salivary gland acini in the adult tick, which reduces the infection rate (139) supporting the idea that other tick factors have no significant role in the survival of the parasite within its vector host. Another difference between sexes in *R. appendiculatus* is that nymphs becoming males ingest more blood than those becoming females, although this does not have a direct effect on the infection rates.

Again, host resistance to ticks is likely to have a large effect on the transmission efficiency of the feeding ticks. It has been shown by Fivaz et al. (35) and Leitch (55) that tick-resistant cattle are capable of limiting the level of *T. parva* transmission when compared to tick-susceptible cattle.

Conclusions

Although our understanding of the life cycle of *Theileria* parasites within their tick vector hosts has increased significantly during the past 20 years, there are still many fundamental questions that remain to be answered if we are to understand more fully the factors involved in the transmission of *Theileria*. In particular, much of the available information on the biology of *Theileria* in ticks is concerned with the parasite in adult ticks and with nymph/adult transmission. In the field there are many more larval and nymphal ticks than adults that feed to repletion (87) and therefore comparable data on the biology of larval/nymphal transmission is required before we can fully assess the roles of the different tick instars in the transmission of *Theileria*.

While the presence of a sexual cycle in *Theileria* has been known for some time, the detailed changes occurring during the differentiation of the gametocytes within the tick gut and the physiological triggers involved in initiating these changes are poorly understood. Progress has been slow in this area because of the lack of a culture system that will support gametogenesis. However, the sexual stages in the tick gut are important subjects for intensive study since sexual recombination between strains may have important effects on the control of theileriosis. In particular, work is required to determine the factor(s) that influence the differences in the extent to which genetic recombination occurs.

The mechanics of entry of the zygote into the gut epithelium is not known nor is the method of exit of the kinete into the hemolymph and the process of entry and establishment of the kinete in the salivary glands.

An understanding of factors that control the transmission of *Theileria* is of fundamental importance in understanding the epidemiology of theileriosis. These are multifactorial and appear to be interrelated to each other so that it is difficult to separate them. New techniques such as the membrane feeding of ixodid ticks that are being perfected at present (118) will have to be applied to transmission research to help untangle these factors.

The effect of diapause in ixodid tick vector populations on the transmission of *Theileria* requires further research. For example, in nondiapausing *R. appendiculatus* from East Africa *T. parva* would die out under conditions prevailing in central and southern Africa where *T. parva* survives in the diapausing tick.

An important aspect of the development of our knowledge of the transmission biology of the *Theileria* parasite is that it will enable us to model the tick and *Theileria* parasite population dynamics with the resultant development of epidemiological models (14, 66). These epidemiology models, when fully developed, will play an increasingly important role in the effective control of theileriosis in the future.

It is expected that all these areas will receive intensified research input in the next few years and that transmission studies on the genus *Theileria* will significantly contribute to the understanding and control of *Theileria* parasites in the future.

Acknowledgments. The subject of this review is a very important part of our work and we would like to thank ILRAD staff, particularly Drs. T.T.

Dolan, S.P. Morzaria, A.J. Musoke, B.D. Perry, R.P. Bishop, Messers F.N. Mwakima, and S.N. Mwaura for their input into this area of research. Outside ILRAD, Prof. G. Gettinby, Prof. L. Tilney, and Drs. D. Berkvens, R. Pegram, R.A.I. Norval, and S.M. Waladde have been particular important as thought stimulators. We would also like to thank Ms. Elizabeth Cushny for her patience in typing the manuscript.

References

1. Aikawa, M. 1966. The fine structure of the erythrocytic stages of three avian malarial parasites, *Plasmodium fallax*, *P. lophurae* and *P. cathemerium*. *Am. J. Trop Med. Hyg.* **15**:449–471.
2. Aikawa, M., Hepler, P.K., Huff. C.G., and Sprintz, H. 1966. The feeding mechanism of avian malarial parasites. *J. Cell. Biol.* **28**:355–365.
3. Barnett, S.F. 1968. Theileriosis. In M. Ristic and D. Weinman, eds., *Infectious Blood Diseases of Man and Animals*, Vol. 2. Academic Press, New York, pp. 269–328.
4. Barnett, S.F., and Bailey, K.P. 1955. *East African Veterinary Research Organization, Annual Report 1952–53, 1954–55*. East African High Commission, Nairobi, Kenya, pp. 51–74.
5. Barta, J.R. 1991. The Dactylosomatidae. *Adv. Parasitol.* **30**:1–37.
6. Bell, L.J. 1980. Organ culture of *Rhipicephalus appendiculatus* with maturation of *Theileria parva* in the tick salivary glands *in vitro*. *Acta Trop.* **37**:319–325.
7. Berkvens D.L. 1990. A study on the ecology of the *Rhipicephalus appendiculatus* complex with special reference to the Eastern Province of Zambia. Ph.D. Thesis, Brunel University.
8. Binnington, K.C. 1978. Sequential changes in the salivary gland structure during attachment and feeding of the cattle tick *Boophilus decoloratus*. *Int. J. Parasitol.* **8**:97–115.
9. Binnington, K.C., Young, A.S., and Obenchain, F.D. 1983. Morphology of normal and *Theileria* infected salivary glands of *Rhipicephalus appendiculatus*. *J. Parasitol.* **69**:421–442.
10. Bishop, R., Sohanpal, B., Kariuki, D.P., Young, A.S., Nene, V., Baylis, H., Allsopp, B.A., Spooner, P.R., Dolan, T.T., and Morzaria, S.P. 1992. Detection of a carrier state in *Theileria parva* infected cattle using the polymerase chain reaction. *Parasitology* **104**:215–232.
11. Blouin, E.F., and Stoltsz, W.H. 1989. Comparative infection rates of *Theileria parva lawrencei* in the salivary glands of *Rhipicephalus appendiculatus* and *Rhipicephalis zambeziensis*. *Onderst. J. Vet. Res.* **54**:211–213.
12. Blouin, E.F., and Stoltsz, W.H. 1989. Development of *Theileria mutans* in the salivary glands of *Amblyomma hebraeum*. Abstracts of papers presented at the 70th Annual Meeting of the Conference of Research Workers in Animal Disease, Chicago, November 6–7, p. 66.
13. Branagan, D. 1974. The feeding performance of the ixodid tick *Rhipicephalus appendiculatus* (Neumann) on rabbits, cattle and other hosts. *Bull. Entomol. Res.* **63**:155–168.

14. Byrom, W. 1990. Simulation models for investigating East Coast fever and other parasitic diseases. Ph.D. Thesis, University of Strathclyde.
15. Cheira, J.W., Newson, R.M., and Cunningham, M.P. 1985. Cumulative effects of host resistance on *Rhipicephalus appendiculatus* Neumann (Acarina: Ixodidae) in the laboratory. *Parasitology* **90**:401–408.
16. Chen, P.P., Conrad, P.A., Ole Moi Yoi, O.K., Brown, W.C., and Dolan, T.T. 1991. DNA probes detect *Theileria parva* in salivary glands of *Rhipicephalus appendiculatus* ticks. *Parasitol. Res.* **77**:590–594.
17. Colborne, J.R.A. 1988. The role of wild hosts in maintaining tick populations on cattle in the southeastern lowveld of Zimbabwe, M. Phil. thesis, University of Zimbabawe.
18. Conrad, P.A., Denham, D., and Brown C.G.D. 1986. Intraerythrocytic multiplication of *Theileria parva in vitro*: An ultrastructural study. *Int. J. Parasitol.* **16**:223–230.
19. Conrad, P.A., Stagg, D.A., Grootenhuis, J.G., Irvin, A.D., Newson, J., Njammungeh, R.E.G., Rossiter, P.B., and Young, A.S. 1987. Isolation of *Theileria* from Afican buffalo (*Syncerus caffer*) and characterization with antischizont monoclonal antibodies. *Parasitology* **14**:413–423.
20. Conrad, P.A., Ole MoiYoi, O.K., Baldwin, C.L., Dolan, T.T., O'Callaghan C.J., Njamunggeh, R.E.G., Grootenhius, J.G., Stagg, D.A., Leitch, B.L., and Young, A.S. 1989. Characterization of buffalo-derived theilerial parasites with monoclonal antibodies and DNA probes. *Parasitology* **98**:179–188.
21. Cowdry, E.V., and Ham, A.W. 1932. Studies on East Coast fever I. The life cycle of the parasite in ticks. *Parasitology* **24**:1–49.
22. de Castro, J.J., Dransfield, R.D., Cunningham, M.P., Dolan, T.T., Newson, R.M., and Young, A.S. 1985. The effects of natural tick infestation on cattle immunized against theileriosis in an endemic area of Kenya. *Res. Vet. Sci.* **39**:279–288.
23. Dolan, T.T. 1989. Theileriosis: A comprehensive review. *Rev. Sci. Tech. Off. Int. Epizoot.* **8**:11–36.
24. Dschunkowsky, E.P., and Luhs, I. 1909. Entwicklungsformen von piroplasmen in Zecken. In *The Proceedings of the 9th International Veterinary Congress*, Den Haag, pp. 51–61.
25. Epstein, H. 1971. *The Origin of the Domestic Animals*, Vol. 1. Africana Pulishing Corporation, New York.
26. Eyles, D.E. 1951. Studies on *Plasmodium gallinaceum* 1. Characteristics of the infection in mosquito *Aedes aegypti. Am. J. Trop. Med. Hyg.* **54**: 101–112.
27. Fawcett, D.W., Doxsey, S., and Büscher, G. 1981. Salivary gland of the tick vector of East Coast fever: I. Ultrastructure of type III acinus. *Tissue Cell* **13**:209–230.
28. Fawcett, D.W., Doxsey, S., and Büscher, G. 1981. Salivary glands in the tick vector of East Coast fever: II. Cellular basis for fluid secretion in type III acinus. *Tissue Cell* **14**:231–251.
29. Fawcett, D.W., Doxsey, S., and Büscher, G. 1982. Salivary glands of the tick vector of East Coast fever: III. Ultrastructure of sporogony in *Theileria parva. Tissue Cell* **14**:183–206.

30. Fawcett, D.W., Doxsey, S., and Büscher, G. 1982. Salivary glands of the tick vector of East Coast fever: IV. Cell type selectivity and host responsiveness to *Theileria parva*. *Tissue Cell* **14**:397–414.
31. Fawcett, D.W., Doxsey, S., Stagg, D.A., and Young, A.S. 1982. The entry of sporozoites of *Theileria parva* into bovine lymphocytes *in vitro*. Electron microscopic observations. *Eur. J. Cell. Biol.* **27**:10–21.
32. Fawcett, D.W., Young, A.S., and Leitch, B.L. 1985. Sporogony in *Theileria* (Apicomplexa: Piroplasmida), A comparative ultrastructural study. *J. Submicro. Cytol.* **17**:643–655.
33. Fawcett, D.W., Conrad, P.A., Grootenhuis, J.G., and Morzaria, S.P. 1987. Ultrastructure of the intra-erythrocytic stage of *Theileria* species from cattle and waterbuck. *Tissue Cell* **19**:643–655.
34. Fivaz, B.H., and Norval, R.A.I. 1990. Immunity of the ox to the brown ear tick *Rhipicephalus appendiculatus* (Neumann), *Exp. Appl. Acarol.* **8**: 51–63.
35. Fivaz, B.H., Norval, R.A.I., and Lawrence, J.A. 1989. Transmission of *Theileria parva bovis* to cattle resistant to the brown ear tick *Rhipicephalus appendiculatus* (Neumann). *Trop. Anim. Hlth Prod.* **21**:129–134.
36. Friedhoff, K.T. 1988. Transmission of *Babesia*. In M. Ristic, ed., *Babesiosis of Domestic Animals and Man*. CRC Press, Boca Raton, FL, PP. 23–53.
37. Gettinby, G., and Byrom, W. 1989. The dynamics of East Coast fever: A modelling perspective for integration of knowledge. *Parasitol. Today* **5**:68–73.
38. Gill, H.G., and Walker, A.R. 1987. The salivary glands of *Hyalomma anatolicum*: Structural changes during attachment and feeding. *Int. J. Parasitol.* **17**:1381–1392.
39. Gonder, R. 1910. Die Entwicklung von *Theileria parva* dem Erreger des Küstenfieber der Rinder in Afrika, Teil 1. *Arch. Protistkde* **22**:143–164.
40. Grootenhuis, J.G., Leitch, B.L., Stagg, D.A., Dolan T.T., and Young, A.S. 1987. Experimental induction of *Theileria parva lawrencei* carrier state in an African buffalo (*Syncerus caffer*). *Parasitology* **94**:424–432.
41. Hazen-Karr, C.G., Kocan, A.A., Kocan, K.M., and Hair, J.A. 1987. The ultrastructure of sporogony in *Theileria cervi* (Bettencourt et al., 1907) in salivary gland of female *Amblyomma americanium* (L) ticks. *J. Parasitol.* **73**:1182–1188.
42. Hepler, P.K., Huff, C.G., and Sprinz, H. 1966. The fine structure of the exoerythrocytic cycle of *Plasmodium fallax*. *J. Cell. Biol.* **30**:333–358.
43. Hulliger, L., Wilde, J.K.H., Brown, C.G.D., and Turner, L. 1964. Mode of multiplication of *Theileria* in culture of bovine lymphocytes. *Nature (London)* **203**:728–730.
44. Irvin, A.D. 1987. Characterization of species and strains of *Theileria*. *Adv. Parasitol.* **26**:145–197.
45. Irvin, A.D., and Boarer, C.D.H. 1980. Some implications of a sexual cycle in *Theileria*. *Parasitology* **80**:571–579.
46. Irvin, A.D., Ocama, J.C.R., and Spooner, P.R. 1982. Cycle of bovine lymphoblastoid cells parasitized by *Theileria parva*. *Res. Vet. Sci.* **33**:298–304.
47. Irvin, A.D., Morzaria, S.P., Munatswa, F.A., and Norval, R.A.I. 1989. Immunization of cattle with a *Theileria parva bovis* stock from Zimbabwe

protects against challenge with virulent *T. p. parva* and *T. p. lawrencei* stocks from Kenya. *Vet. Parasitol.* **32**:271–278.
48. Kaiser, M.M., Sutherst, R.W., and Bourne, A.S. 1982. Relationship between ticks and zebu cattle in Southern Uganda. *Trop. Anim. Hlth Prod.* **14**:63–74.
49. Kaiser, M.M., Sutherst, R.W., Bourne, A.S., Gorissen, L., and Floyd, R.B. 1988. Population dynamics of ticks on Ankole cattle in five ecological zones in Burundi and strategies for their control. *Prevent. Vet. Med.* **6**:199–222.
50. Katende, J.M., Goddeeris, B.M., Morzaria, S.P., Nkonge, C.G., and Musoke, A.J. 1990. Identification of a *Theileria mutans* specific antigen for use in an antibody and antigen detection ELISA. *Parasite Immunol.* **12**:419–433.
51. Kaufman, R.W. 1989. Tick host interaction: A synthesis of current concepts. *Parasitol Today* **5**:47–56.
52. Kubasu, S.S. 1991. Biological diversity of *Rhipicephalus appendiculatus* and transmission of *Theileria parva*. M.Sc. Thesis, University of Nairobi.
53. Koch, R. 1906. Beitrage zur Entwicklungsgeschichte dur Piroplasmen. *Z. Hyg. Infektkrankheit.* **54**:1–9.
54. Lawrence, J.A., Norval, R.A.I., and Uilenberg, G. 1983. *Rhipicephalus zambeziensis* as a vector of bovine Theileriidae. *Trop. Anim. Hlth Prod.* **15**:39–42.
55. Leitch. B.L. 1989. Acquired resistance to *Rhipicephalus appendiculatus* Neumann, 1901 and its effect on transmission of *Theileria parva* to cattle. M.Sc. Thesis, University of Salford.
56. Lewis. A.R. 1981. The pathology of *Rhipicephalus appendiculatus* infection of eland *Taurotragus oryx*. In G.B. Whitehead and J.D. Gibson, eds., *Tick Biology and Control, Proceedings of an International Conference held in Grahamstown, 27–29 January 1981*. Tick Research Unit, Rhodes University, Grahamstown, pp. 15–20.
57. Lewis, E.A. 1950. Conditions affecting the East Coast fever in ticks and in cattle. *E. Afr. Agric J.* **16**:65–83.
58. Lewis, E.A., and Fotheringham, W. 1941. The transmission of *Theileria parva* by ticks. *Parasitology* **33**:251–277.
59. Lightfoot, C.J., and Norval, R.A.I. 1981. Tick problems in wildlife in Zimbabwe, I, the effects of tick parasites on wild ungulates. *S. Afri. J. Wild. Res.* **11**:41–45.
60. MacMillan, W.G., Brocklesby, D.W., and Purnell, R.E. 1971. The fine structure of tick stages of *Theileria parva*. *J. Parasitol* **57**:1128–1129.
61. McMullen, H.L., Sauer, J.R., and Burton, R.L. 1976. Possible role in uptake of water vapour by ticks salivary glands. *Insect Physiol.* **22**:1281–1285.
62. Maritim, A.C., Kariuki, D.P., Young, A.S., and Mutugi, J.J. 1989. The importance of the carrier state of *Theileria parva* in the epidemiology of theileriosis and its control by immunization. In T.T. Dolan, ed., *Theileriosis in Eastern, Central and Southern Africa*, International Laboratory for Research on Animal Diseases, Nairobi, pp. 121–128.
63. Maritim, A.C., Young, A.S., Mutugi, J.J., and Stagg, D.A. 1989. *Theileria* parasites isolated from carrier cattle after immunization with *Theileria parva* by infection and treatment. *Parasitology* **99**:139–147.

64. Martin, H.M., Barnett, S.F., and Vidler, B.O. 1964. Cyclic development and longevity of *Theileria parva* in the tick, *Rhipicephalus appendiculatus*. *Exp. Parasitol.* **15**:527–555.
65. Maule, J.P. 1990. *The Cattle of the Tropics*. University of Edinburgh Press, Edinburgh.
66. Medley, G., Perry, B.D., and Young, A.S. 1993. Preliminary analysis of the transmission dynamics of East Coast fever in eastern Africa. *Parasitology*. **106**:251–264.
67. Mehlhorn, H., and Schein, E. 1976. Elektronenmikroskopiche Untersuchungen an Entwicklungs stadien von *Theileria parva* in Darm de Überträgersecke *Hyalomma anatolium excavatun*. *Z. Trop. Med. Parasitol.* **27**:182–191.
68. Mehlhorn, H., and Schein, E. 1984. The piroplasms: Life cycle and sexual stages. *Adv. Parasitol.* **23**:37–103.
69. Mehlhorn, H., Weber, G., Schein, E., and Büscher, G. 1975. Electronenmikroskopische Üntersuchangen an Entwicklungsstadien von *Theileria annulata* im Darm und in der Hämolymphe von *Hylomma anatolium excavatum*. *Z. Parasitkde* **48**:137–150.
70. Mehlhorn, H., Schein, E., and Warnecke, M. 1978. Electron microscopic studies on the development of the kinetes of *Theileria parva* in gut of the vector tick *Rhipicephalus appendiculatus*. *Acta Trop.* **35**:123–136.
71. Mehlhorn, H., Schein, E., and Warnecke, M. 1979. Electron microscopic studies on *Theileria ovis*: Development of kinetes in the gut of the vector tick, *Rhipicephalus evertsi evertsi* and their transformation within cells of the salivary gland. *J. Protozool.* **26**:377–385.
72. Mitchell G.H., and Bannister, L.H. 1988. Malaria parasite invasion: Interactions with the red cell membrane. *CRC Crit. Rev. Oncol. Hematol.* **8**:255–310.
73. Moll, G., Lohding, A., Young, A.S., and Leitch, B.L. 1986. Epidemiology of theileriosis in calves in an endemic area of Kenya. *Vet. Parasitol.* **19**:255–273.
74. Morzaria, S.P., Irvin, A.D., Wathanga, J., d'Souza, D., Katende J., Young, A.S., Scott, J., and Gettinby, G. 1988. The effect of East Coast fever immunization and different acaricidal treatments on the productivity of beef cattle. *Vet. Rec.* **123**:313–320.
75. Morzaria, S., Young, J., Bishop, R., Young, A., Dolan, T., and Mehlhorn, H. 1992. Evidence for a sexual cycle in *Theileria parva*. In *Annual Scientific Report 1991*. International Laboratory for Research on Animal Diseases, pp. 13–14.
76. Mugera, G.M., and Munyua, W.K. 1973. A study of development stages of *Theileria parva* by electron microscope. *Bull. Epizoot. Dis. Afr.* **11**:51–66.
77. Mukhebi, A.W. 1992. Economic impact of theileriosis and its control in Africa. In R.A.I. Norval, B.D. Perry, and A.S. Young, eds., *The Epidemiology of Theileriosis in Africa*. Academic Press, San Diego, pp. 379–403.
78. Mukhebi, A.W., Perry, B.D., and Kruska, R. 1992. Estimating economic cost of theileriosis control in Africa. *Prevent, Vet. Med.* **12**:73–85.
79. Mwangi, E.M., Rinkunya, F.G.R., Ongare, J.O., and Malonze, M.M. 1983. Patterns of engorgement and the estimation of numbers of *Rhipicephalus*

appendiculatus female ticks which complete engorgement on the host. *Kenya Vet.* **7**:13–15.

80. Needham, G.R., and Coons, L.B. 1984. Ultrastructural changes in type I alveoli of the salivary glands from hydrating and desiccating lone star ticks. In D.A. Griffiths and C.E. Bowman, eds., *Acarology VI*, Vol. 1. Ellis Harwood, Chichester, pp. 366–373.
81. Neitz, W.O. 1957. Theileriosis, gonderioses and cytauxzoonoses: A review. *Onderstep. J. Vet. Res.* **27**:275–430.
82. Newson, R.M., and Cheira, J.W. 1989. Development of resistance in calves to nymphs of *Rhipicephalus appendiculatus* (Acarina: Ixodidae) during test feeds. *Exp. Appl. Acarol.* **6**:19–27.
83. Newson, R.M., Cheira, J.W., Young, A.S., Dolan, T.T., Cunningham, M.P., and Radley, D.E. 1984. Survival of *Rhipicephalus appendiculatus* (Acarina: Ixodidae) and presence of *Theileria parva* (Apicomplexa: Theileriidae) in the field. *Int. J. Parasitol.* **14**:483–489.
84. Norval, R.A.I., and Lightfoot, C.J. 1982. Tick problems in wildlife in Zimbabwe: Factor influencing the occurrence and abundance of *Rhipicephalus appendiculatus. Zimbabwe Vet. J.* **13**:11–20.
85. Norval, R.A.I., and Rechav, Y. 1979. An assembly pheromone and its preception in the tick *Amblyomma variegatum. J. Med. Entomol.* **15**:507–511.
86. Norval, R.A.I., Lawrence, J.A., Young, A.S., Perry, B.D., Dolan, T.T., and Scott, J. 1991. *Theileria parva*: Influence of vector, parasite and host relationship on the epidemiology of theileriosis in southern Africa. *Parasitology* **102**:347–356.
87. Norval, R.A.I., Perry, B.D., and Young, A.S. 1992. *The Epidemiology of Theileriosis in Africa.* Academic Press, San Diego.
88. Norval, R.A.I., Sutherst, R.W., Kurki, J. Gibson, J.D., and Kerr, J.D. 1988. The effect of the brown ear tick *Rhipicephalus appendiculatus* on the growth of Sanga and European breed cattle. *Vet. Parasitol.* **30**:140–164.
89. Ochanda, H., Young, A.S., Mutugi, J.J., Mumo, J.M., and Omwoyo, P.L. 1988. The effect of temperature on the rate of transmission of *Theileria parva* infection to cattle by its tick vector *Rhipicephalus appendiculatus. Parasitology* **97**:239–245.
90. ole MoiYoi, O.K. 1989, *Theileria parva*: An intracellular protozoan parasite that induces reversible lymphocyte transformation. *Exp. Parasitol.* **69**:204–210.
91. Pegram, R.G., Lemche, J., Chizyuka, H.G.B., Sutherst, R.W., Floyd, R.B., Kerr, J.D., and McCosker, P.J. 1989. The effect of tick control on live weight gain of cattle in Central Zambia. *Vet. Med. Entomol.* **33**:313–320.
92. Pinder, M., Kar, S.K., Mayor-Withey, K.S., Lundin L.B., and Roelants, G.E. 1981. Proliferation and lymphocyte stimulatory capacity of *Theileria*-infected lymphoblastoid cells before and after the elimination of intracellular parasites. *Immunology* **44**:51–60.
93. Punyua, D.K., Yonow, T., Newson, R.M., and Gigon, F. 1990. Studies of ticks in different ecological zones of Kenya. In *Seventeenth Annual Report 1989*. The International Centre for Insect Physiology and Ecology, Nairobi, p. 40.

94. Purnell, R.E. 1977. East Coast fever: Some recent research in East Africa. *Adv. Parasitol.* **15**:83–131.
95. Purnell, R.E., and Joyner, L.P. 1968. Development of *Theileria parva* in the salivary glands of the tick *Rhipicephalus appendiculatus*. *Parasitology* **58**:725–732.
96. Purnell, R.E., Young, A.S., Payne, R.C., and Mwangi, J.M. 1975. The development of *Theileria mutans* (Aitong) in the tick *Amblyomma variegatum* compared to *T. parva* (Muguga) in *Rhipicephalus appendiculatus*. *J. Parasitol.* **61**:725–729.
97. Rechav, Y. 1981. Ecological factors affecting seasonal activity of the brown ear tick *Rhipicephalus appendiculatus*. In G.B. Whitehead and J.D. Gibson, eds., *Tick Biology and Control*. Proceeding of an International Conference held in Grahamstown, Rhodes University, pp. 187–192.
98. Rechav, Y. 1982. Dynamics of tick populations (Acari: Ixodidae) in East Cape Province of South Africa. *J. Med. Entomol.* **19**:679–700.
99. Samish, M. 1977. Infective *Theileria annulata* in the ticks without a blood meal stimulus. *Nature* (*London*) **270**:51–52.
100. Schein, E. 1975. On the life cycle of *Theileria annulata* in the midgut and hemolymph of *Hyalomma anatolicum excavatum*. *Z. Parasitkde* **47**:165–167.
101. Schein, E. 1988. Equine babesiosis. In M. Ristic, ed., *Babesiosis of Domestic Animals and Man*. CRC Press, Boca Raton, FL, pp. 197–208.
102. Schein, E., and Friedhoff, K.T. 1978. Lichtmikroskopische Untersuchugen über die Entwicklung von *Theileria annulata* in *Hyalomma anaticum excavatum*. II. Die Entwicklung in Hämolymphe und Speicheldrüsen. *Z. Parasitkde* **56**:287–303.
103. Schein, E., Büscher, G., and Friedhoff, K.T. 1975. Lichtmikroskopische Untersuchungen über die Entwicklung von *Theileria annulata* (Dschunkowsky und Luhs, 1905) im *Hyalomma anatolicum excavatum* (Koch, 1844): I. Die Entwicklung im Darm vollgesogener Nymphen. *Z. Parasitkde* **48**:123–136.
104. Schein, E., Mehlhorn, H., and Warnecke M. 1977. Development of *Theileria parva* in the gut of *Rhipicephalus appendiculatus*. *Parasitology* **75**:309–316.
105. Scholtyseck, E. 1979. *Fine Structure of Parasitic Protozoa*, Springer-Verlag, New York.
106. Shaw, M.K., Tilney, L.G., and Musoke, A.J. 1991. The entry of *Theileria parva* sporozoite in bovine lymphocytes: Evidence for MHC class 1 involvement. *J. Cell. Biol.* **113**:87–101.
107. Short, N.J., and Norval, R.A.I. 1981. The seasonal activity of *Rhipicephalus appendiculatus*, Neumann, 1901 (Acari: Ixodidae) in the highveld of Zimbabawe/Rhodesia. *J. Parasitol.* **67**:77–84.
108. Sinden, R.E. 1983. The cell biology of sexual development of *Plasmodium*. *Parasitology* **86**:7–28.
109. Sinden, R.E. 1983. Sexual development of malarial parasites. *Adv. Parasitol.* **22**:153–216.
110. Sinden, R.E., Hartley, R.H., and Winger, L. 1985. The development of *Plasmodium* ookinetes *in vitro*: An ultrastructural study including a description of the meiotic division. *Parasitology* **91**:227–244.

111. Stagg, D.A., Young, A.S., Leitch, B.L., Grootenhuis, J.G., and Dolan, T.T. 1983. Infection of mammalian cells with *Theileria* species. *Parasitology* **86**:243–254.
112. Tatchell, R.J. 1969. The ionic regulatory role of the salivary secretion of the cattle tick *Boophilus microplus*. *J. Insect. Physiol.* **15**:1421–1430.
113. Tatchell, R.J., Chimawmi, D., Chirchir, S.J., Onga're, J.O., Mwangi, E., Rinkanya, F., and Whittington, D. 1986. A study of the justification for intensive tick control in Kenyan rangelands. *Vet. Rec.* **119**:401–403.
114. Theiler, A. 1906, Report of the Government Veterinary Bacteriologist 1904–1905. In *Transvaal Department of Agriculture, Annual Report of the Director of Agriculture, 1906*. Government Printing and Stationery Office, Pretoria, pp. 75–187.
115. Theiler, A. 1909. Transmission des spirelles and des piroplasmes par differentes éspecies de tiques. *Bull. Soc. Path. Exot.* **2**:293–294.
116. Theiler, A. 1911. Some observations concerning the transmission of East Coast fever by ticks. In *Report of Director of Veterinary Research, Department of Agriculture. Union of South Africa*. Government Printing and Stationary Office, Pretoria, pp. 47–207.
117. Uilenberg, G. 1981. *Theileria* species of domestic animals. In A.D. Irvin, M.P. Cunningham, and A.S. Young, eds., *Advances in the Control of Theileriosis*. Martinus Nijhoff Publishers, The Hague, pp. 4–37.
118. Waladde, S.M., Ochieng, S.A., and Gichuhi, P.M. 1991. Artificial-membrane feeding of the ixodid tick, *Rhipicephalus appendiculatus*, to repletion. *Exp. Appl. Acarol.* **11**:297–306.
119. Walliker, D., Quarkyi, I.A., Willems, T.E., McCulchan, T.F., Szartnman, A., London. W.T., Corcoran, L.M., Burkot, T.R., and Carter, R. 1987. Genetic analysis of the human malaria parasite *Plasmodium falciparum. Science* **236**:1661–1666.
120. Walker, A.R. 1990. Parasitic adaptions in the transmission of *Theileria* by ticks, a review. *Trop. Anim. Hlth Prod.* **22**:23–33.
121. Walker, A.R., and Fletcher, J.D. 1987. Histology of digestion in nymphs of *Rhipicephalus appendiculatus* fed on rabbits and cattle naive and resistant to ticks. *Int. J. Parasitol.* **17**:1393–1411.
122. Walker, A.R., Fletcher, J.D., and Gill, H.S. 1985. Structural and histological changes in the salivary gland of *Rhipicephalus appendiculatus* during feeding. *Int. J. Parasitol.* **15**:81–100.
123. Walker, A.R., Latif, A.A., Morzaria, S.P., and Jongejan, F. 1983. Natural infection rate of *Hyalomma anatolicum* with *Theileria* in Sudan. *Res. Vet. Sci*: **35**:87–90.
124. Walker, J.B. 1974. *The Ixodid Ticks of Kenya*. Commonwealth Institute of Entomology, London.
125. Walker, J.B. 1991. A review of the ixodid ticks (Acari: Ixodidae) occurring in southern African. *Onderstep. J. Vet. Res.* **58**:81–105.
126. Warnecke, M., Schein, E., Voigt, W.P., and Uilenberg, G. 1979. On the life cycle of *Theileria velifera* (Uilenberg, 1964) in the gut and haemolymph of the tick vector *Amblyomma variegatum* (Fabricus, 1794). *Z. Trop. Med. Parasitol.* **30**:318–322.
127. Warnecke, M., Schein, E., Voigt, W.P., Uilenberg, G., and Young, A.S. 1980. Development of *Theileria mutans* (Theiler, 1906) in the gut and

haemolymph of the tick *Amblyomma variegatum* (Fabricus, 1794). *Z. Trop. Med. Parasitol.* **30**:318–322.
128. Wharton, R.H., and Utech, K.B.W. 1970. The relationship between engorgement and dropping of *Boophilus microplus* (Canestini) (Ixodidae) to the assessment of tick numbers on cattle. *J. Aust. Entomol. Soc.* **9**:171–182.
129. Wilson, R.J.M., Gardner, M.J., Feagin, J.E., and Williamson, D.H. 1991. Have malaria parasites three genomes? *Parasitol. Today* **7**:134–136.
130. Yeoman, G.H. 1966. Field vector studies of epizootic East Coast fever. I. A quantitative relationship between *Rhipicephalus appendiculatus* and the epizooticity of East Coast fever. *Bull. Epizoot. Dis. Afr.* **14**:5–27.
131. Yeoman, G.H., and Walker, J.B. 1967. *The Ixodid ticks of Tanzania.* Commonwealth Institute of Entomology, London.
132. Young, A.S. 1981. The epidemiology of theileriosis in East Africa. In A.D. Irvin, M.P. Cunningham, and A.S. Young, eds., *Advances in the Control of Theileriosis.* Martinus Nijhoff Publisher, The Hague, pp. 38–55.
133. Young, A.S., and Leitch, B.L. 1980. A probable relationship between the development of *Theileria* species and the ecdysis of their tick host. *J. Parasitol.* **66**:356–359.
134. Young, A.S., and Leitch, B.L. 1981. Production of *Rhipicephalus appendiculatus* ticks with high infection of *Theileria parva. J. Parasitol.* **67**:751–752.
135. Young, A.S., and Leitch, B.L. 1981. Epidemiology of East Coast fever: Some effects of temperature on the development of *Theileria parva* in the tick vector, *Rhipicephalus appendiculatus. Parasitology* **83**:199–211.
136. Young, A.S., and Morzaria, S.P. 1986. Biology of *Babesia. Parasitol. Today* **2**:211–219.
137. Young, A.S., Leitch, B.L., and Omwoyo, P.L. 1979. The induction of infective stages of *Theileria parva* by exposure of host ticks to high temperature. *Vet. Rec.* **105**:531–533.
138. Young, A.S., Grootenhuis, J.G., Leitch, B.L., and Schein, E. 1980. The development of *Theileria = Cytauxzoon taurotragi* (Martin and Brocklesby, 1960) from eland in its tick vector, *Rhipicephalus appendiculatus. Parasitology* **81**:129–144.
139. Young, A.S., Leitch, B.L., Irvin, A.D., and Dobbelaere, D.A.E. 1981. The effect of irradiation on the susceptibility of *Rhipicephalus appendiculatus* ticks to *Theileria parva* infection. *Parasitology* **82**:473–479.
140. Young, A.S., Leitch, B.L., Dolan, T.T., Newson, R.M., Ngumi, P.N., and Omwoyo, P.L. 1983. Transmission of *Theileria parva* by a population of *Rhipicephalus appendiculatus* under simulated natural conditions. *Parasitology* **86**:255–267.
141. Young, A.S., Leitch, B.L., Stagg, D.A., and Dolan, T.T. 1983. Identification of *Theileria* infection in living salivary glands of ticks. *Parasitology* **86**:519–528.
142. Young, A.S., Leitch, B.L., and Mutugi, J.J. 1984. Some factors controlling the stimulation of sporogony of *Theileria parva* in its tick vector *Rhipicephalus appendiculatus. Int. J. Parasitol.* **14**:97–102.
143. Young, A.S., Leitch, B.L., Newson, R.M., and Cunningham, M.P. 1986. Maintenance of *Theileria parva* infection in an endemic area of Kenya. *Parasitology* **93**:9–16.

144. Young, A.S., Leitch, B.L., Morzaria, S.P., Irvin, A.D., Omwoyo, P.L., and de Castro, J.J. 1987. Development and survival of *Theileria parva* in *Rhipicephalus appendiculatus* exposed in the Trans-Mara, Kenya, *Parasitology* **96**:403–432.
145. Young, A.S., Mutugi, J.J., Kariuki, D.P., Lampard, D., Maritim, A.C., Ngumi, P.N., Linyonyi, A., Leitch, B.L., Ndungu, S.G., Lesan, A.D., Mining, S.K., Grootenhuis, J.G., Orinda, G.O., and Wesonga, D. 1992. Immunization of cattle against theileriosis in Nakuru District of Kenya by infection and treatment and the introduction of unconventional tick control. *Vet. Parasitol.* **42**:225–240.

3
A Model for Relationships among the Tick-Borne Encephalitis Virus, Its Main Vectors, and Hosts

Edward I. Korenberg and Yurii V. Kovalevskii

Introduction

The tick-borne encephalitis (TBE) virus was discovered in 1937 in the Russian Far East by a combined expedition headed by Professor L.A. Zil'ber. According to modern views, this virus is a widely distributed polytypic species characterized by a considerable geographic variation with respect to several antigenic and biological parameters. A major part of the virus range is located within forest zones of Russia (40, 63, 67). Here TBE has the most epidemic importance and has been studied comprehensively for more than 50 years.

Several thousand articles dealing with various aspects of TBE have been published, most of them in Russian (97). The latter is a major obstacle for many researchers, preventing them from making detailed analyses of these papers. This is also true of our views on the circulation of the virus in nature, which were published in Russian as early as 1977 (66). Discussions with many of our colleagues from America and Western Europe proved that after 15 years it would be appropriate to present a revised version of this concept. We are most grateful to Professor Kerry F. Harris for including this chapter in his series, *Advances in Disease Vector Research*.

By and large TBE has been studied reasonably well and can be taken as an example for understanding principles of natural circulation of other transmissible pathogens, specifically those ecologically associated with ticks. Now it is particularly important in the context of the Lyme

---

Edward I. Korenberg, Gamaleya Institute for Epidemiology and Microbiology, Russian Academy of Medical Sciences, 18 Gamaleya Street, Moscow 123098, Russia.
Yurii V. Kovalevskii, Gamaleya Institute for Epidemiology and Microbiology, Russian Academy of Medical Sciences, 18 Gamaleya Street, Moscow 123098, Russia.

borreliosis problem. Therefore, in describing factual data, we would highlight some points that could be missed despite a long history of studies into the pathogen ecology.

Thus, numerous facts concerning TBE epizootiology are in complete agreement with the concept of natural focality of this infectious disease, formulated during the first years of studies into this problem and described in detail in reviews of that time (109, 148). Moreover, all these data make possible the creation of a quantitative model of the virus' circulation in nature.

Basic Principles

Considerable attention has been given to various groups of hematophagous arthropods as probable TBE virus vectors. Isolation of the virus from fleas, gamasid, and trombiculid mites, mosquitoes, and other bloodsuckers obtained in natural foci has been described. Experimental studies have been undertaken concerning the capability of these arthropods to acquire, maintain, and transmit the virus. Most of these data have been summarized in several reviews (20, 87, 116, 125, 136, 140, 145). The analysis of these papers shows that the possibility of virus penetration into various parasitic organisms together with host's blood is indisputable. However, according to our concept (61), which is in agreement with views of other researchers (28, 30, 36, 48, 87, 101) , there is no conclusive evidence that any other group of arthropods except ixodid ticks plays any significant role in TBE virus circulation. In most cases when gamasid mites, fleas, mosquitoes, or other nonixodid obligatory or facultative hematophagous arthropods are capable of transmitting the virus, its dose is so insignificant as to induce no more than a latent immune response in vertebrates (117, 125, 146). We think that the latter determines the probable epizootic significance of these arthropods. Only the presence of ixodid ticks with which the virus is inextricably associated ecologically makes possible its long-term survival in a particular territory.

Natural infection with TBE virus was determined with different degrees of reliability in 18 species of ixodid ticks inhabiting different regions (61). However, only *Ixodes persulcatus* and *I. ricinus* are the principal vectors and reservoirs of the virus, thus ensuring their paramount importance in the epizootiology and epidemiology of tick-borne encephalitis. In some regions of Siberia, under certain conditions, the tick *Haemaphysalis concinna* also may play an important role in the maintenance of natural TBE foci (143). Other species should be regarded as secondary, accessory, or accidental vectors (103, 104). The TBE pathogen is a typical arbovirus of the temperate zone, which, due to several ecological factors, is found predominantly within ranges of its principal vectors (87).

TBE virus is maintained either in starved or engorged *I. persulcatus* and *I. ricinus* throughout the tick's life, even during winter. These species are capable of transmitting the virus to their progeny through eggs and from one stage to another (15, 18, 25–27, 75, 84, 85, 102, 104, 123, 124, 131), thus making them not only vectors but also long-term natural reservoirs of the virus.

Numerous data allow one to determine the individual infection rates for adult *I. persulcatus* and *I. ricinus* ticks in different parts of the TBE virus range (67, 100). A characteristic general trend is that in most regions of the Russian Far East, in all of East Siberia, in some parts of Middle and West Siberia, in Altai, mountain regions of Middle Asia, and Kazakhstan, in the Urals, as well as in the whole European part of the TBE range, the individual infection rate of ticks is similar and, as a rule, does not exceed 1 to 3%. Only in several relatively small regions of the Far East, Middle and West Siberia, and near Urals can this rate attain 10 to 15%. Higher rates of tick infection are an exception. This is a strong evidence that, despite annual, landscape, and other variations of tick infection rates (73, 142), a relatively low infection rate of adults is generally prevalent in different years and in most regions of the virus' range. On this basis we can draw two very important conclusions:

- TBE virus, as every other species, has a certain "survival norm" (or "mortality norm"); in this case, such a "norm" is not high and remains relatively stable.
- It is most probable that the virus population has a similar (i.e., evolutionarily established) basic scheme of reproduction throughout its range.

The data accumulated during more than 50 years of studies into the natural focality of TBE are conclusive evidence for the intimate association between the "life scheme" of the virus and ixodid ticks, its main vectors. Characteristic of these arthropods are a complex cycle of development, polyhostality, and obligatory host changes during metamorphosis (36). If the long-term maintenance of the virus population could be achieved at the expense of any group of host animals and any stage of tick development (i.e., if virus reproduction sufficient to provide for the very existence of its population could proceed by any of several channels functioning simultaneously or consecutively), individual tick infection rate would vary greatly. Relatively stable indices of the infection rate of adults suggest that reproduction of the virus is associated very closely with a certain stage of tick development and, thus, with animal hosts of this stage.

In this connection it would be useful to follow the relationship between infection rate in ticks at the initial and subsequent stages of metamorphosis (114; Table 3.1). The proportion of infected larvae practically does not affect the infection rate at the following stages,

TABLE 3.1. Infection rate of ticks at the initial and subsequent stages of their metamorphosis in the eastern part of Eastern Europe.[a]

Stage of metamorphosis	Year	Positive tests (%)	Stage of metamorphosis	Year	Positive tests (%)	Stage of metamorphosis	Year	Positive tests (%)
L	1961	2.3	N	1962	6.0	Ad	1963	1.0
L	1962	8.0	N	1963	7.4	Ad	1964	2.0
L	1963	0	N	1964	—	Ad	1965	1.0
N	1961	3.3	Ad	1962	0	L	1963	0
N	1962	6.0	Ad	1963	1.0	L	1964	1.3
N	1963	7.4	Ad	1964	2.0	L	1965	—
Ad	1961	7.1	L	1962	8.0	N	1963	7.4
Ad	1962	0	L	1963	0	N	1964	—
Ad	1963	1.0	L	1964	1.3	N	1965	—

[a] From Kucheruk and Korenberg (unpublished).

whereas that of infected nymphs has a noticeable effect on the percentage of resulting infected adult ticks. At the same time, the infection rate of adults affects considerably the proportion of infected ticks at stages of metamorphosis nearest to the adult state. This confirms the concept that the main natural hosts of TBE virus are vertebrates which provide nutrition for the adult ixodid ticks (59, 61); thus, the adult vector is a critically important component for virus circulation.

Analysis of Former Schemes for TBE Virus Circulation

Numerous attempts have been made to construct a scheme for TBE virus circulation in natural foci, in other words, a "life scheme" of the virus (17, 36, 47, 109–113, 120). Several such schemes were included in textbooks and became commonly accepted. Closer to the heart of the matter are those emphasizing intimate association of the very existence of the virus with the life scheme of its main vector. Particularly representative in this respect is the scheme shown in Fig. 3.1.

A major drawback of such schemes, as well as of most schemes proposed for circulation of infections with natural focality, is the lack of a quantitative approach to the analysis of trends characteristic of death and reproduction of the pathogen at various links in the epizootic chain. In such schemes, this process is portrayed as a kind of perpetual mobile, whereas, in fact, it is a complex interaction of populations in a parasitic system (13) that is based on the ecological regularities of their dynamics. We regard the epizootic process as a continuous interaction of the pathogen population with its natural hosts, vectors, and environment providing for the existence of a pathogen (64). Only if fluctuations of the pathogen population size are taken into account, can any scheme provide the bases for estimating the significance of different links in the epizootic chain and for modeling the epizootioly and forecasting its intensity. To create such a scheme for obligatorily transmitted zoonoses is even more difficult, since the "life scheme" of the pathogen appears to be inserted into that of its main vector. The existence of the vector (and, eventually, survival of the pathogen) is provided for by its hosts. Pathogen population size can vary during vector metamorphosis and, moreover, is determined by the availability of susceptible animal hosts for the vector (30). More important, in this case, is pathogen transmission from an infected arthropod specimen to numerous noninfected ones through a susceptible animal. In addition, recent experiments show that TBE virus, similarly to other arboviruses (45), can apparently be transmitted from infected to normal ticks during their simultaneous feeding on vertebrates, even when viremia is insignificant or absent (1, 2, 37).

With respect to TBE virus, the above data suggest that the scheme of its circulation should reflect fluctuations of its population size resulting from the mortality rates of ticks, *I. persulcatus* or *I. ricinus*, at different stages of metamorphosis, the frequency (probability) of transovarial and transstadial virus transmission, as well as of virus reproduction or death in ticks, and, finally, the probability of virus transmission through vertebrates at different stages of vector metamorphosis. Despite a long history of TBE studies, the foregoing parameters cannot all be properly evaluated due to the lack of corresponding data. Hence, the concepts

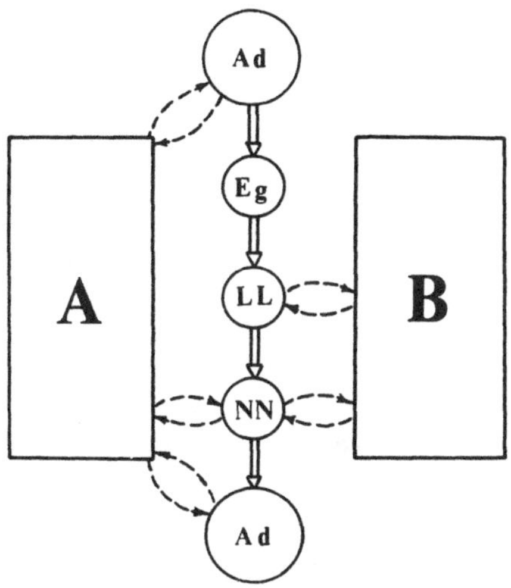

FIGURE 3.1. A scheme for TBE virus transmission during tick development (from Petrishcheva, 1958; see ref. 115). Ad, adult ticks; Eg, eggs; LL, larvae; NN, nymphs. A, hosts of adult ticks and (partly) nymphs (large animals); B, hosts of larvae and (partly) nymphs (rodents and other small mammals, birds). Solid arrows show vertical virus transmission and dashed arrows show virus transmission through hosts.

presented below, although based on factual materials, are, to a considerable extent, speculative.

Generation Balance of the Main Virus Vector

General Remarks

Special studies and reviews (7, 35, 36, 39, 86, 88, 89, 139, 147) provided a basis for the current comprehensive concept of the perennial developmental cycles of *I. persulcatus* and *I. ricinus*. It is not our purpose to describe these cycles. Instead, we will just consider mortality or, more precisely, survival rate of ticks at different stages of their developmental cycle. Such parameters as survival rate of fed females, female productivity, survival of egg layings, and the number of resulting larvae, survival rate of starved and fed larvae or nymphs, as well as starved imago, are critical for the evaluation of general quantitative principles of population reproduction. Certainly, all these parameters are variable, and the mechanisms responsible for changes in a tick population density in different years are based on this variance. However, we think that corresponding mean values can be used for overall qualitative characterization of tick life cycles.

The Number of Surviving Fed Females

The mean values of survival rate are 70 to 90% for fed *I. persulcatus* females (7, 45, 80, 128, 141, 147) and 60 to 87% for *I. ricinus* females (23, 24). Sometimes parasite invasion (5) or other factors result in death of up to a half of fed adult ticks, but such phenomena hardly affect any widespread area at a time. Mortality is the most significant in females fed during the late summer (35, 45). These results, as well as most data presented below, were obtained mostly during studies on the fate of ticks or their eggs deliberately placed into the ground to a certain depth. Apparently, the values obtained are somewhat higher than the real ones, since it was impossible to take into account natural death of engorged females that get into an adverse environment after separating from the host. Tentatively we assume the survival rate of engorged females in nature to be 75%. Thus, if the number of the fed females is n [1],* oviposition would be observed for 0.75 of n [2].

---

*Here and below the numerals in square brackets denote the number of corresponding stage (from the top) in Fig. 3.2.

Egg Production

An *I. persulcatus* female is capable of laying 350 to 5500 eggs (54, 95, 118, 119, 128). Approximately the same number is laid by *I. ricinus* (10, 52, 53, 88). Egg production depends on the weight of the engorged female, i.e., on the amount of blood consumed (10, 36, 118) and several other factors. On the basis of published data, we assume that the average number of eggs within an egg laying is 2500. This value corresponds to the number of eggs laid by a fed female of average weight (10, 36). Hence, the overall egg production of a given population would be

$$2500n \times 0.75 = 1875n \qquad [3]$$

The Number of Surviving Eggs

Due to certain factors, a part of egg layings perishes completely. In *I. persulcatus*, this is observed more often with eggs laid in late summer that die during wintering (36). As a whole, from 70 to 94% of either *I. persulcatus* or *I. ricinus* egg layings are retained (7, 23, 53, 54, 148). In further calculations we assume that the average rate of survival of egg layings is 85%. Therefore, the number of eggs survived after total death of a certain part of egg layings would be equal to

$$1875n \times 0.85 = 1600n \qquad [4]$$

The Number of Unfed Larvae

Apparently, larvae do not develop from all the eggs of the surviving egg layings, since some of remaining eggs should be dead or unviable, According to some observations, from 3 to 5% of eggs can be lost from egg layings during no more than 1 month (77). In *I. ricinus*, larvae emerge from 85–93% of eggs in egg layings (24). More extensive data for estimating the extent of partial loss of eggs in surviving batches are not available at present. Being aware of possible overestimates, we, nevertheless, assume that larvae emerge from 90% of eggs. Consequently, all fed would produce approximately the following number of unfed larvae:

$$600n \times 0.9 = 1440n \qquad [5]$$

The Number of Fed Larvae

It is known that unfed larvae represent the most vulnerable active stage of tick development. Part of them already dies during wintering (7). The main bulk of unfed larvae becomes active in the first half of the following summer. By that time about 300 to 600 larvae, on the average, appear at the site of *I. persulcatus*' egg laying; larvae that cannot find a host survive for no more than a few days (80). In *I. ricinus*, 51 to 92% of unfed larvae are thus lost (24, 139). According to Z.M. Zhmaeva (personal communication), no more than 10% of all larvae succeed in bloodsucking under natural conditions. With such a survival rate, the total number of fed larvae would be

$$1440n \times 0.1 = 144n \qquad [6]$$

The Number of Unfed Nymphs

Engorged larvae can accomplish molting into the next stage during the feeding season. The resulting nymphs, as a rule, enter wintering being unfed. However, larvae fed during late summer usually enter diapause and molt into nymphs only a year later (36, 39, 72). The proportion of larvae developing without diapause varies considerably but most often it is 45 to 70% of the total (45, 98). On average, half of the larvae develop without and half with diapause (7). Mortality rates in these groups are different. In *I. persulcatus*, the proportion of surviving larvae which develop without diapause is 14 to 86% (7, 141), with the average value close to 55%. The more abundant data on the mortality of *I. persulcatus* larvae that enter diapause suggest that their survival rate is even less, i.e., 20 to 35% (7, 98, 129, 141, 147). This parameter for fed *I. ricinus* larvae wintering in diapause under "mild" conditions is about 40% (34), whereas, in a severe climate, almost all larvae die (6, 79). In terms of our general scheme it can be assumed that no more than 30% of larvae after diapause would molt into nymphs. Thus, the total number of unfed nymphs developing from all fed larvae can be expressed as follows:

$$144n \times 0.5 \times 0.55 + 144n \times 0.5 \times 0.3 = 60n \qquad [7]$$

The Number of Fed Nymphs

The mortality rate of unfed nymphs remains unknown. According to common views, they are significantly more durable and viable than larvae. However, mortality of nymphs due to various reasons remains significant. On this basis we assume that an average of 20% succeed in bloodsucking and their total number is

$$60n \times 0.2 = 12n \qquad [8]$$

The Number of Unfed Adult Ticks

Depending on the time of feeding, nymphs, as well as larvae, can enter diapause or not during their subsequent development. Under various conditions, their development in 20 to 90% of cases proceeds without diapause (6, 7, 98, 141). On the average, similar to larvae, about half of nymphs do not enter diapause and about 80% of them usually survive (42, 141, 147). Mortality during diapause and immediately after it is significantly higher, and the survival rate does not exceed 25% (7, 129). Such data suggest that the approximate number of unfed adult ticks developing from fed nymphs would be as follows:

$$12n \times 0.5 \times 0.8 + 12n \times 0.5 \times 0.25 = 6n \qquad [9]$$

The Number of Adult Ticks Giving Rise to a New Generation

It is known that 42 to 80% freshly molted *I. persulcatus* adults and 50 to 85% of those of *I. ricinus* survive after wintering (6, 34, 45). These values, however, reflect natural mortality of unfed adults only partly, since in summer most surviving ticks die, not finding a host. Apparently, no more than 30% of imagos survive and give rise to a new generation i.e.,

$$6n \times 0.3 = 2n \qquad [10]$$

The number of males and females of primary TBE virus vectors, as in many other ixodid tick species (60), is similar (8, 36, 65). Therefore, a

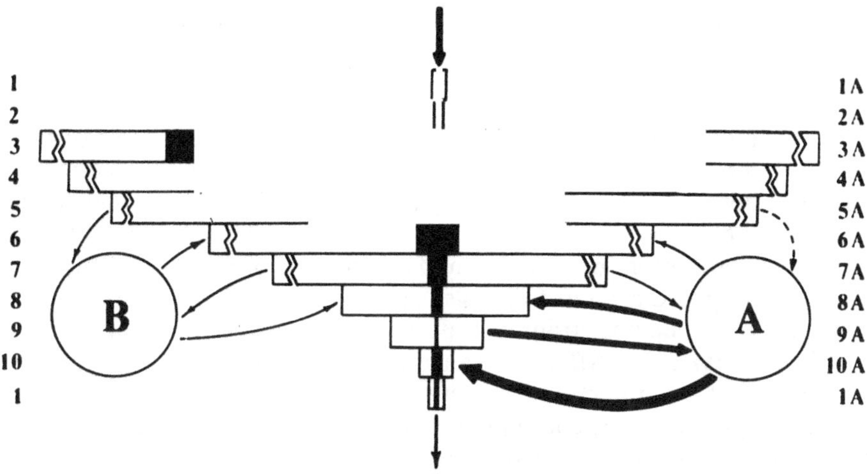

FIGURE 3.2. A scheme of changes in the number of infected ticks during their metamorphosis. Rectangles show population size of the main vector generation at different stages of its development: 1, fed females; 2, fed females after egg laying; 3, the number of eggs (egg production of a given generation); 4, the number of eggs after partial loss of egg layings; 5, unfed larvae; 6, fed larvae; 7, unfed nymphs; 8, fed nymphs; 9, unfed adults; 10, adults giving rise to new tick generation. Black parts of rectangles (1A–10A) show proportion of infected specimens at the same stages. Circles show hosts of both adult ticks and immatures (A) or only immatures (B). Arrows show directions and probable intensity of virus transmission.

new generation would originate from females whose number (n_1) in our calculations is equal to n. Such a situation is quite feasible, although the above data mostly concern only general quantitative parameters of *I. persulcatus* and *I. ricinus* reproduction (Fig. 3.2). In fact, n_1 and n values differ from one another in natural populations, due to variability of the mortality rate at different stages of tick development. However, the difference should be relatively small, since sharp (by dozens and hundreds times) annual fluctuations of adult population size are unusual in these species (36).

Unfortunately, numerous data about the abundance of unfed larvae, nymphs and imagos are, as a rule, hard to compare due to fundamental differences in techniques used for its determination. According to results obtained in forests of the Far East, the density of unfed adults is only 0.6% that of unfed larvae (81). This is in general agreement with the described generation balance and confirms its reliability. Apparently, less than 0.1% of each generation as a whole usually survives under natural conditions (89).

The Virus Population Balance in Relation to the Development of the Main Vector Generation

General Remarks

The next purpose is to follow changes in the number of infected ticks of a certain generation during metamorphosis. However, quantitative characteristics of transovarial and transstadial virus transmission still remain to be studied, and experimental results obtained by different researchers are often inconclusive and can be interpreted differently. The concepts proposed are most divergent and sometimes incompatible, e.g., ranging from denying transovarial virus transmission to accepting it absolutely (25, 115). After analyzing these facts, we suggested that the probability of tick infection, the rate and degree of virus propagation in the tick organism, and, consequently, the possibility of transovarial virus transmission depend on the physical condition and physiological age of these arthropods (59). This hypothesis was confirmed in experiments (58, 93). In addition, it has been shown that the intensity of TBE virus reproduction and death markedly depends on the temperature and relative humidity of the environment (91, 92). Since experimental conditions in different studies were not standard, diversity and incompatibility of the results obtained can be easily explained but their comprehensive evaluation in retrospect seems impossible. Hence, our further calculations will be based on only those experimentally obtained indices that we consider equivalent to natural circumstances.

The Number of Infected Fed Females

On the basis of the above data, we arbitrarily assume that in unfed adult ticks the individual infection rate under natural conditions is 3%. Taking into account reported indices of abundance of adult ticks on their hosts, population serological studies providing evidence concerning probable infection rates of animal hosts (69, 76), and the resulting concepts about the role of large animals in TBE epizootiology (59, 61), we suppose that, due to pathogen transmission through vertebrate hosts, the infection rate of fed adult ticks is 10 times as high, i.e., 30%. The Validity of such an assumption will be discussed later. Thus, if our quantitative analysis of the tick life scheme was initially dealing with n fed females [1], then $0.3n$ of them should be viruliferous [1A], and eggs will be laid by $0.3n \times 0.75 = 0.22n$ [2A] of the latter.

The Number of Infected Eggs

It is doubtless that TBE virus can be transmitted transovarially (102). However, this process can be completed only if the concentration of the pathogen in the tick's hemolymph is sufficient to penetrate the "ovary barrier" (30, 124), as has been shown for some other arboviruses and rickettsia (12, 19, 132). Depending on the dose of the virus acquired by a female tick, the incidence of transovarial virus transmission under experimental conditions can vary from 0 to 100% (55–57). However, even when females receive a high dose of the virus, its transovarial transmission appears to be realized only if a certain period of time passes between acquisition of the virus and egg laying. This period is necessary for accumulation of the virus in the tick's hemolymph and reproductive organs; its minimal duration remains unknown One of the mechanisms responsible for synchronization of the complex life schemes of *I. persulcatus* and *I. ricinus* is that eggs are always laid during the same season (36). This is achieved due to the highly variable duration of oogenesis (10) which in *I. persulcatus* can vary from 2–3 to 40–50 days, depending on the date of the bloodmeal (36, 147). Thus, it can be suggested that the probability of transovarial virus transmission in ticks infected during feeding in early summer (with long period of oogenesis) is higher than in ticks having late bloodmeals (59, 61). Virus amplification in ticks and its transmission to their progeny are more successful when the virus enters the vector organism at the beginning of feeding (3). It is also possible that parameters of transovarial (as well as transstadial; see below) virus transmission depend on the nature of a given virus strain (31) and its clonal structure (138). These factors can account for inconsistent experimental data on the frequency of transovarial transmission that were obtained by different authors by studying eggs of infected females (15, 134) or by testing unfed larvae (33, 38, 43). After analyzing corresponding papers, we assumed in our calculations that transovarial virus transmission under natural conditions is observed with no more than 25% of the infected females; moreover, only 2 to 45% of eggs within an egg laying are infected (11, 134). Furthermore, according to experiments with Powassan virus (21, 22), the pathogen can get on the egg surface immediately before its laying, together with the substance covering the egg and secreted by Gene's organ. Such virus, detectable by common virological testing, seems incapable of penetrating into developing larvae. It is probable that each infected egg laying contains about 25% infected eggs. Therefore, their total number would be

$$0.22n \times 2500 \times 0.25 \times 0.25 = 34.4n \qquad [3A]$$

since a part of egg layings does not survive, the number of viable infected eggs would be

$$34.4n \times 0.85 = 29.2n \qquad [4A]$$

The Number of Infected Unfed Larvae

According to several authors, embryogenesis and larval development are periods adverse for perpetuation of the virus (55, 57). Indeed, some eggs become free of the virus; however, the extent of this process could not be determined, since there is no conclusive data about the infection rate of the eggs at different times after laying. Nor can the results of virological analyses of unfed larvae obtained from infected females (38, 43, 49, 144) provide any positive impression of their individual infection rates, since the studied pools were not standard. Under natural conditions, the virus is found in 4 to 81% of such pools formed by progeny of the same female on different days, along with activation of larvae. This percentage does depend on the total bulk of the virus found in all larvae developing from a given egg laying (82). The data available suggest that the virus is retained and transmitted to larvae in approximately 60% infected eggs. Therefore, the sum total of infected unfed larvae produced by fed infected females would be

$$29.2n \times 0.9 \times 0.6 = 15.8n \qquad [5A]$$

This is in complete agreement with factual data and concepts of other authors, which suggest that the individual infection rate of larvae is from fractions of 1% to several percent (43, 102, 135).

The Number of Infected Fed Larvae

Each infected unfed larvae contains a very small dose of the virus, as has been shown by pool titration experiments (57, 82). It has also been demonstrated that ticks at the larval stages poorly accept the virus and often get rid of it (3, 36, 78). Apparently, virus survival in larvae is the most critical period of its "life scheme." Hence, taking into account the probable indices of abundance of larvae on small mammals, we suggest that the infection rate of larvae hardly increases by a factor of more than 1.5 as a result of virus transmission through hosts. Consequently, the number of infected fed larvae would be

$$15.8n \times 0.1 \times 1.5 = 2.4n \qquad [6A]$$

The Number of Infected Unfed Nymphs

The regularity of transstadial TBE virus transmission in ticks is indisputable. During studies performed with tick pools of different size, virus transmission from larvae to nymphs was registered in 40 to 100% of cases (15, 16, 30, 78, 124). Apparently, we can assume that 75% of fed, infected larvae are capable of transmitting the virus to nymphs. Thus, the number of unfed infected nymphs would be as follows:

$$2.4n \times 0.5 \times 0.55 \times 0.75 + 2.4n \times 0.5 \times 0.3 \times 0.75 = 0.77n \qquad [7A]$$

The Number of Infected Fed Nymphs

It appears that *I. persulcatus* and *I. ricinus* nymphs have the widest range of hosts. They feed on animals of most different size (36). A significant number of nymphs feed on animals of intermediate and large size, together with adult ticks (4, 41, 69, 90, 94, 120, 121, 127). It is very important for the epizootic process that the virus can be transmitted through vertebrate hosts from nymph to nymph, as well as, with similar or even greater probability, from adults to nymphs (69). Hence, taking into account the rate of nymph infestation of medium and large animals, the duration of bloodsucking, and the amount of blood ingested, we believe that infection of nymphs during feeding should take place far more regularly than that of larvae. In our calculations we assume that the number of infected fed nymphs, as compared to that of infected unfed nymphs which found their hosts, increases three to four times:

$$0.77n \times 0.2 \times 3.5 = 0.54n \qquad [8A]$$

The Number of Infected Adult Ticks

According to experimental data, the average incidence rate of transstadial virus transmission from nymphs to adults is about 75% (15, 30, 78). Consequently, under natural conditions all the fed infected nymphs can develop into the following number of unfed infected adult ticks:

$$0.54n \times 0.5 \times 0.8 \times 0.75 + 0.54n \times 0.5 \times 0.25 \times 0.75 = 0.2n \qquad [9A]$$

A significant proportion of such adults dies. However, those that find a host will transmit the virus through it to a certain number of noninfected adult ticks. The resulting number of infected adult ticks would be

$$0.2n \times 0.3 \times 10 = 0.6n \quad [10A]$$

Half of them are females from which starts a new sequence of similar events in the virus life scheme.

Major Ecological Aspects of the TBE Virus Life Scheme

It is known that transovarial virus transmission is accompanied by a sharp decrease in virus titer (15, 55, 57). The concentration of the virus in eggs is extremely low. The virus can often be detected not in eggs but only in developing unfed larvae where the virus titer can be higher (15, 33, 144). These facts, as well as several direct observations (92), suggest that virus amplification, to a certain extent, takes place already during larval development after ecdysis. It is also observed at each stage of development after bloodmeals (15, 131). Although transstadial molting is accompanied by a decrease in virus titer (15), the period of metamorphosis as a whole ensues in amplification of the virus (15, 49). These experimental results are in good agreement with the results of virus titration in ticks collected in natural foci. As a rule, maximum titers and the total virus content in unfed and fed adult ticks are significantly higher than in unfed nymphs, and higher in nymphs than in unfed larvae (68, 70, 74, 82, 83).

Thus, two natural processes can be distinguished with opposite effects on the virus population: during the developmental cycle of each tick generation, the absolute number of infected ticks gradually decreases (Fig. 3.2); however, in an individual organism developing from larva to fed adult tick, the virus titer shows gradual increase (59, 61).

In any natural TBE focus, both infected and not infected unfed ticks are found that can feed either together or separately. Ticks of different generations and stages of development feed on vertebrate hosts without viremia or with various intensities of it. Eventually, this determines virus titers immediately after feeding in ticks of all developmental stages (68). The second group of factors affecting the infection rate is associated with the developmental cycle of ticks, as well as with quantitative aspects of transovarial and transstadial virus transmission. These questions were discussed in detail previously. Ticks can acquire the virus at different developmental stages. During the following developmental cycle (with or without diapause), the virus content in different ticks of the same generation is subject to change. Possible individual differences among ticks of a given population in the ability to acquire and transmit the virus during metamorphosis cannot be excluded either. In nature these factors are closely related. Consequently, actual interaction of host, vector, and virus populations in the natural focus involves a complex of conditions affecting the possibility of virus survival and maintenance in a given vector. As a result, the virus concentration in a tick organism at any stage of development can vary considerably (68, 70, 74). For example, the content of the virus in one adult unfed *I. persulcatus* can be from 2 to 9.7 $\log_{10}$ PFU (70). Thus, such a phenomenon as different individual infection rates in vectors of the same population should be considered quite natural.

Annual changes in the relation between the number of ticks with different virus content within a given population are also characteristic. It has been shown that the overall virus population size in the focus depends mainly on highly infected ticks. Their absolute and relative number varies, but such ticks are found every year (70, 71). In some years the total number of infected ticks increases. Most of them, however, carry a small dose of the virus, and such an increase in numbers hardly affects the total bulk of the virus maintained in ticks. Apparently, this is a mechanism responsible for stability of the virus population and its relative independence from changes in population size of vectors and carriers, i.e., other components of a parasitic system. Such independence is critically important for the very existence of TBE, an obligatorily transmitted virus closely associated with the developmental cycle of ticks. Such a mechanism also provides for the possibility of periodic wide dissemination of the virus in the natural focus (64, 70). We think that all these facts are important for understanding the ecological background of functioning inherent in the quantitative model proposed here.

Analysis of the Model

Let us consider this simplified model of the virus' life scheme (Fig. 3.2). We omit our calculations, which were made using elementary arithmetic operations, but they suggest that the accepted parameters of the model provide for its considerable stability, i.e., for the possibility of long-term maintenance and transmission of the pathogen, even if the initial infection rate of the fed adult ticks is significantly lower. At the same time, one of prerequisites for the survival of the pathogen population is its transmission through a vertebrate host from one adult tick to another. If, according to some authors (17, 36, 38, 96, 113, 130), we assume the absence of such transmission (hence, of the transmission from adults to nymphs as well), the infection rate of fed larvae, compared to unfed ones, should increase at least 35 to 40 times, which is improbable. If we exclude virus transmission through vertebrates to preimaginal stages, the model would be stable only under the condition that the infection rate of fed adult ticks is 25 times greater than that of unfed ones, which is unlikely as well. Therefore, taking into account the actual infection rates of the fed adults, the number of infected ticks should increase about 5-fold during at least one of the preimaginal stages of development. As we have already shown, the principal role here is probably played by nymphs.

It is apparent that, in contrast to the previous concept (25), long-term maintenance of TBE virus population via only transovarial and transstadial virus transmission is impossible (14, 36, 102, 104). In principle, a vector, by itself, can retain a pathogen only under the conditions of highly effective transstadial and transovarial transmission in conjunction with transspermal transmission and several other unlikely circumstances (122). According to some data, sexual transmission of TBE virus in ticks is a fairly common phenomenon (32). However, the index of transovarial transmission is very low (see above).

Without periodic extensive contribution to the population of infected ticks, the natural circulation of TBE virus would cease. Such a contribution is made by regular tick infection during their feeding on vertebrate hosts (Fig. 3.2). According to recent experimental data (1, 2, 37, 44), high viremia of hosts is not a prerequisite in this case. Therefore, common views on the role of different vertebrates in horizontal TBE virus transmission that are based on the probability and intensity of viremia in these animals (36, 62, 99, 105) should now be accepted cautiously.

We attach much importance to transmission of TBE virus to hosts of the adult ticks (59, 61, 69) and share the opinion about the specific evolutionary significance of such hosts for establishment of virus specificity in relation to the vertebrate host (29). Those animals that serve as proper hosts for either adult ticks or nymphs, i.e., hedgehogs, hares, roe, deer, etc., appear particularly important for the maintenance of the epizootic process. Apparently, they provide a shorter TBE transmission pathway in addition to the conventional vertical (transovarial and transstadial) ones, involving pathogen transmission from infected adult ticks to nymphs and vice versa (17, 69, 115, 120). Probably, larvae can be infected in the same way while feeding on a vertebrate host together with nymphs, adult ticks, or both.

We deliberately simplified the scheme shown in Fig. 3.2 to make this picture more clear. It reflects the fate of the virus relative to the fate of a single tick generation. In fact, ticks of various stages feeding together on the same host usually belong to different generations. Hence, one tick generation acquires the virus from another one not only through transovarial transmission but also through vertebrate hosts. The latter provide an especially effective pathway of pathogen transmission, bringing together vector generations of at least three successive years. Hence, Fig. 3.2 shows only one module of the epizootic chain. Its more detailed "block-model" should include at least two similar modules showing changes in the infection rate in generations produced by ticks of the preceding and the following year.

Principles of Dynamics of the Epizootic Process

We already noted while describing the life schemes of the virus and its vectors that the abundance of ticks and their infection rate are most dependent on a particular variant according to which the cycle of tick development is realized. This aspect is of critical importance for understanding the principles and mechanisms determining the dynamics of the epizootic process and should be discussed in more detail. *I. persulcatus* and *I. ricinus* populations usually consist of ticks belonging to different generations (7, 9, 10, 36, 60, 108). This is due to the fact that ticks developing from the same egg laying, depending on a complex of factors, can reach the adult stage within an interval of several years. The mortality rate of a given generation is strictly dependent on the ratio of larvae and nymphs developing without diapause to those entering diapause. In complex, all these factors result in considerable annual variability of the abundance and proportion of different age groups within

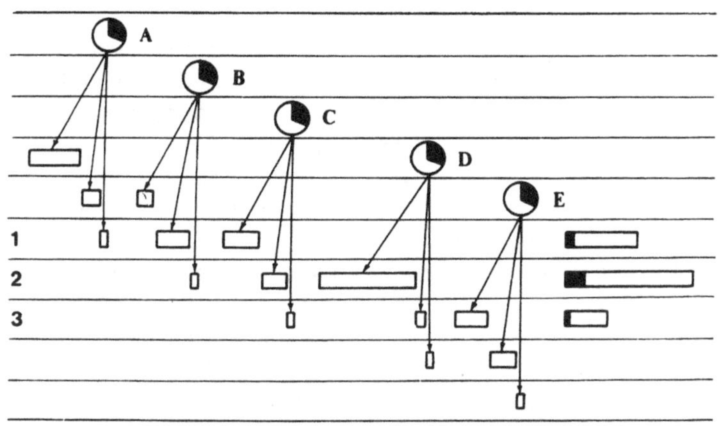

FIGURE 3.3. Adult population size and time of appearance of adult ticks in successive generations of the main TBE virus vector. Horizontal lines separate different years; circles, fed females; rectangles, offspring produced by these females and the overall adult population size during three successive years (1–3); black sectors, the number of infected ticks (A–E) Different ways of development of tick generation: (A) most fed larvae develop without diapause, half of resulting nymphs also develop without diapause; (B) half of larvae develop without diapause, most resulting nymphs enter diapause; (C) most larvae enter diapause, most nymphs develop without diapause; (D) most larvae and nymphs develop without diapause; (E) most larvae develop without diapause, most resulting nymphs enter diapause.

tick populations. A scheme of such a phenomenon underlying the dynamics of the epizootic process in the case of tick-borne encephalitis (60) is shown in Fig. 3.3.

Progeny of fed females can develop in a number of ways (36). Five of them, implying that development is completed during the most probable period of 3 to 5 years, are shown in Fig. 3.3(A–E).

Whatever is the way of development, the number of unfed adult ticks produced by a given number of fed females 3, 4, and 5 years later will be different (7, 147). The size of their population consisting of at least three tick generations simultaneously developed to the adult stage, as well as the number of ticks infected with TBE virus in this population, is different each year (Fig. 3.3, 1–3). Under natural conditions the infection rate characteristic of adult ticks depends on population size and the infection rate of preceding tick generations and on the dynamics of this infection rate during development of each generation. The concept of the close relationship between the age structure of a vector population and annual changes in its size and the intensity of the epizootic process (60) is confirmed by many years of observations in nature.

Conclusions

The data presented here show that, in the case of tick-borne encephalitis (as well as with other infections ecologically linked to ixodid ticks, e.g., Lyme borreliosis), the problems concerning the dynamics and prognosis of intensity of the epizootic process cannot be reduced to only discovering correlations between the population size of a certain group of vertebrate hosts, the number of immune animals or any other factor valid in the preceding year, and the infection rate characteristic of ticks in the current season. The development of principles and methods for such a prognosis should be based on quantitative parameters characterizing virus circulation in the nature. We regard the proposed scheme as an ecologically sound basis for computer modeling of the TBE epizootic process. The resulting model, in turn, can be a component of a more advanced model for epidemic manifestations in the natural focus.

We think that one more remark based on the presented material would be appropriate. The TBE virus population in any year is not homogeneous with respect to its origin. Hence follows the necessity for studies of the ecological polymorphism of the virus population (59, 61, 64). This aspect of TBE epizootiology, as is the case with other infections having natural focality, still remains to be studied.

Acknowledgments. This study was supported in part by FIRCA Grant 00097 (with Professor Andrew Spielman of the Harvard School of Public

Health, Boston, MA) awarded by the Fogarty Center of the National Institutes of Health, U.S.A.

References

1. Alekseev, A.N., and Chunikhin, S.P. 1990. Exchange of tick-borne encephalitis virus between ixodid ticks jointly feeding on animals with subthreshold level of viremia. *Med. Parazitol. Parazitar. Bolezni* No. 2:48–50 (in Russian).
2. Alekseev, A.N., and Chunikhin, S.P. 1991. Exchange of the virus between feeding ticks in the absence of viremia in vertebrate host (distant transfer). *Med. Parazitol. Parazitar. Bolezni* No. 2:50–54 (in Russian).
3. Alekseev, A.N., and Kondrashova, Z.N. 1985. *Organism of Arthropods as the Environment for Pathogens*. Sverdlovsk, 180 p. (in Russian).
4. Arzamasov, I.T., et al. 1969. *Insectivora and Their Parasites on Byelorussian Territory*. Nauka i Tekhnika, Minsk, 174 p. (in Russian).
5. Babenko, L.V. 1969. Parasitic insects as a mortality factor in ixodid ticks. In *Natural Focal Diseases of the Urals, Siberia and Far East*. Sverdlovsk, pp. 84–85.
6. Babenko, L.V., and Rubina, M.A. 1963. A simplified method for predicting the abundance of ticks belonging to the genus Ixodes and some data on their biology. *Med. Parazitol. Parazitar. Bolezni* No. 1:13–18.
7. Babenko, L.V., and Rubina, M.A. 1968. Characteristics of the development of the taiga tick in the region of Kemchug station. In *Problems in Epidemiology of Tick-Born Encephalitis and Biological Characteristics of Its Natural Foci*. Meditsina, Moscow, pp. 138–168 (in Russian).
8. Babenko, L.V., et al. 1977. About sex ratio in natural adult populations of *Ixodes ricinus* L. and *Ixodes persulcatus* P.Sch. (Ixodoidea, Ixodidae). *Med. Parazitol. Parazit. Bolezni* No. 3:294–301.
9. Balashov, Yu.S. 1961. Dynamics of the storage nutrition compounds and determination of age in unfed ixodid ticks. *Zool. Zh.* **40**:1354–1363.
10. Balashov, Yu.S. 1967. *Blood-Sucking Ticks (Ixodoidea) as Vectors of Human and Animal Diseases*. Nauka, Leningrad (in Russian).
11. Balashov, Yu.S. 1984. The role of morphophysiological characteristics of bloodsucking arthropods in pathogen transmission. In *Parazitologicheskii. Sbornik*, Vol. 32. Nauka, Leningrad, pp. 23–42.
12. Balashov, Yu.S., and Daiter, A.B. 1963. On the role of ticks of the superfamily Ixodoidea in Q-rickettsiosis. 5. Localization and dissemination of the Burnett's rickettsia in the organism of the tick *Hyaloma asiaticum* P.Sch. et E.Schl. *Trudy Leningrad. Pasteur Inst. Epidemiol. Microbiol.* **25**:135–153.
13. Beklemishev, Yu.S. 1956. Pathogens as members of biocenoses. *Zool. Zh.* **35**:1765–1779.
14. Beklemishev, V.N. 1959. Some aspects of epidemiology and epizootology of the tick-borne encephalitis. *Med. Parazitol. Parazit. Bolezni* No. 3:310–318.
15. Benda, R. 1958. A common tick *Ixodes ricinus* L. as the virus reservoir and vector of the tick-borne encephalitis. I. Survival of the virus (B_3 strain) during tick development under laboratory conditions. *Zh. Gigieny, Epidemiol., Immunol. (Praha)* **2**:447–466.

16. Benda, R. 1958. A common tick *Ixodes ricinus* L. as the virus reservoir and vector of the tick-borne encephalitis. II. Experimental transmission of tickborne encephalitis to laboratory animals by ticks of different developmental stages. *Zh. Gigieny, Epidemiol. Immunol. (Praha)* **2**:467–480.
17. Blaskovic, D. 1966. *Virus Circulation*. Academia i Artia, Praha (in Russian).
18. Blaskovic, D., and Rehacek, J. 1962. Ticks as virus vectors in Eastern Europe. In *Biological Transmission of Disease Agents*. Academic Press, New York, pp. 135–157.
19. Burgdorfer, W., and Varma, M.T.R. 1967. Trans-stadial and transovarial development of disease agent in arthropods. *Annu. Rev. Entomol.* **12**:347–376.
20. Burlakov, S.A., and Pautov, E.N. 1975. *Mosquitoes and Ticks as Pathogen Vectors in Human Viral Diseases and Rickettsioses*. Meditsina, Moscow (in Russian).
21. Chernesky, M.A. 1969. Powassan virus transmission by ixodid ticks infected after feeding on viremic rabbits injected intravenously. *Can. J. Microbiol.* **15**: 521–530.
22. Chernesky, M.A., and McLean, D.M. 1969. Localization of Powassan virus in *Dermacentor andersoni* ticks by immunofluorescence. *Can. J. Microbiol.* **15**: 1399–1408.
23. Cerny, V. 1959. Viznam keru pro prezivani nasatyck samic klistce obecheho (*Ixodes ricinus* L.). *Vet. casopis (Praha)* **8**:455–460.
24. Cerny, V., et al. 1973. Some factors affecting the horizontal distribution of ticks. In *1 Internationales Arbeitskollokvium uber Naturherve von Infektionskrankheinten in Zentraleurope*, Illmitz und Graz, pp. 116–118.
25. Chumakov, M.P. 1944. Studies into ultraviral encephalites. Communication 6. Transmission of the tick-borne encephalitis virus to progeny in ticks and the problem of natural foci of this infection. *Med. Parazitol. Parazit. Bolezni* No. 6:38–41.
26. Chumakov, M.P., and Naidenova, G.A. 1944. The tick *Ixodes ricinus* as a vector of the tick-borne (spring-summer) encephalitis. *Med. Parazitol. Parazit. Bolezni* No. 4:89–93.
27. Chumakov, M.P., and Zeitlenok, N.A. 1939. Tick-borne spring-summer encephalitis in the Urals and adjacent regions. *Arkhiv Biol. Nauk* **56**:112–120.
28. Chunikhin, S.P. 1973. Ecological analysis of transmission cycles of arboviruses. *Med. Parazitol. Parazit. Bolezni* No. 6:730–735.
29. Chunikhin, S.P. 1975. Major ecological groups of avian arboviruses. In *Proceedings of the All-Union Conference on Bird Migrations*, Vol. 2, Moscow, pp. 151–154 (in Russian).
30. Chunikhin, S.P., and Leonova, G.N. 1985. *Ecology and Geographic Distribution of Arboviruses*. Meditsina, Moscow (in Russian).
31. Chunikhin, S.P., et al. 1979. Studies into characteristics of transstadial and transovarial transmission of the tick-borne encephalitis virus with different degree of pathogenicity for mice. *Med. Parazitol. Parazit. Bolezni* No. 2:61–85.
32. Chunikhin, S.P., et al. 1983. Sexual transmission of the tick-borne encephalitis virus in ixodid ticks. *Parazitologiya* **17**:214–217.

33. Churilova, A.A., Yagodinskii, V.N., and Aleksandrov, Yu.V. 1963. About transovarial virus transmission in ticks. In *Tick-Born Encephalitis and Viral Hemorrhagic Fevers*. Omsk, pp. 243–244 (in Russian).
34. Daniel, M., Honzakova, E., and Cerny, V. 1972. Overwintering of the tick *Ixodes ricinus* L. under conditions of a field experiment. *Folia Parasitol.* **19**:305–314.
35. Filipova, N.A. 1977. Ixodid ticks of the subfamily Ixodinae. In *Fauna of the USSR. Arachnoidea*, Vol. 4. Nauka, Leningrad (in Russian).
36. Filipova, N.A. (ed.). 1985. *Taiga Tick Ixodes persulcatus Schulze (Acarina, Ixodidae). Morphology, Systematics, Ecology, and Medical Significance.* Nauka, Leningrad (in Russian).
37. Galimov, V.R., et al. 1989. Transmission of the tick-borne encephalitis virus by adult taiga ticks in the absence of viremia in their hosts. In *12 All-Union Conference on Natural Focality of Diseases*, Moscow, pp. 43–44 (in Russian).
38. Gorozhankina, T.S., and Smorodintsev, A.A. 1967. About transovarial transmission of the two-wave meningoencephalitis virus in ticks *Ixodes persulcatus. Trudy Inst. Eksperim. Med. Akad. Med. Nauk SSSR* **69**: 106–111.
39. Gray, J.S. 1991. The development and seasonal activity of the tick *Ixodes ricinus*: A vector of Lyme borreliosis. *Rev. Med. Vet. Entomol.* **79**:323–333.
40. Gresicova, M., and Calisher, C.H. 1988. Tick-borne encephalitis. In T.P. Monath, ed., *Epidemiology and Ecology*, Vol. 4. CRC Press Inc., Boca Raton, FL, pp. 177–202.
41. Gusev, V.F. 1955. Phenology and hosts of ticks *Ixodes ricinus* and *Dermacentor pictus*. In *Selected Works of the Leningrad Institute for Advancement of Physicians*, Vol. 10, pp. 5–16 (in Russian).
42. Honzakova, E. 1973. Survival and development of ticks *Ixodes ricinus*, *Ixodes hexagonus* and *Dermacentor marginatus* in the period of maximum summer temperatures. *Folia Parasitol.* **20**:263–266.
43. Ilyenko, V.I., Gorozhankina, T.S., and Smorodintsev, A.A. 1970. Major trends in transovarial tick-borne encephalitis virus transmission by tick vectors. *Med. Parazitol. Parazit. Bolezni* No. 3:263–269.
44. Jones, L.D., Hodgson, E., and Nuttall, P.A. 1989. Enhancement of virus transmission by tick salivary glands. *J. Gen. Virol.* **70**:1895–1898.
45. Kachanko, N.I. 1978. Development if ixodid ticks at Northern boundaries of their ranges in the Amur region. *Parazitologiya* **12**:218–225.
46. Karpov, S.P. 1976. Virus formation and circulation in the tick-borne encephalitis foci inhabited by man. *Zh. Mikrobiol., Epidemiol. Immunobiol.* No. 3:3–8.
47. Karpov, S.P., and Fedorov, Yu.V. 1963. *Epidemiology and Prevention of Tick-Borne Encephalitis*. Tomsk (in Russian).
48. Katin, A.A. 1983. Spatial-temporal characteristic of natural tick-borne encephalitis foci in relation to prediction of their intensity. In *Tick-Borne Encephalitis and Omsk Hemorrhagic Fever in the Tyumen' Region*. Omsk, pp. 22–42 (in Russian).
49. Katin, A.A., Dubov, A.V., and Gorozhankina, T.S. 1969. Relationships between different ixodid tick species and tick-born encephalitis virus. In *Problems of Regional Infection Pathology*. Tyumen', pp. 29–35 (in Russian).

50. Katin, A.A., and Pustovalov, I.I. 1983. The type of relationships between tick and virus populations in tick-borne encephalitis foci in relation to the problem of their prognosis. In *Siberian and Far Eastern Regions of Industrial Development*. Omsk, pp. 41–47 (in Russian).
51. Katin, A.A., Vorobyeva, A.M., and Gurbo, G.D. 1969. About certain characteristics of transovarial and transstakial transmission of tick-borne encephalitis virus in ixodid ticks. In *Problems of Regional Infection Pathology*. Tyumen', pp. 30–31 (in Russian).
52. Kheisin, E.M., and Lebesheva, M.A. 1955. Egg laying and development of *Ixodes ricinus* and *Ixodes persulcatus* at different temperatures and humidity of the environment. In *Selected Works of the Karelian-Finnish State University*, Vol. 6. Petrozavodsk, pp. 5–27 (in Russian).
53. Kheisin, E.M., et al. 1955. Duration of the developmental cycle of *Ixodes persulcatus* P.Sch. under environmental conditions of Karelian-Finnish SSR. In *Selected Works of the Karelian-Finnish State University*, Vol. 6. Petrozavodsk, pp. 102–123 (in Russian).
54. Kheisin, E.M., et al. 1955. Egg laying and development of *Ixodes ricinus* L. under environmental conditions of Karelian-Finnish SSR. In *Selected Works of the Karelian-Finnish State University*. Vol. 6. Petrozavodsk, pp. 47–51 (in Russian).
55. Kondrashova, Z.N. 1969. Comparative study into specificity of relationships between viruses of the tick-borne encephalitis complex and the tick *Ixodes persulcatus* P.Sch. In *Proceedings of the 16 Scientific Conference of the Institute of Poliomyelitis and Viral Encephalitis*, Vol. 2, Moscow, pp. 16–17 (in Russian).
56. Kondrashova, Z.N. 1970. Studies into factors affecting the behavior of the tick-borne encephalitis virus in the tick *Ixodes persulcatus*. In *Tick-Borne Encephalitis and Other Natural Focal Infections*. Sverdlovsk, pp. 22–27 (in Russian).
57. Kondrashova, Z.N., and Filipovets, R.V. 1970. Infection rate of ticks *Ixodes persulcatus* and some problems of transovarial transmission during tick infection with TBE virus at fixed doses. *Vopr. Virusol.* No. 6:703–708.
58. Kondrashova, Z.N., and Kotel'nikova, G.M. 1975. Behavior of the tick-borne encephalitis virus in ticks *Ixodes persulcatus* of different origin. *Med. Parazitol. Parazit. Bolezni* No. 1:25–29.
59. Korenberg, E.I. 1974. About certain general principles of tick-borne-encephalitis epizootology. In *Natural Focal Infections and Invasions*. Vilnius, pp. 62–67 (in Russian).
60. Korenberg, E.I. 1974. Some problems of population ecology of ixodid ticks. *Zool. Zh.* **53**:165–178.
61. Korenberg, E.I. 1976. Some current aspects of natural focality and epidemiology of tick-borne encephalitis. *Folia Parasitol.* **23**:357–366.
62. Korenberg, E.I. 1979. Small mammals and the problem of the natural tocality of tick-borne encephalitis. *Zool. Zh.* **58**:542–552.
63. Korenberg, E.I. 1989. Tick-borne encephalitis. In *Arboviruses and Arbovirus Infections*. Meditsina, Moscow, pp. 256–264.
64. Korenberg, E.I. 1989. Population principles in research into natural focality of zoonoses. *Sov. Sci. Rev. F. Physiol. Gen. Biol.*, Vol. 3. Harwood Acad. Publ., pp. 303–351.

65. Korenberg, E.I., Baranovskii, P.M., and Vinokurova, N.S. 1981. Evaluation of the results obtained during single scoring on collecting grounds of ticks *Ixodes persulcatus* and some ecological characteristics of this species (Ixodidae). *Parazitologiya* **15**:454–458.
66. Korenberg, E.I., and Kovalevskii, Yu.V. 1977. A general scheme for the tick borne encephalitis virus circulation. *Zool. Zh.* **56**:1467–1478.
67. Korenberg, E.I., and Kovalevskii, Yu.V. 1981. Regionalization of the tick-borne encephalitis area. In *Advances in Science and Technology. Medicinal Geography*, Vol. 11. VINITI, Moscow (in Russian).
68. Korenberg, E.I., and Pchelkina, A.A. 1984. Titers of the tick-borne encephalitis virus in engorged ticks *Ixodes persulcatus*. *Parazitologiya* **18**:123–127.
69. Korenberg, E.I., et al. 1975. Studies of deer *Cervus elaphus* in linked foci of tick-borne encephalitis, rickettsioses, leptospiroses and tularemia. *Zool. Zh.* **54**:1057–1065.
70. Korenberg, E.I., et al. 1988. Differences within population in the infection rate of adult ticks *Ixodes persulcatus* P.Sch. with the tick-borne encephalitis virus and evaluation of its total content in ticks. *Vopr. Virusol.* No. 4:456–461.
71. Korenberg, E.I., et al. 1992. Probability models of the rate of infection with tick-borne encephalitis virus in *Ixodes persulcatus* ticks. *Folia Parasitol.* **39**:85–92.
72. Korotkov, Yu.S., and Kislenko, G.S. 1991. Morphogenetic diapause in taiga tick and methods of its quantitative evaluation under the conditions of field experiment. *Parazitologiya* **25**:494–503.
73. Kovalevskii, Yu.V., et al. 1988. Factors determining possibility of infection with tick-borne encephalitis. 2. Infection rate of the vector in mid-taiga forests of the Khabarovsk region. *Med. Parazitol. Parazit. Bolezni* No. 3:22–27.
74. Kozuch, O., Mayer, V., and Nosek, J. 1970. Quantitative study of the tick-borne encephalitis virus in ticks *Ixodes ricinus*. *Acta Virol.* **14**:53–58.
75. Kozuch, O., et al. 1966. Uberleben des Zeckenencephalitisvirus in der Zecke *Ixodes ricinus* und die Ubertragung dieses Virus auf den Zgel (*Erinaceus roumanicus*). *Zentralb. Bakteriol. Parasitenk. Infektion. Hyg. 1 Orig.* **199**:152-151.
76. Kozuch, O., et al. 1978. Surveillance on mosquito-borne natural focus in Zahorsva lowland. In *2 Internationales Arbeitskolloquium uber Naturherde von Infektionskrankheiten in Zentraleuropa*, Graz, pp. 115–118.
77. Krivolutskii, D.A. 1963. Extermination of *Ixodes persulcatus* larvae and nymphs by predators. In *Tick-Borne Encephalitis and Virus Hemorrhagic Fevers*. Omsk, pp. 187–188 (in Russian).
78. Kurenkov, V.B., et al. 1981. Experimental characterization of the taiga tick (*Ixodes persulcatus* Schulze, 1930) as a tick-born encephalitis vector. *Med. Parazitol. Parazit. Bolezni* No. 1:53–58.
79. Kuznetsov, P.K. 1955. About wintering of ticks *Ixodes ricinus* under environmental conditions of the Voronezh region. *Zool. Zh.* **34**:469–470.
80. Levin, M.L. 1987. Biology of unfed larvae of the taiga tick (*Ixodes persulcatus*) under natural conditions. *Zool. Zh.* **66**:348–359.

81. Levin, M.L. 1990. Proportion of unfed larvae, nymphs and adult ticks in *Ixodes persulcatus* population of mid-taiga forests of the Khabarovsk region. In *Sixth All-Union Conference on Problems of Theoretical and Applied Acarology*, Leningrad, pp. 76–77 (in Russian).
82. Levin, M.L., Bannova, G.G., and Karavanov, A.S. 1987. Studies into the rate of infection with the tick-borne encephalitis virus of unfed *Ixodes persulcatus* P.Sch. larvae and nymphs. In *Natural Focal Infections in the Industrially Developing Region of the Baikal-Amur Railway*. Moscow, pp. 73–84 (in Russian).
83. Levin, M.L., et al. 1989. Approaches to estimating the size of the component of the tick-borne encephalitis pathogen population associated with larvae and nymphs of the main vector. In *12th All-Union Conference of Natural Focality of Diseases*, Moscow, pp. 95–96 (in Russian).
84. Levkovich, E.N. 1943. Experimental evidence for epidemiological role of certain tick species in spring-summer tick-borne encephalitis. *Zh. Mikrobiol. Epidemiol. Immunobiol.* No. 10–11:49–54.
85. Levkovich, E.N., and Skrynnik, A.N. 1940. About maintenance of spring-summer encephalitis virus in ticks after wintering. *Arkhiv Biol. Nauk* **59**:118–121.
86. Lutta, A.S., Kheisin, E.N., and Shulman, R.E. 1959. *Ixodid Ticks of the KASSR and Measures for Their Extermination.* Gosizdat KASSR, Petrozavodsk (in Russian).
87. L'vov, D.K., and Lebedev, A.D. 1974. *Ecology of Arboviruses*. Meditsina, Moscow (in Russian).
88. MacLeod, J. 1935. *Ixodes ricinus* in relation to its physical environment. III. Climate and reproduction. *Parasitology* **27**:489–500.
89. MacLeod, J. 1936. *Ixodes ricinus* in relation to its physical environment. IV. An analysis of the ecological complexes controlling distribution and activities. *Parasitology* **28**:259–319.
90. Mel'nikova, T.G. 1953. Ixodid ticks of wild and domestic animals in the Crimea reserve. *Sool. Zh.* **32**:422–434.
91. Mishaeva, N.P. 1974. Experimental study of arboviruses in arthropods. 2. Maintenance of the tick-borne encephalitis virus in ixodid ticks under the conditions of decreasing relative humidity. In *Ecology of Viruses Associated with Birds*. Minsk, pp. 25–27.
92. Mishaeva, N.P., Khod'ko, L.P., and Kalenchuk, V.U. 1974. Experimental study of arboviruses in arthropods. 2. The effect of temperature on reproduction of the tick-borne encephalitis virus in certain species of ixodid ticks. In *Ecology of Viruses Associated with Birds*. Minsk, pp. 27–30.
93. Mishaeva, N.P., et al. 1974. Experimental study of arboviruses in arthropods. The significance of the standard experimental conditions for studies on ticks. In *Ecology of Viruses*, Vol. 2. Moscow, pp. 23–28 (in Russian).
94. Mishin, A.V. 1964. Living conditions of the forest tick in the Udmurt ASSR. *Selected Works Izhevsk Med. Inst.* **20**:66–73.
95. Moiseenko, N.M. 1957. About life cycle of the tick Ixodes persulcatus in different parts of its range. In *Geographic Problems of the Far East*, Vol. 3. Khabarovsk, pp. 157–162 (in Russian).

96. Morozov, Yu.V. 1961. About species composition of animal hosts involved in circulation of the tick-borne encephalitis virus. *Byul. Moskovskogo Obshch. Ispytatelei Prirody, Biol.* No. 3:5–19.
97. *Natural Focality of the Tick-Borne Encephalitis* (bibliography of papers published in Russian), Moscow, 1972.
98. Naumov, R.L. 1975. Distribution of ticks *Ixodes persulcatus* P.Sch. on the northern slope of the Western Sayan and factors determining this distribution. 3. Development of larvae and nymphs and proportion of ticks belonging to different generations in the adult tick population. *Med. Parazitol. Parazit. Bolezni* No. 1:10–16.
99. Naumov, R.L., Chunikhin, S.P., and Gutova, V.P. 1984. Experimental study into relationship between vertebrates and the tick-borne encephalitis virus. 2. Small mammals. *Med. Parazitol. Parazit. Bolezni* No. 2:83–86.
100. Naumov, R.L., and Gutova, V.P. 1977. Geographic and annual variability of the rate of tick infection with the tick-borne encephalitis virus. *Med. Parazitol. Parazit. Bolezni* No. 3:346–355.
101. Naumov, R.L., and Gutova, V.P. 1984. Experimental study into the involvement of gamasid ticks and fleas in circulation of the tick-borne encephalitis virus. A review. *Parazitologiya* **18**:106–115.
102. Naumov, R.L., Gutova, V.P., and Chunikhin, S.P. 1980. Ixodid ticks and the tick-borne encephalitis pathogen. 1. Relationship between the virus and ticks of the genus *Ixodes. Med. Parazitol. Parazit. Bolezni* No. 2:17–23.
103. Naumov, R.L., Gutova, V.P., and Chunikhin, S.P. 1980. Ixodid ticks and the tick-borne encephalitis pathogen. 2. Genera *Dermacentor* and *Haemaphysalis. Med. Parazitol. Parazit. Bolezni* No. 3:66–69.
104. Naumov, R.L., Gutova, V.P., and Chunikhin, S.P. 1981. Ixodid ticks and the tick-borne encephalitis pathogen. 3. Summary and certain conclusions. *Med. Parazitol. Parazit. Bolezni* No. 1:58–61.
105. Naumov, R.L., Gutova, V.P., and Chunikhin, S.P. 1983. Experimental study into relationship between the tick-borne encephalitis virus and vertebrates. 1. Small and medium-size mammals (a literature review). *Med. Parazitol. Parazit. Bolezni* No. 3:78–83.
106. Naumov, R.L., et al. 1976. About periodic oscillations of the rate of tick infection with the tick-borne encephalitis virus. In *Ecology of Viruses*. Baku, pp. 109–110 (in Russian).
107. Nikiforov, L.P. 1968. The role of vertebrates in the tick-borne encephalitis focus located in the Kozul'skii region (Krasnoyarsk province). In *Problems of Tick-Borne Encephalitis Epidemiology and Biological Parameters of Its Natural Foci*. Meditsina, Moscow, pp. 38–42 (in Russian).
108. Pavlovskii, E.N. 1947. Ixodid ticks of the Far East. In *Parazitology of the Far East*. Medgiz, Leningrad, pp. 160–211 (in Russian).
109. Pavlovskii, E.N. 1947. Ticks and tick-borne encephalitis. In *Parazitology of the Far East*. Medgiz, Leningrad, pp. 212–264.
110. Pavlovskii, E.N. 1963. What is the basis for existence of the natural focus of disease and the territory appropriate for it. *Zool. Zh.* **42**:321–325.
111. Pavlovskii, E.N. 1964. *Natural Focality of Transmissive Diseases in Relation to the Landscape Epidemiology of Antropozoonoses*. Nauka, Moscow-Leningrad (in Russian).

112. Pavlovskii, E.N., Krol', N.B., and Smorodintsev, A.A. 1940. *A Short Account of the Tick-Borne (Spring-Summer) Encephalitis*. Medgiz, Moscow-Leningrad (in Russian).
113. Pavlovskii, E.N., et al. 1960. Tick-borne encephalitis. In *Human Diseases with Natural Focality*. Medgiz, Moscow, pp. 41–90.
114. Pchelkina, A.A., et al. 1970. Studies into the infection rate of *Ixodes persulcatus* P.Sch. in foci of tick-borne encephalitis in East-European southern taiga forests. In *The Second Acarological Conference*, Vol. 2. Naukova Dumka, Kiev, pp. 96–97 (in Russian).
115. Petrishcheva, P.A. 1958. *Tick-Borne Encephalitis*. Institute of Sanitary Education, Moscow (in Russian).
116. Petrishcheva, P.A. 1967. Mosquitoes Culicidae and viruses. In *Biological Relationships between Disease Vectors and Pathogens*. Meditsina, Moscow, pp. 17–31 (in Russian).
117. Petrishcheva, P.A., Pchelkina, A.A., and Seledtsov, I.I. 1964. Blood-sucking mosquitoes as a probable link in circulation of the tick-borne encephalitis virus. *Med. Parazitol. Parazit. Bolezni* No. 2:132–135.
118. Popov, V.M. 1959. *A Vector of Tick-Borne Encephalitis*. Medgiz, Moscow (in Russian).
119. Popov, V.M. 1962. *Ixodid Ticks of the Western Siberia*. Tomsk Univ. Publ. House, Tomsk (in Russian).
120. Pretzmann, G., Loew, J., and Radda, A. 1963. Untersuchungen in einem naturherd der fruhsommer-meningoencephalitis (FSME) in Niederosterreich. 3. Mitteilung: versuch einer gesamt darstellung des zyclus der FSME in naturherde. *Zentralb. Bakteriol. Parasitenk. Infection. Hyg.* **190**:299–312.
121. Radda, A.G. 1973. Geoecologische gesichtspunkte beim vorkommen der fruhsommer-meningoencephalitis, geographische zeitschrift. *Beihefte: Erdkundl Wissen*, Heft 35, *Fortschritte der Geomedizinischen Forschungen*. Weisbaden, S. 62–75.
122. Rasnitsyn, S.P. 1976. Analysis of the role of transovarial and transstadial transmission in the maintenance of the pathogen population. *Med. Parazitol. Parazit. Bolezni* No. 3:269–274.
123. Rehacek, J. 1960. Experimental wintering of the tick-borne encephalitis virus in fed *I. ricinus* L. larvae. *Acta Virol.* 4:106–109.
124. Rehacek, J. 1962. Transovarial transmission of the tick-borne encephalitis virus in ticks. *Acta Virol.* **6**:220–226.
125. Sazonova, O.N. 1962. Blood-sucking mosquitoes (Diptera, Culicinae). In *Pathogen Vectors of Natural Focal Diseases*. Medgiz, Moscow, pp. 9–63 (in Russian).
126. Serdyukova, G.V. 1951. Wintering of *I. ricinus* L. eggs in physical environment of the Karelian Isthmus. *Dokl. Akad. Nauk SSSR* **81**:1171–1173.
127. Sharipova, R.R. 1960. On parasitism of taiga ticks *Ixodes persulcatus* P.Sch. on wild animals in natural tick-borne encephalitis foci of the Kalinin region. *Med. Parazitol. Parazit. Bolezni* No. 3:268–270.
128. Shikharbeev, B.V. 1962. About studies into duration of different stages of *Ixodes persulcatus* P.Sch. life cycle in the tick-borne encephalitis focus located in the south-east of the Irkutsk region. In *Selected Works of the*

Irkutsk Research Institute of Epidemiology, Microbiology and Hygiene, Vol. 7, pp. 74–85 (in Russian).

129. Shikharbeev, B.V. 1965. About time of molting of Ixodes persulcatus P.Sch. larvae and nymphs in tick-borne encephalitis foci located in the south-east of the Irkutsk region. In *Selected Works of the Irkutsk Research Institute of Epidemiology, Microbiology and Hygiene*, Vol. 8, pp. 235–261 (in Russian).
130. Shilova, S.A. 1960. About biological validation of epidemiological prognoses of tick-borne encephalitis. *Byul. Moskovskogo Obshch. Ispytatelei Prirody, Biol.* No. 1:37–47.
131. Shubladze, A.K., and Serdyukova, G.V. 1939. The tick *Ixodes persulcatus* as a vector of the spring-summer encephalitis. *Arkhiv Biol. Nauk* **56**: 121–131.
132. Singh, K.R.T., Goverdhan, M.K., and Bhat, H.R. 1968. Transovarial transmission of Kyasanur forest disease virus by *Ixodes petanristae. Indian J. Med. Res.* **56**:628.
133. Solovyev, V.D. 1944. *Summer-Spring Tick-Borne Encephalitis*. Sovetskaya Nauka, Moscow (in Russian).
134. Starodubtseva, D.I., and Korzukhina, L.F. 1967. Studies into the virus content in the eggs of spontaneously infected ticks. In *Selected Works of the Perm' Medical Institute*, Vol. 68, pp. 83–86 (in Russian).
135. Starodubtseva, G.I., Korzukhina, L.F., and Minaeva, V.M. 1972. Studies using hemagglutination test on several problems of transovarial transmission of the tick-borne encephalitis virus in ixodid ticks. In *Current Problems of Virology and Prevention of Viral Diseases*. Moscow, pp. 282–283 (in Russian).
136. Tagil'tsev, A.A., and Tarasevich, L.N. 1982. *Arthropods of the "Shelter Complex" in Natural Foci of Arbovirus Infections*. Nauka, Novosibirsk (in Russian).
137. Tongern, von, H.A.E. 1959. Central European encephalitis—epidemiology and vectors. In *Proceedings of the Sixth International Congress of Tropical Medicine and Malaria*, Vol. 5, Lisbon, pp. 174–179.
138. Tsilinskii, Ya.Ya. 1988. *Population Structure and Evolution of Arboviruses*. Meditsina, Moscow (in Russian).
139. Uspenskaya, I.G. 1987. *Ixodid Ticks of the Interamnian Dnestr-Prut Region*. Shtiintsa, Kishinev (in Russian).
140. Vashchenok, V.S. 1988. *Fleas (Siphonaptera) as Pathogen Vectors in Human and Animal Diseases*. Nauka, Leningrad (in Russian).
141. Vavilova, V.E., Dorokhova, V.S., and Kirilova, M.A. 1965. Some data about duration of development and time of molting of the tick *Ixodes persulcatus* in Primorsky region. In *Selected Works of the Vladivostok Research Institute of Epidemiology, Microbiology and Hygiene*, Vol. 3, pp. 73–78 (in Russian).
142. Vereta, L.A. 1975. *Principles of Predicting Incidence Rate of Tick-Borne Encephalitis*. Meditsina, Moscow (in Russian).
143. Voshchakina, N.V., and Davydova, M.S. 1954. A focus of tick-borne encephalitis with *Haemaphysalis concinna* as a main vector under physical environment of the Krasnoyarsk region. In *Selected Works of Omsk State Research Institute of Epidemiology, Microbiology and Hygiene*, Vol. 2, pp. 48–51 (in Russian).

144. Yagodinskii, V.N., and Aleksandrov, Yu.V. 1954. About two specific features of transovarial tick-borne encephalitis virus transmission. In *Proceedings of the 11th Scientific Conference of the Institute of Poliomyelitis and Viral Encephalites*. USSR Academy of Medical Sciences, Moscow, pp. 209–210 (in Russian).
145. Zemskaya, A.A. 1973. *Parasitic Gamasid Ticks and Their Medical Significance*. Meditsina, Moscow (in Russian).
146. Zemskaya, A.A., and Pchelkina, A.A. 1974. About the role of gamasid ticks in foci of tick-borne encephalitis. *Med. Parazitol. Parazit. Bolezni* No. 4:405–406.
147. Zhmaeva, Z.M. 1969. About the development of *Ixodes persulcatus* P.Sch. in the European southern taiga forests. In *Tick-Borne Encephalitis in Udmurtia and Adjacent Regions*. Izhevsk, pp. 118–141 (in Russian).
148. Zil'ber, L.A. 1945. *Epidemic Encephalites*. Medgiz, Moscow (in Russian).

4
Genetics of Insect Vector Competence for Arboviruses

Walter J. Tabachnick

Introduction and Background

Ever since arthropods were first incriminated as vectors of disease, medical and veterinary entomologists have recognized that identification of the particular arthropod species transmitting a disease agent is critical for understanding disease epidemiology and for effective disease control. As a result of this recognition, between or interspecies variation in vector ability is well accepted. As a first step, it remains critically important to identify those species that are vectors of different pathogens.

This chapter is concerned with the recognition that individual insects, within a known vector species, also show variability as vectors of particular pathogens, and there is a need to characterize the causes of this variation. This realization has become increasingly important in interpreting and understanding vector-borne diseases. Many studies demonstrate heterogeneity between individuals and populations of a vector species for traits directly involved in vector ability to transmit a pathogen.

The reader is directed to several important reviews that discuss the vector ability of arthropods (9, 13, 24, 26, 54, 65). I will not attempt to review this literature in its entirety, citing only limited examples; the reader may consult these reviews for more detail. I will use standard definitions where vector capacity is defined as the broad suite of traits associated with the basic reproductive rate of a vector–pathogen. Such traits include host preference, biting or feeding rates, population densities, vector longevity, and those traits associated with the general term vector competence. Vector competence includes traits that are directly associated with the interaction between the pathogen and the

Walter J. Tabachnick, Arthropod-Borne Animal Diseases Research Laboratory, USDA-ARS, University Station, P.O. Box 3965, Laramie, Wyoming 82071, USA.

vector. These include susceptibility of the arthropod to infection with the pathogen after an incubation period, the ability of the arthropod to transmit the pathogen to a suitable host, and the ability of the arthropod to transmit the pathogen to its progeny (transovarial).

The recognition of intraspecies variability in the suite of characters associated with vector capacity has only recently begun to influence our view of the role of arthropod vectors in disease transmission. Although intraspecies variability has been extensively studied using morphologic, behavioral, and physiologic characters, isozymes, and more recently nucleotide sequence divergence, acceptance that individuals and populations of a known vector may differ in vector capacity traits has not easily replaced a stereotypic view of vector species with a more appropriate population concept. The stereotypic view of arthropod vectors, by virtually ignoring population variation, sees all individuals and populations of the species as equally dangerous. The population concept views variation in the species as critically important to the role of the vector in disease epidemiology. The population view provides the raison d'être for characterizing vector variation to enable more efficient predictability of the conditions and the particular populations that pose a danger. The population view also provides the basis for utilizing known genetic variation in vector capacity traits for novel methods aimed at interrupting and controlling vector populations and their influence in transmitting disease.

The focus of this chapter is insect vectors of arboviruses and the use of genetics to understand the role of insects in arboviral disease epidemiology. I will discuss one particular aspect of vector competence, susceptibility of a vector to infection with an arbovirus. I have chosen this emphasis because more work has been done on this than other components of vector competence and there are several fundamental differences between the history and state of knowledge of arboviral insect vectors compared to arthropod vectors of other pathogens, i.e., the anopheline vectors of *Plasmodia*, culicine vectors of filaria, *Glossina* vectors of trypanosomes, and *Ixodes* vectors of spirochetes. Although the discussion is relevant to these vectors, their pathogens, and diseases, there are differences that need to be considered. The major differences are that the genetics, in the form of genetic maps and markers of many of the insect vectors of arboviruses, are considerably more advanced (50, 60, 61), although this is rapidly changing due to the development of molecularly based genetic maps and markers (4, 92). The population genetics of arboviral vectors are well characterized (5, 6, 36–38, 49, 58, 59, 62, 73, 75) and a large literature is available that characterizes variation in one aspect of vector competence, susceptibility to infection (21, 24, 26, 54). The epidemiologic histories of many of the arboviral diseases, a necessary prerequisite for relating vector and disease variation, are well documented, i.e., yellow fever (57), dengue (19), western equine encephalitis (65), and the veterinary disease, bluetongue (88). One final difference is that studies of the genetic and physiologic mechanisms controlling vector susceptibility to arboviruses have been difficult to interpret. Several studies have reported single loci controlling susceptibility to infection of both *Anopheles* to *Plasmodia* and *Culex* to filaria. For example, Gubler et al. (21) list five separate studies showing the mode of inheritance for vector competence of malaria parasites and eight for filarial infections. Several similar studies of arboviral vectors have failed to identify, in equally convincing fashion, the underlying genetic control of susceptibility to infection (21). It is apparent that there are fundamental differences between arboviral and parasite vectors that result in a more complex mechanism of genetic control for the former.

Knowledge of the genetic factors controlling vector competence, and the environmental factors that influence these factors, will provide the basis for predictability of where and when there is danger of endemic or epidemic disease (75). The genetic characterization of presumed vector populations and identification of the appropriate environmental factors will allow efficient use of resources for insect control and provide the basis for predicting the danger to any region of the importation of exotic viruses. Finally, this knowledge will provide novel methods to interrupt transmission cycles through genetic or environmental manipulation of dangerous vector populations.

Vector Competence

The series of events that must occur within the insect for it to be a capable vector of an arbovirus are reviewed elsewhere (9, 13, 24, 26, 54). Briefly, after ingestion of the blood meal containing virus, the cells in the posterior part of the insect midgut become infected. The virus multiplies in the midgut cells and passes into the body cavity through the basal lamina. The virus is transported by the hemolymph to secondary target tissues, one of which is the salivary gland. The virus must pass into the salivary gland where it replicates and is eventually transferred to a susceptible host when the insect takes a subsequent blood meal.

A variety of mechanisms have been studied, each of which may contribute to the susceptibility of an insect to become infected and transmit an arbovirus. These include virus inactivation by digestive enzymes, permeability of the peritrophic membrane surrounding the blood meal, barriers that may prevent the virus from passing through a variety of membranes, i.e., the basal lamina, gut or mesenteronal infection and escape barriers, and the salivary gland.

The physiologic aspects of vector competence have been explored in a select number of species. The most detailed knowledge has been obtained for midgut barriers to western equine encephalitis infection in *Culex tarsalis* (27, 31, 42). However, the genetic basis of any aspect of the vector competence phenotype remains largely unknown, i.e., resistance to infection due to a gut barrier, replication of virus in target tissue, transmission efficiency due to salivary gland barriers, or infection capability of the salivary gland cells.

Genetics of Vector Competence

Evidence for Genetic Control

Many studies have inferred genetic control of insect susceptibility to infection with an arbovirus by characterizing geographic variation among populations or strains. This has been shown for *Aedes aegypti* for yellow fever (77) and dengue viruses (20, 78), *Ae. albopictus* for dengue (22) and chikungunya viruses (82), *Ae. triseriatus* for La Crosse virus (18), *Cx. pipiens* for St. Louis encephalitis and Rocio viruses (55, 56), *Cx. tritaeniorhynchus* for West Nile virus (29), *Cx. tarsalis* for western equine encephalitis virus (28), *Cx. annulirostris* for Murray Valley encephalitis virus (39), and *Culicoides variipennis* for bluetongue virus (34), among others.

There have been numerous studies employing a variety of selection schemes that have provided refractory and highly susceptible strains. These strains have been used in genetic mating experiments to determine the actual mode of genetic control. Highly susceptible and resistant strains have been selected in *Ae. aegypti* for yellow fever (53, 87) and dengue viruses (79), *Cx. tarsalis* for western equine encephalitis virus (24, 25, 27), and *C. variipennis* for bluetongue virus (33, 74).

The majority of studies to date have not identified a single locus or the genetic mechanism involved in the control of per os susceptibility of any arboviral vector. Polygenic inheritance has been inferred in some studies, while one or two loci have been inferred with different dominant–recessive relationships hypothesized between resistant and susceptible alleles, in others. In almost all cases these hypotheses have been based on limited genetic data. Only in *C. variipennis* has a convincing genetic mechanism been postulated for controlling per os susceptibility for bluetongue virus (74). The single locus controlling *C. variipennis* susceptibility to bluetongue virus infection functions by a maternally inherited factor where the paternal gene is dominant in female offspring. Since resistance has been hypothesized as dominant in some studies of per os susceptibility, while susceptibility has been thought the dominant allele in others, it is interesting that in *C. variipennis* either allele can be dominant to the other, if paternal in origin.

The difficulty encountered in determining single genes controlling vector susceptibility to arbovirus infection is not unexpected considering the long sequence of physiologic events that must occur for an insect to become infected. However, this sequence must also occur in the infection of *Anopheles* or *Culex* mosquitoes when they encounter their respective pathogens plasmodia and filaria. Yet, single locus controls have not been as elusive in these cases (10, 21, 85, 86). It is still unclear even in these cases if other genetic or environmental factors cause quantitative differences in addition to the identified single locus. It is possible that arboviral–vector interactions are more usually influenced by polygenes, or genes that each has a small effect on the phenotype, whereas the variation in the interactions of larger pathogens with insect tissues is controlled by genes having major effect. If this is indeed the case, the reasons for the difference will provide insight into the different modes of interaction between micro (viruses) and macro (plasmodia, filaria) parasites and their vectors. The critical goal, however, remains to identify the number of major and minor loci, to localize and map the principal genetic controlling factors, and to determine the degree of environmental control. As the genetic mechanisms controlling vector capacity traits are identified, it will be interesting to compare the findings in parasite and arboviral vector species. It will also be of interest to investigate whether similar genetic mechanisms predominate in different populations and between different species.

Evidence for Environmental Effects

Identifying the genetic controlling elements for vector capacity traits will be a first step in understanding and predicting the vector potential of individuals and populations of an insect vector species. It will be important to determine the effects of important environmental parameters on vector capacity genotypes. The range of phenotypes, elicited from a genotype under varying environmental parameters, is termed the "norm of reaction" for the particular genotype.

The temperature of extrinsic incubation is known to influence the rate of arboviral replication and, hence, the susceptibility of insects to arboviral infection. Kramer et al. (43) demonstrated that higher incubation temperatures greatly enhanced *Cx. tarsalis* susceptibility to western equine encephalitis virus. This has also been demonstrated for several species of *Aedes* and *Cx. pipiens* for Rift Valley fever virus (83, 84). B.A. Mullens et al. (personal communication, 1992) have shown that a *C. variipennis* strain, homozygous for genes that confer high susceptibility to infection with BLU virus, also has a norm of reaction in varying temperatures of extrinsic incubation. *Culicoides variipennis* of this genotype, maintained at 32°C, became infected and transmitted virus in as little as 5 days after an infectious blood meal. This is significantly faster than transmission by genetically identical insects at 21 and 15°C.

Other environmental factors are capable of influencing vector ability and these must also be studied in relation to the controlling genetic factors. It must be realized that any nongenetic factors that interact with the vector are part of the insect's environment. Larval nutritional effects on vector susceptibility have been reported (17, 40) and must be considered in norms of reaction studies. Environmental effects include variation in the arbovirus and variation in the host. The effect of viral variation on vector competence has been studied (53, 55) and has been observed in the variation in *C. variipennis* susceptibility to infection by different serotypes of bluetongue virus (J.O. Mecham et al., personal communication, 1992). Beaty et al. (2) found that variation in a segment of La Crosse encephalitis virus has a significant effect on the susceptibility of the primary North American vector, *Ae. triseriatus*. Similar studies of how host variation influences vector ability need to be done and these parameters must be explored in their relation to specific vector competence genotypes.

It is likely that vector competence genotypes will differ, relative to one another, in their response to the same environmental variations. A usually competent genotype may be relatively incompetent in a specific environment, while a usually incompetent genotype can become highly competent in particular situations. Laboratory experiments on vector competence, which do not account for the possible complex interactions between genotype and environment, are often impossible to interpret and apply to field conditions because of a limited understanding of the insect genetic material used and the effect of the laboratory environment (76).

Molecular Biology and Genetics of Vector Competence

The task for the future will be to take advantage of new methodologies to identify the genetic controlling elements of all aspects of vector competence, and eventually the norms of reaction in different insect vector species. The available information suggests that genetic control of insect vector competence, particularly susceptibility to infection with arboviruses, is complex in nature and may often be controlled by several different genes or polygenes. Loci controlling traits that are quantitative and influenced by other loci (polygenic), i.e., susceptibility to infection with an arbovirus, have been termed quantitative trait loci or "QTLs." Molecular biologic technology is currently being applied in the form of DNA genetic markers that can be used to identify and localize, by genetic and physical mapping methods, the QTLs of complex phenotypes.

The traditional approach used to identify the genetic locations of QTLs requires measurement of recombination frequencies between single genetic markers and individual QTLs. This procedure usually underestimates the number of QTLs and is not efficient in assigning their location because distant linkage cannot be distinguished from a small phenotypic effect (45). Another disadvantage of traditional mapping is the large number of progeny required for a given level of confidence (70). An alternative approach (44) combines the advantages of interval mapping with selective genotyping to reduce the number of progeny to be scored while increasing the probability of identifying the number and relative importance of QTLs.

A prerequisite for interval mapping is the availability of abundant, highly polymorphic genetic markers distributed evenly throughout the genome. DNA restriction fragment length polymorphisms (RFLPs) have been recognized as useful genetic markers and have recently been reported in insect disease vectors (67, 68, D. Severson, personal communication, 1992). RFLPs are caused by differences in DNA sequences that result in different fragment lengths when the DNA is digested with a particular DNA restriction endonuclease. Specific probes of cloned genomic DNA fragments are used to detect RFLPs. RFLPs have proven useful in mapping genes in interspecies crosses and have been identified in a variety of species, particularly in plants (63, 64, 71). Nevertheless, RFLPs have limitations: (i) polymorphism is relatively rarer in inbred laboratory strains than between species, (ii) identification of RFLPs is time consuming, and (iii) typing RFLPs of individual flies generated in crosses requires many individuals, because usually only a few probes can be hybridized to the DNA of a single individual.

Genetic polymorphisms are available that take advantage of the polymerase chain reaction (PCR) (16) to amplify small amounts of DNA enabling easy detection. This is critical for small insects to ensure that single individuals can easily be genetically characterized for numerous loci. Random amplified polymorphic DNA (RAPDs) are detected using short primers of 10 oligonucleotides for the PCR reaction (91). Random sequences of these short primers amplify random fragments of DNA, many of which show polymorphism between individuals. The technique has proven useful in *Ae. aegypti* (1), in *Ae. scutellaris* species (36), and in *C. variipennis* (W.J. Tabachnick and M.A. Robertson, unpublished data). Analysis of three *Culicoides variipennis sonorensis* laboratory colonies and a field population of *Culicoides variipennis occidentalis* using eight different primers detected numerous polymorphisms within and among populations, several of which were unique for each subspecies (M.A. Robertson and W.J. Tabachnick, unpublished data). Hundreds of separate PCR reactions can be done on any individual insect. Although useful for discriminating populations, RAPDs are not ideal for mapping because most are dominant and require backcrosses between homozygous genotypes to identify heterozygotes.

An alternative type of DNA polymorphism is based on variation in the length of simple sequence repeats (SSRs) or microsatellites (3, 90). SSRs occur frequently in eukaryotic genomes (23, 72, 81), they are usually present about every 10 kb of DNA sequence (80), and they are easily typed using PCR and primers that flank the SSR (3). SSRs have shown extremely high levels of polymorphism in both humans (89, 90) and mice (11, 30, 48), and have been detected in insects (23, 41; L. Zhang, personal communication, 1992; W.C. Black, personal communication 1992; M.A. Robertson and W.J. Tabachnick, unpublished data). Single sequence length polymorphisms (SSLPs) have been identified in the mouse (14) and have been used in mapping QTLs in the rat (32). The conserved nature of SSRs and the high frequency of SSLPs have been attributed to internal genomic mechanisms regulating gene expression as hot spots for recombination and/or targets for mutagens and carcinogens(23, 80).

The procedures for detecting SSLPs are straightforward (14, 32, 69, 90). Random DNA clones, containing SSRs, are identified and isolated from an M13 library of genomic DNA containing small inserts (200–500 bp) using the SSR as a probe. Inserts containing an SSR are sequenced (one pass per fragment) and PCR primers are designed for each using available computer programs to generate PCR assays that work under a single uniform set of conditions. Dietrich et al. (14) designed 455 primer pairs based on sequences of random clones containing CA or GT repeats from mouse genomic DNA; 77% yielded SSLPs between any two mouse subspecies and 50% were polymorphic between two inbred laboratory strains. The mouse SSLPs were used to type 46 F_2 intercross progeny to produce a map with an average spacing of 4.3 cM providing linkage to 99% of the mouse genome (14). A similar protocol in the rat, using 112 SSLPs, located QTLs associated with hypertension (32).

Genetic maps based on SSLPs are being constructed for the insect vectors *An. gambiae* (L. Zhang, personal communication, 1992), *Ae. aegypti* (W.C. Black, personal communication, 1992), and *C. variipennis* (M.A. Robertson and W.J. Tabachnick, unpublished data). *Culicoides variipennis* possess abundant repeats of CACA . . . distributed throughout the genome (M.A. Robertson and W.J. Tabachnick, unpublished data), which are being used to detect 150–250 SSLPs, or about 60–70 for each *C. variipennis* chromosome. The initial work will use markers to determine linkage groups and to estimate the numbers of loci involved in per os susceptibility to infection with bluetongue virus. Ultimately, these maps will be used to identify the number and locations of a variety of other vector capacity and vector competence genotypes using computer mapping algorithms (46, 47).

Once the number of loci are identified and their general linkage assignments known, it will be necessary to use other molecular biological methods to isolate and clone these genes. For example, *Ae. aegypti* has a genome of 8×10^8 bp and, therefore, about 800 markers would be required to place any vector competence locus within 1 million bases of any marker, a distance appropriate for current techniques of chromosome walking and gene cloning. I expect that within the next few years new molecular techniques will become available that will allow for more rapid gene cloning. Already, physical maps of organisms are being constructed directly using clones labeled with fluorescent dyes and physically ordered by fluorescence *in situ* hybridization (FISH) onto the chromosomes (66). The technique is being applied to order cloned DNA on the chromosomes of *Ae. aegypti* and already such markers are available on each of the three chromosomes (D.L. Knudson, personal communication, 1992). The ability to use homologous probes between species will allow vector competence genes, i.e., isolated in *Ae. aegypti*, to be screened in other vector species, i.e., *Ae. albopictus*, *C. variipennis*, and in the macroparasite vectors *An. gambiae* and *Cx. pipiens*.

To fully exploit the use of molecular biologic tools, vector biologists will require the appropriate insect genetic stocks. For example, successful cloning of specific *Drosophila* genes that control mutant phenotypes was made possible by the use of *Drosophila* stocks containing deletions in the chromosome region of interest. This allows for deletion mapping to focus cloning strategies on targeted small portions of the genome. The availability of deletion stocks, chromosomal balancing systems, and translocation stocks would greatly improve the ability to identify and clone critical vector competence loci.

Conclusions

The genetic mechanisms controlling insect vector competence for arboviruses remain largely unknown. I anticipate that the application of modern molecular biologic techniques will provide detailed genetic and physical maps for many insect vectors. These maps will provide the tools for the identification of the major genes controlling insect vector competence. The identification of important vector competence genotypes will allow work to proceed to determine the norms of reaction for each genotype for different environmental parameters and lead the way for an improved understanding of vector–virus interactions. With the ability to characterize individual vectors and their populations for vector competence, vector capacity genotypes, and with characterized norms of reaction, will come the ability to make accurate assessments of the danger posed by any vector. This predictability is vital in today's world where there is danger of importing exotic pathogens into currently disease free regions. Without the ability to predict vector capacity, resident insect species pose unknown danger. For example, African horse sickness virus has not entered the United States, although it has caused periodic epidemics outside Africa, as has occurred in Spain during 1988–1991 (8, 51). *Culicoides variipennis* from the United States transmit African horse sickness virus in the laboratory (7, 35), however, with no understanding of environmental or genetic influences, it is impossible to predict with assurance the danger posed by U.S. *C. variipennis* populations. Predictability will enable more effective targeting of control efforts to specific vector populations prior to their becoming dangerous. New control strategies will be developed that take advantage of genetic and environmental variation to reduce the effect of dangerous vectors. Already, methods are being explored to create transgenic mosquitoes with the goal of producing mosquitoes that are less efficient as vectors of disease (12, 15, 52). Identification of the controlling genes is critical to this goal. Control strategies that take advantage of modifications of environmental parameters, i.e., vector nutrition, to reduce vector efficiency remain to be explored. The molecularly based research on vector competence and capacity that I have described offers tremendous challenges for the future. I believe this is the road that will bring the dream of controlling arboviral disease, by manipulating and changing vector populations, to fruition.

Acknowledgments. I want to thank M.A. Robertson, K.E. Murphy, W.C. Black, B.R. Miller, and F.R. Holbrook for helpful discussions on vector competence genetics and their constructive comments on this manuscript. I also thank O.L. Roney for her assistance in preparing the manuscript.

References

1. Ballinger-Crabtree, M.E., Black, W.C., and Miller, B.R. 1992. Use of genetic polymorphism detected by RAPD-PCR for differentiation and identification of *Aedes aegypti* subspecies and populations. *Am. J. Trop. Med. Hyg.* **47**:893–901.
2. Beaty, B.J., Holterman, M., Tabachnick, W.J., Shope, R.E., Rozhon, E.J., and Bishop, D.H.L. 1981. Molecular basis of bunyavirus transmission by mosquitoes: Role of the middle-sized RNA segment. *Science* **211**:1433–1435.
3. Beckmann, J.S., and Soller, M. 1990. Toward a unified approach to genetic mapping of eukaryotes based on sequence tagged microsatellite sites. *Biotechnology* **8**:930–932.
4. Besansky, N.J., Finnerty, V., and Collins, F.H. 1992. Molecular perspectives on the genetics of mosquitoes. *Adv. Genet.* **30**:123–184.
5. Black, W.C. IV, Ferrari, J.A., Rai, K.S., and Sprenger, D.A. 1988. Breading structure of a colonizing species: *Aedes albopictus* in the United States. *Heredity* **60**:173–181.
6. Black, W.C. IV, McLain, D.K., and Rai, K.S. 1989. Patterns of variation in the rDNA cistron within and among world populations of a mosquito. *Aedes albopictus* (Skuse). *Genetics* **121**:539–550.
7. Boorman, J., Mellor, P.S., Penn, M., and Jennings, M. 1975. The growth of African horse sickness in embryonated hen eggs and the transmission by

Culicoides variipennis (Coquillett) (Diptera: Ceratopogonidae). *Arch. Virol.* **47**:343–349.
8. Brown, C.C., and Dardiri, A.H. 1990. African horse sickness: A continuing menace. *J. Am. Vet. Med. Assoc.* **196**:2019–2020.
9. Chamberlain, R.W., and Sudia, W.D. 1961. Mechanism of transmission of viruses by mosquitoes. *Annu. Rev. Entomol.* **6**:371–390.
10. Collins, F.H., Sakai, R.H., Vernick, K.D., Paskewitz, S., Seeley, D.C., Miller, L.H., Collins, W.E., Campbell, C.C., and Gwadz, R.W. 1986. Genetic selection of a *Plasmodium-refractory* strain of the malaria vector *Anopheles gambiae*. *Science* **234**:607–610.
11. Cornall, R.J., Aitman, T.J., Hearne, C.M., and Todd, J.A. 1991. The generation of a library of PCR-analyzed microsatellite variants for genetic mapping of the mouse genome. *Genomics* **10**:874–881.
12. Crampton, J., Morris, A., Lycett, G., Warren, A., and Eggleston, P. 1990. Transgenic mosquitoes: A future vector control strategy? *Parasitol. Today* **6**:31–36.
13. DeFoliart, G.R., Grimstad, P.R., and Watts, D.M. 1987. Advances in mosquito-borne arbovirus/vector research. *Annu. Rev. Entomol.* **32**:479–505.
14. Dietrich, W., Katz, H., Lincoln, S.E., Shin, H.S., Friedman, J., Dracopoli, N., and Lander, E.S. 1992. A genetic map of the mouse suitable for typing intraspecific crosses. *Genetics* **131**:423–447.
15. Eggleston, P. 1991. The control of insect-borne disease through recombinant DNA technology. *Heredity* **66**:161–172.
16. Erlich, H.A. 1989. Polymerase chain reaction. *J. Clin. Immunol.* **9**:437–447.
17. Grimstad, P.R., and Haramis, L.D. 1984. *Aedes triseriatus* (Diptera: Culicidae) and La Crosse virus. III. Enhanced oral transmission by nutrition-deprived mosquitoes. *J. Med. Entomol.* **22**:447–453.
18. Grimstad, P.R., Craig, G.B. Jr., Ross, Q.E., and Yuill, T.M. 1977. *Aedes triseriatus* and La Crosse virus: Geographic variation in vector susceptibility and ability to transmit. *Am. J. Trop. Med. Hyg.* **26**:990–996.
19. Gubler, D.J. 1988. Dengue. In T.P. Monath, ed., *The Arboviruses: Epidemiology and Ecology*. CRC Press, Boca Raton, FL, pp. 223–260.
20. Gubler, D.J., Nalim, S., Tan, R., Saipan, H., and Sulianti-Saroso, J. 1979. Variation in susceptibility to oral infection with dengue viruses among geographic strains of *Aedes aegypti*. *Am. J. Trop. Med. Hyg.* **28**:1045–1052.
21. Gubler, D.J., Novak, R., and Mitchell, C.J. 1982. Arthropod vector competence—epidemiological, genetic and biological observations. In W.W.M. Steiner, W.J. Tabachnick, K.S. Rai, and S. Narang, eds., *Recent Developments in the Genetics of Insect Disease Vectors*. Stipes, Champaign, IL, pp. 343–378.
22. Gubler, D.J., and Rosen, L. 1976. Variation among geographic strains of *Aedes albopictus* in susceptibility to infection with dengue viruses. *Am. J. Trop. Med. Hyg.* **25**:318–325.
23. Hamada, H., Petrino, M.G., and Takunaga, T. 1982. A novel repeated element with z-DNA forming potential is widely found in evolutionary diverse eukaryotic genomes. *Proc. Natl. Acad. Sci. U.S.A.* **79**:6465–6469.
24. Hardy, J.L. 1988. Susceptibility and resistance of vector mosquitoes. In T.P. Monath, ed., *The Arboviruses: Epidemiology and Ecology*. CRC Press, Boca Raton, FL, pp. 87–126.

25. Hardy, J.L., Apperson, G., Asman, S.M., and Reeves, W.C. 1978. Selection of a strain of *Culex tarsalis* highly resistant to infection following ingestion of western equine encephalomyelitis virus. *Am. J. Trop. Med. Hyg.* **27**:313–321.
26. Hardy, J.L., Houk, E.J., Kramer, L.D., and Reeves, W.C. 1983. Intrinsic factors affecting vector competence of mosquitoes for arboviruses. *Annu. Rev. Entomol.* **28**:229–262.
27. Hardy, J.L., and Reeves, W.C. 1990. Experimental studies on infection in vectors. In W.C. Reeves, ed., *Epidemiology and Control of Mosquito-borne Arboviruses in California, 1943–1987.* California Mosquito and Vector Control Association, Inc., Sacramento, CA, pp. 145–253.
28. Hardy, J.L., Reeves, W.C., and Sjogren, R.D. 1976. Variations in the susceptibility of field and laboratory populations of *Culex tarsalis* to experimental infection with western equine encephalomyelitis virus. *Am. J. Epidemiol.* **103**:498–505.
29. Hayes, C.G., Baker, R.H., Baqar, S., and Ahmed, T. 1984. Genetic variation for West Nile virus susceptibility in *Culex tritaeniorhynchus*. *Am. J. Trop. Med. Hyg.* **33**:715–724.
30. Hearne, C.M., McAleer, M.A., Love, J.M., Aitman, T.J., Cornall, R.J., Ghosh, S., Knight, A.M., Prins, J.-B., and Todd, J.T. 1991. Additional microsatellite markers for mouse genome mapping. *Mammalian Genome* **1**:273–282.
31. Houk, E.J., Kramer, L.D., Hardy, J.L., and Presser, S.B. 1986. An interspecific mosquito model for the mesenteronal infection barrier to western equine encephalomyelitis virus (*Culex tarsalis* and *Culex pipiens*). *Am. J. Trop. Med. Hyg.* **35**:190–197.
32. Jacob, H.J., Lindpainter, K., Lincoln, S.E., Kusumi, K., Bunker, R.K., Mao, Y-P, Ganten, D., Dzau, V.J., and Lander, E.S. 1991. Genetic mapping of a gene causing hypertension in the stroke-prone spontaneously hypertensive rat. *Cell 67*:213–224.
33. Jones, R.H., and Foster, N.M. 1974. Oral infection of *Culicoides variipennis* with bluetongue virus: development of susceptible and resistant lines from a colony population. *J. Med. Entomol.* **11**:316–323.
34. Jones, R.H., and Foster, N.M. 1978. Heterogeneity of *Culicoides variipennis* to oral infection with bluetongue virus. *Am. J. Trop. Med. Hyg.* **27**:178–183.
35. Kamal, M. 1971. Biological studies of some midges and their relation to disease transmission, particularly horse sickness. *Bull. Soc. 1st Entomol. XXXII* **1948**:97–121.
36. Kambhampati, S., Black, W.C. IV, and Rai, K.S. 1991. Random amplified polymorphic DNA of mosquito species and populations (Diptera: Culicidae): Techniques, statistical analysis, and applications. *J. Med. Entomol.* **29**:939–945.
37. Kambhampati, S., Black, W.C. IV, Rai, K.S., and Sprenger, D. 1989. Temporal variation in the genetic structure of a colonizing species: *Aedes albopictus* (Skuse) in the United States. *Heredity* **64**:282–287.
38. Kambhampati, S., and Rai, K.S. 1991. Mitochondrial DNA variation within and among populations of the mosquito, *Aedes albopictus*. *Genome* **34**:293–297.
39. Kay, B.H., Edman, J.D., Fanning, I.D., and Mottram, P. 1989. The vector competence of *Culex annulirostris*, *Aedes sagax* and *Aedes alboannulatus* for

Murray Valley encephalitis virus at different temperatures. *Med. Vet. Entomol.* **3**:107–112.

40. Kay, B.H., Edman, J.D., Fanning, I.D., and Mottram, P. 1989. Larval diet and the vector competence of Culex *annulirostris* (Diptera: Culicidae) for Murray Valley Encephalitis virus. *J. Med. Entomol.* **26**:487–488.
41. Kirschoff, C. 1988. GATA tandem repeats detect minisatellite regions in blowfly DNA (Diptera: Calliphoridae). *Chromosoma* **96**:107–11.
42. Kramer, L.D., Hardy, J.L., Houk, E.J., and Presser, S.B. 1989. Characterization of the mesenteronal infection with western equine encephalomyelitis virus in an incompetent strain of *Culex tarsalis*. *Am. J. Trop. Med. Hyg.* **41**:241–250.
43. Kramer, L.D., Hardy, J.L., and Presser, S.B. 1983. Effect of temperature of extrinsic incubation on the vector competence of *Culex tarsalis* for western equine encephalomyelitis virus. *Am. J. Trop. Med. Hyg.* **32**:1130–1139.
44. Lander, E.S., and Botstein, D. 1986. Mapping complex genetic traits in humans: New methods using a complete RFLP linkage map. *Cold Spring Harbor Symp. Quant. Biol.* **51**:49–62.
45. Lander, E.S., and Botstein, D. 1989. Mapping Mendelian factors underlying quantitative traits using RFLP linkage maps. *Genetics* **121**:185–199.
46. Lander, E.S., Green, P., Abrahamson, J., Barlow, A., Daly, M.J., and Lincoln, S.E. 1987. MAPMAKER: An interactive computer package for constructing primary genetic linkage maps of experimental and natural populations. *Genomics* **1**:174–181.
47. Lincoln, S.E., and Lander, E.S., 1987. Constructing genetic linkage maps with MAPMAKER: A tutorial and reference manual. *Whitehead Institute Technical Report.*
48. Love, J.M., Knight, A.M., McAleer, M.A., and Todd, J.A. 1990. Towards construction of a high resolution map of the mouse genome using PCR analyzed microsatellites. *Nucl. Acids Res.* **18**:4123–4130.
49. Matthews, T.C., and Craig, G.B., Jr. 1980. Genetic heterozygosity in natural populations of the tree-hole mosquito, *Aedes triseriatus*. *Ann. Entomol. Soc. Am.* **73**:739–743.
50. Matthews, T.C., and Munstermann, L.E. 1990. Linkage maps for 20 enzyme loci in *Aedes triseriatus*. *J. Hered.* **81**:101–106.
51. Mellor, P.S., Boned, J., Hamblin, C., and Graham, S. 1990. Isolations of African horse sickness virus from insects made during the 1988 epizootic in Spain. *Epidemiol. Infect.* **105**:447–454.
52. Meredith, S.E.O., and James, A.A. 1990. Biotechnology as applied to vectors and vector control. *Ann. Parasit. Humaine Comparee* **65**:113–118.
53. Miller, B.R., and Mitchell, C.J. 1991. Genetic selection of a flavivirus-refractory strain of the yellow fever mosquito *Aedes aegypti*. *Am. J. Trop. Med. Hyg.* **45**:399–407.
54. Mitchell, C.J. 1983. Mosquito vector competence and arboviruses. In K.F. Harris, ed., *Current Topics in Vector Research*, Vol., 1, Praeger, New York, pp. 63–92.
55. Mitchell, C.J., Gubler, D.J., and Monath, T.P. 1983. Variation in infectivity of St. Louis encephalitis virus strains for *Culex pipiens quinquefasciatus* (Diptera: Culicidae). *J. Med. Entomol.* **20**:526–533.

56. Mitchell, C.J., Monath, T.P., and Cropp, C.B. 1981. Experimental transmission of Rocio virus by mosquitoes. *Am. J. Trop. Med. Hyg.* **30**:465–472.
57. Monath, T.P. 1989. Yellow fever. In T.P. Monath, ed., *The Arboviruses: Epidemiology and Ecology*. CRC Press, Boca Raton, FL, pp. 139–231.
58. Munstermann, L.E. 1980. Distinguishing geographic strains of the *Aedes altropalpus* group (Diptera: Culicidae) by analysis of enzyme variation. *Ann. Entomol. Soc. Am.* **73**:699–704.
59. Munstermann, L.E. 1985. Geographic patterns of genetic variation in the treehole mosquito *Aedes triseriatus*. In L.P. Lounibos, et al., eds., *Ecology of Mosquitoes*. Florida Medical Entomology Laboratory, Vero Beach, FL, pp. 327–343.
60. Munstermann, L.E. 1990. Gene map of the yellow fever mosquito *Aedes* (Stegomyia) *aegypti* (2N = 6). In S.J. O'Brien, ed., *Genetic Maps: Locus Maps of Complex Genomes*, Cold Spring Harbor Press, New York, pp. 3.179–3.183.
61. Munstermann, L.E. 1990. Gene map of the eastern North American tree hole mosquito, *Aedes triseriatus* (2N = 6). In S.J. O'Brien, ed., *Genetic Maps: Locus Maps of Complex Genomes*. Cold Spring Harbor Press, New York, pp. 3.184–3.187
62. Munstermann, L.E., Taylor, D.B., and Matthews, T.C. 1982. Population genetics and speciation in the Aedes triseriatus group. In W.M. Steiner, W.J. Tabachnick, K.S. Rai, and S. Narang, eds., *Recent Developments in the Genetics of Insect Disease Vectors*. Stipes, Champaign, IL, pp. 433–453.
63. Paterson, A.H., Damon, S., Hewitt, J.D., Zamir, D., Rabinowitch, H.D., Lincoln, S.E., Lander, E.S., and Tanksley, S.D. 1991. Mendelian factors underlying quantitative traits in tomato: Comparison across species, generations, environments. *Genetics* **127**:181–197.
64. Paterson, A.H., Lander, E.S., Hewitt, J.D., Peterson, S., Lincoln, S.E., and Tanksley, S.D. 1988. Resolution of quantitative traits into Mendelian factors using a complete map of restriction fragment length polymorphisms. *Nature (London)* **335**:721–725.
65. Reeves, W.C. 1990. *Epidemiology and Control of Mosquito-Borne Arboviruses in California, 1943–1987*. California Mosquito and Vector Control Association, Sacramento, CA.
66. Ried, T., Baldini, A., Rand, T.C., and Ward, D.C. 1992. Simultaneous visualization of seven different DNA probes by *in situ* hybridization using combinatorial fluorescence and digital imaging microscopy. *Proc. Natl. Acad. Sci. U.S.A.* **89**:1388–1392.
67. Robertson, M.A., and Tabachnick, W.J. 1992. Molecular genetic approaches to *Culicoides variipennis* vector competence for bluetongue viruses. In T.E. Walton, and B.I. Osburn, eds., *Bluetongue, African Horse Sickness, and Related Orbiviruses*. CRC Press, Boca Raton, FL, pp. 271–277.
68. Romans, P., Seely, D.C., Kew, Y., and Gwadz, R.W. 1991. Use of a restriction fragment length polymorphism (RFLP) a genetic marker in crosses of *Anopheles gambiae* (Diptera: Culicidae): Independent assortment of a diphenol oxidase and an esterase locus. *J. Med. Entomol.* **28**: 147–151.
69. Serikawa, T., Kuramoto, T., Hilbert, P., Mori, M., Yamada, J., Dubay, C.J., Lindpainter, K., Ganten, D., Guenet, J.-L., Lathrop, G.M., and

Beckmann, J.S. 1992. Rat gene mapping using PCR-analyzed microsatellites. *Genetics* **131**:701–721.
70. Simpson, S.P. 1989. Detection of linkage between quantitative trait loci and restriction fragment length polymorphisms using inbred lines. *Theor. Appl. Genet.* **77**:815–819.
71. Song, K., Osborn, T.C., and Williams, P.H. 1990. Brassica taxonomy based on nuclear restriction fragment length polymorphisms (RFLPs). 3. Genome relationships in *Brassica* and related genera and the origin of B-Oleracea and B-Rapa (SynCampestris). *Theor. Appl. Genet.* **79**:497–506.
72. Stallings, R.L., Ford, A.F., Nelson, D., Torney, D.C., Hildebrand, C.E., and Moyzis, R.K. 1991. Evolution and distribution of $(GT)_n$ repetitive sequences in mammalian genomes. *Genomics* **10**:807–815.
73. Tabachnick, W.J. 1991. Evolutionary genetics and arthropod-borne disease. The yellow fever mosquito. *Am. Entomol.* **37**:14–24.
74. Tabachnick, W.J. 1991. Genetic control of oral susceptibility to infection of *Culicoides variipennis* for bluetongue virus. *Am. J. Trop. Med. Hyg.* **45**:666–671.
75. Tabachnick, W.J. 1992. Genetic differentiation among North American populations of *Culicoides variipennis*. *Ann. Entomol. Soc. Am.* **85**:140–147.
76. Tabachnick, W.J. 1992. Genetics, population genetics, and evolution of *Culicoides variipennis*: Implications for bluetongue virus transmission in the USA and its international impact. In T.E. Walton and B.I. Osburn, eds., *Bluetongue, African Horse Sickness, and Related Orbiviruses*. CRC Press, Boca Raton, FL, pp. 262–271
77. Tabachnick, W.J., Wallis, G.P., Aitken, T.H.G., Miller, B.R., Amato, G.D., Lorenz, L., Powell, J.R., and Beaty, B.J. 1985. Oral infection of *Aedes aegypti* with yellow fever virus: Geographic variation and genetic considerations. *Am. J. Trop. Med. Hyg.* **34**:1219–1224.
78. Tardieux, I., Poupel, O., Lapchin, L., and Rodhain, F. 1990. Variation among strains of *Aedes aegypti* in susceptibility to oral infection with dengue type 2. *Am. J. Trop. Med. Hyg.* **43**:308–313.
79. Tardieux, I., Poupel, O., Lapchin, L., and Rodhain, F. 1991. Analysis of inheritance of oral susceptibility of *Aedes aegypti* (Diptera: Culicidae) to dengue-2 virus using isofemale lines. *J. Med. Entomol.* **28**:518–521.
80. Tautz, D. 1989. Hypervariability of simple sequences as a general source for polymorphic DNA markers. *Nucl. Acids Res.* **17**:6463–6471.
81. Tautz, D., and Renz, M. 1984. Simple sequences are ubiquitous repetitive components of eucaryotic genomes. *Nucl. Acids Res.* **12**:4127–4138.
82. Tesh, R.B., Gubler, D.J., and Rosen, L. 1976. Variation among geographic strains of *Aedes albopictus* in susceptibility to infection with chikungunya virus. *Am. J. Trop. Med. Hyg.* **25**:326–333.
83. Turrell, M.J. 1989. Effect of environmental temperature on the vector competence of *Aedes fowleri* for Rift Valley fever virus. *Res. Virol.* **140**: 147–154.
84. Turrell, M.J., Rossi, C.A., and Bailey, C.L. 1985. Effect of extrinsic incubation on the ability of *Aedes tritaeniorhynchus* and *Culex pipiens* to transmit Rift Valley fever virus. *Am. J. Trop. Med. Hyg.* **34**:1211–1218.
85. Vernick, K.D., and Collins, F.H. 1989. Association of a *Plasmodium*-refractory phenotype with an esterase locus in *Anophles gambiae*. *Am. J. Trop. Med. Hyg.* **40**:593–597.

86. Vernick, K.D., Collins, F.H., and Gwadz, R.W. 1989. A general system of resistance to malaria infection in *Anopheles gambiae* controlled by two main genetic loci. *Am. J. Trop. Med. Hyg.* **40**:585–592.
87. Wallis, G.P., Aitken, T.H.G., Beaty, B.J., Lorenz, L., Amato, G.D., and Tabachnick, W.J. 1985. Selection for susceptibility and refractoriness of *Aedes aegypti* to oral infection with yellow fever virus. *Am. J. Trop. Med. Hyg.* **34**:1225–1231.
88. Walton, T.E., and Osburn, B.I. 1992. *Bluetongue, African Horse Sickness, and Related Orbiviruses*. CRC Press, Boca Raton, FL.
89. Weber, J.L. 1990. Informativeness of human (dC-dA)n(dG-dT)n polymorphisms. *Genomics* **7**:524–530.
90. Weber, J.L., and May, P.E. 1989. Abundant class of human DNA polymorphisms which can be typed using the polymerase chain reaction. *Am. J. Hum. Genet.* **44**:388–396.
91. Williams, J.G.K., Kubelik, A.R., Livak, K.J., Rafalski, A., Tingey, S.V. 1990. DNA polymorphisms amplified by arbitrary primers are useful as genetic markers. *Nucl. Acids Res.* **18**:6531–6536.
92. Zheng, L., Saunders, R.D.C., Fortini, D., Torre, A.D., Coluzzi, M., Glover, D.M., and Kafatos, F.C. 1991. Low-resolution genome map of the malaria mosquito, *Anopheles gambiae*. *Proc. Natl. Acad. Sci. U.S.A.* **88**:11187–11191.

5
The Endocrinology of the Adult Female Mosquito

Henry H. Hagedorn

Introduction

Because of its medical importance, the mosquito is one of the most intensively studied insects. The reproductive endocrinology of the adult female is an important topic of current research in this area. This is so not only because of the medical importance of the animal, but also because mosquitoes provide many technical advantages for such research. The ease with which mosquitoes can be reared in large numbers, the cyclic nature of reproduction, and the simplicity of dissection all have contributed to the vast amount of information available on this topic. Thus despite its small size, the mosquito can hold its own against larger model insects such as the tobacco hornworm, *Manduca sexta*.

In this review the known physiological roles of each of the known and inferred hormones will be discussed in turn. To set the stage, an overview of the endocrinology of the adult female *Aedes aegypti* will be presented. As shown in Fig. 5.1, the titers of juvenile hormone (JH) and 20-hydroxyecdysone (20-HE) are low in the newly emerged adult female. During the first 2 days of adult life the level of JH rises, perhaps in response to peptide hormones from the brain such as the allatotropins, known from other insects, but not yet studied in the mosquito. The level of JH remains high but gradually declines in the aging female. However, if the female is given a blood meal, the levels of JH drop sharply and remain low for 36 hr. After a blood meal a peptide hormone, the egg development neurosecretory hormone (EDNH) that stimulates the production of ecdysone by the ovary, is released. The rise in ecdysteroid titer begins within a few hours after the blood meal, and reaches a peak by 18–24 hr. As the level of ecdysteriods falls, JH titers begin to rise again.

Henry H. Hagedorn, Department of Entomology and Center for Insect Science, The University of Arizona, 430 Forbes Building #36, Tucson, Arizona 85721, USA.

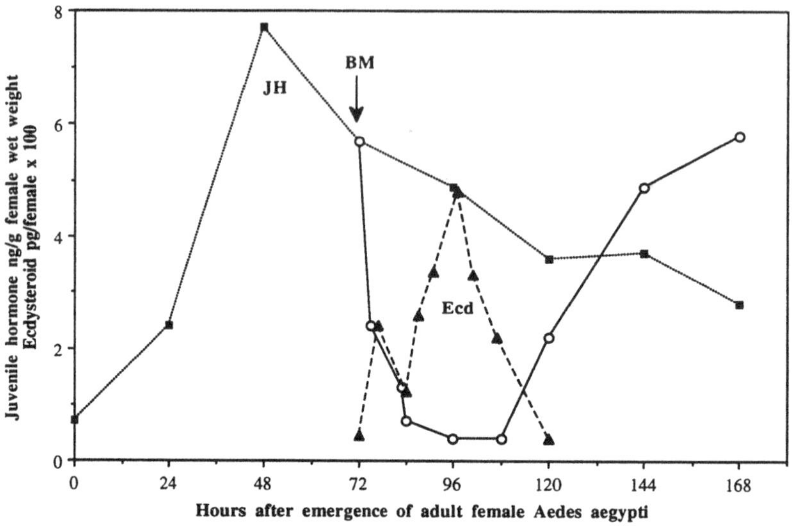

FIGURE 5.1. Juvenile hormone (JH) and ecdysteroid (Ecd) levels in female *Aedes aegypti* after emergence and after a blood meal (BM). The juvenile hormone levels are shown for both the non-blood-fed female (solid squares, dotted line) and the blood-fed female (open circles, solid line). Ecdysteroid levels (triangles, dashed line) are shown after a blood meal. Juvenile hormone data are from Shapiro et al. (161) and ecdysteroid data are from an experiment done on 6/10/80 by G.D. Wheelock (unpublished) chosen to show an experiment in which the early peak was evident. The ecdysteroid levels were determined by radioimmunoassay of whole animals extracted in 50% methanol (169).

It is within the context of these changes in hormone titers that the physiological events that occur during reproduction in the adult female can be placed (Fig. 5.2). Juvenile hormone has been shown to affect several aspects of reproductive biology after eclosion of the adult female. The ovary grows to the resting stage (58, 67), the endocytotic complex develops (144), and the ovary becomes competent to respond to EDNH by producing ecdysone (160). The fat body undergoes an increase in ploidy (36), the synthesis of ribosomes is stimulated (145), and the fat body becomes competent to respond to ecdysone (42). In addition, the behavior of the female changes; she becomes receptive to the male (60) and, in some species, she becomes avid for the blood meal (124). The release of EDNH after a blood meal stimulates the production of ecdysone by the ovary (69). The ecdysone converted to 20-HE and affects the separation of new follicles (8), and expression of several genes including vitellogenin (52, 139), a vitelline envelope protein (107) and a serine carboxypeptidase (31). It also stimulates trypsin (26) and dopa decarboxylase (157) enzyme activites, and may affect the behavior of the female (25). There is some evidence for factors that regulate the activity

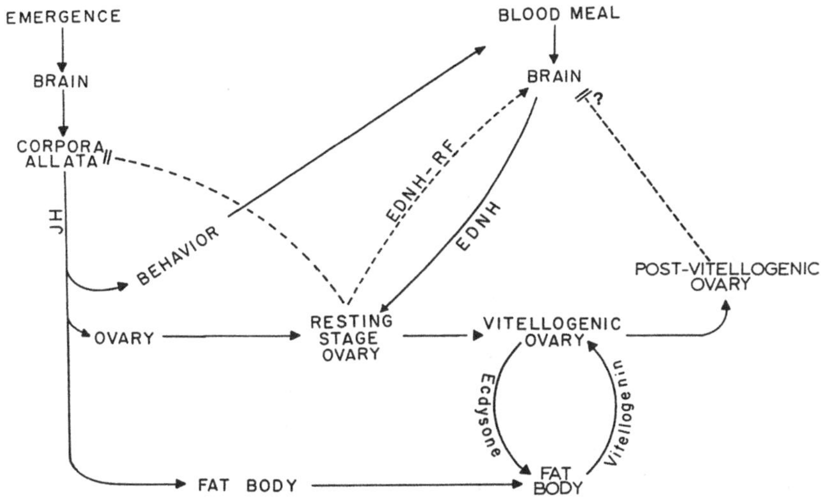

FIGURE 5.2. Physiology of egg development in the mosquito, *Aedes aegypti*. Dotted lines indicate proposed pathways.

of the corpora allata (152), allow the release of EDNH (12, 104), and inhibit the further development of eggs when mature eggs are present in the ovary (38, 123). Details of these events are discussed below.

The female mosquito can take a number of blood meals and there is some, incomplete, evidence that juvenile hormone, EDNH, and 20-HE levels rise and fall for each batch of eggs (100, 161). The hormonal changes are coordinated by eclosion and by the taking of blood meals as is shown if Fig. 5.3.

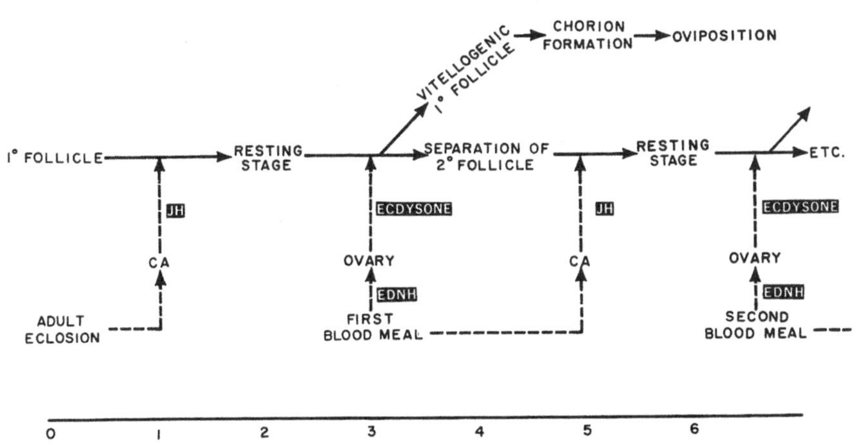

FIGURE 5.3. Repetitive egg development in the mosquito, *Aedes aegypti*.

Juvenile Hormone

JH Changes in the Adult Female?

JH Levels in the Mosquito

As is the case in most insects, JH III is the only JH present in larval and adult *A. aegypti* (3). Changes in amounts of JH in whole body extracts of *A. aegypti* were followed by Shapiro et al. (161). JH levels in whole body extracts can presumably be related to physiologically relevant hormone titers as JH is known to be released from the corpora allata immediately after synthesis and there is no evidence for the storage of JH in mosquitoes. Storage is known in a few other insects, such as the accessory glands of the male *Cecropia* moth (162). As shown in Fig. 5.1, the amount of JH in whole body extracts rises after eclosion until day 2, declining gradually thereafter to day 7. If a blood meal is given on day 3, JH levels decline rapidly during the first 12 hr, remain low for another 24 hr, and then rise again between 36 and 96 hr.

The activity of JH degrading enzymes has been shown to be important in the regulation of JH levels in the mosquito. Shapiro et al. (161) found that JH esterase activity rose slowly for the first 12 hr after a blood meal, and then increased more rapidly to a peak at 42 hr. The highest levels of esterases were seen when JH levels were lowest. The importance of the JH esterase was shown by use of a JH esterase inhibitor, BPAT, which, when topically applied, caused abnormal hatching of the egg (161). A JH epoxide-hydrolase, also present since the final metabolite of JH was the acid-diol (161), has a secondary role in JH degradation (22), but the levels of JH epoxide hydrolase activity are high in the adult female *A. aegypti* (L. Harshman and B. Hammock, personal communication).

Measuring Activity of the Corpora Allata

The synthetic activity of the corpora allata has been followed from eclosion through a blood meal in female *Culex pipiens* by Readio et al. (149) who used an assay based on the method of Pratt et al. (136) in which the corpora allata are incubated with [^{3}H]methionine to label newly synthesized juvenile hormone. Their results generally confirmed the pattern seen by Shapiro et al. (161) in that the corpora allata was found to be synthetically active after eclosion, JH synthesis rapidly declined after a blood meal, and was restored by 3 days after the blood meal. The activity of the corpora allata did not change from eclosion through day 8 if the female was fed sugar, whereas Shapiro et al. (161) found that the amounts of JH gradually declined from day 2 to 7 in the sugar-fed female. These differences could be due to more rapid elimination of JH in the older female, or the use of different mosquito species.

Borovsky and Carlson (22) developed a new assay for JH synthesis in which the corpora allata are left *in situ* but exposed to a bathing medium containing [12-^{3}H]methyl farnesoate. It differs from the assay of Pratt et al. (136) in that the corpora allata are left intact, and the label used is a metabolic intermediate that measures only the last step of JH production rather than [^{3}H]methionine, which acts as a methyl donor and labels overall JH synthesis. The advantages of this assay are the ease in preparation of the tissue and the use of a labeled compound that avoids the labeling of proteins. The disadvantages are first, that fluctuations in the pool of unlabeled methyl farnesoate could cause changes in the specific activity of the labeled methyl farnesoate, and second, that the presence or absence of degradative enzymes, particularly JH esterases, could bias the results. Finally, measuring the last step of the synthetic pathway ignores the possibility that earlier steps are rate limiting, as has indeed been found to be the case in the locust (165). All are potentially serious sources of error and cannot be controlled in the *in situ* assay. Using *A. aegypti*, Borovsky et al. (22) compared the results from this assay to an *in vivo* assay in which labeled methyl farnesoate was topically applied. The *in vivo* assay gave results very similar to those obtained by Readio et al. (149) in that rates of JH labeling were low shortly after a blood meal, and rose again 48 hr later. However, the assay using exposed corpora allata showed a rise in methyl farnesoate epoxidation during the first 5 hr after a blood meal. It is difficult to interpret these results without detailed information on whether the *in situ* corpora allata assay is a valid measure of JH synthesis.

Regulation of Corpora Allata Activity

It has been suggested that the ovary controls the activity of the corpora allata at two stages of egg development, after the ovary reaches the resting stage, and when it is full of mature oocytes ready to be oviposited. Rossignol et al. (152) showed that ovaries removed from pupae would grow to the resting stage after implantation into sugar-fed females only if the host ovaries had been removed. They suggested that the resting stage ovary inhibits the activity of the corpora allata, since growth to the resting stage is JH dependent (58, 67). Similar results were obtained by Meola and Readio (125), and the decline in JH amounts in the sugar-fed female reported by Shapiro et al. (161) is consistent with this interpretation. However, the effect of the resting stage ovary cannot be directly on the corpora allata since Readio et al. (149) found that the activity of the corpora allata did not change after the primary follicles reached the resting stage. Furthermore, Shapiro et al. (161) found no change in JH esterases during this period. Thus the effect of the resting stage ovary on the growth of implanted preresting stage ovaries remains unexplained.

An effect of the mature ovary on the activity of the corpora allata has been shown by Readio et al. (149), who examined the effect of oviposition on JH synthesis by the corpora allata. If mature eggs were present the activity of the corpora allata was depressed, but activity was restored after the eggs were laid. Ovaries at this stage of development are known to produce an unidentified "oostatic hormone" that inhibits development of a second batch of eggs (see below), and this factor is therefore a candidate for having an allatostatic activity.

These results demonstrate that the activity of the corpora allata is regulated by a number of endogenous and exogenous factors. Recent work on the control of the corpora allata in other insects has led to the isolation and sequencing of peptides that have allatostatic and allatotropic effects (82, 95, 137, 138, 173). It is likely that related peptides regulate the activity of the corpora allata in mosquitoes and investigations into the control of the corpora allata of mosquitoes should be rewarding.

Effect of JH on Nutritional Reserves

The diet of the anautogenous adult female mosquito varies considerably. After eclosion, sugar, presumably from nectar, is the major nutrient. In mosquitoes that have been starved as larvae, which may include most mosquitoes in the wild (41), a sugar meal may be required to ensure previtellogenic development of the oocytes. The blood meal presents an entirely different diet with protein being the major nutrient. Most of the reserves in either a meal of sugar or protein are stored as triglycerides. Van Handel and Lea (167) have shown that a hormone from neurosecretory cells of the brain regulates the production of triglycerides from a sugar meal, but has no effect on the production of triglycerides from a protein meal. Removal of the corpora allata had no effect on metabolism of a sugar or protein meal.

However, the nutritional status of the female seems to have an important effect on the activity of the corpora allata. For example, Feinsod and Spielman (41) found that oocytes of small adult females, derived from larvae that were starved during larval development, failed to undergo normal previtellogenic growth, which has been demonstrated to be under the control of JH (see below). Feeding on blood or sucrose stimulated normal previtellogenic growth, but the blood meal in these small females failed to cause vitellogenic growth. Topical applications of JH had the same effect as the sucrose or blood meal. This suggests that the corpora allata were not active after eclosion in the small female. Furthermore, the corpora allata appear to have been nutritionally inactivated since feeding restored function.

The same conclusion can be drawn from experiments where normally sized females were starved after eclosion (41). Klowden and Chambers (92) showed that such starved females failed to develop eggs after a small blood enema, but egg development was restored by feeding a sugar meal or topically applying JH. They further showed that JH stimulated the utilization of stored carbohydrate reserves. It is possible, therefore, that one of the effects of JH is to stimulate the utilization of stored carbohydrate and they postulate that the normal role of JH after eclosion is to provide nutrients to the growing oocytes.

The endocrinology of small females in which a blood meal failed to stimulate vitellogenic growth is unknown, and it would be interesting to discover whether 20-HE levels rise after a blood meal in these animals. It is also not known if the sucrose meal stimulates production of JH by the corpora allata in the female starved after eclosion.

JH and Previtellogenic Oocyte Growth

Previtellogenic growth of the primary oocyte occurs after eclosion, and of secondary, and succeeding, oocytes after a blood meal. Several patterns of growth of the secondary oocyte have been described. The previtellogenic growth of the secondary oocyte to the resting stage in *A. aegypti* begins 36 hr after the blood meal (67) and requires JH (40). Feinsod and Spielman (40) predicted that JH levels would be low during the first day afer a blood meal, and would rise by the third day, as was indeed found by Shapiro et al. (161). Thus in this mosquito, growth of the secondary oocyte is delayed until the second rise in JH that occurs between the second and third days after the blood meal. In *Anopheline* mosquitoes, growth of the secondary oocyte occurs concomitantly with the primary oocyte (*Anopheles stephensi*, 150; *Anopheles albimanus*, 109). Redfern (150) suggested that the growth of the secondary oocyte in *A. stephensi* is controlled by 20-HE rather than JH. However, it is more likely that Redfern was unable to distinguish between induction of follicle separation after a blood meal that is under the control of 20-HE (8) from further growth to the resting stage that is stimulated by the second rise in JH (Fig. 5.1). It would be difficult to distinguish these in *Anopheline* mosquitoes without knowing when JH titers rose after a blood meal. A different pattern of secondary oocyte growth is seen in *C. pipiens* where the follicles show an initial period of growth 2 to 5 days after the blood meal and a second growth phase after oviposition (148). Allatectomy inhibited the second growth phase, and a second blood meal was ineffective in stimulating egg development in the allatectomized female, demonstrating that the second period of oocyte growth was essential. The effect of oviposition on growth of the secondary follicle has not been investigated in *A. aegypti*.

JH and the Development of Competence

JH has been implicated in the development of competence, in particular the ability of the fat body to respond to 20-HE and the ovary to respond to EDNH (see below). The fat body is not capable of responding to

20-HE immediately after eclosion, but becomes competent to respond during the first 2 days as JH titers rise (42). The implication that JH was involved was shown by topical application of methoprene to isolated abdomens. A similar role for JH was suggested by Shapiro and Hagedorn (160) who found that the ovary was not competent to respond to EDNH immediately after eclosion. Competence to respond was induced by the topical application of JH I or methoprene. In both cases the biochemical basis of competence is still not known, although an increase in hormone receptors and changes in the structure of the DNA are some of the more obvious possibilities.

JH and Ribosomal RNA Synthesis

The dramatic change in total RNA in the fat body of the adult female was evident in some of the first biochemical and ultrastructural studies of the reproductive cycle in mosquitoes. A modest increase in free ribosomes and endoplasmic reticulum occurred after emergence, but a 3-fold increase in total RNA was seen after a blood meal that was correlated with a large increase in rough endoplasmic reticulum in the fat body cells (9, 11, 65). Raikhel and Lea (141, 143) pursued these ultrastructural changes in considerably more detail. Immunocytochemistry was used to follow the secretory pathway of vitellogenin from rough endoplasmic reticulum, through Golgi complexes to secretory granules (143). A striking feature of these data was the buildup of a vast amount of protein synthetic machinery by 30 hr after the blood meal followed by its degradation in autophagic vacuoles by 48 hr.

It was clear from the ultrastructural work that ribosomal RNA synthesis had to be coordinated with other events during egg development. Park and Fallon (130) measured copy number of the ribosomal genes during development of *A. aegypti* and found that copy number increased during metamorphosis from 400 to 1200 copies per haploid genome. No further changes were seen in the adult fat body after a blood meal. Hotchkin and Fallon (75) measured changes in ribosomal RNA after a blood meal, and Johnson and Fallon (79) characterized the ribosomal proteins of *A. aegypti*. No gross qualitative effects of 20-HE were seen on phosphorylation of ribosomal proteins. Raikhel and Lea (145) examined the ultrastructure of the nucleoli, the cellular source of ribosomes, and found that they increased in size during the previtellogenic phase of oogenesis. They further found that allatectomy decreased the size of the nucleoli and topical application of JH, or 7-S-methoprene, restored the normal amount of ribosomal RNA and ribosomal RNA synthesis. Thus it appears that previtellogenic proliferation of ribosomes is regulated by JH. They found no evidence that JH regulates the quantitatively much more significant increase in ribosomal RNA after the blood meal, and the role of hormones in this part of the process is not understood. The cyclic construction and destruction of large numbers of ribosomes, without cell death, presents a number of interesting problems to the biochemist. For example, how is the rate of turnover of ribosomes regulated? Fritz et al. (45) found little evidence for changes in general ribonuclease activity in fat body during a reproductive cycle, but the possibility of a specific ribonuclease was not ruled out.

JH and Development of the Follicle Cells

Laurence and Simpson (98) have shown that the follicle cells of the ovary undergo distinct cycles of replication after emergence and after a blood meal in several mosquitoes, including *A. aegypti*, *C. pipiens molestus*, *Aedes togoi*, and *Mansonia uniformis*. In *A. aegypti*, total cell number per follicle increased from 18 to 214 by 96 hr after eclosion and then doubled by 10 hr after a blood meal. Raikhel and Lea (146) showed that many aspects of growth and development of the follicle cells after eclosion are under the control of JH, including an increase in free ribosomes, rough endoplasmic reticulum, and Golgi complexes. Although they did not examine the effect of JH on follicle cell number, it is likely that JH regulates this process after eclosion. As mentioned above, Laurence and Simpson (98) found that follicle cell number also increased after a blood meal. A peak of thymidine incorporation into follicle cell nuclei was found at 4 hr after feeding. It is not clear what controls this second cycle of cell replication.

JH and Ploidy of the Fat Body

Dittmann et al. (36) showed that the fat body cells of *A. aegypti* undergo an increase in ploidy, from $2n$ to $4n$, after emergence and again after a blood meal (from $4n$ to $8n$). These ploidy changes were correlated with increases in JH. Topical application of JH III or methoprene to abdomens isolated at emergence stimulated increases in ploidy. Synthesis of DNA (36) was apparent within 12 hr after emergence and after a blood meal; an increase in total DNA was also apparent within this time frame (Hagedorn, unpublished observations). However, the increase in ploidy was not evident until several days later. For example, after a blood meal, DNA synthesis rose to a peak at 12 hr and then declined to background levels by 24 hr. But, microspectrophotometry failed to show an increase in ploidy until 72 hr after the meal, which is when JH levels rise after a blood meal. Dittmann et al. (36) reviewed the evidence that newly synthesized DNA takes several days to become part of the chromosomal, high molecular weight, DNA, which may account for the delay in the ploidy increase. Thus, the evidence suggests that JH regulates a step in this longer term process, rather than DNA synthesis, but the nature of the processes being regulated are not understood, nor is it clear how DNA synthesis is controlled. This appears to be an excellent system for studying the effects of hormones on ploidy changes.

JH and the Development of the Endocytic Complex

Roth and Porter (153) described the presence of unique structures on the surface of the mosquito oocyte, called coated pits, that they surmised were involved in the uptake of vitellogenin. These structures have been since investigated in many systems and have been found to be present in cells actively taking up macromolecules by a process called receptor-mediated endocytosis. The process has been intensively investigated in mosquitoes (2, 140, 144) and reviewed by Raikhel (141). Raikhel and Lea (144) have shown that the development of the endocytic complex, which they define as the specialized oocyte cortex found in the resting stage oocyte that consists of microvilli and coated pits, occurs after adult eclosion, and is prevented from forming by removal of the corpora allata. Reimplantation of corpora allata, or topical application of JH III restored development of the complex. There is no evidence that uptake of vitellogenin by the endocytic complex after a blood meal is under hormonal control (147).

JH, Biting Behavior, and Sexual Receptivity

In some mosquito species, preoviposition behaviors are regulated, in part, by JH. In *C. pipiens*, JH appears to stimulate biting behavior; allatectomy at emergence blocked the development of biting behavior (124). This effect is not evident in Aedine or Anopheline mosquitoes (24). In *A. aegypti*, JH induces the female to become receptive to mating after emergence (60).

Ecdysteroids

What Kinds of Ecdysteroids Are Present?

The original descriptions of the kinds of ecdysteroids found in *A. aegypti* after a blood meal indicated that incubated ovaries released ecdysone into the medium, while whole animal extracts, prepared 14–26 hr after the blood meal, contained only 20-hydroxyecdysone (20-HE) (66). These analyses were done after separation of ecdysteroids by sequential use of continuous flow thin-layer chromatography, reverse-phase HPLC, and gas chromatography of trimethylsilyimidazole derivatives. The ecdysone RIA was used to follow active fractions through the separation procedure. These data suggested that ecdysone produced by the ovary is rapidly converted to 20-HE. Similar results were obtained by Masler et al. (117) who examined *Aedes atropalpus*.

However, other reports have suggested that significant amounts of ecdysone are present in whole animal extracts. For example, in *A. stephensi*, Redfern (150) found that 70% of the total ecdysteroid found in the blood fed female 30–32 hr after a blood meal was ecdysone. The amount of 20-HE in this mosquito was not different from *A. aegypti*. Borovsky et al. (20) found that 70% of total ecdysteroids was ecdysone in whole *A. aegypti* extracts taken 24 hr after a blood meal, however, most was in the ovary. Ecdysteroids have been found in the ovaries of many other insects, where they may be involved in regulating embryonic molting and reinitiation of meiosis (74, 96). These ovarian ecdysteroids are typically conjugates, which may or may not be detected by the radioimmunoassays used in these studies. It may be that the differences seen in these studies reflect differences in the timing of movement of steroids into the ovary and their conjugation. Indeed, Lu and Hagedorn (109) found that in whole animal extracts of *A. albimanus*, free ecdysteroids became undetectable between 36 and 42 hr after a blood meal, while conjugated ecdysteroids rose thereafter. The activity of the conjugating enzyme, 20-HE sulfotransferase, has been followed in *A. togoi*, but the blood-fed adult was not examined (159). Generally, therefore, the identification of the steroids present in mosquitoes may be in question. The use of multiple methods of separation and derivitization of the steroids should result in reliable identification, but mass spectroscopy will be necessary for definitive identification, and this has never been done for steroids found in mosquitoes.

Smith and Mitchell (163) followed ecdysone 20-monooxygenase activity (which converts ecdysone to 20-hydroxyecdysone) after a blood meal. They found that the apparent activity of this enzyme was high in the gut, Malpighian tubules, and ovaries of the non-blood-fed female. Enzyme activity was high immediately after the blood meal, but then fell to low levels in these tissues by 20 hr. In contrast, enzyme activity in the fat body was low in the non-blood-fed and newly fed females, but increased 20–24 hr after a blood meal. These data support the report (62) that the conversion of ecdysone to 20-HE by fat body is 2-fold greater after a blood meal. The fact that Smith and Mitchell (163) found significant enzyme activity in the ovary is quite surprising as only ecdysone was found after incubation of ovaries (66).

The suggestion that ecdysone is not active after injection into non-blood-fed females (49) was not confirmed by Lea (103), who found that ecdysone and 20-HE were about equally effective in stimulating yolk deposition as would be expected if ecdysone were rapidly converted to 20-HE.

The amounts of ecdysteroids present in non-blood-fed female *A. aegypti* have been variously reported as high as 80–100 pg/female (20,

66), and as low as 27 ± 6 pg/female (56, 139). This is most likely to be due to differences between strains. The physiological meaning of the presence of 20-HE prior to the blood meal is unknown.

Calculations of the half-life of ecdysteroids in the mosquito have been made by following the decline in RIA positive material after injection or hormone (61, 62) or after ovarietomy (20). The calculated half-life from either method is on the order of several hours. However, what is needed is an examination of the decline in 20-HE titers, not total immunoreactive material, and in the hemolymph, not in whole body extracts. The actual half-life is likely to be much shorter when measured under these conditions.

In conclusion, therefore, there are inconsistencies in the data available on the steroids present in the adult mosquito and this remains a fertile area for investigation.

Effects of 20-HE on Egg Development

The observations of Spielman et al. (164) in 1971 that injected or ingested 20-HE stimulated egg development in *A. aegypti* had little impact because they did not fit into the prevailing paradigm that ecdysone was absent in adult insects. A few years later, however, experiments on the role of dopa decarboxylase in tanning of the eggshell (47, 157) and on vitellogenin synthesis by the fat body (39, 64) provided two physiological systems in which 20-HE apparently played an essential role. The discovery that the ovary was the source of 20-HE in the adult (66) completed the physiological argument. Since then a number of effects have been ascribed to the presence of 20-HE in the adult female including various aspects of egg development, behavior, and parasites as discussed below. The hypothesis that 20-HE stimulated vitellogenin synthesis was debated vigorously and has been the topic of several reviews (13, 48, 61).

Effects of 20-HE *In Vivo*

The Isolated Abdomen

One aspect of the debate on the role of 20-HE on egg development concerned the large (5–10 μg) doses of 20-HE that were required to obtain an effect *in vivo*, and the small amounts of vitellogenin made in responsse to injections of 20-HE (62). The nature of the argument was considerably altered when Borovsky et al. (19) showed that topical applications of small amounts (25 pg) of a JH analog, methoprene, reduced the amounts of 20-HE required to stimulate normal levels of vitellogenin in isolated abdomens from micrograms to nanograms.

It would appear that the experiments of Borovsky et al. (19) changed the problem from one of the role of 20-HE to the role of JH (or methoprene?). However, the role of JH in modulating vitellogenin synthesis is ambiguous and probably an artifact. First, removal of the corpora allata after the initial rise in JH titers after eclosion does not affect egg development. Lea (99, 101, 102) showed that removal of the corpora allata 3 days after eclosion did not prevent egg development following a blood meal in *A. aegypti*, *Aedes sollicitans*, *Aedes taeniorhynchus*, *C. pipiens*, and *Culex nigripalpus*. The effect varied between species. For example, 80–90% of the allatectomized *A. sollicitans*, *A. taeniorhynchus*, *C. pipiens*, and *C. nigripalpus* developed eggs after a blood meal, while 50% of the *A. aegypti* did so. Also, diet had an effect as only 16% of *A. sollicitans* that were starved as larvae responded (suggesting again that the corpora allata were not active after emergence in starved animals). In one experiment when Lea (99) allatectomized *A. sollicitans* 3 days after eclosion and then waited 12 days to feed them a blood meal, 50% of the animals still developed eggs after a blood meal. Borovsky et al. (19) questioned these data because the numbers of maturing oocytes was not reported.

Second, JH titers drop after a blood meal. As shown in Fig. 5.1, JH titers are high in the non-blood-fed female, and that they decline during the first 12 hr after a blood meal (161). The isolated abdomens used by Borovsky et al. (19) may or may not have experienced the same decline in JH titer as would have occurred in the whole animal, depending on the effect of decapitation on JH esterases. The topical application of 25 pg of methoprene would change only the rate of decline. However, it is known that topical application of methoprene to decapitated mosquitoes or isolated abdomens causes a rise in ecdysteroid titers in both *A. aegypti* and *A. atropalpus* (10, 19). This effect is also known in *D. melanogaster* (158).

Third, determination of 20-HE titers after a blood meal in several mosquitoes often (but not always) shows two peaks (Fig. 5.1 and 56, 66, 118, 150, 157). When it is detected, the first of these peaks occurs in the first 10 hr and the second always appears at 24–30 hr. Two doses of 20-HE were also shown to be necessary to stimulate normal separation of new follicles (8). Martinez and Hagedorn (116) showed that two doses of 20-HE (50 pg followed by 30 ng, 18 hr later) had the same effect as 50 pg of methoprene followed by 50 ng of 20-HE, 18 hr later.

In all of the experiments of Borovsky et al. (19) the abdomens were treated with methoprene immediately after ligation, and 18–24 hr later injected with 20-HE, or implanted with ecdysone-secreting ovaries. If one effect of methoprene is to cause a transient rise in 20-HE production by the ovaries, then the ultimate effect of the treatment with both hormones is to recreate the titer changes that are seen *in vivo* after a blood meal. Finally, it should be noted that methoprene was used in all of these experiments, and it has not been shown how much JH III is needed to replicate these results. Thus the low doses of methoprene used may be misleading. It is also possible that the effects of methoprene are different from JH III (92).

A quite different argument was made by Klowden and Chambers (92). They found that application of JH III caused the mobilization of carbohydrate reserves. This may be significant for nutritionally deprived mosquitoes, and may also be an important factor in experiments involving isolated abdomens, and perhaps in experiments with incubated fat body preparations. But as Klowden and Chambers (92) state, the fact that JH has stimulatory effects in certain experimental systems does not mean that the hormone is normally involved.

Thus, it appears to this reviewer that the weight of the evidence available now does not support a role for JH after a blood meal. However, given the evidence for regulatory effects of declining titers of hormones in insect systems, this possibility should be investigated in *A. aegypti*.

The Infused Mosquito

One paper deserves mention because it provides a potentially useful model system for future investigations. Lea (103) infused live animals with small amounts of 20-HE that were considered to replicate titers measured by Hagedorn et al. (66). Of 27 that were infused, 2 showed an increase in yolk. Although Lea (103) did not regard these as sufficient to postulate a role for 20-HE in egg development, since two of the infused mosquitoes did respond the question becomes under what conditions does infusion cause the majority to respond? How much 20-HE has to be infused to cause a significant increase in vitellogenin titers? Can normal vitellogenesis ever be achieved? How would isolated abdomens respond to infusion of 20-HE? This system seems ideal to investigate the question of whether two doses of 20-HE are more effective than one large dose (116). Answers to these questions could lead to a better understanding of the role of 20-HE in both vitellogenin synthesis and vitellogenesis. Thus the infused mosquito, in combination with an assay for vitellogenin, represents a model system that could be used to investigate the factors that regulate vitellogenin synthesis and vitellogenesis.

Effects of 20-HE *In Vitro*

The stimulation of vitellogenin synthesis by 20-HE in fat body preparations has been described by a number of groups (62, 63, 147). The initial observations (39, 66) were clouded by controversy regarding the assays used (17, 48, 61, 68). At least in the mind of the present reviewer, the technical questions regarding the radioimmunoprecipitation assay initially used have been settled (68) and several variations of this assay have been successfully used by others (30, 35, 50, 51, 147). The dynamic nature of egg development in the mosquito presents major problems to researchers desiring to measure changes in synthesis of macromolecules. Because digestion of the blood meal results in rapidly changing titers of amino acids, fatty acids, and nucleic acids, the investigator should be cautious in the use of radioactive tracers, particularly when they are injected into the live animal (17). Results of such *in vivo* assays should be regarded only as gross indications of changes, not as precise measures of synthetic rate. These problems can be reduced by either using *in vitro* assays for synthesis of macromolecules (11), where the concentration of substrates can be more easily controlled, or by directly measuring the amount of vitellogenin using an immunological technique.

The development of a new assay based on an ELISA and a monoclonal antibody raised against the vitellin of *A. aegypti* (113) provided a way to reinvestigate the effects of 20-HE without using radioactive tracers. Ma et al. (114, 115) confirmed the basic observations including the effect of 20-HE on vitellogenin synthesis, the timing of competence of the fat body to respond, and the role of JH in the development of competence. They further confirmed the hypothesis that JH regulated the recovery of responsiveness after the blood meal (115). Bohm et al. (11) found that the fat body lost its capacity to respond to 20-HE between 24 and 48 hr after a blood meal even in an *in vitro* system. If hormone was removed at 24 hr and then replaced 12 hr later, synthesis of vitellogenin declined and then recovered, but it did not recover if the hormone was removed at 48 hr. Ma et al. (114) also found that the response to hormone declined after 24 hr even in the continuous presence of 20-HE. However, they additionally showed that after 6 hr of exposure to 20-HE, the fat body produced vitellogenin for 24 hr regardless of the presence or absence of hormone, whereas in the experiments by Bohm et al. (11), the synthetic rate declined when the hormone was removed before 24 hr. This novel result provides further evidence that the response of the fat body to 20-HE is programmed after the initial exposure to hormone, but it is not clear why the timing of the effect was so different between the experiments of Bohm et al. (11) and Ma et al. (114).

20-HE and Gene Expression

Vitellogenin

The vitellogenin gene has been cloned and the regulation of its expression examined. Using a restriction fragment of the cloned gene as a probe, mRNA coding for vitellogenin was detected in total RNA from blood-fed females but not in RNA from sugar-fed females. This result eliminated the suggestion (based on the use of RNA inhibitors) that vitellogenin synthesis was regulated by 20-HE at the translational level, and that juvenile hormone stimulated transcription of the vitellogenin gene (43,

44, 81). Vitellogenin gene transcript was found in fat body cells, but in contrast to *D. melanogaster*, vitellogenin transcript was not found in follicle cells of the ovary (107). This evidence suggests that vitellogenin synthesis by follicle cells is a unique feature of the Cyclorraphid Diptera. Injection of 20-HE (5 μg) into decapitated sugar-fed females resulted in low levels of vitellogenin transcript, but injection into females fed blood and immediately decapitated resulted in much higher levels of transcript (52, 139), demonstrating once again the importance of factors from the blood meal, either nutritional or hormonal. A detailed examination of vitellogenin gene expression by incubated fat body revealed a clear effect of the JH analog, methoprene. Methoprene by itself had no effect, but the response was greater when combined with 20-HE than 20-HE by itself (115, 139). This rather surprising effect may be related to the effect of JH on isolated abdomens seen by Borovsky et al. (19) and Martinez and Hagedorn (116) discussed above, but could also be a result of the effect of JH on other factors such as the concentration of 20-HE receptors in fat body cells, or recruitment of nutritional reserves within the fat body cells (92). A quite different approach, which is the focus of current studies, involves a search for gene sequences responsible for interaction with hormone–receptor complexes and other transcription factors. This approach may help sort out the relative roles of JH and 20-HE in regulating vitellogenin gene expression.

Vitelline Envelope Proteins

Anderson and Spielman (2) and Raikhel and Lea (142, 146) described the formation of a membrane, or envelope, by the fusion of "plaques" that begin to appear between the follicular epithelium and the oocyte about 8–10 hr after the blood meal that initiates vitellogenic growth of the oocyte. These plaques increase greatly in size between 18 and 30 hr and by 36 hr the plaques have fused into a continuous membrane approximately 1.2 μm thick, effectively sealing the oocyte from the overlying follicle cells and preventing further uptake of yolk proteins. A few hours later the deposition of the chorion begins. Raikhel and Lea (146) have shown that the formation of the vitelline "envelope" is induced in decapitated females given 20-HE by injection, or in isolated abdomens given blood by enema.

A gene (originally designated as 15a) cloned from an *A. aegypti* genomic library by Gemmill et al. (52) has been found to code for proteins related to the vitellin membrane of *D. melanogaster* (107). Expression of the 15a gene has been found to be under the control of 20-HE. The timing of expression was found to be very similar to that of vitellogenin, but the tissue of expression was the follicle cells of the ovary as determined by *in situ* hybridization. It is likely, therefore, that 15a codes for a protein that forms part of the vitelline envelope as described from ultrastructural observations of the ovary. In parallel with studies on the vitellogenin gene, this gene provides an interesting contrast in that it is also under the control of 20-HE but expressed in the follicle cells of the ovary rather than the fat body.

53-kDa Serine Carboxypeptidase

Hays and Raikhel (72) and Cho et al. (31) described a 53-kDa serine carboxypeptidase that is synthesized in the fat body after a blood meal and sequestered by the ovaries where it becomes part of the yolk. The 53-kDa protein is an inactive precursor that is activated at the beginning of embryonic development and is present throughout development of the embryo where it is thought to be involved in processing vitellogenin. Synthesis of the protease by incubated fat body can be stimulated by $10^{-6} M$ 20-HE. A cDNA clone coding for this protease has been sequenced and found to contain two out of three conserved domains present in animal and plant carboxypeptidases (31).

Effects of 20-HE on Gut Trypsin

Digestion of blood is mainly accomplished by trypsins that appear in the gut after a blood meal. Trypsin activity is reduced by half after removal of either the medial neurosecretory cells or the ovaries (26). The activity can be restored by reimplantation of ovaries or the injection of 20-HE. At least two trypsin genes are expressed at different times after the blood meal; early (0–8 h) and late (6–48 h) (55). Barillas-Mury et al. (5) have reported the isolation of a cDNA clone coding for late trypsin, and an early trypsin gene has also been cloned (F.G. Noriega and M.A. Wells, personal communication). A stored form of the early mRNA is present in the gut cells prior to a blood meal. The kinetics of appearance of this early mRNA suggests that JH may be involved in its regulation. Following a blood meal the early mRNA is translated. The resulting products of digestion activate transcription of the late trypsin gene. Transcription of the late trypsin gene is prevented by inhibitors of early trypsin activity (i.e. soybean trypsin inhibitor) or by cycloheximide, a protein synthesis inhibitor, suggesting a complex control mechanism.

An observation that may be related to blood digestion was made by Cole and Gillett (33) and Rosenberg (151) who showed that retention of the blood meal in the gut is promoted by the presence of the brain or 20-HE.

20-HE and Dopa Decarboxylase

The egg of *A. aegypti* turns black within a few hours after deposition, suggesting a sclerotization process similar to that seen in the cuticle after molting. Schlaeger and Fuchs (156) showed that dopa decarboxylase was involved in the sclerotization of the egg, as it is in the sclerotization of the cuticle. Schlaeger and Fuchs (156) and Schlaeger et al. (157) then showed that injection of 20-HE into adult females stimulates dopa decarboxylase activity in the ovary. The effect of the hormone was not seen until 24 hr after an injection, and then rose to a peak by 72 hr that had the same kinetics but was 100-fold smaller than after a blood meal. Schlaeger et al. (157) were the first to measure ecdysteroid levels in mosquitoes after a blood meal and concluded that the induction of dopa decarboxylase is one of a number of events induced by 20-HE.

Behavioral Effects of 20-HE

Host-seeking behavior by *A. aegypti* is terminated by a blood meal, but the mechanisms involved are complex. Klowden (88, 91) reviewed the evidence for two separate physiological systems involved in terminating host-seeking behavior, the first being induced by distention of the gut and the second coming from the maturing ovary. Distention-induced termination lasts only as long as the midgut contains blood and appears to be mediated by stretch receptors. The maturing ovary induces a longer term inhibition and the factor from the ovary may be 20-HE as suggested by Bowen and Loess-Perez (25). The ovarian factor stimulates the release of a second factor that is thought to cause the behavioral inhibition (89). M.J. Klowden (personal communication) believes that the second factor could be the *Aedes* head peptide I isolated by Matsumoto et al. (121). Oviposition terminates the inhibition (87).

Inhibitory Effects of 20-HE and JH

JH titers rise in the normal newly eclosed female during the first two days (Fig. 5.1). Growth of the oocytes to the resting stage is promoted by JH (58, 67). Hagedorn et al. (67) found that topical application of JH I to isolated abdomens promoted growth to the resting stage, however, if 20-HE was experimentally applied at the same time, previtellogenic egg development was blocked suggesting that 20-HE must be absent during the rise in JH titers.

Similarly, if JH is applied after a blood meal, normal development of the eggs is blocked (80, 131, 161). Judson and de Lumen (80) found that topical application of JH I or methoprene (in oil) 72 hr prior to a blood meal, immediately after a blood meal or 24 hr after the blood meal, blocked growth of the oocytes, usually at stage III–IVa, at the end of the vitellogenic stage of oocyte development before the deposition of the chorion. These follicles also showed a failure of the normal degeneration of the nurse and follicle cells. Patterson (132) found that topical application of JH II or JH analogs (in methyl ethyl ketone) during a short period 32–36 hr after the blood meal caused a failure of the eggs to darken normally, and subsequent failure of hatching. Shapiro et al. (161) also found that application of methoprene (in acetone), or inhibition of JH esterases, resulted in reduced egg hatch. These experiments suggest that certain stages of egg development, most likely the deposition of a normal chorion, require the absence of JH. It is interesting to note the possible connection with dopa decarboxylase, which causes the darkening of the egg and requires 20-HE as described above. The timing of the effect of 20-HE on dopa decarboxylase is about the same as the critical period for JH application. It may be, therefore, that one of the events disturbed by JH application after a blood meal is the normal appearance of dopa decarboxylase. The results discussed here are quite different from those described by Borovsky et al. (19) and Martinez and Hagedorn (116) where small doses of methoprene potentiated the effect of 20-HE in isolated abdomens. The assay used in the latter two reports was the size of the oocytes 66 hr after ligation, not oviposition of normal eggs.

Peptides

We know from other model systems that much of the physiological life of animals is regulated by peptides, yet we know very little about the peptide hormones of insects. The reasons for this are mainly technical. The small size of insects coupled with the need for large sample sizes for classical methods of separation and analysis has slowed the pace of discovery down considerably. New advances in technology, which have reduced the limits of detection down to femtomoles, and the sample size down to insect-sized levels, have resulted in a rapid increase in the publication of peptide sequences. One of the current problems is that many of these molecules are known chemically in great detail, but their physiological roles remain unknown, providing an exciting challenge to future researchers.

The Egg Development Neurosecretory Hormone, EDNH

The egg development neurosecretory hormone (EDNH) that stimulates the production of ecdysone by the ovary (69) is produced by medial neurosecretory cells of the brain (100) and released from a neurohemal site behind the brain distinct from the location of the intrinsic cells of the corpora cardiaca (122). Peptides isolated from mosquito heads have been shown to stimulate production of ecdysone by the incubated ovary and vitellogenesis in the injected female. Attempts to purify peptides with EDNH activity have been reported by several groups (18, 51, 70, 119, 120, 171, 172). It is quite evident from this work that there is a family of peptides that can stimulate egg development in blood-fed decapitated females and/or ecdysone production by incubated ovaries. The precise source of these peptides is not clear, and although Hagedorn et al. (69) suggested that it may be the median neurosecretory cells identified by Lea (100) on the basis of crude dissections, it is not certain that the molecules being isolated are EDNH as defined by Lea. Matsumoto et al. (120) have isolated one of these peptides in sufficient quantities to obtain an amino acid composition, but despite much effort the sequence is still unknown. Several of the characteristics of this group of peptides, including their size and behavior during isolation, have suggested that they are related to the insulin family of peptides (171), and therefore to Bombyxin (earlier known as 4K PTTH). Bombyxins are the insulin-related peptides isolated from *B. mori* that stimulate ecdysone production by prothoracic glands of *Samia cynthia ricini* (78, 126), but whose main physiological role may be to stimulate ecdysone production by the ovary of the silkworm (77, 155). Given the failure of more traditional approaches to isolation of EDNH, an approach using the techniques of molecular biology may be more successful.

Injection of EDNH into decapitated non-blood-fed females has no effect. EDNH has a small effect when injected into blood-fed females that are immediately decapitated and injected (51), but the effect is much increased if a delay of 2 hr is allowed to elapse between feeding and decapitation and injection (169). Isolated ovaries, however, do respond to EDNH (64). Borovsky et al. (19) used isolated abdomens prepared immediately after the blood meal and found that topical application of methoprene was required to obtain an effect of injected EDNH. These experiments suggest that during the 2-hr delay, events occur that are induced by juvenile hormone. Since the titer of JH-III is high before the blood meal and drops thereafter (161) it is puzzling that an additional application of a JH analog (methoprene) would have a beneficial effect.

Factors Promoting the Release of EDNH

The control over the release of EDNH is a question that has attracted considerable attention. It is one of the most obvious remaining black boxes in the pathway from the blood meal to oviposition. Chang and Judson (29) devised an ingenious assay for the release of EDNH and used it to suggest that peptides released by the digestion of gamma globulins were the most likely primary factors causing the release of EDNH. They argued that since soybean trypsin inhibitor eliminated the response, peptides released by the endopeptidase action of trysin were the causative factor. They also showed, however, that injection of amino acids induced a small, but significant, level of EDNH release. Uchida et al. (166) provided convincing evidence that slow infusion of amino acids could stimulate significant levels of egg development. Slow infusion also eliminated stretching as a factor. The results of Uchida et al. (166) suggested that the infused amino acids stimulated the release of EDNH as the critical time for infusion was similar to the critical time for decapitation, but they did not directly examine this point.

Borovsky (12) and Lea and Van Handel (104) used different approaches to provide evidence suggesting that a factor from the ovary was necessary to stimulate the release of EDNH from the corpora cardiaca. The most significant results of these experiments were that EDNH was not released in the absence of resting stage ovaries. Thus, factors from the blood meal, such as peptides or amino acids, could not, by themselves, act to cause the release of EDNH; a factor from the ovary is also required. The nature of the factor from the ovary, and how it acts is not clear, but since an implanted ovary is effective, a nervous signal is not involved. This factor has some of the properties of a classical releasing factor and has been called EDNH-RF in Fig. 5.2.

Factors with Oostatic Activity

Meola and Lea (123) and Else and Judson (38) described an inhibitory effect of ovaries containing mature oocytes on development of eggs after a second blood meal. The "oostatic" effect had been described earlier in houseflies by Adams et al. (1) who used a 95% ethanol extract of ovaries. Kelly et al. (85, 86) showed that saline extracts with oostatic effects could be prepared from *Musca domestica*, *D. melanogaster*, and *A. atropalpus*. They further showed that the oostatic factor appeared in the ovary of *A. atropalpus* between 32 and 72 hr after emergence, which is after the ecdysteroid titers decline in this autogenous species (10). The oostatic factor was shown to decrease the production of ecdysone by the ovary *in vitro* (86). Borovsky (14) partially purified a 2.2-kDa peptide from ovaries of *A. aegypti* that had the properties of an oostatic hormone as defined by earlier investigators (38, 123) and did not prevent the release of EDNH.

Borovsky et al. (21) later isolated a factor from mature ovaries that apparently inhibits synthesis of trypsin-like enzymes by the gut. The latter was initially called an oostatic hormone but has more recently been called the trypsin-modulating oostatic factor (TMOF, 21). TMOF has been described as terminating the synthesis of midgut proteases, specifically trypsin (16). It is active when injected in high doses (2 nmol, equivalent to about 2 μg, cause about a 50% inhibition), but not when ingested (16). TMOF has been purified, sequenced, and synthesized. It is a decapeptide with the sequence NH_2-Tyr-Asp-Pro-Ala-Pro-Pro-Pro-Pro-Pro-Pro-COOH. Not surprisingly, sequence similarities are seen between this molecule and other proteins containing a polyproline sequence (21), but the importance of these similarities is highly questionable as they are based almost entirely on the relatively informationless polyproline sequence.

Attempts to homologize TMOF with the oostatic hormone (16, 21) are unconvincing. First, most of the work on mosquito oostatic hormone by Kelly was done using the autogenous mosquito, *A. atropalpus*, in which egg development occurs without a blood meal (85, 86), and therefore an effect on digestion of the blood meal is clearly irrelevant. Second, the oostatic hormone partially purified by Borovsky (14) is 100 times more active than purified TMOF (21); 3.5 ng vs 0.5 μg for about 50% inhibition. Third, the oostatic hormone has a mass of 2.2 vs 1.047 kDa for TMOF. Interactions with column material and differences in standards could cause errors in mass determination of this magnitude. Coelution of these two factors on a single column has not been demonstrated. Finally, the proposed physiological roles of these two factors, on ecdysone production and on trypsin synthesis, are quite different.

Lin et al. (107) cloned several genes coding for proteins similar to the vitelline membrane proteins of *Drosophila*. One of these genes contains a sequence at the 5′ end of the coding region identical to TMOF except for the reversal of the first two amino acids. It is possible that the peptide isolated by Borovsky as TMOF is a proteolytic product of the vitelline membrane produced during homogenization of the ovary. This supposition is strengthened by the presence of a possible proteolytic cleavage site at the N-terminal end of the decapeptide sequence. Recently, Borovsky et al. (23) described the development of an RIA and ELISA that were used to demonstrate accumulation of TMOF in the ovary after a blood meal, at the same time that the vitelline membrane is forming (107, 146). In view of the near identity between TMOF and the vitelline membrane protein it is not surprising that an ELISA would indicate accumulation of TMOF in the ovary. It remains to be demonstrated that TMOF has the properties of a hormone (i.e., that it is normally present in the hemolymph about 48 hr after a blood meal in concentrations sufficient to inhibit trypsin synthesis, and that receptors for TMOF can be found in the target tissue).

Peptides from the Accessory Gland

The accessory glands of male mosquitoes can have a number of effects on the female after mating. Craig (34) showed that compounds from the accessory gland can induce females to reject further matings, thus essentially inducing monogamy, others however, have shown that the effect wears off in females that develop a batch of eggs (174, 175). Fuchs et al. (46) isolated peptides of 30–60 kDa from extracts of whole *A. aegypti* that induced monogamy when injected. This activity was called matrone. More recent work by Young and Downe (176) has shown that extracts of accessory glands of *Culex tarsalis* contain factors with matrone-like activity that have a molecular weight near 2 kDa. Molecules in a similar size range with matrone-like activity have been isolated from *D. melanogaster*, and glands from the latter insect are active in *A. aegypti* (106), but the cross-reactivity of accessory gland components varies considerably among species (91). Young and Downe (176) showed that active material isolated from *C. tarsalis* would migrate as a much larger molecule when isolated using the methods of Fuchs, suggesting that it has the unfortunate property of being "sticky." It is, therefore, likely that the accessory glands of mosquitoes contain peptides similar to those in *D. melanogaster* that cause the female to reject further matings, and it is possible that an approach to the identification of these molecules using DNA probes based on *D. melanogaster* peptides would be successful.

Klowden and Chambers (93) found a stimulatory effect of male accessory gland components on egg development in starved females that could not be attributed to a nutritional effect, but rather seemed to be acting to stimulate the mobilization of resources toward egg development rather than survival of the nutritionally marginal animal.

O'Meara and Evans (128) discovered that autogeny in *A. taeniorhynchus* was stimulated by mating. They later reported (129) that this effect was due to accessory gland fluid, which they suggested stimulated the release of the egg development neurosecretory hormone. Borovsky (15) followed up on these observations by showing that accessory gland extracts were not effective in decapitated females unless the decapitation was delayed for 12 hr. These extracts were also not effective in causing the release of ENDH in ovariectomized females. Borovsky (15) suggested that in this mosquito a factor from the accessory gland stimulated the release of the EDNH-releasing factor from the ovary (12, 104). There is no direct evidence supporting this hypothesis (i.e., do incubated ovaries release EDNH-RF in response to accessory gland extracts?).

Diuretic Hormones

The production of a copious urine after a blood meal, and after eclosion, is a characteristic feature of mosquito life. Nijhout and Carrow (127) were the first to show that diuresis after a blood meal is under the control of a factor from the head. Gillett (54) demonstrated the same for the diuresis that occurs after eclosion. Three factors that stimulate fluid production and changes in transepithelial voltage by the Malpighian tubule were identified by Petzel et al. (133, 134). The effect of each of these factors on incubated Malpighian tubules and after injection *in vivo* was examined by Wheelock et al. (170) who showed that factor I caused urine excretion but not fluid secretion, while factors II and III affected both fluid secretion and urine excretion. It was suggested that factor I may act on the rectum. Factor III has been the subject of further investigation as it causes its effects via cAMP. Petzel et al. (135) showed that cAMP levels increase in the Malpighian tubule after a blood meal and in the isolated Malpighian tubule after exposure to factor III. The effects of ions, forskolin, theophylline, ouabain, SITS, and bumetanide on the Malpighian tubule have been used to design model for how factor III affects fluid production by the tubule (73).

Hayes et al. (71) found that the leucokinins, peptides isolated from the cockroach *Leucophaea maderae*, had diuretic effects on the mosquito tubule. Hayes (personal communication) has recently isolated several peptides related to the leucokinins from *Culex salinarius* that also have effects on the Malpighian tubule. Similar effects have been described for 5-HT by Veenstra (168).

Evidence for Other Factors from the Head

Two peptides related to the RFamide family were isolated from mosquito heads (27, 121). *Aedes* head peptide I is a decapeptide while *Aedes* head peptide II is a tripeptide. The physiological function of these peptides is unknown, although as mentioned above, *Aedes* head peptide I may be involved in regulating biting behavior.

There is increasing evidence that multiple factors, presumably peptides released from the head after a blood meal, are important in egg development. Kohler and Raikhel (94) showed that factors that regulate protein synthesis in the ovary, and vitellogenin uptake are released in two pulses. The first occurs within minutes of beginning a blood meal and the second occurs 8 to 16 hr later. Both require the act of blood feeding, as an enema has no effect. Greenplate et al. (56) measured the effect of decapitation on ecdysteroid and vitellogenin levels after a blood meal and presented evidence that EDNH is also released in two pulses, the first before 8 hr and the second after 8 hr. Similar results were obtained by Lu and Hagedorn (109) with *A. albimanus*. Greenplate et al. (56) also found evidence for a factor from the head that regulates the total number of oocytes developed, as was suggested by the earlier work of Lea et al. (105). The data from these investigations support the concept of Gillett (53) that growth of the oocytes is divided into two phases: initiation and promotion. However, contrary to Gillett's hypothesis, both phases require factors from the head.

Several investigations (56, 94) found that giving an enema to a normal female did not stimulate protein synthesis, vitellogenin uptake or ecdysone production by the ovary. Thus it seems absolutely essential for the blood to pass through the mouth of the mosquito for some events to occur normally. This fact should be noted by researchers who routinely use enemas to supply blood meals.

Hormonal Regulation of Egg Development in Mosquitoes Other Than *Aedes aegypti*

In general, the basic pattern of the regulation of egg development by juvenile hormone and 20-hydroxyecdysone has been found in several mosquito species as is detailed below. However, there are major differences between mosquito species. For example, the effect of JH on biting behavior shown in *Culex pipiens* is not apparent in *A. aegypti*. Also, the quantities and kinds of ecdysteroids present during vitellogensis differ in several species, and perhaps strains, of mosquitoes. Factors found in one species are sometimes not active in another. For example, *Anopheles* ovaries do not replace *Aedes* ovaries as a source of EDNH-RF. It is not valid to conclude, therefore, that results obtained from a model insect such as *A. aegypti* will necessarily pertain to all other mosquitoes

Aedes atropalpus and Other *Aedes* Species

A. atropalpus is especially interesting because it is autogenous, that is, it does not require a blood meal to develop the first batch of eggs. A series of papers emanating from the laboratory of M.S. Fuchs has examined the endocrinology of egg development in some detail. These data are of interest for what they tell us about the difference between autogenous and anautogenous mosquitoes, and shed some light on the interpretation of what is known about *A. aegypti*.

Vitellogenic oocyte growth in *A. atropalpus* begins 18–24 hr after eclosion of the adult. The appearance of vitellogenin in the hemolymph at about 18 hr after eclosion (113) is coincident with increasing amounts of 20-HE in whole body extracts (117). Immediately after eclosion, the decapitated female is not sensitive to injected 20-HE, but becomes sensitive by 12 hr after eclosion (83). Competence to respond to 20-HE is under JH control (84). The amount of 20-HE required to stimulate egg development is remarkably low (4.8 ng), compared to *A. aegypti*. Isolated abdomens prepared shortly after eclosion require much higher dose of 20-HE (4.8 μg) to stimulate egg development, but after an application of JH I (0.5 ng) they become sensitive to the dose of 4.8 ng of 20-HE. The requirement for JH can be overridden by an injection of a much larger dose of 20-HE (4.8 μg). Conversely, a larger dose of JH I (0.5 μg) can completely eliminate the need for injecting 20-HE (84). However, the large dose of JH I causes an increase 20-HE levels (10), and the effect of JH on the isolated abdomen may be to create a small peak of 20-HE, which, when followed by the injection of 20-HE, stimulates egg development, as has been discussed above for the case of *A. aegypti*.

JH is known to regulate competence to respond to 20-HE in both *A. aegypti* and *A. atropalpus*. In addition, JH can stimulate the ovaries to produce 20-HE in both insects. In the case of *A. aegypti* the effect of JH on competence occurs prior to the blood meal, and is, therefore, separable from any effect JH might have when applied to isolated abdomens after a blood meal. However, in the isolated abdomen of *A. atropalpus* prepared immediately after eclosion, applied JH could be having two effects, that of stimulating competence and of causing a rise in

20-HE titers. Given what we know about *A. aegypti*, both effects are probably relevant to the acquired sensitivity to 20-HE.

Guilvard et al. (57) examined the levels of ecdysteroids and juvenile hormones using radioimmunoassays in two autogenous mosquitoes, *Aedes caspius* and *Aedes detritus*. At adult eclosion, the follicles are described as containing a distinct oocyte with nurse cells. By 24 hr previtellogenic oocyte growth has occurred, and by 40 hr vitellogenesis begins. The levels of ecdysteroids rise to a peak at 40 hr in both species, followed by a peak of juvenile hormone immunoreactive material. The changes in hormones resemble those seen after a blood meal in *A. aegypti*, but the previtellogenic rise in JH is not present, yet the previtellogenic growth of the oocyte occurs normally. It appears that the ovary of the newly eclosed *A. caspius* and *A. detritus* contains follicles that are slightly more advanced than is seen in the ovary of *A. aegypti*, and it may be that previtellogenic ovary development is stimulated by events during eclosion.

Culex pipiens

Egg development was examined in *Culex pipiens pallens* in a series of experiments carried out at the Shanghai Institute of Entomology. Ecdysteroid titers were found to rise sharply and remain high between 12 and 36 hr after a blood meal, and injection of 5–10 μg of 20-HE was shown to stimulate the production of vitellogenin (178) and oocyte growth (28). 20-HE was found to have a half-life of about 8 hr after an injection of 5 μg of hormone (177). By 32 hr after the injection, 110 ng of hormone was still present in the mosquito, almost all of which was in the hemolymph.

Baldridge and Feyereisen (4) showed that both 20-HE and ecdysone were present in whole body extracts, and the ratio of the two steroids changed during egg development, suggesting the increasing activity of a 20-monooxygenase. The role of a head factor in egg development was demonstrated by decapitation, but the operation had to be performed within 2–4 min of the beginning of the meal to prevent egg development. This is in contrast to the results of Larsen and Bodenstein (97) who found a critical time of 30 min in *C. pipiens*. However, the timing of the release of EDNH is known to vary considerably among species and strains of mosquitoes (32).

Anopheles stephensi and *Anopheles albimanus*

The pattern of egg development, the effect of decapitation, and titers of ecdysteroids after a blood meal in *Anopheles stephensi* and *Anopheles albimanus* are similar to the results obtained from *A. aegypti*. The amounts of ecdysteroids as measured by RIA were 10-fold greater in *A.*

stephensi (150), and twice as high for *A. albimanus* (109) as has been reported for *A. aegypti* (56, 66). However, most of the excess ecdysteroid in *A. stephensi* was ecdysone. In both mosquitoes, injection of large amounts of 20-HE into resting stage females caused the primary oocyte to degenerate. In *A. albimanus*, injection of 5µg of 20-HE into blood-fed, decapitated females, stimulated nearly normal amounts of vitellogenin synthesis, and some degree of vitellogenesis by the oocyte.

Implications for Vector Biology

Parasites, whether viral, bacterial, or multicellular, must somehow interact with the host. Hormones are a feature of the host milieu that can provide the parasite with important cues as to the status of the host. For example, high ecdysone titers in the adult female mosquito signify the vitellogenic stage of egg development, as high titers of juvenile hormone signify the previtellogenic stage. Brown and Lea (27) discussed the possibility that peptide hormones are also cues for parasites. In addition, hormones can also provide mechanisms for altering the host to suit the parasite. Many cases are known where a parasite changes the hormonal status of the host, thus preventing molting or reproduction (6, 7). Such interactions are so common between parasites and their hosts that their absence would be considered remarkable (7). The mechanisms by which this is achieved can be quite complex, as in parasitoids of lepidoptera that employ viruses, components of the venom, and teratocytes to subdue the host and direct its metabolism to serve the needs of the parasite.

Two parasites that affect mosquitoes have been found to utilize the hormonal system, *Amblyospora* and *Coelomomyces*. These are discussed below. In addition, Ilan et al. (76), found that JH affected the growth of *Crithidia fasciculata*, a trypanosomatid commensal in the mosquito gut. In contrast to these cases growth of malaria parasites in *A. aegypti* was not inhibited by decapitation (154). However, the number of mosquitoes that were infected was reduced by about a third, and the intensity of the infection was also reduced. These were regarded secondary effects of the failure of normal digestion of the meal. Gwadz and Spielman (59) failed to find any effect of decapitation, allatectomy, JH treatment, or injection of 20-HE on development of the filarial nematode, *Brugia pahangi* in *A. aegypti*.

Given the accumulating evidence for physiological interactions between insects and their parasites that involve host hormones, it is very likely that similar events occur in mosquitoes that are vectors of organisms causing disease. Work in this area must be carefully done as the interactions between host and parasite can occur at so many levels that a cursory examination may well miss the event. Parasites that undergo developmental changes during residence in the host are especially likely to be amenable for study. For example, nematodes, which molt in the insect host, are likely to cue into host hormones.

The Case of *Amblyospora*

The life cycle of the microsporidian *Amblyospora* sp. alternates between copepods and mosquitoes. The haploid spores overwinter in copepods and kill the host in the spring causing the release of spores that are eaten by mosquito larvae. Infected larvae develop normally and adults emerge that are infected with binucleate cells of *Amblyospora*. These develop differently in males and females. In males, sporulation kills the host, releasing spores that can be eaten by mosquito larvae. In females, sporulation does not occur until after a blood meal, when binucleate cells develop in the oenocytes and infect the ovaries, resulting in vertical transmission to the next generation of mosquito larvae. Lord and Hall (108) have shown that the post-blood-meal sporulation in female adults was inhibited by decapitation or ovariectomy prior to a blood meal and was stimulated by injection of 20-HE into either normal or ovariectomized females.

The Case of *Coelomomyces*

Coelomomyces stegomyiae is a chytridiomycete fungus that also alternates between mosquitoes and copepods. Infected larval mosquitoes usually die in the fourth instar, but those adults that do survive have fungal hyphae mainly in the ovaries (110). Hyphae continue to grow as the ovary develops during previtellogenesis, but spores are not produced until after a blood meal is taken (112). Injection of 20-HE causes sporulation and therefore appears to be the stimulating factor as is the case in *Amblyospora*. Egg development in infected females is not normal after a blood meal, mature oocytes do not form, and vitellogenin and 20-HE titers are reduced (111). Nevertheless, adults do mate, blood feed, and show oviposition behavior. Instead of ovipositing eggs, however, the females deposit spores, effectively disseminating the fungus.

The Case of *Lambornella*

Lambornella clarki is a tetrahymenid ciliate that infects mosquito larvae, often killing them. Larvae can survive, however, and produce adults (37). In the infected adult female the ciliates move to the ovary from which they are later deposited. Females are capable of multiple ciliate depositions, but they take no blood meals. *Lambornella* is different from *Amblyospora* and *Coelomomyces* in that development of the parasite is not dependent on a blood meal, and, therefore, presumably not dependent on 20-HE either. However, the parasite does invade the ovaries, and parasites are oviposited in place of eggs, as in the case of *Coelomomyces*. It would be interesting to investigate the question of the role of hormones in the movement of the parasites toward the oocytes.

Conclusions

A general point that emerges from these studies, which never ceases to impress this reviewer, is that, when looked at globally, the complexity of mosquito endocrinology is the equal of any animal. It seems that, at the physiological level, the small size of the mosquito does not cause concomitant reduction in the complexity of the physiological systems needed to regulate processes such as reproduction. Furthermore, there is ample evidence of many opportunities for further work from this review, which have been indicated in the text. Certainly the most exciting prospect is that the advent of molecular biology has provided the physiologist with many new tools for investigating complex systems in small animals. These tools are only beginning to be applied to investigations of mosquito endocrinology. The future looks bright for eventually obtaining an in-depth understanding of mosquito reproductive biology.

Acknowledgments. Thanks are due to Carolina Barillas-Mury, Dov Borovsky, Larry Harshman, Mark Klowden, and K. Uchida for providing unpublished data, René Feyereisen for helpful discussions, and Jan Veenstra, Diana Wheeler, and Rolf Ziegler for helpful comments on the manuscript.

References

1. Adams, T.S., Hintz, A.M., and Pomonis, J.G. 1968. Oostatic hormone production in houseflies, *Musca domestica*, with developing oocytes. *J. Insect Physiol.* **14**:983–993.
2. Anderson, W.A., and Spielman, A. 1971. Permeability of the ovarian follicle of *Aedes aegypti* mosquitoes. *J. Cell Biol.* **50**:201–221.
3. Baker, F.C., Hagedorn, H.H., Schooley, D.A., and Wheelock, G. 1983. Mosquito juvenile hormone: Identification and bioassay activity. *J. Insect Physiol.* **29**:465–470.
4. Baldridge, G.D., and Feyereisen, R. 1986. Ecdysteroid titer and oocyte growth in the northern house mosquito, *Culex pipiens*. *Comp. Biochem. Physiol.* **83A**:325–329.
5. Barillas-Mury, C., Graf, R., Hagedorn, H.H., and Wells, M.A. 1991. cDNA and deduced amino acid sequence of the blood meal-induced trypsin from the mosquito, *Aedes aegypti*. *Insect Biochem.* **21**:825–831.

6. Beckage, N. 1990. Parasitic effects on host development. In R. Baker and P.E. Dunn, eds., *New Directions in Biological Control: Alternatives for Suppressing Agricultural Pests and Diseases*. Alan R. Liss, New York, pp. 497–515.
7. Beckage, N.E. 1991. Host-parasite hormonal relationships: A common theme? *Exp. Parasitol.* **72**:332–338.
8. Beckemeyer, E.F., and Lea, A.O. 1980. Induction of follicle separation in the mosquito by physiological amounts of ecdysterone. *Science* **209**:819–821.
9. Behan, M., and Hagedorn, H.H. 1978. Ultrastructural changes in the fat body of adult female *Aedes aegypti* in relationship to vitellogenin synthesis. *Cell Tissue Res.* **186**:499–506.
10. Birnbaum, M.J., Kelly, T.J., Woods, C.W., and Imberski, R.B. 1984. Hormonal regulation of ovarian ecdysteroid production in the autogenous mosquito, *Aedes atropalpus. Gen. Comp. Endocrinol.* **56**:9–18.
11. Bohm, M.K., Behan, M., and Hagedorn, H.H. 1978. Termination of vitellogenin synthesis by mosquito fat body, a programmed response to ecdysterone. *Physiol. Entomol.* **3**:17–25.
12. Borovsky, D. 1982. Release of egg development neurosecretory hormone in *Aedes aegypti* and *Aedes taeniorhynchus* induced by an ovarian factor. *J. Insect Physiol.* **28**:311–316.
13. Borovsky, D. 1984. Control mechanisms for vitellogenin synthesis in mosquitoes. *BioEssays* **1**:264–267.
14. Borovsky, D. 1985. Isolation and characterization of highly purified mosquito oostatic hormone. Arch. Insect Biochem. Physiol. **2**:333–349.
15. Borovsky, D. 1985. The role of the male accessory gland fluid in stimulating vitellogenesis in *Aedes taeniorhynchus. Arch. Insect Biochem. Physiol.* **2**:405–413.
16. Borovsky, D. 1988. Oostatic hormone inhibits biosynthesis of midgut proteolytic enzymes and egg development in mosquitoes. *Arch. Insect Biochem. Physiol.* **7**:187–210.
17. Borovsky, D., and Van Handel, E. 1977. A specific radioimmunoassay for vitellogenesis in mosquitoes. *J. Insect Physiol.* **23**:655–658.
18. Borovsky, D., and Thomas, B.R. 1985. Purification and partial characterization of mosquito egg development neurosecretory hormone: Evidence for gonadotropic and steroidogenic effects. *Arch. Insect Biochem. Physiol.* **2**:265–281.
19. Borovsky, D., Thomas, B.R., Carlson, D.A., Whisenton, L.R., and Fuchs, M.S. 1985. Juvenile hormone and 20-hydroxyecdysone as primary and secondary stimuli of vitellogenesis in *Aedes aegypti. Arch. Insect Biochem. Physiol.* **2**:75–90.
20. Borovsky, D., Whisenton, L.R., Thomas, B.R., and Fuchs, M.S. 1986. Biosynthesis and distribution of ecdysone and 20-hydroxyecdysone in *Aedes aegypti. Arch. Insect Biochem. Physiol.* **3**:19–30.
21. Borovsky, D., Carlson, D.A., Griffin, P.R., Shabanowitz, J., and Hunt, D.F. 1990. Mosquito oostatic factor: A novel decapeptide modulating trypsin-like enzyme biosynthesis in the midgut. *FASEB J.* **4**:3015–3020.
22. Borovsky, D., Carlson, D.A., and Ujvàry, I. 1992. *In vivo* and *in vitro* biosynthesis and metabolism of methyl farnesoate, juvenile hormone III,

and juvenile hormone III acid in the mosquito, *Aedes aegypti. J. Med. Entomol.* **29**:619–629.
23. Borovsky, D., Powell, C.A., and Carlson, D.A. 1992. Development of specific RIA and ELISA to study trypsin modulating oostatic factor in mosquitoes. *Arch. Insect Biochem. Physiol.* **21**:13–22.
24. Bowen, M.F., and Davis, E.E. 1989. The effects of allatectomy and juvenile hormone replacement on the development of host-seeking behavior and lactic acid receptor sensitivity in the mosquito *Aedes aegypti. Med. Vet. Entomol.* **3**:53–60.
25. Bowen, M.F., and Loess-Perez, S. 1989. A re-examination of the role of ecdysteroids in the development of host-seeking inhibition in blood-fed *Aedes aegypti* mosquitoes. In D. Borovsky and A. Spielman, eds., *Host Regulated Developmental Mechanisms in Vector Arthropods.* University of Florida, Vero Beach, pp. 286–291.
26. Breigel, H., and Lea, A.O. 1979. Influence of the endocrine system on tryptic activity in female *Aedes aegypti. J. Insect Physiol.* **25**:227–230.
27. Brown, M.R., and Lea, A.R. 1989. Neuroendocrine and midgut endocrine systems in the adult mosquito. *Adv. Dis. Vector Res.* **6**:29–58.
28. Cao, M., Zhu, X., Chen, Z., Liu, Z., and Jiang, R. 1980. Studies on the induction of the ovarian development in the mosquito, *Culex pipiens pallens. Acta Biol. Exp. Sinica* **13**:99–104.
29. Chang, Y.-y., and Judson, C.L. 1977. Peptides as stimulators of egg development neurosecretory hormone release in the mosquito *Aedes aegypti. Comp. Biochem. Physiol.* **57C**:147–151.
30. Chang, Y.-y., and Judson, C.L. 1977. The role of isoleucine in differential egg production by the mosquito *Aedes aegypti* following feeding on human or guinea pig blood. *Comp. Biochem. Physiol.* **57A**:23–28.
31. Cho, W.L., Deitsch, K.S., and Raikhel, A.S. 1991. An extraovarian protein accumulated in mosquito oocytes is a carboxypeptidase activated in embryos. *Proc. Natl. Acad. Sci. U.S.A.* **88**:10821–10824.
32. Clements, A.N. 1956. Hormonal control of ovary development in mosquitoes. *J. Exp. Biol.* **33**:211–223.
33. Cole, S.J., and Gillett, J.D. 1979. The influence of the brain hormone on retention of blood in the midgut of the mosquito, *Aedes aegypti. Proc. R. Soc. London Ser.* **B 205**:411–422.
34. Craig, G.B., Jr. 1967. Mosquitoes: Female monogamy induced by a male accessory gland substance. *Science* **156**:1499–1501.
35. Dhadialla, T.S., and Raikhel, A.S. 1990. Biosynthesis of mosquito vitellogenin. *J. Biol. Chem.* **265**:9924–9933.
36. Dittmann, F., Kogan, P.H., and Hagedorn, H.H. 1989. Ploidy levels and DNA synthesis in fat body cell of the adult mosquito, *Aedes aegypti*: The role of juvenile hormone. *Arch. Insect Biochem. Physiol.* **12**:133–143.
37. Egerter, D.E., Anderson, J.R., and Washburn, J.O. 1986. Dispersal of the parasitic ciliate *Lambornella clarki*: Implications for ciliates in the biological control of mosquitoes. *Proc. Natl. Acad. Sci. U.S.A.* **83**:7335–7339.
38. Else, J.G., and Judson, C.L. 1972. Enforced egg-retention and its effects on vitellogenesis in the mosquito, *Aedes aegypti. J. Med. Entomol.* **9**: 527–530.

39. Fallon, A.M., Hagedorn, H.H., Wyatt, G.R., and Laufer, H. 1974. Activation of vitellogenin synthesis in the mosquito *Aedes aegypti* by ecdysone. *J. Insect Physiol.* **20**:1815–1823.
40. Feinsod, F.M., and Spielman, A. 1980. Independently regulated juvenile hormone activity and vitellogenesis in mosquitoes. *J. Insect Physiol.* **26**:829–832.
41. Feinsod, F.M., and Spielman, A. 1980. Nutrient mediated juvenile hormone secretion in mosquitoes. *J. Insect Physiol.* **26**:113–117.
42. Flanagan, T.R., and Hagedorn, H.H. 1977. Vitellogenin synthesis in the mosquito: The role of juvenile hormone in the development of responsiveness to ecdysone. *Physiol. Entomol.* **2**:173–178.
43. Fong, W.-F., and Fuchs, M.S. 1976. Studies on the mode of action of ecdysterone in adult female *Aedes aegypti. Mol. Cell. Endocrinol.* **4**:341–351.
44. Fong, W.-F., and Fuchs, M.S. 1976. The long term effect of a-amanitin on RNA synthesis in adult *Aedes aegypti. Insect Biochem.* **6**:123–130.
45. Fritz, M.A., Hotchkin, P.G., and Fallon, A.M. 1986. Changes in ribonuclease activity during development of the mosquito, *Aedes aegypti. Comp. Biochem. Physiol.* **84B**:355–361.
46. Fuchs, M.S., and Hiss, E.A. 1970. The partial purification and separation of the protein components of matrone from *Aedes aegypti. J. Insect Physiol.* **16**:931–939.
47. Fuchs, M.S., and Schlaeger, D.A. 1973. The stimulation of dopa decarboxylase activity by ecdysone and its enhancement by cyclic AMP in adult mosquitoes. *Biochem. Biophys. Res. Commun.* **54**:784–789.
48. Fuchs, M.S., and Kang, S.H. 1981. Ecdysone and mosquito vitellogenesis: A critical appraisal. *Insect Biochem.* **11**:627–633.
49. Fuchs, M.S., Schlaeger, D.A., and Shroyer, C. 1979. α-Ecdysone does not induce ovarian development in adult *Aedes aegypti. J. Exp. Zool.* **207**:153–159.
50. Fuchs, M.S., Sundland, B.R., and Kang, S.-H. 1980. *In vivo* induction of ovarian development in *Aedes atropalpus* by a head extract from *Aedes aegypti. Int. J. Invert. Reprod.* **2**:121–129.
51. Fuchs, M.S., Kang, S.H., Kelly, T.J., Masler, E.P., and Whisenton L.R. 1981. Endocrine control of ovarian development in an autogenous mosquito. In F. Senhal, A. Zabaz, J.J. Menn, and B. Cymborowski, eds., *Insect Development and Behavior.* Wroclaw Technical University Press, Wroclaw, Poland, pp. 569–590.
52. Gemmill, R.M., Hamblin, M., Glaser, R.L., Racioppi, J.V., Marx, J.L., White, B.N., Calvo, J.M., Wolfner, M., and Hagedorn, H.H. 1986. Isolation of mosquito vitellogenin genes and induction by 20-hydroxyecdysone. *Insect Biochem.* **16**:761–774.
53. Gillett, J.D. 1956. Initiation and promotion of ovarian development in the mosquito *Aedes aegypti. Ann. Trop. Med. Parasitol.* **50**:375–380.
54. Gillett, J.D. 1984. The effects of decapitation and the influence of size and sex on diuresis in newly emerged mosquitoes. *Physiol. Entomol.* **9**:139–144.
55. Graf, R., and Briegel, H. 1989. The synthetic pathway of try psin in the mosquito *Aedes aegypti* and *in vitro stimulation in isolated midguts. Insect Biochem.* **19**:129–137.

56. Greenplate, J.T., Glaser R.L., and Hagedorn, H.H. 1985. The role of factors from the head in the regulation of egg development in the mosquito *Aedes aegypti*. *J. Insect Physiol.* **31**:323–329.
57. Guilvard, E., DeReggi, M., and Rioux, J.-A. 1984. Changes in ecdysteroid and juvenile hormone titers correlated to the initiation of vitellogenesis in two *Aedes* species. *Gen. Comp. Endocrinol.* **53**:218–223.
58. Gwadz, R.W., and Spielman, A. 1973. Corpus allatum control of ovarian development in *Aedes aegypti*. *J. Insect Physiol.* **19**:1441–1448.
59. Gwadz, R.W., and Spielman, A. 1974. Development of the filarial nematode, *Brugia pahangi*, in *Aedes aegypti* mosquitoes: Non dependence upon host hormones. *J. Parasitol.* **60**:134–137.
60. Gwadz, R.W., Lounibos, L.P., and Craig, G.B. 1971. Precocious sexual receptivity induced by a juvenile hormone analog in females of the yellow fever mosquito, *Aedes aegypti*. *Gen. Comp. Endocrinol.* **16**:47–51.
61. Hagedorn, H.H. 1983. The role of ecdysteroids in the adult insect. In R.G.H. Downer and H. Laufer, eds., *Invertebrate Endocrinology*, Vol. 1, *Endocrinology of Insects*. Alan R. Liss, New York, pp. 271–304.
62. Hagedorn, H.H. 1985. The role of ecdysteroids in reproduction. In G.A. Kerkut and L.I. Gilbert, eds., *Comprehensive Insect Physiology, Biochemistry and Pharmacology*, Vol. 8. Pergamon Press, Oxford, pp. 205–262.
63. Hagedorn, H.H. 1989. Physiological roles of ecdysteroids in the adult insect. In J. Koolman, ed., *Ecdysone*. Georg Thieme Verlag, Stuttgart, pp. 279–289.
64. Hagedorn, H.H., and Fallon, A.M. 1973. Ovarian control of vitellogenin synthesis by the fat body in *Aedes aegypti*. *Nature* (*London*) **244**:103–105.
65. Hagedorn, H.H., Fallon, A.M., and Laufer, H. 1973. Vitellogenin synthesis by the fat body of the mosquito *Aedes aegypti*: Evidence for transcriptional control. *Dev. Biol.* **31**:285–294.
66. Hagedorn, H.H., O'Connor, J.D., Fuchs, M.S., Sage, B., Schlaeger, D.A., and Bohm, M.K. 1975. The ovary as a source of alpha ecdysone in an adult mosquito. *Proc. Natl. Acad. Sci. U.S.A.* **72**:3255–3259.
67. Hagedorn, H.H., Turner, S., Hagedorn, E.A., Pontecorvo, D., Greenbaum P., Pfieffer, D., Wheelock, G., and Flannagan, T.R. 1977. Postemergence growth of the ovarian follicles of *Aedes aegypti*. *J. Insect Physiol.* **23**:203–206.
68. Hagedorn, H.H., Kunkel, J.G., and Wheelock, G. 1978. The specificity of an antiserum against mosquito vitellogenin and its use in radioimmunological precipitation assay for protein synthesis. *J. Insect Physiol.* **24**:481–489.
69. Hagedorn, H.H., Shapiro, J.P., and Hanoaka, K. 1979. Ovarian ecdysone secretion is controlled by a brain hormone in an adult mosquito. *Nature (London)* **282**:92–94.
70. Hanaoka, K., and Hagedorn, H.H. 1980. Brain hormone control of ecdysone secretion by the ovary in a mosquito. In J.A. Hoffmann, ed., *Progress in Ecdysone Research*. Elsevier/North-Holland, Amsterdam, pp. 467–480.
71. Hayes, T.K., Pannabecker, T., Hinckley, D., Holman, G., Nachman, R., Petzel, D., and Beyenbach, K.W. 1989. Leucokinins, a new family of ion

transport stimulators and inhibitors in insect Malpighian tubules. *Life Sci.* **44**:1259–1266.
72. Hays, A.R., and Raikhel, A.S. 1990. A novel protein produced by the vitellogenic fat body and accumulated in mosquito oocytes. *Roux's Arch. Dev. Biol.* **199**:114–121.
73. Hegarty, J.L., Zhang, B., Pannabecker, T.L., Petzel, D.H., Baustian, M.D., and Beyenbach, K.W. 1991. Dibutyryl cAMP activates bumetanide-sensitive electrolyte transport in Malpighian tubules. *Am. J. Physiol.* **261**:C521–C529.
74. Hoffmann, J.A., and Lagueux, M. 1985. Endocrine aspects of embryonic development in insects. In G.A. Kerkut and L.I. Gilbert, eds., *Comprehensive Insect Physiology, Biochemistry and Pharmacology*, Vol. 1. Pergamon Press, Oxford, pp. 435–460.
75. Hotchkin, P.G., and Fallon, A.M. 1987. Ribosome metabolism during the vitellogenic cycle of the mosquito *Aedes aegypti. Biochim. Biophys. Acta* **924**:352–359.
76. Ilan, J., Ilan, J., and Ricklis, S. 1969. Inhibition of juvenile hormone of growth of *Crithidia fasciculata* in culture. *Nature (London)* **224**:179–180.
77. Ishizaki, H., and Suzuki, A. 1984. Prothoracic hormone of *Bombyx mori*. In J. Hoffmann and M. Porchet, eds., *Biosynthesis, Metabolism and Mode of Action of Invertebrate Hormones*. Springer-Verlag, Berlin, pp. 63–77.
78. Iwami, M., Kawakami, A., Ishizaki, H., Takahashi, S.Y., Adachi T., Yoshiaki, Suzuki, Y., Nagasawa, H., and Suzuki, A. 1989. Cloning of a gene encoding bombyxin, an insulin-like brain secretory peptide of the silkmoth *Bombyx mori* with prothoracicotropic activity. *Dev. Growth Differ.* **31**:31–37.
79. Johnson, A.M., and Fallon, A.M. 1985. Characterization of the ribosomal proteins from mosquito (*Aedes albopictus*) cells. *Eur. J. Biochem.* **150**:507–515.
80. Judson, C.L., and de Lumen, H.Z. 1976. Some effects of juvenile hormone and analogs on ovarian follicles of the mosquito, *Aedes aegypti. J. Med. Entomol.* **13**:197–201.
81. Kaczor, W.J., and Hagedorn, H.H. 1980. The effects of alpha-amanitin and cordycepin on vitellogenin synthesis by mosquito fat body. *J. Exp. Zool.* **214**:229–233.
82. Kataoka, H., Toschi, A., Li, J.P., Carney, R.L., Schooley, D.A., and Kramer, S.J. 1989. Identification of an allatotropin from adult *Manduca sexta. Science* **243**:1481–1483.
83. Kelly, T.J., and Fuchs, M.S. 1980. *In vivo* induction of ovarian development in decapitated *Aedes atropalpus* by physiological levels of 20-hydroxyecdysone. *J. Exp. Zool.* **213**:25–32.
84. Kelly, T.J., Fuchs, M.S., and Kang, S.-H. 1981. Induction of ovarian development in autogenous *Aedes atropalpus* by juvenile hormone and 20-hydroxyecdysone. *Int. J. Invert. Reprod.* **3**:101–112.
85. Kelly, T.J., Birnbaum, M.J., Woods, C.W., and Borkovec, A.B. 1984. Effects of house fly oostatic hormone on egg development neurosecretory hormone action in *Aedes atropalpus. J. Exp. Zool.* **229**:491–496.
86. Kelly, T.J., Masler, E.P., Schwartz, M.B., and Haught, S.B. 1986. Inhibitory effects of oostatic hormone on ovarian maturation and ecdysteroid production in Diptera. *Insect Biochem.* **16**:273–279.

87. Klowden, M.J. 1981. Initiation and termination of host-seeking inhibition in *Aedes aegypti* during oocyte maturation. *J. Insect Physiol.* **27**:799–803.
88. Klowden, M.J. 1983. The physiological control of mosquito host-seeking behavior. In K.F. Harris, ed., *Current Topics in Vector Research*, Vol. 1. Praeger, New York, pp. 93–116.
89. Klowden, M.J. 1987. Distention-mediated egg maturation in the mosquito. *Aedes aegypti. J. Insect Physiol.* **33**:83–88.
90. Klowden, M.J. 1989. Influence of the ovaries and fat body on the initiation and termination of pre-oviposition behavior in the mosquito, *Aedes aegypti. J. Insect Physiol.* **35**:567–570.
91. Klowden, M.J. 1990. The endogenous regulation of mosquito reproductive behavior. *Experentia* **46**:660–670.
92. Klowden, M.J., and Chambers, G.M. 1989. Ovarian development and adult mortality in *Aedes aegypti* treated with sucrose, juvenile hormone and methoprene. *J. Insect Physiol.* **35**:513–517.
93. Klowden, M.J., and Chambers, G.M. 1991. Male accessory gland substances activate egg development in nutritionally stressed *Aedes aegypti* mosquitoes. *J. Insect Physiol.* **37**:721–726.
94. Kohler, C.N., and Raikhel, A.S. 1991. Initiation of vitellogenin uptake and protein synthesis in the mosquito (*Aedes aegypti*) ovary in response to a blood meal. *J. Insect Physiol.* **37**:703–711.
95. Kramer, S.J., Toschi, A., Miller, C.A., Kataoka, H., Quistad, G.B., Li, J.P., Carney, R.L., and Schooley, D.A. 1991. Identification of an allatostatin from the tobacco hornworm *Manduca sexta. Proc. Natl. Acad. Sci. U.S.A.* **88**:9458–9462.
96. Lanot, R., Dorn, A., Günster, B., Thiebold, J., Lagueux, M., and Hoffmann, J.A. 1989. Functions of ecdysteroids in oocyte maturation and embryonic development of insects. In J. Koolman, ed., *Ecdysone, from Chemistry to Mode of Action*. Georg Thieme Verlag, Stuttgart, pp. 262–270.
97. Larsen, J.R., and Bodenstein, D. 1959. The humoral control of egg maturation in the mosquito. *J. Exp. Zool.* **140**:343–382.
98. Laurence, B.R., and Simpson, M.G. 1974. Cell replication in the follicular epithelium of the adult mosquito. *J. Insect Physiol.* **20**:703–715.
99. Lea, A.O. 1963. Some relationships between environment, corpora allata, and egg maturation in Aedine mosquitoes. *J. Insect Physiol.* **9**:793–809.
100. Lea, A.O. 1967. The median neurosecretory cells and egg maturation in mosquitoes. *J. Insect Physiol.* **13**:419–429.
101. Lea, A.O. 1969. Egg maturation in mosquitoes not regulated by the corpora allata. *J. Insect Physiol.* **15**:537–541.
102. Lea, A.O. 1970. Endocrinology of egg maturation in autogenous and anautogenous *Aedes taeniorhynchus. J. Insect Physiol.* **16**:1689–1696.
103. Lea, A.O. 1982. Artifactual stimulation of vitellogenesis in *Aedes aegypti* by 20-hydroxyecdysone. *J. Insect Physiol.* **28**:173–176.
104. Lea, A.O., and Van Handel, E. 1982. A neurosecretory hormone-releasing factor from ovaries of mosquitoes fed blood. *J. Insect Physiol.* **28**:503–508.
105. Lea, A.O., Briegel, H., and Lea, H.M. 1978. Arrest, resorption, or maturation of oocytes in *Aedes aegypti*: Dependence on the quantity of blood and the interval between bloodmeals. *Physiol. Entomol.* **3**:309–316.
106. Leahy, M.G. 1967. Non-specificity of the male factor enhancing egg-laying in Diptera. *J. Insect Physiol.* **13**:1283–1292.

107. Lin, Y., Hamblin, M.T., Kanost, M.R., Knipple, D.C., Wolfner, M.F., and Hagedorn, H.H. 1993. Structure, expression and hormonal control of genes from the mosquito, *Aedes aegypti*, which encode proteins similar to the vitelline membrane proteins of *Drosophila melanogaster. Dev. Biol.* **155**:558–568.
108. Lord, J.C., and Hall, D.W. 1983. Sporulation of *Amblyospora* (Microspora) in female *Culex salinarius*: Induction by 20-hydroxyecdysone. *Parasitology* **87**:377–383.
109. Lu, Y.H., and Hagedorn, H.H. 1986. Regulation of egg development in the mosquito *Anopheles albimanus. Int. J. Invert. Reprod. Dev.* **9**:79–94.
110. Lucarotti, C. 1987. *Coelomomyces stegomyiae* infection in adult *Aedes aegypti. Mycologia* **79**:362–369.
111. Lucarotti, C. 1992. Invasion of *Aedes aegypti* ovaries by *Coelomomyces stegomyiae. J. Invert. Pathol.* **60**:176–184.
112. Lucarotti, C.J., and Klein, M.B. 1988. Pathology of *Coelomomyces stegomiae* in adult *Aedes aegypti* ovaries. *Can. J. Bot.* **66**:877–884.
113. Ma, M., Newton, P.B., He, G., Kelly, T.J., Hus, H.T., Masler, E.P., and Borkovec, A.B. 1984. Development of monoclonal antibodies for monitoring *Aedes atropalpus* vitellogenesis. *J. Insect Physiol.* **30**:529–536.
114. Ma, M., Gong, H., Zhang, J.-Z., and Gwadz, R. 1987. Response of cultured *Aedes aegypti* fat bodies to 20-hydroxyecdysone. *J. Insect Physiol.* **33**:89–93.
115. Ma, M., Zhang, J.-Z., Gong, H., and Gwadz, R. 1988. Permissive action of juvenile hormone on vitellogenin production by the mosquito, *Aedes aegypti. J. Insect Physiol.* **34**:593–596.
116. Martinez, T., and Hagedorn, H.H. 1987. Development of responsiveness to hormones after a blood meal in the mosquito, *Aedes aegypti. Insect Biochem.* **17**:1095–1098.
117. Masler, E.P., Fuchs, M.S., Sage, B., and O'Connor, J.D. 1980. Endocrine regulation of ovarian development in the autogenous mosquito, *Aedes atropalpus. Gen. Comp. Endocrinol.* **241**:250–259.
118. Masler, E.P., Fuchs, M.S., Sage, B., and O'Connor, J.D. 1981. A positive correlation between oocyte production and ecdysteroid levels in adult *Aedes. Physiol. Entomol.* **6**:45–49.
119. Masler, E.P., Hagedorn, H.H., Petzel, D., and Borkovec, A.B. 1983. Partial purification of egg development neurosecretory hormone and reverse-phase liquid chromatographic techniques. *Life Sci.* **33**:1925–1931.
120. Matsumoto, S., Brown, M.R., Suzuki, A., and Lea, A.O. 1989. Isolation and characterization of ovarian ecdysteroidogenic hormones from the mosquito, *Aedes aegypti. Insect Biochem.* **19**:651–656.
121. Matsumoto, S., Brown, M.R., Crim, J.W., Vigna, S.R., and Lea, A.O. 1989. Isolation an primary structure of neuropeptides from the mosquito, *Aedes aegypti*, immunoreactive to FMRFamide antiserum. *Insect Biochem.* **19**:277–283.
122. Meola, R.W., and Lea, A.O. 1972. The ultrastructure of the corpus cardiacum of *Aedes sollicitans* and the histology of the cerebral neurosecretory system of mosquitoes. *Gen. Comp. Endocrinol.* **18**:210–234.
123. Meola, R.W., and Lea, A.O. 1972. Humoral inhibition of egg development in mosquitoes. *J. Med. Entomol.* **9**:99–103.

124. Meola, R.W., and Petralia, R.S. 1980. Juvenile hormone induction of biting behavior in *Culex* mosquitoes. *Science* **209**:1548–1550.
125. Meola, R.W., and Readio, J. 1988. Juvenile hormone regulation of biting behavior and egg development in mosquitoes. *Adv. Dis. Vector Res.* **5**:2–24.
126. Nagasawa, H., Kataoka, H., Isogai, A., Tamura, A., Suzuki, A., Mizoguchi, A., Fujiwara, Y., Suzuki, A., Takahashi, S., and Ishizaki, H. 1986. Amino acid sequence of a prothoracicotropic hormone of the silkworm *Bombyx mori. Proc. Natl. Acad. Sci. U.S.A.* **83**:5840–5843.
127. Nijhout, H.F., and Carrow, G.M. 1978. Diuresis after a bloodmeal in female *Anopheles freeborni. J. Insect Physiol.* **24**:293–298.
128. O'Meara, G.F., and Evans, D.G. 1976. The influence of mating on autogenous egg development in the mosquito, *Aedes taeniorhynchus. J. Insect Physiol.* **22**:613–617.
129. O'Meara, G.F., and Evans, D.G. 1977. Autogeny in saltmarsh mosquitoes induced by a substance from the male accessory gland. *Nature (London)* **267**:342–343.
130. Park, Y.-J., and Fallon, A.M. 1990. Mosquito ribosomal genes: Characterization of gene structure and evidence for changes in copy number during development. *Insect Biochem.* **20**:1–11.
131. Patterson, J.W. 1971. Critical sensitivity of the ovary of *Aedes aegypti* adults to sterilization by juvenile hormone mimics. *Nature New Biol.* **233**:176–177.
132. Patterson, J.W. 1974. A comparison of the morphogenetic and sterilizing activities of juvenile hormone mimics on *Aedes aegypti. J. Insect Physiol.* **20**:2095–2106.
133. Petzel, D.H., Hagedorn, H.H., and Beyenbach, K.W. 1985. Preliminary isolation of mosquito natriuretic factor. *Am. J. Physiol.* **249**:R379–R386.
134. Petzel, D.H., Hagedorn, H.H., and Beyenbach, K.W. 1986. Peptide nature of two mosquito natriuretic factors. *Am. J. Physiol.* **250**:R328–R332.
135. Petzel, D.H., Berg, M., and Beyenbach, K.W. 1987. Hormone-controlled cAMP mediated fluid secretion in the yellow fever mosquito. *Am. J. Physiol.* **253**:R701–R711.
136. Pratt, G.E., Tobe, S.S., Weaver, R.J., and Finney, J.R. 1975. Spontaneous synthesis and release of C_{16} juvenile hormone by isolated corpora allata of female locust *Schistocerca gregaria* and female cockroach *Periplaneta americana. Gen. Comp. Endocrinol.* **26**:478–484.
137. Pratt, G.E., Farnsworth, D.E., Siegel, N.R., Fok, K.F., and Feyereisen, R. 1989. Identification of an allatostatin from adult *Diploptera punctata. Biochem. Biophys. Res. Commun.* **163**:1243–1247.
138. Pratt, G.E., Farnsworth, D.E., Fok, K.F., Siegel, N.R., McCormack, A.L., Shabanowitz, J., Hunt, D.F., and Feyereisen, R. 1991. Identity of a second type of allatostatin from cockroach brains: An octadecapeptide amide with a tyrosine-rich address sequence. *Proc. Natl. Acad. Sci. U.S.A.* **88**:2412–2416.
139. Racioppi, J.V., Gemmill, R.M., Kogan, P.K., Calvo, J.M., and Hagedorn, H.H. 1986. Expression and regulation of vitellogenin messenger RNA in the mosquito *Aedes aegypti. Insect Biochem.* **16**:255–262.
140. Raikhel, A.S. 1984. Accumulations of membrane-free clathrin-like lattices in the mosquito oocyte. *Eur. J. Cell Biol.* **35**:279–283.
141. Raikhel, A.S. 1992. Vitellogenesis in mosquitoes. *Adv. Dis. Vector Res.* **9**:1–39.

142. Raikhel, A.S., and Lea, A.O. 1982. Abnormal vitelline envelope induced by unphysiological doses of ecdysterone in *Aedes aegypti. Physiol. Entomol.* **7**:55–64.
143. Raikhel, A.S., and Lea, A.O. 1983. Previtellogenic development and vitellogenin synthesis in the fat body of a mosquito: An ultrastructural and immunocytochemical study. *Tissue Cell* **15**:281–300.
144. Raikhel, A.S., and Lea, A.O. 1985. Hormone-mediated formation of the endocytotic complex in mosquito oocytes. *Gen. Comp. Endocrinol.* **57**:422–433.
145. Raikhel, A.S., and Lea, A.O. 1990. Juvenile hormone controls previtellogenic proliferation of ribosomal RNA in the mosquito fat body. *Gen. Comp. Endocrinol.* **77**:423–434.
146. Raikhel, A.S., and Lea, A.O. 1991. Control of follicular epithelium development and vitelline envelope formation in the mosquito: Role of juvenile hormone and 20-hydroxyecdysone. *Tissue Cell* **23**:577–591.
147. Raikhel, A.S., and Dhadialla, T.S. 1992. Accumulation of yolk proteins in insect oocytes. *Annu. Rev. Entomol.* **37**:217–251.
148. Readio, J., and Meola, R. 1985. Two stages of juvenile hormone-mediated growth of secondary follicles in *Culex pipiens. J. Insect Physiol.* **7**:559–562.
149. Readio, J., Peck, K., Meola, R., and Dahm, K.H. 1988. Corpus allatum activity (in vitro) in female *Culex pipiens. J. Insect Physiol.* **34**:131–135.
150. Redfern, C.P.F. 1982. 20-Hydroxyecdysone and ovarian development in *Anopheles stephensi. J. Insect Physiol.* 2:97–109.
151. Rosenberg, R. 1980. Ovarian control of blood meal retention in the mosquito *Anopheles freeborni. J. Insect Physiol.* **26**:477–480.
152. Rossignol, P.A., Feinsod, F.M., and Spielman, A. 1981. Inhibitory regulation of corpus allatum activity in mosquitoes. *J. Insect Physiol.* **27**:651–654.
153. Roth, T.F., and Porter, K.R. 1964. Yolk protein uptake in the oocyte of the mosquito *Aedes aegypti. J. Cell. Biol.* **20**:313–332.
154. Rozeboom, L.E. 1961. The effect of the gonadotropic hormone cycle of the adult mosquito on development of the malaria parasite. *J. Parasitol.* **47**:597–599.
155. Saegusa, H., Mizoguchi, A., Nagasawa, H., Suzuki, A., and Ishizaki, H. 1989. Changes in hemolymph titer of bombyxin during postembryonic development of *Bombyx mori. Zool. Sci.* **6**:1182 (abstract).
156. Schlaeger, D.A., and Fuchs, M.S. 1974. Dopa decarboxylase activity in *Aedes aegypti*: A preadult profile and its subsequent correlation with ovarian development. *Dev. Biol.* **38**:209–219.
157. Schlaeger, D.A., Fuchs, M.S., and Kang, S.H. 1974. Ecdysone mediated stimulation of dopa decarboxylase activity and its relationship to ovarian development in *Aedes aegypti. J. Cell Biol.* **61**:454–465.
158. Schwartz, M.B., Kelly, T.J., Imberski, R.B., and Rubenstein, E. 1985. The effects of nutrition and methoprene treatment on ovarian ecdysteroid synthesis in *Drosophila melanogaster. J. Insect Physiol.* **31**:947–957.
159. Shampengtong, L., and Wong, K.P. 1989. An in vitro assay of 20-hydroxyecdysone sulfotransferase in the mosquito, *Aedes togi. Insect Biochem.* **19**:191–196.

160. Shapiro, J.P., and Hagedorn, H.H. 1982. Juvenile hormone and the development of ovarian responsiveness to brain hormone in the mosquito *Aedes aegypti*. *Gen. Comp. Endocrinol.* **46**:176–183.
161. Shapiro, A.B., Wheelock, G.D., Hagedorn, H.H., Baker, F.C., Tsai, L.W., and Schooley, D.A. 1986. Juvenile hormone and juvenile hormone esterase in adult females of the mosquito *Aedes aegypti*. *J. Insect Physiol.* **32**:867–877.
162. Shirk, P.D., Dahm, K.H., and Röller, H. 1976. The accessory sex glands as the repository for juvenile hormone in male *Cecropia* moths. *Z. Naturf.* **31c**:199–200.
163. Smith, S.L., and Mitchell, M.J. 1986. Ecdysone 20-monooxygenase systems in a larval and an adult dipteran. *Insect Biochem.* **16**:49–55.
164. Spielman, A., Gwadz, R.W., and Anderson, W.A. 1971. Ecdysone-initiated ovarian development in mosquitoes. *J. Insect Physiol.* **17**:1807–1814.
165. Tobe, S.S., and Pratt, G.E. 1975. The synthetic activity and glandular volume of the corpus allatum during ovarian maturation in the desert locust. *Schistocerca gregaria*. *Life Sci.* **17**:417–422.
166. Uchida, K., Ohmori, D., Yamakura, F., and Suzuki, K. 1992. Mosquito egg development induced by infusion of amino acids into the hemocoel. *J. Insect Physiol.* **38**:953–959.
167. Van Handel, E., and Lea, A.O. 1970. Control of glycogen and fat metabolism in the mosquito. *Gen. Comp. Endocrinol.* **14**:381–384.
168. Veenstra, J.A. 1988. Effects of 5-hydroxytryptamine on the malpighian tubules of *Aedes aegypti*. *J. Insect Physiol.* **34**:299–304.
169. Wheelock, G.D., and Hagedorn, H.H. 1985. Egg maturation and ecdysiotropic activity in extracts of mosquito (*Aedes aegypti*) heads. *Gen. Comp. Endocrinol.* **60**:196–203.
170. Wheelock, G.H., Petzel, D.H., Gillett, J.D., Beyenbach, K.W., and Hagedorn, H.H. 1988. Evidence for hormonal control of diuresis after a blood meal in the mosquito *Aedes aegypti*. *Arch. Insect Biochem. Physiol.* **7**:75–89.
171. Wheelock, G.H., Sieber, K.P., and Hagedorn, H.H. 1991. Rapid isolation of a nurohormone from mosquito heads by high-performance liquid chromatography. *J. Chromatogr.* **54**:508–514.
172. Whisenton, L.R., Kelly, T.J., and Bollenbacher, W.E. 1987. Multiple forms of cerebral peptides with steroidogenic function in pupal and adult brains of the yellow fever mosquito, *Aedes aegypti*. *Mol. Cell. Endocrinol.* **50**:3–14.
173. Woodhead, A.P., Stay, B., Seidel, S.L., Khan, M.A., and Tobe, S.S. 1989. Primary structure of four allatostatins: Neuropeptide inhibitors of juvenile hormone synthesis. *Proc. Natl. Acad. Sci. U.S.A.* **86**:5997–6001.
174. Young, A.D.M., and Downe, A.E.R. 1982. Renewal of sexual receptivity in mated female mosquitoes, *Aedes aegypti*. *Physiol. Entomol.* **7**:467–471.
175. Young, A.D.M., and Downe, A.E.R. 1983. Influence of mating on sexual receptivity and oviposition in the mosquito, *Culex tarsalis*. *Physiol. Entomol.* **8**:213–217.
176. Young, A.D.M., and Downe, A.E.R. 1987. Male accessory gland substances and the control of sexual receptivity in female *Culex tarsalis*. *Physiol. Entomol.* **12**:233–239.

177. Zhu, X.X., and Chen, Z.-f. 1985. Kinetics of disappearance of exogenous 20-hydroxyecdysone in adult females of the mosquito, *Culex pipiens pallens*. *Cont. Shanghai Inst. Entomol.* **5**:33–39.
178. Zhu, X.X., Chen, Z.-f., and Cao, M.-x. 1980. Endogenous molting hormone level and vitellogenin synthesis during adult stage of the mosquito, *Culex pipiens*. *Cont. Shanghai Inst. Entomol.* **1**:63–68.

Note added in Proof

Two papers have been published on ecdysone during larval and pupal development in mosquitoes. Whisenton et al. (1989) have followed ecdysteroid levels in pupal *A. aegypti* during pharate adult development until eclosion. They found differences in the timing of ecdysteroid peaks between male and female pupae that correlated with the shorter developmental period of males. Jenkins et al. (1992) have shown that the prothoracic gland in larval and pupal mosquitoes does not secrete ecdysone. The latter appears to be produced instead by unidentified tissues in the thorax and abdomen.

Jenkins, S.P., Brown, M.R., and Lea, A.O. 1992. Inactive prothoracic glands in larvae and pupae of *Aedes aegypti*: ecdysteroid release by tissues in the thorax and abdomen. *Insect Biochem. Molec. Biol.* **22**:553–559.
Whisenton, L.R., Warren, J.T., Manning, M.K., and Bollenbacher, W.E. 1989. Ecdysteroid titers during pupal-adult development of *Aedes aegypti*: basis for a sexual dimorphism in the rate of development, *J. Insect Physiol.* **35**:67–73.

6
Human Malaria Transmission: Reconciling Field and Laboratory Data

Thomas R. Burkot and Patricia M. Graves

Introduction

The human malaria parasite life cycle appears deceptively simple because it involves only one species of vertebrate host (humans) and anopheline mosquitoes to transmit the pathogen. However, a number of host, vector, and pathogen factors have evolved that interact with ecological and logistic considerations to determine whether transmission of the parasite from a mosquito to a human, or the converse, will successfully occur.

Our understanding of the transmission of human malaria is shaped by the ethical and logistic constraints in the design of experiments and observations involving a human parasite. Consequently, animal models have been used to circumvent some of these concerns. Results obtained using animal models carry with them the need for caution in extrapolating the results to humans. Caution must also be used in applying laboratory results to the interpretation of field observations as well as in extrapolating field results from one epidemiological situation to another.

Our comprehension of human malaria transmission is also biased by the technology used to measure the basic parameters involved. The increased sophistication of biotechnology offers more sensitive and elegant methods for measuring some interactions involved in human malaria transmission. However, there is a danger that attention may be focused on minor aspects of the transmission of human malaria while important factors needing research are ignored because new methods have not evolved for their study, so they are not fashionable enough to attract the attention of researchers and funding agencies.

Thomas R. Burkot, Division of Vector-Borne Infectious Diseases, National Center for Infectious Diseases, Centers for Disease Control, Foothills Campus, P.O. Box 2087, Fort Collins, Colorado 80522, USA.
Patricia M. Graves, Queensland Institute of Medical Research, 300 Herston Road, Brisbane Q 4029, Australia.

Before examining the individual factors affecting transmission, we will review the basic strategies for measuring malaria transmission. Then we will discuss the multitude of factors that determine whether malaria is successfully transmitted. The factors involved in this process will be discussed in chronological order starting with an established blood infection of asexual parasites. This will be followed by an examination of the factors involved in the establishment of the infection in the mosquito by the ingestion of the sexual stages of the parasite. The factors that interact during the extrinsic incubation phase of the life cycle in which a mosquito goes from being infected to becoming infectious will be reviewed, culminating in the process of an infectious mosquito passing sporozoites to a susceptible host while blood feeding. When examining the factors involved in human malaria transmission, it is necessary to be aware of the limitations of the techniques used to detect malaria parasites, and the methods by which the vector and host populations are measured. Brief discussions of new techniques for identifying and measuring parasite and vector populations will therefore be included in the relevant sections.

Measuring Malaria Transmission

In the simplest view, measuring malaria transmission can be seen as comparing measurements of parasite populations in humans and vectors together with the rate of contact between vector and human populations. Malaria transmission can be estimated from three different approaches: entomological, parasitological, or theoretical, and the measurement may be either direct or indirect. For example, estimating entomological transmission of malaria involves measuring the sporozoite inoculation rate (sporozoite rate multiplied by the man-biting rate). Similarly, direct measurement of malaria transmission by parasitology involves monitoring individuals for the presence of blood stage parasites to record the incidence of malaria. This is usually done by microscopic examination of Giemsa-stained thick blood films. In theory, transmission should be estimated by the infant conversion rate in highly endemic areas to avoid the problem of immunity depressing the malaria incidence. However, low infant numbers in small villages coupled with the presence of maternal immunity often makes this impractical. Conversion rates in adults can be used in low endemic or epidemic situations where few persons are immune to malaria.

The corresponding indirect measure of transmission involves recording the rate at which antibody negative individuals (for malaria antigens) become antibody positive. Serology will most accurately reflect transmission in areas where the transmission level is low, the antibody half-life is short, the ability of the human host to respond to the antigen of interest is not genetically restricted, and the epitope to which the immune response is being monitored is unique to a particular species.

Recently, monitoring the human immune response to the *Plasmodium falciparum* and *P. vivax* circumsporozoite (CS) proteins has been proposed as a potentially useful method for indirectly measuring malaria transmission (13, 58, 64, 110, 159, 164). Studies in Thailand have shown that all individuals tested, whether immune or nonimmune, responded to the *P. falciparum* CS protein and the antibody half-life was 28 days (159). In *P. vivax* antibody-positive individuals, the antibodies had a similar half-life, but only 16 of 26 persons experiencing their first episode of *P. vivax* malaria developed antibodies to the CS protein (13). Studies from Africa suggested that levels and prevalence of antibodies against the CS protein reflected malaria transmission (64). However, other studies have not been as supportive of the CS protein as a universally suitable measure of malaria transmission. Monitoring antibody levels of individuals after they left malaria-endemic areas indicated an anti-CS antibody half life of several years (58). In addition, the problems of similar epitopes within a malarial antigen (31) or on different antigens in different stages of the parasite's life cycle (41, 49, 107, 114) giving rise to cross-reacting antibodies must also be taken into consideration (86).

Two studies in a hyperendemic area of Papua New Guinea failed to find a significant association between sporozoite inoculation rates and antibody prevalence or titers to the CS proteins of *P. falciparum* or *P. vivax* (18, 23). Anti-CS antibodies may serve as an adequate indirect measure of transmission in low or seasonal transmission zones but may not be sensitive enough to reflect minor changes in transmission rates. The African studies in which antibodies reflected transmission were done in areas where large seasonal fluctuations in transmission occur, while the Papua New Guinean study was in an area of perennial transmission where there are minor but consistent differences in transmission among the villages studied. Serology was inadequate as a means for measuring transmission in the area of perennial transmission.

The theoretical or mathematical approach to malaria transmission is best exemplified by the concept of vectorial capacity in which one attempts to estimate the potential for transmission by mathematically relating the factors that affect malaria transmission independently of any measurements of malaria parasites (71). Vectorial capacity was intended to be used to predict the potential for transmission should the parasite be introduced into a nonendemic area and also to predict the degree of malaria control required to decrease the prevalence of the disease (110).

Factors Influencing Malaria Transmission

Asexual Blood Stages

Obviously an asexual infection is a prerequisite for a person to be infectious to mosquitoes, and the prevalence and density of the asexual infection will be factors contributing to the prevalence and density of gametocytes. The genetic and immune factors controlling asexual parasitemia levels are beyond the scope of this review (see 2, 73, 78, 108 for reviews). Cultural and environmental factors including the availability and correct use of antimalarials will also affect parasite incidence.

Malaria parasites are most frequently detected in humans by microscopic examination of thick blood smears stained with Giemsa. Accuracy of identification depends on the quality of the microscope, training of the microscopist, the cross-checking system, and the number of thick film fields examined (115). DNA probes for *P. falciparum* and *P. vivax* with sensitivities comparable to routine microscopy have been developed (4, 133). Probes that detect the species-specific regions of ribosomal RNA have been proposed as potentially sensitive tools for malaria diagnosis because of the large number of such molecules in the cell (158). Such probes have yet to be tested on patients or residents of malarious areas.

Mosquito Infection Probability

The mosquito infection probability is defined here as the probability of a mosquito becoming infected with malaria after feeding on a human. It is not identical to the proportion of persons infectious to mosquitoes (sometimes called "reservoir of infection") since not all mosquitoes that feed on an infected person become infected. For a person to be infectious to mosquitoes, gametocytes must be circulating in the peripheral blood and ingested by susceptible mosquitoes. Mosquito infection probability is therefore a function of factors affecting the gametocyte rate, the infectivity of gametocytes to mosquitoes, and the susceptibility of mosquitoes to infection with malaria.

Mosquito infection probability may be estimated in three ways. The first involves feeding laboratory-reared mosquitoes at random on people to determine the proportion of the population that is infectious; the second involves feeding mosquitoes on gametocyte carriers and combining the information obtained with the gametocyte rates in the population. The third method uses measurements of infection rates in the wild mosquito population to determine the force of infection.

Until the last few years, field measurements of mosquito infection probability were scarce. Recently, attempts to measure it have increased, prompted partly by interest in the effects of transmission-blocking vaccines. One of the earliest attempts (117) used the first method of random mosquito feeding to estimate that 10–11% of a West African village population was infectious to mosquitoes. After allowing for the proportion of mosquitoes infected, the infection probability was actually about 2%. A slightly lower figure of 1.3% for *Anopheles farauti* was obtained in recent studies in Papua New Guinea (75). A very similar estimate was obtained in the same study area by feeding *A. farauti* on *P. falciparum*, *P. vivax*, and *P. malariae* gametocyte carriers (50, 45, and 36% of whom were infectious, respectively, with an overall average of 15% of mosquitoes infected). Using the prevalence of gametocyte carriers to extrapolate the results to the whole population gave an estimate of 1.2% for mosquito infection probability (75). A study in Bukina Faso obtained a mosquito infection probability of 4% for *P. falciparum* in *An. gambiae* (10), however, this is an overestimate since feeds were only performed on "heavy" gametocyte carriers (>50 gametocytes/mm^3) who infected 37.2% of mosquitoes fed on them. The second method of estimating mosquito infection probability was also applied recently in Sri Lanka, by Gamage-Mendis et al. (69). In the study area *P. vivax* is endemic, with relatively low transmission rates, and *P. falciparum* is epidemic. Feeding mosquitoes on gametocyte-positive persons demonstrated that 47.5% of people infected with *P. vivax* (about 70% of gametocyte carriers) and 7.8% of people infected with *P. falciparum* (about 90% of gametocyte carriers) were infectious to laboratory-reared *A. tessellatus*. Since the infection rate was low, prevalence data were not available, but using the case-incidence rate coupled with an estimate of 5 days for the duration of infectiousness gave a value of 0.05% of people infectious for *P. vivax* and 0.01% infectious for *P. falciparum*. The mosquito infection probability would be even lower since these figures do not account for the proportion of mosquitoes infected.

Recent studies have applied the third method of estimating infection probability, namely the use of infection rates in wild-caught mosquitoes, in Papua New Guinea (20, 76) and Tanzania (100). Although different models were used to analyze the data, the principle underlying the methods was first applied in the 1960s by Gillies and Wilkes (72). Using multiparous age-grading of *A. gambiae* in Tanzania, the infection probability was estimated in the original study to be 6–7% by observing the rate at which the sporozoite rate increased in different age classes. Twenty-five years later, Lines et al. (100) performed a similar study [with age-grading done by the same individual as in the Gillies and Wilkes (72) study], but the model was reformulated in terms of the proportion remaining uninfected, to allow for superinfection. Recalculation of the Gillies and Wilkes (72) data with this model gave a revised estimate of 8%, whereas Lines et al. (100) observed the infection probability to be 20–22%, and convincingly argued that there has been a 2.5-fold increase over the past 25 years. They attributed this to possible genetic selection for increased gametocyte production by chloroquine, although they do not consider the effects of suppression of antigametocyte immunity by use of chloroquine and consequent increase in gametocyte rates (relative to asexual stages) on emergence of chloroquine resistance. Gametocyte rates at the two time periods would be of interest, although they might not resolve the cause of the putative increase.

Mosquito infection rates were also used to estimate infection probability in Papua New Guinea (20, 76). The model used (149) differed from that of Lines et al. (100) in that no age grading was required; the method involved holding some of the wild-caught mosquitoes in an insectary to obtain "delayed infection rates," which were compared with infection rates determined immediately after capture. The estimates obtained were in the range of 4.9–17.4% (20, 76) depending on mosquito species and village, which are much higher than the rates obtained by direct feeding studies (see above) in the same area. In view of this discrepancy and the unknown factors in direct feeding studies (which include susceptibility of laboratory colonies and possible nonrandom feeding behaviour), it is regarded as more desirable to estimate mosquito infection probability from wild-caught mosquitoes by the methods of Saul et al. (149) and Lines et al. (100), where epidemiological circumstance make this possible.

Having described estimates of mosquito infection probability, we will now consider the biological factors that affect its magnitude. Obviously, the infection probability cannot be higher than the gametocyte rate, unless there is marked nonrandom feeding on gametocyte-positive individuals (90). However, the infection probability is usually lower than the gametocyte rate, because of the numerous factors affecting gametocyte infectivity.

Gametocyte Rates

Gametocyte rates in a population will be determined by asexual parasite rates and the immune status of the population. The ratio of gametocytes to asexual parasites is not fixed. Seasonal as well as spatial variations for *P. falciparum* gametocyte rates were observed (10, 139). Gametocyte rates of *P. falciparum* usually range from 5 to 30% of asexual infection rates in immune populations (69, 75, 115). The proportion of infections with *P. vivax* gametocytes is generally much higher, averaging 40–75% (69, 75). Both species in endemic areas have a marked age-dependent variation in the ratio between asexual stages and gametocytes. This effect of immunity was demonstrated in Papua New Guinea, where gametocyte rates for both *P. falciparum* and *P. vivax* declined more rapidly with age than did asexual parasite rates (75). Further evidence for the specific immune suppression of *P. falciparum* gametocytes, independent of the immune control of asexual stages, was found by comparing gametocyte rates in immune and non-immune populations in Irian Jaya (3). Malaria control programs may successfully reduce malaria transmission and consequently lower immunity in the population, with effects on gametocyte rates which may be apparent after the cessation of control. Such was the case after a DDT spraying program in Irian Jaya, when the proportion of gametocyte-positive infections (in 3 to 8 year olds) increased from 7 to 40% for *P. falciparum* and 9 to 50% for *P. vivax* (113).

Gametocyte Infectivity

Several recent reviews have outlined the course of gametocytemia during infections as well as the effects of gametocyte numbers, sex ratio, and drugs on infectivity (32, 33, 121). This discussion will focus mainly on new information.

Numerous authors have commented on the fact that persons with high numbers of gametocytes are often completely noninfectious, whereas persons with low or even nondetectable gametocyte densities may infect a large proportion of mosquitoes. Although many individual studies have failed to demonstrate statistically significant positive correlations between gametocyte density and either oocyst numbers or frequency of mosquito infection (69, 75, 147), consideration of groups of studies together does reveal a trend toward higher infectivity at higher gametocyte densities, although noninfectious persons occur at all levels of gametocytemia (32). Gametocyte density will determine the maximum possible infection rate in mosquitoes, which is rarely achieved due to the influence of numerous factors, including gametocyte sex ratio (27), antimalarial drugs (94, 95), and antigamete immunity (reviewed in 34).

In humans, transmission-blocking antigamete antibodies are induced by natural infections of *P. vivax* (111, 132) and *P. falciparum* (74). These antibodies may have a regulatory effect on malaria transmission in certain epidemiological situations (56). However, the same antibodies can be responsible for both transmission blocking and enhancing effects in *P. vivax* (124). At high titers, antibodies block the infection of mosquitoes, while at low titers, it is postulated that these same antibodies facilitate fertilization resulting in higher mosquito infection rates and higher densities of oocysts. In the monkey model *P. cynomolgi*, factors in the serum enhanced infectivity 2- to 3-fold (relative to nonimmune monkey serum) during the early patent period, while serum taken postpeak of parasitemia suppressed infectivity (119). Over the next 3 to 4 months suppression waned until the serum again enhanced infectivity, and the enhancement and suppression effects correlated with low and high antibody titers in the serum, respectively. If applicable to human *P. vivax*, these studies demonstrate the importance of enhancing factors in the serum on infectivity in the critical early period of the infection. Other immune factors besides antibodies have recently been demonstrated to be of importance in infectivity. In the "crisis" of *P. cynomolgi* and the clinical paroxysm of *P. vivax*, two cytokines (tumor necrosis factor and interferon-γ), together with other unidentified factors, mediate killing of intraerythrocyic parasites including gametocytes (112, 118). These transient but dramatic affects appear to be more common in nonimmune patients with no previous history of malaria than in those from endemic regions (112). The complex interplay between transmission-enhancing and blocking antibodies and cytokines may be responsible for the interesting observations of Gamage-Mendis et al. (69) that infectivity "per gametocyte" of *P. vivax* increased markedly with age, while gametocyte density decreased.

Mosquito Susceptibility

Following ingestion, macrogametocytes are fertilized by the microgametes in the mosquito gut to form a zygote. The ookinete that develops must penetrate the gut wall to form an oocyst, inside which sporozoites will form. Genetic variation exists in anopheline susceptibility to malaria infection [reviewed by Curtis and Graves (51)]. In general, malaria strains were more compatible with anophelines from the same geographic area. While size of the female mosquito had no influence on susceptibility (157), larger *An. dirus* become more heavily infected with oocysts (91). From laboratory colonies of *An. gambiae* and *An. stephensi*, susceptible and refractory lines for malaria infection have been established (44, 65). Although the *An. gambiae* colony was selected against *P. cynomolgi*, the refractory line was fully refractory to simian malarias (*P. gonderi*, *P. inui*, and *P. knowlesi*), a rodent malaria, *P. berghei*, and an avian malaria, *P. gallinaceum*. In addition, this line was highly refractory to the human malarias, *P. falciparum*, *P. vivax*, and *P. ovale*. Analysis of the mechanism for resistance to infection indicated that two unlinked genetic loci controlled expression of the susceptible and refractory phenotypes through physiologically independent means (156). *Anopheles gambiae* in Kenya had a significant association between inversion karyotypes on chromosome 2 and malaria infection rates (126). However, it was unclear whether the difference in infection rates was a function of a difference in malaria susceptibility or a result of mosquito behavioral differences.

In susceptible mosquito species, variation in susceptibility is related to the speed of digestion of the blood meal and formation of the peritrophic membrane (127). In populations of *An. gambiae* and *An. stephensi*, individual mosquitoes with slower rates of digestion had significantly higher *P. falciparum* oocyst infections.

The Components of Vectorial Capacity

The classic definition of vectorial capacity (71) is the daily rate at which future inoculations arise from a currently infective case. It recognizes three basic factors that should affect the transmission potential of any arthropod-borne human disease: the density of vectors relative to man, the man-biting habit (frequency of feeding on man), and vector survivorship through the extrinsic incubation period (the time from when a mosquito becomes infected until it becomes infectious).

The third term is expressed as a function of the daily survival rate. However, the daily survival rate is impossible to measure in practice. Instead, what is usually measured is the survival per feeding cycle, which is converted to daily survival using a "guesstimate" of the number of days comprising the feeding cycle. This introduces unnecessary error into vectorial capacity. More recent formulations have expressed vectorial capacity in terms of survival through the feeding cycle (149). Similarly, it is simpler to express and measure mosquito density and man-biting habit as the human biting rate (bites/person/night) and human blood index (proportion of meals taken on humans).

Furthermore, Garrett-Jones and Shidrawi (71) did not include in their formulation any factors for the probability of a mosquito becoming infected after feeding on a host with infective parasites, or for the probability that an infectious mosquito transmits an infection while biting a susceptible host, perhaps because they were assumed to be equal to 1. While this is obviously not the case, we also prefer to exclude these parameters from new expressions of vectorial capacity and include them in the mosquito and human infection probabilities, respectively, since they are to some extent influenced by the immune status of the human population.

Although vectorial capacity is a useful qualitative concept, Dye (60, 61) has argued convincingly that it is inestimable in practice, because the inaccuracies involved in measuring each of the separate parameters are compounded when multipled together. He suggests using indices (e.g., biting rate, human blood index) that change in direct proportion to the vectorial capacity as a means of making comparative rather than absolute estimates, for example, of the effects of different control strategies (61). He argues that even though measurements of many parameters are inevitably biased, the crucial question is not the extent of the bias, but whether the bias is constant in changing circumstances, such as before and after control measures are instituted (60).

The Man-Biting Rate

Determination of the man-biting rate of anopheline vectors is one of the most basic parameters involved in the transmission of any vector-borne disease. In an analysis of data from the Garki Project (115), the variation in malaria sporozoite inoculation rates was mostly explained by the variation in the biting rates (59). It is one of the first parameters that entomologists and malariologists have tried to measure, and it is one that is probably the least accurately measured. Traditionally, the density of the biting vector population is measured as a function of the human population (by either the man-biting rate or by a variety of traps including the exit trap on experimental huts or by bednet traps).

The man-biting or man-landing rate is determined by collectors catching mosquitoes landing on exposed parts of their bodies. This approach to estimating man-biting rates grossly overestimates the true mosquito attack rate, particularly in areas with high vector populations. The error results from several factors including host defensive behavior and cultural factors including the use of bednets, house construction, repellents, etc., which would lessen the true feeding success rate of mosquitoes. Using mark–release–recapture date to estimate the size of the entire anopheline population in a Papua New Guinean village, Saul (148) was able to estimate the true man-biting rate in that village by taking into account the human blood index and the number of villagers. In this village with a high domestic animal population and virtually 100% bednet usage, the true man-biting rate was estimated at 23 bites/night, compared to an estimate of over 500 by landing catches.

While landing catches estimate the relative numbers of anophelines seeking a blood meal, exit traps in experimental huts, bednet traps, and permethrin spray trapping attempt to measure the number of mosquitoes after they have taken a blood meal. All these measurements tend to underestimate the feeding mosquito population due to the inefficiency of the trapping method. Light traps have also been used to estimate anopheline density (97, 105), and good correlations between this measure and the landing rate enable it to be used as a relative estimate of the true mosquito biting rates (99).

All of these estimates of vector populations yield relative, not absolute, values. Comparing relative changes in attack rates in a village with time or among villages with similar epidemiological and cultural characteristics in an area yields measures of changes in transmission useful for evaluating the efficacy of control measures.

Studies on host defensive behavior using a variety of culicines and nonhuman vertebrate hosts demonstrated an inverse relationship between mosquito biting numbers and feeding success [reviewed by Edman et al. (63)]. Differences in engorgement success rates were seen among host animals of the same species, and evidence for host–species selection based on defensive behavior was found.

Human host selection by anophelines in malaria-endemic areas has been studied by examining engorged anophelines for various markers characteristic of individuals in a household. These markers have included white blood cell ratios (39), ABO blood types (15, 21), and haplotypes (9). In an alternate approach, evidence for preferential anopheline attack rates on adults were inferred from the higher prevalence of antibodies against the *P. falciparum* CS protein in adults compared with children following an epidemic of *P. falciparum* in an area without *P. falciparum* in the previous decade (110). Adults were preferred to infants and children for *An. gambiae* (7, 15, 40, 129) and *An. albimanus* (116), but not in the members of the *An. punctulatus* complex (21). However, all of these studies found large variations in the attractiveness of individuals.

Differences in amounts of defensive behavior and relative attractiveness among individuals result in nonhomogeneous mixing of mosquito and human populations. In addition, mark–recapture studies in Sabah, Malaysia, demonstrated that sympatric *An. balabacensis* preferentially return for a second blood meal on the same host species as their first blood meal (81). Recent models have shown that this always causes a relative increase in vectorial capacity (62). Models based on rodent malaria studies have also explored the consequences of mosquitoes preferentially selecting infected individuals (90).

Studies have shown that the feeding success rate on mice infected with *P. yoeli*, *P. berghei*, and *P. chabaudi* by mosquitoes is greatest at the time of maximum gametocytemia (52, 53). In addition, mosquito probing time was decreased on mice infected with *P. chabaudi* compared to uninfected mice (144). Caution must be exercised when extrapolating the results of rodent malaria studies to humans, as a number of significant differences exist. In the rodent studies cited, parasitemias up to 45% were recorded with a concomitant drop in both body temperature (of up to 6°C) and activity patterns (54). In contrast, human malarias in endemic areas consist predominantly of asymptomatic infections [reviewed by Covell (50)], with the vast majority of parasitemias well below 0.1%. Correlations between human malaria infection and fever, if they exist, are weak (115). Interestingly, depressed body temperatures have been reported in young children with high malaria parasite densities (125).

Most probably, the increased feeding success rate in rodents was due to the change in defensive behavior and not to any increased attractiveness to mosquitoes due to the presence of circulating malaria parasites. In the only comparable study examining human host selection in a malaria-endemic area, significant differences were not seen in host selection by *An. punctulatus* for malaria parasitemic or aparasitemic individuals (25). In that study, significant differences in relative attractiveness of individuals to anophelines were seen but were unrelated to parasite status or host size. This study occurred in a hyperendemic area where most infections were asymptomatic, with fever cases not significantly related to malaria parasitemias. The possibility exists in low endemic areas or during epidemics that stronger associations between clinical and parasitological cases of malaria could exist. In such a scenario, infection might lower the defensive behavior of infected individuals, resulting in an increased feeding success rate.

The Human Blood Index

The human blood index (HBI) affects vectorial capacity in a linear fashion. Anthropophilic mosquitoes have greater opportunities to ingest gametocytes and become infected and subsequently to pass on sporozoites to a susceptible human.

In determining the HBI, engorged resting mosquitoes are analyzed for host blood source. The best estimates of HBI for exophilic mosquitoes are obtained from the analyses of outdoor resting mosquitoes. In this population, mosquitoes that have fed both indoors and outdoors will be sampled. The difficulties of representative sampling of resting mosquitoes have been reviewed by Garrett-Jones (70). Logistic problems in finding outdoor resting engorged mosquitoes for exophilic mosquitoes have necessitated workers using bednet traps enclosing two host species to measure relative attractiveness by identifying the proportion of mosquitoes caught inside the net that had fed on each host.

Mosquitoes can be classified as "opportunistic" or "fixed" as regards host preference (16). Anophelines with fixed feeding preferences include *An. gambiae*, which is highly anthropophilic, and *An. quadriannulatus*, which is highly zoophilic (46). Opportunistic or catholic feeders exhibit a wide range of host selection depending on the epidemiological circumstances that determine relative host availability. Within a 20-km radius in Papua New Guinea, *An. farauti* No. 1 had an HBI determined from ELISA analysis of engorged outdoor resting mosquitoes that ranged from 7 to 93% (21). Sociological factors including clothing, house construction, animal husbandry practices, bednet usage, and relative numbers of humans and animals all help determine the relative availability of host species to mosquitoes [reviewed in (103, 146)]. A significant increase in the HBI for *An. farauti* No. 1 resulted from penning pigs that had previously been allowed to range freely through a village (17).

A significant number of anophelines are interrupted in the process of engorging (8, 16, 21). The process of being interrupted leads to increased probing with mixed blood meals. Mixed meals may involve two hosts of the same species (cryptic mixed meals) or different species (patent mixed meals). Mixed meals increase the classic definition of vectorial capacity, but may in fact diminish the probability that a mosquito feeding on a gametocytemic human becomes infected, as the total number of gametocytes ingested will be less for a partial meal. Since only a small proportion of the human population will be gametocytemic at any one time, it is unlikely that the mosquito will complete the blood meal on a second gametocytemic individual. A study of *P. falciparum* oocyst densities in different size classes of *An. dirus* found that significantly more oocysts developed in the larger females that had imbibed larger blood meals (91). The implications of probing on multiple hosts for malaria transmission to humans will be discussed later.

Survivorship

Of the components affecting malaria transmission, changes in mosquito survivorship were judged to have the largest impact (104). Since vectorial capacity expresses survivorship from when the mosquito became infected to when it became infectious (the extrinsic incubation period), the daily estimate of survivorship is raised to a power equal to the number of days in the extrinsic incubation period (generally estimated in the range of 9–14 days for *P. vivax* and *P. falciparum*). Similarly, estimates of survivorship per feeding cycle are raised to a power equal to the number of feeding cycles per extrinsic incubation period. Small changes in estimates of survivorship per day or feeding cycle thus translate into large changes in survivorship over the extrinsic incubation period.

Vector survival rates have been estimated by parous rates (85, 87, 98, 115, 140), mark–release–recapture experiments (37, 80, 148), multiparous age grading (36, 72, 100), and from mosquito infection rates (20, 76, 149). All of these methods estimate survival rate per feeding cycle. Estimates by all four approaches are usually in the range of 0.4–0.6. The length of the feeding cycle will determine the number of feeding cycles that occur in the extrinsic incubation period, and can be determined from cross-correlations on time-series parous rate data (85) or from mark–release–recapture experiments. The length of the extrinsic incubation period, however, is usually derived from an educated guess. The advantage of estimating survival from mosquito infection rates is that survival per extrinsic incubation period can be estimated without knowledge of the duration of the extrinsic incubation period (20, 76, 149).

This technique relies on two estimates of malaria infection rates in mosquitoes. These may be either sporozoite rates from landing catches or total infection rates (oocysts and/or sporozoites) in mosquitoes caught in resting or landing catches and held in the laboratory long enough for all infections to be detectable ("delayed infecton rates"). Comparisons of the infection rates in any two of the three mosquito samples allow one to calculate survivorship per feeding cycle. Survival per extrinsic incubation period may be obtained from sporozoite rates and one other delayed infection rate together with knowledge of the HBI. The disadvantage of the technique is that it can be applied only to epidemiological situations where large numbers of biting and resting anophelines can be collected. The approach is also logistically constrained by the need to capture adequate numbers of infected mosquitoes, thereby virtually ensuring that it will only be attempted in highly endemic areas.

When applied to the mosquito populations in the hyperendemic Madang area of Papua New Guinea where transmission occurs throughout the year, some unexpected findings resulted. Comparing survivorship estimates per feeding cycle yielded no significant differences at different times of the year in a village (despite significant temporal fluctuations in malaria transmission) (20) or among different villages (20, 76). These survivorship estimates were consistent with estimates obtained using parity rates (35, 36). However, significant differences in survivorship per extrinsic incubation period were found among the mosquito populations in the different villages (76) and at different times of the year within a village (20). It appears that survivorship per extrinsic incubation period is dependent on the number of feeding cycles, and that this number may vary, giving rise to significant differences in survivorship for the extrinsic incubation period. In Papua New Guinea, correlations between sporozoite inoculation rates and parasite prevalences in children in different villages could be found (19). Sporozoite rates, in turn, were significantly correlated to the HBI (17). Reanalysis of the same data revealed a correlation between parasite prevalence in children and the HBI and vector biting rate (61). In this hyperendemic area with perennial transmission, fluctuations in survivorship per feeding cycle were not responsible for differences in transmission in different villages (20). This contrasts with a hyperendemic area of Thailand where seasonal differences in transmission could be related to differences in survivorship as determined by parity dissections (140). In Kenya, large seasonal differences in survivorship per gonatropic cycle were reported for *An. arabiensis* (87). However, as confidence limits on the data were not given, the significance of these differences is unknown. In The Gambia, a fall in parity rates from 0.58 in the wet to 0.41 in the dry season combined with a drop in the HBI and possible direct effects of high temperature on the parasite were suggested as the reasons for a low prevalence of malaria morbidity in the dry season (98).

Vector–Parasite Interactions

During the time that sporozoites are developing in the oocysts, parasitism may exert effects on the mosquito. Studies by Klein et al. (92) using *An. dirus A* infected with *P. cynomolgi* demonstrated an increased mortality in mosquitoes with more than 10 oocysts. Mosquitoes with greater than 71 oocysts were noted to have extensive deterioration of the gut and salivary glands by bacterial contamination. However, wild-caught infected anophelines have significantly fewer oocysts than this. In Papua New Guinea, the geometric mean number of oocysts seen in the *An. punctulatus* complex was 3.7 with a mode of 1 (76). In the Gambia, similar results were found with 40% of *An. gambiae* having 1 oocyst (45). In Thailand, 32 and 39% of infected *An. dirus* and *An. minimus*, respectively, had a single oocyst, and median numbers of oocysts were 3.5 for *An. dirus* and 3.0 for *An. minimus* (140). Studies involving *An. gambiae* infected with *P. falciparum* from gametocytemic individuals were unable to demonstrate an increased mortality in malaria-infected mosquitoes (38, 136). From these studies it appears unlikely that malaria will significantly decrease mosquito survivorship for most infections. In a study of wild-caught anophelines in a hyperendemic area, an increased mortality was noted in mosquitoes infected with both malaria and filarial worms, *Wuchereria bancrofti* (24). However, the increase in mortality was related to the stage of the filarial parasite and not the malarial infection.

In a matter of some controversy, researchers have debated about the impact on malaria infections of anophelines ingesting antibodies against CS proteins in a blood meal. It is well established that immunoglobulins cannot only pass into the hemocoel of a number of hematophagous arthropods including *An. stephensi*, *An. gambiae*, and *An. albimanus* [154; reviewed in (84)] but that such antibodies can still recognize their target epitopes. One group demonstrated that ingestion of antibodies against the *P. falciparum* CS protein by *An. stephensi* 5 days after ingestion of gametocytes resulted in a significant increase in the number of sporozoites in the salivary glands produced per oocyst (155). A second study argued that any increase in *P. falciparum* sporozoite numbers is a function of a general nutritional effect from the second blood meal and that antibodies against the CS protein have no direct effect on the numbers of sporozoites which result (128). A third study found no evidence for an increase in sporozoite numbers per oocyst following the ingestion of antisporozoite antibodies in a blood meal following an infectious *P. vivax* meal but noted that the resulting salivary glands had greater than three times as many sporozoites as mosquitoes that had not received a second blood meal (141). Finally, a fourth study found that ingestion of anti-*P. falciparum* sporozoite antibodies by *P. falciparum* oocyst infected *An. gambiae* did not affect oocyst numbers, sporozoite rates, or sporozoite loads (6).

A high proportion of naturally infected *An. gambiae* and *An. funestus* in western Kenya transmit sporozoites bound with human IgG acquired during a previous blood meal (6). A previous laboratory study demonstrated that *P. falciparum* sporozoites from *An. stephensi* that had ingested anti-*P. falciparum* CS antibodies could not be neutralized by sera from humans immunized with an anti-*P. falciparum* sporozoite vaccine (57). The implication is that increases in the level of humoral immunity to the CS proteins in a population (through increased transmission or by vaccination) may facilitate sporozoite transmission.

Infection of mosquitoes at the oocyst level and survival through the period of the extrinsic incubation period do not necessarily guarantee that the mosquito will become a vector of malaria. In experiments in which oocysts were transplanted from one species of anopheline to another, Rosenberg (138) was able to demonstrate that even though a mosquito may support the growth of oocysts to maturation with release of sporozoites into the hemocoel, the sporozoites may not be able to penetrate the salivary glands. In a separate study, Rosenberg and Rungsiwongse (141) estimated that only 20% of *P. vivax* sporozoites released from oocysts subsequently invade the salivary glands of *An. dirus*. Using a CS antigen-capture ELISA (discussed below), workers in Brazil found antigen-positive species not previously incriminated as malaria vectors (55). Feeding laboratory-reared F1 progeny of the suspected vector species (as defined by ELISA) on gametocytemic individuals, Klein et al. (93) demonstrated a wide range in susceptibility to oocyst infection. Even more importantly, differences in the numbers of sporozoites were found in the salivary glands with some CS antigen and oocyst-positive species failing to become infected with sporozoites in the glands.

Vectors

The quantitative study of malaria transmission from mosquitoes to humans requires two basic hurdles to be overcome: vector incrimination and vector identification. Incrimination of a mosquito as a human malaria vector requires the identification of the sporozoite stage of a human malaria in the salivary glands of that mosquito (discussed below). The second aspect of vector incrimination involves the species identification of the mosquito. Human malarias are transmitted only by anopheline mosquitoes. In areas where malaria has been studied in detail, transmission of human malaria is usually by a complex of species that are oftentimes morphologically indistinguishable. The separation of vector species from their isomorphic nonvector sibling species has attracted considerable attention from mosquito taxonomists.

A number of methods for distinguishing sibling species in anopheline complexes have been developed. The first method is by examining the gonads of the offspring of cross-matings between standard colony mosquitoes and the offspring of unidentified parents. This method is obviously time-consuming and labor-intensive with only a few specimens being identified at one time. The second identification technique is to examine the banding patterns of polytene chromosomes. In the *An. punctulatus* complex, polytene chromosomes can be examined only from preparations of the salivary glands of fourth instar larvae (13). Lack of a suitable preservative also necessitates the use of fresh material for identification and therefore renders this approach unsuitable for large-scale field work. A third approach, isoenzyme analysis, requires fresh material or material stored at −70°C or below. Limited freezer facilities in many malaria-endemic areas limits the application of this technique. Immunological identification of species in the *An. gambiae* complex has been developed based on the use of monoclonal antibodies against vitellogenin (102). As most anophelines are sampled in the host-seeking stage and are therefore without large amounts of vitellogenin, it is unlikely that this technique will find many applications in transmission studies.

The development of DNA probes has facilitated taxonomic work on many anopheline complexes including the *An. gambiae* complex (42, 43, 67, 68, 82), the *An. dirus* complex (123), the *An. quadrimaculatus* complex (129), and the *An. punctulatus* complex (48). Using this approach, mosquitoes may be collected and stored in a variety of conditions: frozen, dried at room temperature, or fixed in alcohol for later analyses. Detection systems in use include isotopic, chromogenic, and chemiluminescent probes, which thereby allow their use in field stations and laboratories in underdeveloped areas. The extreme sensitivity of the DNA probes allows species to be identified with only a small fraction of the mosquito; the remainder of the mosquito can then be analyzed for malaria parasites or aspects of mosquito behavior, including survivorship measurements, host blood meal source identification, etc.

Sporozoite versus Sporozoite Antigen

A number of reviews have examined the technology of malaria parasite detection in vector populations (28, 160). The development of stage-specific immunoassays for the detection of circumsporozoite (CS) proteins has allowed field workers to circumvent the problems previously associated with identifying sporozoites in the salivary glands of mosquitoes (primarily parasite species identification and the logistics of dissecting thousands of individual mosquitoes) (26, 30, 161, 163, 164). These assays have a sensitivity of approximately 25 sporozoites. Thus far, DNA and RNA detection systems have not been able to duplicate reliably the sensitivity of the ELISAs for sporozoite detection, nor are they as cost-effective. When interpreting ELISA-based sporozoite rates reported in the literature, one must remember that the ELISAs detect sporozoite antigen, whether in oocysts, salivary glands, or shed in the hemocoel of the mosquito (26, 30, 101, 137, 161). The presence of CS antigen is not the same as visual confirmation of sporozoites in the salivary glands. However, workers in The Gambia (45), Papua New Guinea (162), and Kenya (1) have found good correlations between detection of CS protein and sporozoite rates in the salivary glands. The discovery of *P. vivax* polymorphs (142) and the development of ELISAs to differentiate them (165) now enables even more detailed studies on the transmission of human malaria to be undertaken (29).

Human Infection Probability

The human infection probability may be defined as the probability of a malaria parasite infection being successfully established by the bite of a mosquito. This depends on the sporozoite infection rate in the mosquito population, the numbers of sporozoites inoculated, the immune status of the person, the genetic susceptibility of the person to malaria infection, and the presence of antimalarials in the person.

Sporozoite Inocula

Successful invasion of a sporozoite into the liver depends on the number of sporozoites inoculated and the route of inoculation. Studies on the number of sporozoites regurgitated into capillary tubes have shown that although sporozoite load in the salivary glands was correlated with sporozoite number ejected, for most infected mosquitoes, the number of sporozoites ejected and counted in the capillary tube was surprisingly small (median number, 15; range, 0–978) (143). Applying the methods of Rosenberg et al. (143) to *An. gambiae* naturally infected with *P. falciparum* gave nearly identical results: geometric mean ejected between 2.3 and 3.8 (5). However, unpublished work of Spielman suggests that the number of sporozoites ejected may be 10-fold more than found in the capillary tubes. Working with *P. gallanaceum*-infected *Aedes aegypti*, 10-fold more sporozoites were reingested into the ventral diverticulum when feeding from a capillary tube (Spielman, personal communication).

In an earlier study, the number of viable *P. berghei* sporozoites inoculated by *An. stephensi* was determined by counting the average number of exoerythrocytic forms in the liver of mice and rats per fed mosquito (153). The range of sporozoites found (5.6 to 266) was similar to the work of Rosenberg et al. (143). Circumstantial epidemiological evidence for greater numbers of sporozoites being inoculated than indicated in the capillary feeding experiments may be found from serological investigations. The development of antibodies against the *P. falciparum* and *P. vivax* CS proteins following first exposures in Thailand (13, 159), Brazil (66), and Sri Lanka (110) suggests that significantly more than the median number of 15 estimated by Rosenberg et al. (143) or the geometric mean of 2.3–3.8 reported by Beier et al. (5) are ejected into humans in endemic areas.

The attention focused on determining the exact size of the sporozoite inoculum stems, in part, from the failures of the first generation anti-sporozoite vaccines to provide sterile immunity (79). Vaccine developers rationalize continued support by citing studies that claim that the dose of sporozoites inoculated has clinical reprecussions for the length of the prepatent period and incubation period as well as the length and severity of the clinical stages of the infection. Close examination of the studies cited reveals disparities with data from field work in the dose of the inocula used. Infections initiated by "lightly" infected mosquitoes with less than 50 *P. vivax* oocysts were compared with infections resulting from "heavily" infected mosquitoes with more than 100 oocysts (11). The often cited work of Jeffery et al. (88) compared *P. falciparum* sporozoite challenges from the bites of an average of 5.4 to 7.1 infected mosquitoes with those from a high challenge of 25.3 to 44.8 bites. Prepatent periods differed from 1.0 to 2.8 days for the three *P. falciparum* strains tested, with incubation periods varying by 0.8 to 2.4 days. Alternatively, needle-inoculated challenges compared inocula of less than 100 to greater than 10 million sporozoites. Mean days to patency for *P. cynomolgi* sporozoite inocula of less than 100 (17.7 ± 7.2) were within one standard deviation of inocula exceeding 10,000 (13.8 ± 4.1) (150). Although the mean level of parasitemia on day 6 postinjection rose significantly with increasing numbers of *P. gallinaceum*, the minimum number of sporozoites injected was 500 (131). None of these challenge inocula described is epidemiologically relevant because the median oocyst burden for mosquitoes naturally infected with human malaria is between 3 and 5 with a mode of 1. Furthermore, although daily sporozoite inoculation rates greater than 3 have been reported (22), these inoculation rates are based on landing catches as an approximation of the true mosquito attack rate. The true feeding success rate may be less than 1/25 of that reported (148). The relevance of sporozoite inocula to clinical outcome must await experiments in which the sporozoite challenge approximates that in endemic areas: the bite of a single mosquito with a sporozoite load resulting from less than 10 oocysts.

Although the absolute numbers of sporozoites inoculated may still be debated, Rosenberg et al. (143) found that it was proportional to the number of sporozoites in the salivary glands. This aspect of the laboratory results is consistent with field observations. Using antigen-capture ELISAs to estimate sporozoite numbers in naturally infected Papua New Guinea mosquitoes, geometric means of 2300 for *P. falciparum* and 200 for *P. vivax* were obtained (22). Comparing relative inoculation rates by the anophelines in villages to the parasite prevalences in children in the same village for these two species of human malaria revealed that *P. falciparum* was at least twice as efficient at establishing infections than *P. vivax* (19). The increased efficiency of transmission of *P. falciparum* was hypothesized to be due to a greater number of sporozoites inoculated by the more heavily infected *P. falciparum*-infected mosquitoes (19).

The invasion of mosquito salivary glands by malaria sporozoites induces pathological changes resulting in behavior changes. Sporozoites of *P. gallinaceum* impair the mosquito's apyrase secretion (which inhibits aggregation of host platelets (134)) without reducing saliva volume from the salivary glands (145). The reduced apyrase flow inhibits the mosquito's ability to inhibit vertebrate platelet formation, thereby interfering with the mosquito's ability to locate blood, resulting in increased probing. It was hypothesized that the increased probing would result in increased numbers of sporozoites inoculated as well as more interrupted blood meals. The result would be increased transmission of sporozoites to a greater number of hosts.

Site of Inoculation

The site of sporozoite inoculation may influence the success of establishing an infection in a susceptible human. Sporozoite inoculation directly into a venule would enhance the prospect of hepatocyte invasion while sporozoite deposition into the skin might lessen the chances of infection. Studies using *P. vivax* sporozoites inoculated directly into a vein demonstrated that as few as 10 sporozoites of one strain were sufficient to establish malaria infections in 100% of susceptible humans (152). Experiments using infected mosquitoes gave variable results, with some experiments establishing *P. falciparum* infections using lightly infected mosquitoes (79) while other more heavily infected mosquitoes were less successful in establishing human infections (88). When one or two *An. stephensi* infected with *P. falciparum* sporozoites were fed on human volunteers, only 50% of volunteers became infected (135). In contrast, an epidemiological study in Madagascar estimated that almost all mosquito inoculations of *P. falciparum* resulted in blood stage infections (96).

Studies with Rift Valley Fever-infected *Culex pipiens* feeding on suckling mice suggest that pathogens are inoculated extravascularly by feeding mosquitoes rather than directly into the vascular system (151). Extrapolating these results to sporozoite-infected anophelines, it appears that most sporozoites would be ejected while probing prior to cannulation of a blood vessel (134). Although most sporozoites would be released prior to blood vessel location in the first saliva droplets (143), the early work of Boyd and Stratman-Thomas (12) demonstrated that *P. vivax* sporozoites can penetrate tissues to establish infections. In the experiment, sporozoites from three mosquito bites were inoculated into an induced blister. A malaria infection was verified 15 days later, compared with a prepatent period of 13 days for two controls. This early experiment is supported by an analysis of allelic serotype frequencies of polymorphic malaria blood stage proteins in Gambian households. The significant clustering of serotypes in households compared with their frequency in the population suggested that interrupted feeding by mosquitoes occurs and accounts for multiple infections during a single blood meal (47). These field observations are supported by laboratory results which show that a single mosquito can transmit the bird malaria *P. gallinaceum* to three hosts within a 5-min period, allowing the mosquito just 10 sec per host for probing (89).

Infection in the Human Host

The susceptibility of the human host to establishment of a malaria infection depends on the innate genetic susceptibility of the person and the immune status induced by exposure to malaria. Repeated exposure of humans in endemic areas to the bites of sporozoite-infected mosquitoes induces humoral (120) and cellular immune responses to the sporozoite stage (106). While the possible protective effects of such naturally induced exposure are still a matter of debate, laboratory-based experiments suggest that the humoral immune response to sporozoites is only partially protective (109), confiming the observations from endemic areas that most adults still are susceptible to malaria. More than 99% of Liberian adults studied had detectable malaria parasites at some time over the course of a year (125). Antibodies recognizing the CS proteins may exert an effect not only on the sporozoite stage, but on the exoerythrocytic stages. Laboratory studies using the rodent malaria, *P. yoelii*, showed that high anti-CS antibody titers inhibited exoerythrocytic development whereas low titers resulted in an enhancement of up to 150% more maturing liver schizonts than controls (122). Furthermore, *P. falciparum* sporozoites are unable to be neutralized by immune sera containing antibodies against the *P. falciparum* CS protein if anti-*P. falciparum* CS sera are ingested by the mosquito during sporogony (57).

Although induced immune responses against both sporozoites and blood stages may not sterilely protect against malaria, they may depress the density of parasites to such a level that some infections are undetectable by the presently available methods, thereby giving the impression of a reduced transmission rate from mosquito to man. The hypothesis that parasite density and the associated severity of illness may be directly related to the number of sporozoites inoculated (11, 78) awaits validation by experiments using sporozoite inocula comparable to those naturally transmitted in endemic areas. The possibility that severe illness may be associated with superinfection or the inoculation of new parasite strains must also be considered. In The Gambia, where malaria transmission is seasonal, there is no significant drop in parasite rates in the human population in the dry season. However, the fact that all malaria deaths and a high proportion of the febrile episodes occur at the end of the rainy season (77) supports the hypothesis that transmission of new malaria strains results in clinical cases. The situation is complicated by the finding that an HLA antigen and an HLA haplotype are associated with protection against severe malaria (83). However, in most malaria-endemic areas, even in areas of low endemicity, most malaria parasite infections are asymptomatic (50).

Conclusions

Much of this chapter has been concerned with factors affecting transmission at the level of an individual infection in either the mosquito or the human host. However, only when a factor exerts a significant influence at the population level is an impact on transmission seen. Studies of rare events or those involving animal models may not be epidemiologically relevant to understanding human malaria transmission or for predicting the impact of control strategies on transmission. It must also be emphasized that measurements of field parameters involved in transmission are, in many cases, relative, and not absolute measurements. Sporozoite antigen infection rate determinations carry errors in their estimation. However, if the error is constant, one can use the CS antigen rates to estimate relative changes in transmission.

In contrast, the precision of laboratory-based experiments generates absolute numbers. To be epidemiologically relevant, laboratory work must be related to field observations. The effects of the malaria parasite on the vector must be ascertained at densities seen in malaria-endemic areas. Parasite densities and clinical effects in animal models used in the laboratory are often vastly different from those in infected humans. Extrapolations of animal model observations to human malarias must be done with great caution.

Despite many of the fine details that we now know about factors affecting malaria transmission at the molecular level, there are still "great neglected research areas" that have been ignored for a variety of reasons. Chief among these areas is our inability to accurately measure one of the most basic parameters involved in the transmission of all mosquito-borne human pathogens, the man-biting rate. Precise measurement of many parameters is not important for practical applications if the errors in measurement are constant. When errors vary in a density-dependent manner, more accuracy is needed to ascertain how changes in one parameter relate to changes in another. The man-biting rate is one such parameter that will vary with defensive behavior (including bednets) and availability of alternate hosts. Comparisons of biting and inoculation rates between areas or after certain interventions cannot be done with any certainty until more accurate ways of measuring biting rates are determined.

References

1. Adungo, N.I., Mahadevan, S., Mulaya, N.L., Situbi, A.P., and Githure, J.I. 1991. Comparative determination of *Plasmodium falciparum* sporozoite rates in Afrotropical *Anopheles* from Kenya by dissection and ELISA. *Ann. Trop. Med. Parasitol.* **85**:387–394.
2. Allison, A.C. 1988. The role of cell-mediated immune responses in protection against plasmodia and in the pathogenesis of malaria. In W.H. Wernsdorfer and I. McGregor, eds., *Malaria. Principles and Practices in Malariology*, Vol. 1. Churchill Livingstone, Edinburgh, pp. 253–306.
3. Baird, J.K., Jones, T.R., Masbar, P.S., Ratiwayanto, S., and Leksana, B. 1991. Evidence for specific suppression of gametocytemia by *Plasmodium falciparum* in residents of hyperendemic Irian Jaya. *Am. J. Trop. Med. Hyg.* **44**:183–190.
4. Barker, R.H. 1990. DNA probe diagnosis of parasitic infections. *Exp. Parasitol.* **70**:494–499.
5. Beier, J.C., Onyango, F.K., Koros, J.K., Ramadan, M., Ogwang, R., Wirtz, R.A., Koech, D.K., and Roberts, C.R. 1991. Quantitation of malaria sporozoites transmitted by wild Afrotropical *Anopheles*. *Med. Vet. Entomol.* **5**:71–79.

6. Beier, J.C., Oster, C.N., Koros, J.K., Onyango, F.K., Githeko, A.K., Rowton, E., Koech, D.K., and Roberts, C.R. 1989. Effect of human circumsporozoite antibodies in *Plasmodium*-infected *Anopheles* (Diptera: Culicidae). *J. Med. Entomol.* **26**:547–553.
7. Boreham, P.F.L., Chandler, J.A., and Jolly, J. 1978. The incidence of mosquitoes feeding on mothers and babies at Kisumu, Kenya. *J. Trop. Med. Hyg.* **81**:63–67.
8. Boreham, P.F.L., and Garrett-Jones, C. 1973. Prevalence of mixed blood meals and double feeding in a malaria vector (*Anopheles sacharovi* Favre). *Bull. W.H.O.* **48**:605–614.
9. Boreham, P.F.L., and Lenahan, J.K. 1976. Methods for detecting multiple blood meals in mosquitoes (Diptera: Culicidae). *Bull. Entomol. Res.* **66**:671–679.
10. Boudin, C., Lyannaz, J., Bosseno, M.F., Carnevale, P., and Ambroise-Thomas, P. 1991. Epidemiology of *Plasmodium falciparum* in a rice field and a savanna area in Burkina Faso: seasonal fluctuations of gametocytaemia and malaria infectivity. *Ann. Trop. Med. Parasitol.* **85**:377–385.
11. Boyd, M.F. 1940. The influence of sporozoite dosage in vivax malaria. *Am. J. Trop. Med.* **20**:279–286.
12. Boyd, M.F., and Stratman-Thomas, W.K. 1934. Studies on benign tertian malaria. 7. Some observations on inoculation and onset. *Am. J. Hyg.* **20**:488–495.
13. Brown, A.E., Webster, H.K., Krinchai, K., Gordon, M., Wirtz, R.A., and Permpanich, B. 1991. Characteristics of natural antibody responses to the circumsporozoite protein of *Plasmodium vivax*. *Am. J. Trop. Med. Hyg.* **44**:21–27.
14. Bryan, J.H., and Coluzzi, M. 1971. Cytogenetic observations on *Anopheles farauti* Laveran. *Bull. W.H.O.* **45**:266–267.
15. Bryan, J.H., and Smalley, M.E. 1978. The use of ABO blood groups as markers for mosquito biting studies. *Trans. R. Soc. Trop. Med. Hyg.* **72**:357–360.
16. Burkot, T.R. 1988. Non-random host selection by anopheline mosquitoes. *Parasitol. Today* **4**:156–162.
17. Burkot, T.R., Dye, C., and Graves, P.M. 1989. An analysis of some factors affecting the sporozoite rates, human blood indexes and biting rates of the members of the *Anopheles punctulatus* complex in Papua New Guinea. *Am. J. Trop. Med. Hyg.* **40**:229–234.
18. Burkot, T., Garner, P., Paru, R., Dagoro, H., Barnes, A., McDougall, S., Wirtz, R.A., Campbell, G., and Spark, R. 1990. Effects of untreated bednets on the transmission of *Plasmodium falciparum, P. vivax* and *Wuchereria bancrofti* in Papua New Guinea. *Trans. R. Soc. Trop. Med. Hyg.* **84**:773–779.
19. Burkot, T.R., Graves, P.M., Cattani, J.A., Wirtz, R.A., and Gibson, F.D. 1987. The efficiency of sporozoite transmission of the human malarias, *Plasmodium falciparum* and *P. vivax*. *Bull. W.H.O.* **65**:375–380.
20. Burkot, T.R., Graves, P.M., Paru, R., Battistutta, D., Barnes, A., and Saul, A. 1990. Variations in malaria transmission rates are unrelated to the anopheline survivorship per feeding cycle. *Am. J. Trop. Med. Hyg.* **43**:321–327.

21. Burkot, T.R., Graves, P.M., Paru, R., and Lagog, M. 1988. Mixed blood feeding by the malaria vectors in the *Anopheles punctulatus* complex (Diptera: Culicidae). *J. Med. Entomol.* **25**:205–213.
22. Burkot, T.R., Graves, P.M., Paru, R., Wirtz, R.A., and Heywood, P. 1988. Human malaria transmission studies in the *Anopheles punctulatus* complex in Papua New Guinea: Sporozoite rates, inoculation rates and sporozoite densities. *Am. J. Trop. Med. Hyg.* **39**:135–144.
23. Burkot, T.R., Graves, P.M., Wirtz, R.A., Brabin, B.J., Battistutta, D., Cattani, J.A., Maizels, R.M., and Alpers, M.P. 1989. Differential antibody responses to *Plasmodium falciparum* and *P. vivax* circumsporozoite proteins in a human population. *J. Clin. Microbiol.* **27**:1346–1351.
24. Burkot, T.R., Molineaux, L., Graves, P.M., Paru, R., Battistutta, D., Dagoro, H., Barnes, A., Wirtz, R.A., and Garner, P. 1990. The prevalence of naturally acquired multiple infections of *Wuchereria bancrofti* and human malarias in anophelines. *Parasitology* **100**:369–375.
25. Burkot, T.R., Narara, A., Paru, R., Graves, P.M., and Garner, P. 1989. Human host selection by anophelines: No evidence for preferential selection of malaria or microfilariae infected individuals in a hyperendemic area. *Parasitology* **98**:337–342.
26. Burkot, T.R., Williams, J.L., and Schneider, I. 1984. Detection of *Plasmodium falciparum* infected mosquitoes by a double antibody enzyme-linked immunosorbent assay. *Am. J. Trop. Med. Hyg.* **33**:783–788.
27. Burkot, T.R., Williams, J.L., and Schneider, I. 1984. Infectivity to mosquitoes of *Plasmodium falciparum* clones grown in vitro from the same isolate. *Trans. R. Soc. Trop. Med. Hyg.* **78**:339–341.
28. Burkot, T.R., and Wirtz, R.A. 1986. Immunoassays of malaria sporozoites in mosquitoes. *Parasitol. Today* **2**:155–157.
29. Burkot, T.R., Wirtz, R.A., Paru, R., Garner, P., and Alpers, M.P. 1992. The population dynamics in mosquitoes and humans of two *Plasmodium vivax* polymorphs distinguished by different circumsporozoite protein repeat regions. *Am. J. Trop. Med. Hyg.* **47**:778–786.
30. Burkot, T.R., Zavala, F., Gwadz, R.W., Collins, F., Nussenzweig, R.S., and Roberts, D.R. 1984. Identification of malaria-infected mosquitoes by a two-site enzyme linked immunosorbent assay. *Am. J. Trop. Med. Hyg.* **33**:227–231.
31. Burkot, T.R., Zhou, W.D., Geysen, H.M., Wirtz, R.A., and Saul, A. 1991. Fine specificities of monoclonal antibodies against the *Plasmodium falciparum* circumsporozoite protein: Recognition of both repetitive and non-repetitive regions. *Parasite Immunol.* **13**:161–170.
32. Carter, R., and Graves, P.M. 1988. Gametocytes. In W.H. Wernsdorfer and I. McGregor, eds., Malaria. *Principles and Practices in Malariology*, Vol. 1. Churchill Livingstone, Edinburgh, pp. 253–306.
33. Carter, R., and Gwadz, R.W. 1980. Infectiousness and gamete immunization in malaria. In J.P. Kreier, ed., *Malaria*, Volume 3, *Immunology and Immunization*. Academic Press, New York, pp. 263–298.
34. Carter, R., Kumar, N., Quakyi, I., Good, M., Mendis, K., Graves, P.M., and Miller, L. 1988. Immunity to sexual stages of malaria parasites. *Prog. Allergy* **41**:193–214.

35. Charlwood, J.D. 1986. Survival rate variation of *Anopheles farauti* (Diptera: Culicidae) between neighboring villages in coastal Papua New Guinea. *J. Med. Entomol.* **23**:361–365.
36. Charlwood, J.D., Birley, M.H., Dagoro, H., Paru, R., and Holmes, P.R. 1985. Assessing survival rates of *Anopheles farauti* (Diptera: Culicidae) from Papua New Guinea. *J. Anim. Ecol.* **54**:1003–1016.
37. Charlwood, J.D., Graves, P.M., and Birley, M.H. 1986. Capture-recapture studies with mosquitoes of the group of *Anopheles punctulatus* Donitz (Diptera: Culicidae) from Papua New Guinea. *Bull. Entomol. Res.* **76**:211–227.
38. Chege, G.M.M., and Beier, J.C. 1990. Effect of *Plasmodium falciparum* on the survival of naturally infected Afrotropical *Anopheles* (Diptera: Culicidae). *J. Med. Entomol.* **27**:454–458.
39. Clyde, D.F., and Shute, G.T. 1955. A technique for the investigation of mosquito feeding preferences on man. *Trans. R. Soc. Trop. Med. Hyg.* **49**:64–67.
40. Clyde, D.F., and Shute, G.T. 1958. Selective feeding habits of anophelines amongst Africans of different ages. *Am. J. Trop. Med. Hyg.* **7**:543–545.
41. Cochrane, A., Uni, S., Maracic, M., Di Giovanni, L., Aikawa, M., and Nussenzweig, R.S. 1989. A circumsporozoite protein is present in micronemes of mature blood stages of malaria parasites. *Exp. Parasitol.* **69**:351–356.
42. Collins, F.H., Mehaffey, P.C., Rasmussen, M.O., Brandling-Bennett, A.D., Odera, J.S., and Finnerty, V. 1988. Comparison of DNA-probe and isoenzyme methods for differentiating *Anopheles gambiae* and *Anopheles arabiensis* (Diptera: Culicidae). *J. Med. Entomol.* **25**:116–120.
43. Collins, F.H., Petrarca, V., Mpofu, S., Brandling-Bennett, A.D., Were, O., Rasmussen, M.O., and Finnerty, V. 1988. Comparison of DNA probe and cytogenetic methods for identifying field collected *Anopheles gambiae* complex mosquitoes. *Am. J. Trop. Med. Hyg.* **39**:545–550.
44. Collins, F.H., Sakai, R.K., Vernick, K.D., Paskewitz, S., Seeley, D.C., Miller, L.H., Collins, W.E., Campbell, C.C., and Gwadz, R.W. 1986. Genetic selection of a *Plasmodium*-refractory strain of the malaria vector *Anopheles gambiae*. *Science* **234**:607–609.
45. Collins, F.H., Zavala, F., Graves, P.M., Cochrane, A.H., Gwadz, R.W., Akoh, J., and Nussenzweig, R.S. 1984. First field trial of an immunoradiometric assay for the detection of malaria sporozoites in mosquitoes. *Am. J. Trop. Med. Hyg.* **33**:538–543.
46. Coluzzi, M. 1984. Heterogeneities of the malaria vectorial system in tropical Africa and their significance in malaria epidemiology and control. *Bull. W.H.O.* (Suppl.) **62**:107–113.
47. Conway, D.J., and McBride, J.S. 1991. Genetic evidence for the importance of interrupted feeding by mosquitoes in the transmission of malaria. *Trans. R. Soc. Trop. Med. Hyg.* **85**:454–456.
48. Cooper, L., Cooper, R.D., and Burkot, T.R. 1991. The *Anopheles punctulatus* complex: DNA probes for identifying the Australian species using isotopic, chromogenic, and chemiluminescence detection systems. *Exp. Parasitol.* **73**:27–35.

49. Coppel, R.L., Favaloro, J.M., Crewther, P.E., Burkot, T.R., Bianco, A.E., Stahl, H.D., Kemp, D.J., Anders, R.F., and Brown, G.V. 1985. A blood stage antigen of *Plasmodium falciparum* shares determinants with the sporozoite coat protein. *Proc. Natl. Acad. Sci. U.S.A.* **82**:5121–5125.
50. Covell, G. 1960. Relationship between malarial parasitaemia and symptoms of the disease. *Bull. W.H.O.* **22**:605–619.
51. Curtis, C.F., and Graves, P.M. 1983. Genetic variation in the ability of insects to transmit filariae, trypanosomes and malarial parasites. In K.F. Harris, ed., *Current Topics in Vector Research*. Praeger, New York, Vol. 1, pp. 31–62.
52. Day, J.F., Ebert, K.M., and Edman, J.D. 1983. Feeding patterns of mosquitoes (Diptera: Culicidae) simultaneously exposed to malarious and healthy mice, including a method for separating blood meals from conspecific hosts. *J. Med. Entomol.* **20**:120–127.
53. Day, J.F., and Edman, J.D. 1983. Malaria renders mice susceptible to mosquito feeding when gametocytes are most infective. *J. Parasitol.* **69**: 163–170.
54. Day, J.F., and Edman, J.D. 1984. The importance of disease induced changes in mammalian body temperature to mosquito blood feeding. *Comp. Biochem. Physiol.* **77A**:447–452.
55. de Arruda, M., Carvalho, M.B., Nussenzweig, R.S., Maracic, M., Ferreira, A.W., and Cochrane, A.H. 1986. Potential vectors of malaria and their different susceptibility to *Plasmodium falciparum* and *Plasmodium vivax* in northern Brazil. *Am. J. Trop. Med. Hyg.* **35**:873–881.
56. de Zoysa, A.P.K., Mendis, C., Gamage-Mendis, A.C., Weerasinghe, S., Herath, P.R.J., and Mendis, K.N. 1991. A mathematical model for *Plasmodium vivax* malaria transmission: Estimation of the impact of transmission-blocking immunity in an endemic area. *Bull. W.H.O.* **69**: 725–734.
57. do Rosario, V.E., Appiah, A., Vaughan, J.A., and Hollingdale, M.R. 1989. *Plasmocium falciparum*: Administration of anti-sporozoite antibodies during sporogony results in production of sporozoites which are not neutralized by human anti-circumsporozoite protein vaccine sera. *Trans. R. Soc. Trop. Hyg.* **83**:305–307.
58. Druilhe, P., Pradier, O., Marc, J.-P., Miltgen, F., Mazier, D., and Parent, G. 1986. Levels of antibodies to *Plasmodium falciparum* sporozoite surface antigens reflect malaria transmission rates and are persistent in the absence of reinfection. *Infect. Immun.* **53**:393–397.
59. Dye, C. 1986. Vectorial capacity: Must we measure all its components? *Parasitol. Today* **2**:203–209.
60. Dye, C. 1990. Epidemiological significance of vector-parasite interactions. *Parasitology* **101**:409–415.
61. Dye, C. 1992. The analysis of parasite transmission by bloodsucking insects. *Annu. Review Entomol.* **37**:1–19.
62. Dye, C., and Hasibeder, G. 1986. Population dynamics of mosquito-borne disease: Effects of flies which bite some people more frequently than others. *Trans. R. Soc. Trop. Hyg.* **80**:69–77.
63. Edman, J., Day, J., and Walker, E. 1985. Vector-host interplay-factors affecting disease transmission. In L.P. Lounibos, J.R. Rey, and J.H. Frank, eds., *Ecology of Mosquitoes*. Chapman and Hall, New York, pp. 273–285.

64. Esposito, F., Lombardi, S., Modiano, D., Zavala, F., Reeme, J., Lamizana, L., Coluzzi, M., and Nussenzweig, R.S. 1988. Prevalence and levels of antibodies to the circumsporozoite protein of *Plasmodium falciparum* in an endemic area and their relationship to resistance against malaria infection. *Trans. R. Soc. Trop. Hyg.* **82**:827–832.
65. Feldmann, A.M., and Ponnudurai, T. 1989. Selection of *Anopheles stephensi* for refractoriness and susceptibility to *Plasmodium falciparum. Med. Vet Entomol.* **3**:41–52.
66. Fontes, C.C.J., Bathurst, I., and Krettli, A.U. 1991. *Plasmodium vivax* sporozoite antibodies in individuals exposed during a single malaria outbreak in a non-endemic area. *Am. J. Trop. Med. Hyg.* **44**:28–33.
67. Gale, K.R., and Crampton, J.M. 1987. DNA probes for species identification of mosquitoes in the *Anopheles gambiae* complex. *Med. Vet. Entomol.* **1**:127–136.
68. Gale, K.R., and Crampton, J.M. 1987. A DNA probe to distinguish the species *Anopheles quadriannulatus* from other species of the *Anopheles gambiae* complex. *Trans. R. Soc. Trop. Med. Hyg.* **81**:842–846.
69. Gamage-Mendis, A.C., Rajalaruna, J., Carter, R., and Mendis, K.N. 1991. Infectious reservoir of *Plasmodium vivax* and *Plasmodium falciparum* malaria in an endemic region of Sri Lanka. *Am. J. Trop. Med. Hyg.* **45**:479–487.
70. Garrett-Jones, C. 1964. The human blood index of malaria vectors in relation to epidemiological assessment. *Bull. W.H.O.* **30**:241–261.
71. Garrett-Jones, C., and Shidrawi, G.R. 1969. Malaria vectorial capacity of a population of *Anopheles gambiae*. An exercise in epidemiological entomology. *Bull. W.H.O.* **40**:531–545.
72. Gillies, M.T., and Wilkes, T.J. 1965. A study of the age-composition of populations of *Anopheles gambiae* Giles and *A. funestus* Giles in north-eastern Tanzania. *Bull. Entomol. Res.* **56**:237–262.
73. Good, M.F., Saul, A., and Graves, P.M. 1992. Malaria. In *Vaccines: New Approaches to Immunological Problems*. Butterworths, London (in press).
74. Graves, P.M., Carter, R., Burkot, T.R., Quakyi, I.A., and Kumar, N. 1988. Antibodies to *Plasmodium falciparum* gamete surface antigens in Papua New Guinea sera. *Parasite Immunol.* **10**:209–218.
75. Graves, P.M., Burkot, T.R., Carter, R., Cattani, J.A., Lagog, M., Parker, J., Brabin, B.J., Gibson, F.D., Bradley, D.J., and Alpers, M.P. 1988. Measurement of malarial infectivity of human populations to mosquitoes in the Madang area, Papua New Guinea. *Parasitology* **96**:251–263.
76. Graves, P.M., Burkot, T.R., Saul, A., Hayes, R., and Carter, R. 1990. Estimation of anopheline survival rate, vectorial capacity, and mosquito infection probability from malaria vector infection rates in villages near Madang, Papua New Guinea. *J. Appl. Ecol.* **27**:134–147.
77. Greenwood, B.M., Bradley, A.K., Greenwood, A.M., Byass, P., Jammeh, K., Marsh, K., Tulloch, S., Oldfield, F.S.J., and Hayes, R. 1987. Mortality and morbidity from malaria among children in a rural area of The Gambia, West Africa. *Trans. R. Soc. Trop. Med. Hyg.* **81**:478–486.
78. Greenwood, B., Marsh, K., and Snow, R. 1991. Why do some African children develop severe malaria?, *Parasitol. Today* **7**:277–281.
79. Herrington, D.A., Clyde, D.F., Losonsky, G., Cortesia, M., Murphy, J.R., Davis, J., Baqar, S., Felix, A.M., Heimer, E.P., Gillessen, D., Nardin, E.,

Nussenzweig, R.S., Nussenzweig, V., Hollingdale, M.R., and Levine, M.M. 1987. Safety and immunogenicity in man of a synthetic peptide malaria vaccine against *Plasmodium falciparum* sporozoites. *Nature (London)* **328**:257–259.

80. Hii, J.L.K., Birley, M.H., and Sang, V.Y. 1990. Estimation of survival rate and oviposition interval of *Anopheles balabacensis* mosquitoes from mark-recapture experiments in Sabah, Malaysia. *Med. Vet. Entomol.* **4**:135–140.
81. Hii, J.L.K., Chew, M., Sang, V.Y., Munstermann, L.E., Tan, S.G., Panyim, S., and Yasothornsrikul, S. 1991. Population genetic analysis of host seeking and resting behaviors in the malaria vector, *Anopheles balabacensis* (Diptera: Culicidae). *J. Med. Entomol.* **28**:675–684.
82. Hill, S.M., Urwin, R., and Crampton, J.M. 1991. A comparison of non-radioactive labeling and detection systems with synthetic oligonucleotide probes for the species identification of mosquitoes in the *Anopheles gambiae* complex. *Am. J. Trop. Med. Hyg.* **44**:609–622.
83. Hill, A.V.S., Allsopp, C.E.M., Kwiatkowski, D., Anstey, N.M., Twumasi, P., Rowe, P.A., Bennett, S., Brewster, D., McMichael, A.J., and Greenwood, B.M. 1991. Common West African HLA antigens are associated with protection from severe malaria. *Nature (London)* **352**:595–600.
84. Hollingdale, M.R., and do Rosario, V. 1989. Malaria transmission-enhancing activity in mosquitoes by mammalian host anti-sporozoite antibodies. *Exp. Parasitol.* **68**:365–368.
85. Holmes, P.R., and Birley, M.H. 1987. An improved method for survival rate analysis from time series of haematophagous dipteran populations. *J. Anim. Ecol.* **56**:427–440.
86. Hope, I.A., Hall, R., Simmons, D.Ll., Hyde, J.E., and Scaife, J.G. 1984. Evidence for immunological cross-reaction between sporozoites and blood stages of a human malaria parasite. *Nature (London)* **308**:191–194.
87. Ijumba, J.N., Mwangi, R.W., and Beier, J.C. 1990. Malaria transmission potential of *Anopheles* mosquitoes in the Mwea-Tebere irrigation scheme, Kenya. *Med. Vet. Entomol.* **4**:425–432.
88. Jeffery, G.M., Young, M.D., Burgess, R.W., and Eyles, D.E. 1959. Early activity in sporozoite-induced *Plasmodium falciparum* infections. *Ann. Trop. Med. Parasitol.* **54**:51–58.
89. Kelly, R., and Edman, J.D. 1992. Multiple transmission of *Plasmodium gallinaceum* (Eucoccida: Plasmodiidae) during serial probing by *Aedes aegypti* (Diptera: culicidae) on several hosts. *J. Med. Entomol.* **29**:329–331.
90. Kingsolver, J.G. 1987. Mosquito host choice and the epidemiology of malaria. *Am. Nat.* **130**:811:–827.
91. Kitthawee, S., Edman, J.D., and Sattabongkot, J. 1990. Evaluation of survival potential and malaria susceptibility among different size classes of laboratory-reared *Anopheles dirus*. *Am. J. Trop. Med. Hyg.* **43**:328–332.
92. Klein, T.A., Harrison, B.A., Grove, J.S., Dixon, S.V., and Andre, R.G. 1986. Correlation of survival rates of *Anopheles dirus* A (Diptera: Culicidae) with different infection densities of *Plasmodium cynomolgi*. *Bull. W.H.O.* **64**:901–907.
93. Klein, T.A., Lima, J.B.P., Tada, M.S., and Miller, R. 1991. Comparative susceptibility of Anopheline mosquitoes in Rondonia, Brazil to infection by *Plasmodium vivax*. *Am. J. Trop. Med. Hyg.* **45**:463–470.

94. Klein, T.A., Tada, M.S., Lima, J.B.P., and Katsuragawa, T.H. 1991. Infection of *Anopheles darlingi* fed on patients infected with *Plasmodium vivax* before and during treatment with chloroquine in Costa Marques, Rondonia, Brazil. *Am. J. Trop. Med. Hyg.* **45**:471–478.
95. Klein, T.A., Tada, M.S., and Lima, J.B.P. 1991. Infection of *Anopheles darlingi* fed on patients with *Plasmodium falciparum* before and after treatment with quinine or quinine plus tetracycline. *Am. J. Trop. Med. Hyg.* **44**:604–608.
96. Lepers, J.P., Fontenille, D., Rason, M.D., Chougnet, C., Astagneau, P., Coulanges, P., and DeLoron, P. 1991. Transmission and epidemiology of newly transmitted falciparum malaria in the central highland plateaux of Madagascar. *Ann. Trop. Med. Parasitol.* **85**:297–304.
97. Lindsay, S.W., Snow, R.W., Broomfield, G., Semega Janneh, M., Wirtz, R.A., and Greenwood, B.M. 1989. Impact of permethrin treated bednets on malaria transmission in The Gambia. *Med. Vet. Entomol.* **3**:263–271.
98. Lindsay, S.W., Wilkins, H.A., Zieler, H.A., Daly, R.J., Petrarca, V., and Byass, P. 1991. Ability of *Anopheles gambiae* mosquitoes to transmit malaria during the dry and wet seasons in an area of irrigated rice cultivation in The Gambia. *J. Trop. Med. Hyg.* **94**:313–324.
99. Lines, J.D., Curtis, C.F., Wilkes, T.J., and Hjunwa, K.J. 1991. Monitoring human biting mosquitoes in Tanzania with light traps hung beside mosquito nets. *Bull. Entomol. Res.* **81**:77–84.
100. Lines, J.D., Wilkes, T.J., and Lyimo, E.O. 1991. Human malaria infectiousness measured by age-specific sporozoite rates in *Anopheles gambiae* in Tanzania. *Parasitology* **102**:167–177.
101. Lombardi, S., Esposito, F., Zavala, F., Lamizana, L., Rossi, P., Sabatinelli, G., Nussenzweig, R.S., and Coluzzi, M. 1987. Detection and anatomical localization of *Plasmodium falciparum* circumsporozoite protein and sporozoites in the afrotropical malaria vector *Anopheles gambiae* s.l. *Am. J. Trop. Med. Hyg.* **37**:491–494.
102. Ma, M., Beier, J.C., Petrarca, V., Gwadz, R.W., Zhang, J.-Z., Song, Q., and Koech, D.K. 1990. Differentiation of *Anopheles gambiae* and *An. arabiensis* (Diptera: Culicidae) by ELISA using immunoaffinity purified antibodies to vitellogenin. *J. Med. Entomol.* **27**:564–569.
103. MacCormack, C.P. 1984. Human ecology and behaviour in malaria control in tropical Africa. *Bull. W.H.O.* **62** (Suppl.):81–87.
104. Macdonald, G. 1957. *The Epidemiology and Control of Malaria*, Oxford University Press, London.
105. Magesa, S.M., Wilkes, T.J., Mnzava, A.E.P., Njunwa, K.J., Myamba, J., Kivuyo, M.D.P., Hill, N., Lines, J.D., and Curtis, C.F. 1991. Trial of pyrethroid impregnated bednets in an area of Tanzania holoendemic for malaria. Part 2. Effects on the malaria vector population. *Acta Tropica* **49**:97–108.
106. Malik, A., Egan, J.E., Houghten, R.A., Sadoff, J.C., and Hoffman, S.L. 1991. Human cytotoxic T lymphocytes against the *Plasmodium falciparum* circumsporozoite protein. *Proc. Natl. Acad. Sci. U.S.A.* **88**:3300–3304.
107. Mattei, D., Berzins, K., Wahlgren, M., Udomsangpetch, R., Perlmann, P., Griesser, H.W., Scherf, A., Muller-Hill, B., Bonnefoy, S., Guillotte, M., Langsley, G., Pereira da Silva, L., and Mercereau-Puijalon, O. 1989. Cross-

reactive antigenic determinants present on different *Plasmodium falciparum* blood-stage antigens. *Parasite Immunol.* **11**:15–30.
108. McGregor, I.A., and Wilson, R.J.M., 1988. Specific immunity: acquired in man, Wernsdorfer, W.H., and McGregor, I. (eds). In *Malaria. Principles and Practices in Malariology*, Vol. 1. Churchill Livingstone, Edinburgh, pp. 253–306.
109. Mellouk, S., Mazier, D., Druilhe, P., Berbiguier, N., and Danis, M. 1986. In vitro and in vivo results suggesting that anti-sporozoite antibodies do not totally block *Plasmodium falciparum* sporozoite infectivity. *N. Engl. J. Med.* **315**:648.
110. Mendis, C., Del Giudice, G., Gamage-Mendis, A.C., Tougne, C., Pessi, A., Weerasinghe, S., Carter, R., and Mendis, K.N. 1992. Anti-circumsporozoite protein antibodies measure age related exposure to malaria in Kataragama, Sri Lanka. *Parasite Immunol.* **14**:75–86.
111. Mendis, K.N., Munesinghe, Y.D., de Silva, Y.N.Y., Keragalla, I., and Carter, R. 1987. Malaria transmission-blocking immunity induced by natural infections of *Plasmodium vivax* in humans. *Infect. Immun.* **55**:369–372.
112. Mendis, K.N., Naotunne, T., de, S., Karunaweera, N.D., Del Giudice, G., Grau, G.E., and Carter, R. 1990. Anti-parasite effects of cytokines in malaria. *Immunol. Lett.* **25**:217–220.
113. Metselaar, D. 1960. Relative increase in the prevalence of *Plasmodium falciparum* some years after the beginning of a house-spraying campaign in Netherlands New Guinea. *Trans. R. Soc. Trop. Hyg.* **54**:523–528.
114. Moelans, I.I.M.D., Meis, J.F.G.M., Kocken, C., Konings, R.N.H., and Schoenmakers, J.G.G. 1991. A novel protein antigen of the malaria parasite *Plasmodium falciparum*, located on the surface of gametes and sporozoites. *Mol. Biochem. Parasitol.* **45**:193–204.
115. Molineaux, L., and Gramiccia, G. 1980. *The Garki Project: Research on the Epidemiology and Control of Malaria in the Sudan Savanna of West Africa.* W.H.O., Geneva, pp. 255–259.
116. Muirhead-Thompson, R.C. 1951. The distribution of Anopheline mosquito bites among different age groups. *Br. Med. J.* **1**:1114–1117.
117. Muirhead-Thompson, R.C. 1957. The malaria infectivity of an African village population to mosquitoes (*Anopheles gambiae*). *Am. J. Trop. Med. Hyg.* **6**:971–979.
118. Naotunne, T., De, S., Karunaweera, N.D., Del Giudice, G., Kularatne, M.U., Grau, G.E., Carter, R., and Mendis, K.N. 1991. Cytokines kill malaria parasites during infection crisis: extracellular complementary factors are essential. *J. Exp. Med.* **173**:523–529.
119. Naotunne, T., De, S., Rathnayake, K.D.L., Jayasinghe, A., Carter, R., and Mendis, K.N. 1990. *Plasmodium cynomolgi*: Serum-mediated blocking and enhancement of infectivity to mosquitoes during infections in the natural host. *Macaca sinica. Exp. Parasitol.* **71**:305–313.
120. Nardin, E.H., Nussenzweig, R.A., McGregor, I.A., and Bryan, J.H. 1979. Antibodies to sporozoites: Their frequent occurrence in individuals living in an area of hyperendemic malaria. *Science* **206**:597–599.
121. Nedelman, J. 1989. Gametocytemia and infectiousness in falciparum malaria: Observations and models. In K.F. Harris, ed., *Advances in Disease Vector Research*, Vol. 6. Springer-Verlag, New York, pp. 59–89.

122. Nudelman, S., Renia, L., Charoenvit, Y., Yuan, L., Miltgen, F., Beaudoin, R.L., and Mazier, D. 1989. Dual action of anti-sporozoite antibodies in vitro. *J. Immunol.* **143**:996–1000.
123. Panyim, S., Yasothornsrikul, S., Tungprodubkul, S., Baimai, V., Rosenberg, R., Andre, R.G., and Green, C.A. 1988. Identification of isomorphic malaria vectors using a DNA probe. *Am. J. Trop. Med. Hyg.* **38**:47–49.
124. Peiris, J.S.M., Premawansa, S., Ranawaka, M.B.R., Udagama, P.V., Munasinghe, Y.D., Nanayakkara, M.V., Gamage, C.P., Carter, R., David, P.H., and Mendis, K.N. 1988. Monoclonal and polyclonal antibodies both block and enhance transmission of human *Plasmodium vivax* malaria. *Am. J. Trop. Med. Hyg.* **39**:26–32.
125. Petersen, E., Hogh, B., Marbiah, N.T., David, K., and Hanson, A.P. 1991. Development of immunity against *Plasmodium falciparum* malaria: Clinical and parasitologic immunity cannot be separated. *J. Infect. Dis.* **164**:949–953.
126. Petrarca, V., and Beier, J.C. 1992. Intraspecific chromosomal polymorphism in the *Anopheles gambiae* complex as a factor affecting malaria transmission in the Kisumu area of Kenya. *Am. J. Trop. Med. Hyg.* **46**:229–237.
127. Ponnudurai, T., Billingsley, P.F., and Rudin, W. 1988. Differential infectivity of *Plasmodium* for mosquitoes. *Parasitol. Today* **4**:319–321.
128. Ponnudurai, T., Lensen, A.H.W., van Gemert, G.J.A., Bensing, M.P.E., Bolmer, M., and Meuweissen, J.H.E.Th. 1979. Sporozoite load of mosquitoes infected with *Plasmodium falciparum. Trans. R. Soc. Trop. Med. Hyg.* **83**:67–70.
129. Port, G.R., Boreham, P.F.L., and Bryan, J.H. 1980. The relationship of host size to feeding by mosquitoes of the *Anopheles gambiae* Giles complex (Diptera: Culicidae). *Bull. Entomol. Res.* **70**:133–144.
130. Porter, C.H., and Collins, F.H. 1991. Species-diagnostic differences in a ribosomal DNA internal transcribed spacer from the sibling species *Anopheles freeborni* and *Anopheles hermsi* (Diptera: Culicidae). *Am. J. Trop. Med. Hyg.* **45**:271–279.
131. Porter, R.J., Laird, R.L., and Dusseau, E.M. 1954. Studies on malarial sporozoites. II. Effect of age and dosage of sporozoites on their infectiousness. *Exp. Parasitol.* **3**:267–274.
132. Ranawaka, M.B., Munesinghe, Y.D., de Silva, D.M.R., Carter, R., and Mendis, K.N. 1988. Boosting of transmission-blocking immunity during natural *Plasmodium vivax* infections in humans depends upon frequent reinfection. *Infect. Immun.* **56**:1820–1824.
133. Relf, W.A., Pugh, R.E., Tapchaisri, P., Khusmith, S., Healey, A., Upcroft, P., Tharavanij, S., and Kidson, C. 1990. Diagnosis of *Plasmodium vivax* malaria using a specific deoxyribonucleic acid probe. *Trans. R. Soc. Trop. Med. Hyg.* **84**:630–634.
134. Ribeiro, J.M.C., Rossignol, P.A., and Spielman, A. 1985. Salivary gland apyrase determines probing time in anopheline mosquitoes. *J. Insect Physiol.* **31**:689–692.
135. Rickman, L.S., Jones, T.R., Long, G.W., Paparello, S., Schneider, I., Paul, C.F., Beaudoin, R.L., and Hoffman, S.L. 1990. *Plasmodium falciparum*-infected *Anopheles stephensi* inconsistently transmit malaria to humans. *Am. J. Trop. Med. Hyg.* **43**:441–445.

136. Robert, V., Verhave, J.P., and Carnevals, P. 1990. *Plasmodium falciparum* infection does not increase the precocious mortality rate of *Anopheles gambiae*. *Trans. R. Soc. Trop. Med. Hyg.* **84**:346–347.
137. Robert, V., Verhave, J.P., Ponnudurai, T., Louwe, L., Scholtens, P., and Carnevale, P. 1988. Study of the distribution of circumsporozoite antigen in *Anopheles gambiae* infected with *Plasmodium falciparum*, using the enzyme-linked immunosorbent assay. *Trans. R. Soc. Trop. Med. Hyg.* **82**:389–391.
138. Rosenberg, R. 1985. Inability of *Plasmodium knowlesi* sporozoites to invade *Anopheles freeborni* salivary glands. *Am. J. Trop. Med. Hyg.* **34**:687–691.
139. Rosenberg, R., Andre, R.G., and Ketrangsee, S. 1990. Seasonal fluctuation of *Plasmodium falciparum* gametocytaemia. *Trans. R. Soc. Trop. Med. Hyg.* **84**:29–33.
140. Rosenberg, R., Andre, R.G., and Somchit, L. 1990. Highly efficient dry season transmission of malaria in Thailand. *Trans. R. Soc. Trop. Med. Hyg.* **84**:22–28.
141. Rosenberg, R., and Rungsiwongse, J. 1991. The number of sporozoites produced by individual malaria oocysts. *Am. J. Trop. Med. Hyg.* **45**:574–577.
142. Rosenberg, R., Wirtz, R.A., Lanar, D.E., Sattabongkot, J., Hall, T., Waters, A.P., and Prasittisuk, C. 1989. Circumsporozoite protein heterogeneity in the human malaria parasite *Plasmodium vivax*. *Science* **245**:973–976.
143. Rosenberg, R., Wirtz, R.A., Schneider, I., and Burge, R. 1990. An estimation of the number of malaria sporozoites ejected by a feeding mosquito. *Trans. R. Soc. Trop. Med. Hyg.* **84**:209–212.
144. Rossignol, P.A., Ribeiro, J.M.C., Jungery, M., Spielman, A., and Bailey, C.L. 1985. Enhanced mosquito blood-finding success on parasitemic hosts: Evidence for vector-parasite mutualism. *Proc. Natl. Acad. Sci. U.S.A.* **82**:7725–7727.
145. Rossignol, P.A., Ribeiro, J.M.C., and Spielman, A. 1984. Increased intradermal probing time in sporozoite-infected mosquitoes. *Am. J. Trop. Med. Hyg.* **33**:17–20.
146. Rozendaal, J.A. 1989. Impregnated mosquito nets and curtains for self protection and vector control. *Trop. Dis. Bull.* **86**:R1–R41.
147. Sattabongkot, J., Maneechai, N., and Rosenberg, R. 1991. *Plasmodium vivax*: Gametocyte infectivity of naturally infected Thai adults. *Parasitology* **102**:27–31.
148. Saul, A. 1987. Estimation of survival rates and population size from mark-release-recapture experiments of bait-caught haematophagous insects. *Bull. Entomol. Res.* **77**:589–602.
149. Saul, A., Graves, P.M., and Kay, B.H. 1990. A cyclical feeding model for pathogen transmission and its application to determine vectorial capacity from vector infection rates. *J. Appl. Ecol.* **27**:123–133.
150. Schmidt, L.H., Fradkin, R., Genther, C.S., Rossan, R.N., and Squires, W. 1982. I. The characteristics of untreated sporozoite-induced and trophozoite-induced infections. *Am. J. Trop. Med. Hyg.* **31** (Suppl.):612–645.
151. Turrell, M.J., and Spielman, A. 1992. Nonvascular delivery of Rift Valley fever virus by infected mosquitoes. *Am. J. Trop. Med. Hyg.* **47**:190–194.

152. Ungureanu, E., Killick-Kendrick, R., Garnham, P.C.C., Branzei, P., Romanescu, C., and Shute, P.G. 1976. Prepatent periods of a tropical strain of *Plasmodium vivax* after inoculations of tenfold dilutions of sporozoites. *Trans. R. Soc. Trop. Med. Hyg.* **70**:482–483.
153. Vanderberg, J.P. 1977. *Plasmodium berghei*: Quantitation of sporozoites injected by mosquitoes feeding on a rodent host. *Exp. Parasitol.* **42**:169–181.
154. Vaughan, J.A., and Azad, A.F. 1988. Passage of host immunoglobulin G from blood meal into hemolymph of selected mosquito species (Diptera: Culicidae). *J. Med. Entomol.* **25**:472–474.
155. Vaughan, J.A., Do Rosario, V., Leland, P., Adjepong, A., Light, J., Woollett, G.R., Hollingdale, M.R., and Azad, A.F. 1988. *Plasmodium falciparum:* Ingested anti-sporozoite antibodies affect sporogony in *Anopheles stephensi* mosquitoes. *Exp. Parasitol.* **66**:171–182.
156. Vernick, K.D., Collins, F.H., and Gwadz, R.W. 1989. A general system of resistance to malaria infection in *Anopheles gambiae* controlled by two main genetic loci. *Am. J. Trop. Med. Hyg.* **40**:585–592.
157. Ward, R.A. 1963. Genetic aspects of the susceptibility of mosquitoes to malarial infection. *Exp. Parasitol.* **13**:328–341.
158. Waters, A.P., and McCutchan, T.F. 1989. Ribosomal RNA: nature's own polymerase-amplified target for diagnosis. *Parasitol. Today* **6**:56–59.
159. Webster, H.K., Boudreau, E.F., Pang, L.W., Permpanich, B., Sookto, P., and Wirtz, R.A. 1987. Development of immunity in natural *Plasmodium falciparum* malaria: Antibodies to the falciparum sporozoite vaccine 1 antigen (R32tet32). *J. Clin. Microbiol.* **25**:1002–1008.
160. Wirtz, R.A., and Burkot, T.R. 1991. Detection of malarial parasites in mosquitoes. In K.F. Harris, ed. *Advances in Disease Vector Research*, Vol. 8. Springer-Verlag, New York, pp. 77–106.
161. Wirtz, R.A., Burkot, T.R., Andre, R.G., Rosenberg, R.M., Collins, W.E., and Roberts, D.R. 1985. Identification of *Plasmodium vivax* sporozoites in mosquitoes using an enzyme-linked immunosorbent assay. *Am. J. Trop. Med. Hyg.* **34**:1048–1054.
162. Wirtz, R.A., Burkot, T.R., Graves, P.M., and Andre, R.G. 1987. Field evaluation of enzyme-linked immunosorbent assays for *Plasmodium falciparum* and *Plasmodium vivax* sporozoites in mosquitoes (Diptera: Culicidae) from Papua New Guinea. *J. Med. Entomol.* **24**:433–437.
163. Wirtz, R.A., Charoenvit, Y., Burkot, T.R., Esser, K.M., Beaudoin, R.L., Collins, W.E., Rosenberg, R., and Andre, R.G. 1990. Evaluation of monoclonal antibodies against *Plasmodium vivax* sporozoites for ELISA development. *Med. Vet. Entomol.* **5**:17–22.
164. Wirtz, R.A., Duncan, J.F., Njelesani, E.K., Schneider, I., Brown, A.E., Oster, C.N., Were, J.B.O., and Webster, H.K. 1989. ELISA method for detecting *Plasmodium falciparum* circumsporozoite antibody. *Bull. W.H.O.* **67**:535–542.
165. Wirtz, R.A., Sattabongkot, J., Hall, T., Burkot, T.R., and Rosenberg, R. 1992. Development and evaluation of an ELISA for *Plasmodium vivax* variant strain-VK247 sporozoites. *J. Med. Entomol.* **29**:854–857.

166. Wirtz, R.A., Zavala, F., Charoenvit, Y., Campbell, G.H., Burkot, T.R., Schneider, I., Esser, K.M., Beaudoin, R.L., and Andre, R.G. 1987. Comparative testing of *Plasmodium falciparum* sporozoite monoclonal antibodies for ELISA development. *Bull. W.H.O.* **65**:39–45.

7
Factors Affecting Filarial Transmission by Simuliids

Anthony J. Shelley

Introduction

Of the nine filariae commonly transmitted by simuliids one, *Onchocerca volvulus*, is preeminent because of its pathogenic effects to man. Currently 17 million people are infected by this parasite; many have irreversible skin lesions and more than 300,000 are blind (44). The literature built up around both the parasite and vector responsible for transmission of river blindness is thus considerably larger than that associated with the other simuliid-transmitted filariae that are either nonpathogenic to man or affect only animals. Consequently, this review has been structured around *O. volvulus* with references to other filarial species where data are available. Recent advances have been made in various areas including vector–parasite relationships, particularly in the foci of Latin America and with reference to vector cytospecies and parasite strains in Africa, as well as the immunity of simuliids to filaria infections. Underpinning this work is the precise identification of vector and filaria species and considerable effort has been made to achieve this using both traditional morphological and more recent biochemical techniques. The review has been structured to provide the reader with an introduction to the literature on basic facts about the parasites and their simuliid vectors, followed by the natural sequence of parasite transmission and the factors that affect it.

Anthony J. Shelley, Department of Entomology, The Natural History Museum, Cromwell Road, London SW7 5BD, United Kingdom.

The Filarial Parasite

Adult filariae are typically long, filiform, obligate parasitic nematode worms that live in the blood, lymph, and adjacent tissues or body cavities of their vertebrate hosts. Females are generally viviparous, producing larvae or microfilariae, which need to be ingested by a hematophagous arthropod host before further larval development will take place through the L1 and L2 to the infective L3 stage. Once L3 larvae have been introduced to the vertebrate host during blood feeding, development to the L4 larval stage first occurs, followed by ecdysis to male or female worms. Details of morphology and life history patterns of the order Spirurida, which contains the superfamily Filarioidea, are given in Chabaud (11).

The nine filaria species known to be transmitted by simuliids belong to the family Onchocercidae: *Dirofilaria ursi* in bears, *Mansonella ozzardi* in man, *Onchocerca cervipedis* and *O. tarsicola* in deer, *O. dukei*, *O. lienalis*, and *O. ochengi* in cattle, *O. volvulus* in man, and *Splendidofilaria fallisensis* in ducks (Table 7.1). The cattle filaria *O. gutturosa* is considered to be transmitted by *Culicoides* spp. in England and Germany (28, 35) and does not develop when inoculated into *S. soubrense* and *S. yahense* (65). References to simuliids as vectors of *O. gutturosa* in Japan, the Ukraine, and Tanzania (83, 92, 130) therefore require confirmation. Other filaria species suspected of being transmitted by simuliids are *O. alcis*, a parasite of moose in Scandinavia, *O. flexuosa* and *O. tubingensis* in European deer, and *O. armillata* and *O. denkei* in African cattle (28, 118). In South America *Microfilaria bolivarensis* and *Mansonella perstans* and several species of monkey filariae are likely to have a simuliid or ceratopogonid vector.

Other filarial species undoubtedly have simuliids as vectors because unidentified filariae are often found in flies, especially in onchocerciasis investigations in Africa (28, 39, 96, 105, 140). These unidentified species have been assigned to the genus *Agamofilaria* (140). Of the two filarial species that have man as definitive host only *O. volvulus* is seriously pathogenic (94), the relatively mild symptoms attributed to *M. ozzardi* possibly being due to other parasitic infections (145). Little is known about the pathogenic effects of filarial species that infect animals but heavy infections of *O. gutturosa* and *O. lienalis* cause economic damage to cattle hides (83). The main interest in animal filariae is when they occur in areas where human onchocerciasis exists and are transmitted by the same simuliid species, because this can cause overestimation of human onchocerciasis transmission indices. They have also been used as models for human onchocerciasis.

The Filarioidea have a worldwide distribution with species transmitted by simuliids occurring in both tropical and temperate regions. Filariae of medical importance are all tropical, *O. volvulus* being found in Africa, Latin America, and southern Arabia and *M. ozzardi* in the Caribbean and

TABLE 7.1. Species of filarial worms transmitted by simuliid flies.[a]

Filaria species	Vertebrate host	Simuliid species	Development site	Distribution
Dirofilaria ursi	Bears	*S. venustum s.l.*[b]	Malpighian tubules	Canada
Mansonella ozzardi	Man	*S. amazonicum*	Thorax	Brazil, Colombia
		S. argentiscutum		Brazil, Colombia
		S. oyapockense s.l.		Brazil, Colombia, Guyana, Venezuela
Mansonella perstans	Man	*S. oyapockense s.l.*?	?	Guyana, Venezuela
Microfilaria bolivarensis	Man	*S. oyapockense s.l.*?	?	Venezuela
Onchocerca cervipedis	Deer	*Prosimulium impostor*	Thorax	USA
Onchocerca dukei	Cattle	*S. bovis*	Thorax	Cameroon
Onchocerca lienalis	Cattle	*S. ornatum s.l.*	Thorax	UK
		S. reptans		UK
		S. erythrocephalum		USSR
		S. jenningsi s.l.		USA
Onchocerca ochengi	Cattle	*S. sanctipauli* subcomplex and *S. sirbanum*	Thorax	West Africa
Onchocerca tarsicola	Deer	*Prosimulium nigripes*	Thorax	Germany
		S. ornatum s.l.		Germany
Onchocerca volvulus	Man	*S. albivirgulatum*	Thorax	Zaire river basin
		S. callidum		Guatemala
		S. damnosum siblings		
		S. damnosum s. str.		West Africa to Sudan and Uganda
		S. kilibanum		East Africa
		S. mengense		Cameroun
		S. rasyani		Yemen
		S. sanctipauli (+cytoforms)		West Africa
		S. sirbanum		West Africa and Sudan
		S. soubrense (+cytoforms)		West Africa
		S. squamosum		West Africa (+Zaire?)
		S. yahense		West Africa
		Jimma sibling		(Vector in Ethiopia?)
		Kapere sibling		(Vector in Zaire?)
		Ketaketa sibling		(Vector in Tanzania?)
		Nkusi sibling		Tanzania
		S. exiguum complex		
		Cayapa form		Ecuador
		Aguarico form		Ecuador
		Unidentified form		Colombia
		Unidentified form		Venezuela?
		S. gonzalezi s.l.		Guatemala
		S. guianense s.l.		Brazil, Venezuela
		S. haematopotum		Guatemala
		S. metallicum s.l.		Guatemala, Mexico, Venezuela
		S. neavei group		
		S. ethiopiense		Ethiopia
		S. neavei		Uganda, Zaire, previously Kenya
		S. woodi		Tanzania
		S. ochraceum s.l.		Guatemala, Mexico
		S. oyapockense s.l.		Brazil, Venezuela
		S. quadrivittatum		Ecuador
		S. veracruzanum		Guatemala
		S. yarzabali		Brazil, Venezuela
Splendidofilaria fallisensis	Ducks	*S. rugglesi*	Hemocoel	Canada
		S. anatinum		Canada

[a] Adapted from Crosskey (28).
[b] s.l. denotes existence of species complex but vector not identified as sibling.

from Mexico to Argentina. *Splendidofilaria fallisensis* parasitizes wild and domestic ducks in Canada and is probably widespread in other parts of the world and in other avian species; *D. ursi* is found in bears in the USSR, Japan, and North America; *O. cervipedis* affects mule deer and moose in North America; *O. tarsicola* occurs in European cervids; *O. lienalis and O. gutturosa* are cattle parasites worldwide; and *O. ochengi* and *O. dukei* parasitize cattle in Africa.

The taxonomy and identification of filariae are still far from satisfactory because of their parasitic habit and hence relative inaccessibility in their definitive and intermediate hosts. Most species appear to be host specific and do not easily infect small laboratory mammals for life history studies (7) and hence taxonomic characters other than adult morphology (e.g., ecology and larval morphology) are frequently used for identification (90, 98). Various important works contain details of the classification and morphology of the superfamily Filarioidea down to the 100 constituent genera and 500 species (3, 12–14, 23, 88–90, 93, 127, 128).

Two major identification problems that have been recently studied are the separation of infective larvae of animal and human *Onchocerca* species and differentiation of strains within *O. volvulus*. Since no simuliid species is totally anthropophilic sympatric human and animal filariae can be found in the same vector species, which can cause overestimates of vector competence in epidemiological and control programs for *O. volvulus*. The problem was first noted (96) in Uganda where infective larvae of *O. volvulus*, *S. fallisensis*, and an unidentified filarial species infecting *S. neavei s.s.* were distinguished on gut and tail morphology. Using these characters several workers in Africa were able to show that infective larvae other than *O. volvulus* occurred in the main onchocerciasis vector (*S. damnosum s.l.*) in West Africa (39, 59, 105, 140). Similar problems will be encountered as epidemiological studies on human filariasis progress in Latin America because of the sympatry of *O. volvulus* and *O. gutturosa* (119), *M. ozzardi* and *M. perstans* (145), and the omnipresence of monkey filaria species, which probably have simuliids as intermediate hosts. Since the earlier work in Africa morphometric analyses have been used to distinguish *O. volvulus* and *O. ochengi* found sympatrically in *S. damnosum* (probably *S. sirbanum*) in West Africa (80) and to describe *M. ozzardi* in *S. amazonicum*, *S. argentiscutum*, and *S. oyapockense s.l.* (*as* sanguineum *group*) in Brazil and Colombia (123, 133, 134). Scanning electron microscopy has also been used to separate *O. volvulus* from an unknown filaria (54). Identification of L3 larvae in simuliids has been facilitated by a compilation of morphological studies (5) and a review of morphological and chemical techniques (98). Successful attempts have also been made to distinguish L1 larvae of *O. volvulus* from other unidentified filariae by staining of the cephalic armature in infected *S. sanctipauli* of the *S. damnosum* complex in Liberia and in *S. damnosum s.l.* and *S. bovis* in

North Cameroon where *O. ochengi*, *O. armillata*, *O. gutturosa*, *O. dukei*, and *O. hamoni* occur in cattle, antelopes and birds (116). L2 larvae show no useful characters for species recognition.

Nonmorphological methods have been developed to distinguish both species and strains of filariae with varying degrees of success. Isoenzyme pattern variation has been used to separate adult *O. gutturosa*, *O. lienalis*, and *O. gibsoni* all occurring in cattle, but would not separate populations of *O. gutturosa* from the United Kingdom, Australia, and Mexico (52, 79). Enzymatic and acid phosphatase differences can also be used to distinguish unidentified L3 filarial larvae (probably from cattle that are infected by *O. armillata*, *O. dukei*, *O. gutturosa*, *O. ochengi*) and *O. volvulus* in *S. damnosum s.l.* in West Africa (100, 101), but no enzymatic differences were seen between *O. volvulus* and *O. ochengi* (102). The use of monoclonal antibodies to identify L3 larvae of *O. volvulus* has also been investigated (84) but not found to be sufficiently species specific due to cross-reactivity to *O. lienalis*. However, cross-reactivity to *B. pahangi* in patients infected with mansonelliasis in Haiti make this method useful for detection of low level infections in an area where other filarial species are absent (70).

Differences in severity of symptoms and the specificity of parasite populations to simuliid vector species indicated that savanna and forest strains of *O. volvulus* occurred in West Africa (38, 47, 94). Their precise identification was important since the savanna strain was shown to be more pathogenic to man than the forest strain (7) and so control has been principally aimed at the former. Vector incompatibility also suggests that strain differences exist between *O. volvulus* from West Africa and Latin America (42, 74). Various methods have been used to distinguish *O. volvulus* strains in order to understand the complex epidemiology of onchocerciasis in West Africa. Morphological variation may be of limited use to distinguish savanna and forest populations (50) and West African *O. volvulus* from Venezuelan (9). Biochemical, immunochemical, and histochemical methods on both adult worms and microfilariae only function at the group level and not on individual samples (53, 97, 98, 103, 148). Savanna and forest strains of *O. volvulus* also differ in antigenic (77) and isoenzyme patterns (53) and acid phosphatase activity (103). More recent work on DNA sequences has provided a precise method of identification for species as well as intraspecific strains. Probes for separating *O. volvulus* from *O. gibsoni*, *O. gutturosa*, *O. ochengi*, and *O. cervicalis* and two mosquito transmitted filariae (82) as well as savanna from forest strains of *O. volvulus* (51) have now been developed. The "UNNASCH DNA" probe is now in use in the Onchocerciasis Control Programme in West Africa to distinguish nonblinding forest strains from blinding sanvanna strains of *O. volvulus* (61).

Despite the development of the techniques mentioned above microscopic examination of vectors using morphological characters for species, parasite number, and life stage determination is the standard field technique still in use. Its principal drawbacks are that species determination is not always possible in areas where two or more filariae are sympatric or where the parasites are damaged. Potentially, the most sensitive method for identification that can theoretically be used in all life stages without the fear of environmental influence is the measurement of variation in repetitive DNA sequences relative to species or strains (110). These species-specific sequences can be cloned and used as a probe against genomic DNA in unidentified samples. For identification purposes simple field techniques such as dot-blots or squash blots would be ideal. The first generation DNA probes produced by cloning highly repeated noncoding DNA sequences suffered several disadvantages (34). They often cross-hybridized with DNA of other filarial species and were unable to distinguish larval stage and hence gave an overestimate of infectivity rates. This has been eliminated in second generation probes by synthesis of oligonucleotides from portions of the molecule that show greatest interspecific differences. Another problem is the lack of sensitivity of radiolabeled probes in identifying filariae within the vector and in their limitation for field use. To overcome the latter problem nonradioactive labels can be used but these make the probes at least 10 times less sensitive. To increase sensitivity amplification of species-specific DNA segments by the polymerase chain reaction is now being used in nonradioactive probes for *O. volvulus*. The ultimate goal would be to have a selection of probes that can identify vector cytospecies and parasite species or strain in individual specimens.

The Simuliid Vector

The family Simuliidae are stout-bodied, small (mostly less than 5 mm long) flies in the order Diptera. Adults are generally dark colored, but orange and yellow species commonly occur in the Neotropical region. Although some do not bite, females of most species require a blood meal from a warm-blooded vertebrate for egg development and it is during feeding that parasite transmission occurs. The immature stages (larvae and pupae) occur in running water and develop from eggs laid directly in the water or on surrounding moist areas. Full details of the life history of the Simuliidae may be found in Crosskey (28). Only about 10% (150) of the 1500 or so species bite man and his domestic animals (28) and of these less than 50 are serious biting pests or vectors of various protozoal parasites and 9 species of filariae. Table 7.1 indicates the simuliid species that are responsible for their transmission.

Crosskey (25, 28) thoroughly reviews the geographic distribution of simuliids in relation to zoogeographic regions. The family Simuliidae is almost cosmopolitan occurring on all major land masses with the exception of Antarctica. Generally blackflies are absent from areas devoid of running water such as the polar regions and deserts, although they may be found on small islands and at desert oases where running water is limited. Major pest species to man and his domestic stock, including most of the filarial vector species, are generally associated with areas of abundant vegetation and water courses.

The taxonomy of the family, which contains about 1500 species, is relatively well organized and has been comprehensively reviewed (27). Although identification keys are available for species of most world regions a notable exception is South America, even though simuliids are major biting nuisances and vectors of human filariae there. Species determination is often difficult because of the relative morphological homogeneity of the family, requiring the use of characters from more than one stage of the life cycle for species separation. These problems are well illustrated by the *S. amazonicum* and *S. exiguum* species groups (120), which are currently being studied in South America in relation to transmission of human filariae.

Additionally, many of the better studied species that transmit filariae to man (especially *O. volvulus*) are known to be complexes of sibling species, which may be separated using banding patterns of polytene chromosomes from larval salivary glands. Identification to cytospecies of adult females that transmit the parasites is often impossible because polytene chromosomes of Malpighian tubules are generally poorly developed. More advanced DNA, hydrocarbon, and enzyme techniques for adult identification have been developed (109, 137) but none is yet totally cytospecies specific and so a combination of methods has to be used.

Identification to cytospecies level is of crucial importance to fully understand transmission cycles and their subsequent relevance to control programs, because individual cytospecies are unique in their biology and susceptibility to filaria species. *Simulium damnosum* of the Afrotropical region was the first vector complex discovered in East Africa about 30 years ago and is now known to consist of over 40 members (26, 28). Twelve are vectors of *O. volvulus* in Africa and one in the Arabian peninsula. Nine cytospecies (some with various forms) are now known from larval chromosomes from West Africa (109, 110) and these are divided into four subcomplexes: *damnosum* containing *S. damnosum s.s.*, *S. sirbanum*, and *S. dieguerense*, *squamosum* containing *S. squamosum* and *S. yahense*, *sanctipauli* containing *S. sanctipauli*, *S. soubrense*, and *S. soubrense B*, and "*Kibwezi*" containing *S. mengense*. All of these are proven vectors to different degrees except *S. dieguerense*, which is suspected of transmitting only *O. volvulus* (28, 136). No single diagnostic technique is available to distinguish adult females of each cytospecies of the first three more widespread subcomplexes but a combination of methods will separate some. Thus DNA probes will distinguish the *damnosum*, *squamosum*, and *sanctipauli* subcomplexes from one another (110), enzymes distinguish *S. yahense* from *S. squamosum* in the *squamosum* subcomplex, morphology *S. sanctipauli* from *S. yahense*, cuticular hydrocarbons *S. damnosum s.s.*, *S. sirbanum*, *S. sanctipauli*, and *S. yahense* (137) and various morphological features will separate some of the savanna from forest cytospecies (144).

Work on vector sibling species in Latin America is at a more preliminary stage. Of the 11 *O. volvulus* vector species present larval cytotaxonomy has established the presence of species complexes in *S. metallicum* (11 cytotypes), *S. exiguum* (4 cytotypes), *S. ochraceum* (3 cytotypes), *S. gonzalezi* (2 cytotypes), and *S. oyapockense* (2 cytotypes). *Simulium guianense* is also suspected of consisting of at least two sibling species, one entirely zoophilic and the other largely anthropophilic (22, 111, 119, 121). Enzyme studies on the *S. metallicum* complex (137) and more recently on the *S. exiguum* complex in Ecuador (15) have shown important variation but have not yet been developed for separating adult cytospecies. Preliminary studies (111) have also shown that species relationships based on morphology do not agree with those based on cytology and cuticular hydrocarbon profiles in species of the *S. amazonicum* group and *S. limbatum*. Mainly distributional data in relation to cytotaxa and onchocerciasis in Latin America indicate which cytotaxa may be involved in transmission (28, 121). Cytotaxa A, B, and C of the *S. ochraceum* complex have been found in the Soconusco, Oaxaca, and Chamula foci in Mexico, respectively, using larval polytene chromosomes and adult cuticular hydrocarbons for identification (85). Cytotaxa D and E of the *S. metallicum* complex are present in the northern Venezuela foci and cytotaxa A, B, H, T, X in the Central American foci (22). However, the only direct evidence for cytotaxa involvement in transmission comes from Ecuador where all four forms (Cayapa, Aguarico, Quevedo, and Bucay) are able to host the filaria and only two (Cayapa and Aguarico) are present in the main onchocerciasis focus and one (Cayapa) has been confirmed as a natural vector (121, 124).

Filaria Transmission

Basic Cycles

Filariae use a broad spectrum of vertebrates as definitive hosts with the main exception of fish. The known Arthropod intermediate hosts include Copepods, Orthoptera, Diptera, Acari, and Mallophaga. In 1973 life cycles of only 93 filarial species were known, emphasis having been placed on filariae that affect man and his domestic animals with mosquitoes the most studied vectors (115). Previous reviews cover

Spirurid nematodes (11), Filarioidea (68, 73, 93, 115), and onchocerciasis (7, 94, 95).

Development of filariae in the Arthropod host can be in the hemocoel, fat body, thoracic, and other muscles and the Malpighian tubules. In species transmitted by simuliids development always occurs in the thoracic musculature with the exception of *D. ursi* (Malpighian tubules) and *S. fallisensis* (hemocoel). After ingestion by simuliids during blood feeding microfilariae of all species, except *Dirofilaria ursi*, which continue on to the Malpighian tubules, reach the stomach of the fly, bore through its wall using a cephalic hook (4, 57, 68) and enter the hemocoel. Development to the infective larval stage occurs here in *S. fallisensis*, whereas in all other filaria species the parasite seeks out and penetrates the thoracic muscles. At its final site the microfilaria shortens and fattens and primordial cells begin developing to form future organs in the L1 or sausage stage. This undergoes ecdysis to the L2 or preinfective stage, which is quiescent but grows rapidly and shows considerable development of body organs. The second ecdysis then produces the active L3 or infective larva, which migrates from its site of development to the proboscis of the fly ready for transmission to the vertebrate host. The development cycle in the simuliid is temperature dependent, lasting for a minumum of 5 days in tropical conditions. References giving details of life stages and development of the following species are *O. volvulus* in *S. damnosum s.l.* (4, 8), *O. lienalis* (as *gutturosa*) in *S. ornatum* (49), *O. cervipedis* in *Prosimulium impostor* (142), *O. tarsicola* in *S. ornatum* and *Prosimulium nigripes* (118), *O. ochengi* in *S. damnosum s.l.* (33, 102), *O. dukei* in *S. bovis* (141), *D. ursi* in *S. venustum* (1), *Splendidofilaria* (as *Ornithofilaria fallisensis*) in *S. rugglesi* and *S. anatinum* (2), and *M. ozzardi* in *S. amazonicum*, *S. argentiscutum*, and *S. oyapockense s.l.* (123, 133, 134).

Parasite Availability in the Definitive Host

Contact between the simuliid vector and the parasite is initially governed by the availability of the filaria in the definitive host. At the community level the higher the prevalence rates and intensity of infection in the definitive host the greater the chance of the parasite meeting the vector. Of the several factors that affect this availability the behavior of the definitive host has been well studied in *O. volvulus* and *M. ozzardi* (28, 119, 145). Workers in agriculture and fishing associated with vector breeding grounds are the groups at highest risk. Human migratory habits influence transmission in the hypoendemic North Chiapas onchocerciasis focus in Mexico and the lowland areas of the Amazonia focus of Brazil (119). In the former, visits to hyperendemic areas of the South Chiapas focus for coffee berry harvesting when *S. ochraceum s.l.* populations are at their highest causes annual infection of workers. They are then sources of the parasite for local vectors (principally *ochraceum s.l.*) on their return home. Similarly, Yanomami Indians visiting hyperendemic upland foci in Brazil are infected by the efficient vector *S. guianense s.l.* and maintain the lowland hypoendemic focus on their return where the less effective vector *S. oyapockense s.l.* is responsible only for sporadic transmission. The use of items of clothing such as long trousers and long-sleeved shirts can considerably reduce man–vector contact and parasite transmission.

To ensure transmission adult worms need to produce sufficient numbers of microfilariae for a prolonged period to travel from their usually subdermal origin to either the peripheral blood system or skin where they are ingested by the vector. In the case of *O. volvulus* it is estimated that adult females live for up to 16 years (94) and can produce about 1000 microfilariae per day (117). The microfilariae live for 6–24 months and in heavily infected individuals can number 50–200 million (44). The only data available for *M. ozzardi* indicate that microfilariae live for 2–3 years (145).

Microfilariae of filaria species transmitted by simuliids can be divided into two groups based on the principal tissue from which they are transmitted. In the case of *D. ursi*, *M. ozzardi*, and *S. fallisensis* this is in the animal's blood stream whereas in the *Onchocerca* species it is the host's skin. The distribution of microfilariae in the definitive host of *Onchocerca* species is often related to the site of the bite of the vector. This is particularly well illustrated by the filaria species that infect animals, such as *O. tarsicola*, whose adults live on the tendons of the leg joints of cervids but whose microfilariae concentrate mainly in the animal's ear pinnae where the vectors *S. ornatum s.l.* and possibly *Prosimulium nigripes* bite in large numbers (118). Adults of another deer filaria *O. cervipedis* are found in a similar site and the mechanism for transmission is probably similar to *O. tarsicola* by the probable vectors *S. venustum s.l.*, *S. decorum*, and *Prosimulium impostor* (107, 142). Microfilariae of *O. lienalis* adults occur in the gastrosplenic omentum of cattle while microfilariae concentrate in the umbilical region where the vectors *S. jenningsi*, *erythrocephalum*, *ornatum s.l.*, and *reptans* bite (28, 78, 83). The distribution in the skin of microfilariae is also important for transmission of the human filarial parasite *O. volvulus*. In Africa highest densities occur in the lower parts of the body where the vector species *S. damnosum s.l.*, *S. neavei s.l.*, and *S. albivirgulatum* prefer to bite (45, 93, 94). In Latin America one of the factors that determines whether a simuliid species is a primary or secondary vector of *O. volvulus* is its biting site preference in relation to skin microfilarial densities. In all South American foci microfilarial densities are higher in the lower parts of the body where the more efficient primary vector species (*exiguum s.l.*, *guianense s.l.*, *metallicum s.l.*) bite, while secondary vectors (*S. oyapockense s.l.*, *S. yarzabali*) bite in the upper parts of the body (6, 87,

119, 148). In Central America densities are higher in the torso and transmission is principally by *S. ochraceum s.l.*, while *S. metallicum*, *callidum*, *haematopotum*, *gonzalezi*, and *veracruzanum* are secondary vectors (119). The importance of high microfilarial densities in the skin for effective transmission to take place is well illustrated by the use of ivermectin against the parasite in man in Africa where the consequent greatly reduced skin microfilarial densities reduced uptake of microfilariae and ultimately L3 production by *S. yahense* (30).

Daily and seasonal periodicity of microfilariae in the host's tissues may also affect transmission. Diurnal fluctuations in densities of *O. volvulus* in relation to biting times of *S. damnosum s.l.* and *S. woodi* (*S. neavei* group) in Africa and *S. ochraceum s.l.* in Guatemala have been observed by some but not all authors (67, 94) and seasonal variation in parasite densities in relation to vector biting rates has also been observed in *S. damnosum s.l.* in West Africa and *S. ochraceum s.l.* in Guatemala (55, 67). The epidemiological significance of this periodicity is debatable in *O. volvulus* because of the conflicting data, relatively small variations compared with the periodic form of the mosquito transmitted *Wuchereria bancrofti*, and limitations of the biopsy method for accurately recording microfilarial densities. Similar conflicting data exist for daily periodicity in *M. ozzardi* and little work has been done to relate this to vector simuliid and ceratopogonid biting cycles (145).

Other factors that affect the availability of parasites to their vectors is the appearance of host immunity to the parasite in ducks infected with *S. fallisensis* (2), and skin degeneration in individuals heavily infected with *O. volvulus* impeding the intake of microfilariae (37).

Parasite Uptake by the Vector

Having found the host, probably through visual and olfactory detection, the female fly selects a site to probe using sensilla on the legs and palps as tactile mechanoreceptors and chemoreceptors (28). Unlike many blood sucking insects, which ingest their parasites directly from the host's blood stream, simuliids are pool feeders scarifying the host's tissues and engorging on the resultant blood pool from which blood and skin inhabiting filariae are ingested. The position of microfilariae of *O. volvulus* and *M. ozzardi* in the subpapillary layers of the skin and in skin capillaries is such that they come into contact with the short proboscis of simuliids when feeding (86, 95). Adenosine and adenosine phosphates in the host's blood act as phagostimulants. Different fly species engorge at different rates with an average engorgement time of 5–6 min on man, which increases in the case of *S. damnosum s.l.* when feeding on skin that has thickened as a result of onchocerciasis (28). There is a paucity of data for feeding times on hosts other than man, but engorgement time on cattle lasts longer. The saliva of simuliids, which is injected into the bite, contains anticoagulants and agglutinins preventing blood clotting in the wound or mouthparts and causing adhesion of red blood corpuscles in the fly's stomach. In some species of simuliid these substances in the saliva appear to attract microfilariae of *O. volvulus* when the fly is feeding. The original work on *S. damnosum* (probably *S. damnosum s.s.* and/or *S. sirbanum* in savanna and *S. squamosum* and/or *S. yahense* in forest localities) by Duke (37) in West Africa showed that the numbers of microfilariae ingested by flies were not significantly higher than skin densities at the site of the bite. However, other populations of *S. damnosum s.l.* in Africa did show higher microfilarial densities, which in *S. sanctipauli* and *S. yahense* of the damnosum complex were often so high (up to 4000) as to cause the death of the vector (56, 143). Similarly high microfilarial densities were found in three vector species (*S. ochraceum s.l.*, *S. metallicum s.l.*, and *S. callidum*) in Guatemala (10, 66, 74, 99, 129). Cross-infection experiments (42, 74) showed that these three Guatemalan species as well as *S. metallicum s.l.* and *S. exiguum s.l.* in Venezuela ingest the equivalent number of parasites as *S. damnosum s.l.* when fed on forest and savanna strains of *O. volvulus*, but significantly more when ingesting blood from infected volunteers from local foci. Other examples of this concentration effect are seen in *S. oyapockense s.l.* in Brazil (125) and probably in *S. haematopotum* in Guatemala (131) as well as in ticks ingesting microfilariae from infected lizards (94) and some phlebotomine vectors of *Leishmania* (135). Xenodiagnosis of onchocerciasis where low filarial densities occur based on the concentration effect has been used in East Africa but the method was never refined (143). Attempts to use *S. oyapockense s.l.* in Brazil to detect low-density infections of *O. volvulus* in man were curtailed because 2.5% of the wild flies used had partially fed on other individuals before obtaining their full blood meal and reared female flies only reluctantly fed on man (125). Immunodiagnostic techniques currently being developed will provide a sensitive test for low-density infections not detected by skin biopsies and thus obviate the need for xenodiagnosis.

Development of Parasite in the Vector and Transfer to the Definitive Host

Microfilariae ingested with blood during the infective feed pass over the cibarium in the head of the fly and continue to the mid gut. In the gut two factors affect the success of their future development, mechanical damage by the cibarium and mechanical obstruction by the peritrophic membrane of the mid gut. Some vectors of *O. volvulus* in Latin America possess rows of sharp teeth on the cibarium that lacerate the parasite and prevent its further development (99, 126). The exact epidemiological importance of these teeth has not been quantified because transmission models like those used for *S. damnosum s.l.* in Africa (106) have not yet been developed due to the paucity of information on vector bionomics. One authority has dismissed this factor as being epidemiologically insignificant (28) but an examination of experimental infection data for vector species in Latin America clearly shows its importance (121). These species may be divided into two groups according to the presence or absence of cibarial teeth. In those with an armed cibarium (*S. haematopotum*, *S. ochraceum s.l.*, *S. oyapockense s.l.*, *S. quadrivittatum*) less than 2% of ingested microfilariae develop to the infective L3 stage because of damage by the cibarial teeth while up to 60% develop in species with unarmed cibaria (*S. callidum*, *S. exiguum s.l.*, *S. guianense s.l.*, *S. metallicum s.l.*). Also actual numbers of L3 larvae developing from microfilariae are usually higher in species with unarmed cibaria. Cibarial armatures have already been shown to have a similarly lethal effect on *Dirofilaria repens*, *Brugia pahangi*, and *Wuchereria bancrofti* transmitted by species of *Anopheles* and *Culex*, and the absence of teeth in vector *Aedes* in Pacific Islands explains the continued transmission of Bancroftian filariasis in areas of mass DEC chemotherapy (21, 81).

The parasites then pass down the alimentary canal to the mid-gut, where digestive enzymes as well as the chitinous peritrophic membrane are secreted, and become very active in an effort to penetrate the stomach wall and reach the fly's hemocoel. A major constraint here is the morphology and speed of formation of the peritrophic membrane, which once hardened acts as an impenetrable barrier to microfilariae that then disentegrate in the gut (113, 114). The peritrophic membrane forms initially within 30 min of feeding and subsequent layers form during 24 hr. The form of the membrane varies with species being thin and delicate in *S. neavei s.l.* and *S. metallicum s.l.*, thick with up to 15 laminae in *S. damnosum s.l.*, and even variable within a species complex. The presence of higher numbers of infective larvae in forest (*S. yahense*, *S. soubrense*, *S. squamosum*, *S. mengense*) than in savanna cytospecies (*S. damnosum s.s.*, *S. sirbanum*) of the *S. damnosum* complex in West Africa is due to a thicker membrane being present in the former (46, 50, 105). The peritrophic membrane severely restricts migration of microfilariae to the hemocoel with up to 93% of *O. volvulus* in *S. damnosum s.l.* and 75% of *O. lienalis* (as *O. gutturosa*) in *S. ornatum s.l.* being trapped in the blood meal (113). Successful penetration of the stomach by microfilariae therefore has to occur soon after their entry into the gut. In *S. damnosum s.l.* 40–50% of *O. volvulus* have migrated within an hour of feeding (71), while similar rapid migrations have been seen in *O. lienalis* in *S. ornatum s.l.* (49) and *S. fallisensis* in *S. rugglesi* and *S. anatinum* (2). Some loss of *O. volvulus* microfilariae occurs from the fly's gut in drops of fluid expelled through the anus when *S. damnosum s.l.* feeds (75) as occurs in mosquitoes (68), but this phenomenon probably has little epidemiological importance because of the small numbers of parasites involved.

Having penetrated the stomach wall and entered the insect's hemocoel the microfilariae migrate to the thoracic musculature, although in heavy infections they may invade other parts of the body such as the legs and halteres within the first 24 hr of feeding. Development to the L3 stage then occurs if the parasite has been ingested by the appropriate vector. Different degrees of adaptation between the parasite and its vector were first observed 30 years ago in savanna and forested areas of West Africa (47) and later confirmed using experimentally infected flies and inoculations of microfilariae into wild flies (50, 64, 105, 112, 139). These cross-transmission experiments have shown significant differences in the capacity of vector cytospecies to support the development of each parasite strain. It initially appeared that forest strains of *O. volvulus* develop well in forest cytospecies (*S. yahense*) to a variable extent in savanna–forest mosaic cytospecies (*S. sanctipauli*, *S. squamosum*, *S. soubrense*) and poorly in savanna cytospecies (*S. damnosum, S. sirbanum*) and similar degrees of development were thought to occur in savanna strains of the parasite in relation to the vector cytospecies. However, in other areas it is now known that both forest and savanna strains of the parasite inoculated into the vector will develop in savanna populations (*S. damnosum s.s.*, *S. sirbanum*) and savanna–forest mosaic populations (*S. squamosum*) and rainforest populations (*S. damnosum s.s.*, *S. squamosum*, *S. mengense*). These latter findings suggest that the peritrophic membrane rather than factors in the hemolymph or muscles limits development of microfilaria to L3s in naturally infected flies. More work is required to elucidate the parasite–vector interaction in West Africa.

Incompatibility between parasite and vector also occurs in Latin America. Forest and savanna strains of *O. volvulus* from West Africa either developed poorly or failed to develop in Guatemalan (*S. ochraceum s.l.*, *S. metallicum s.l.*, *S. callidum*) and Venezuelan (*S. metallicum s.l.*, *S. exiguum s.l.*) vectors (42, 74) and few microfilariae of the Guatemalan form of the parasite fully developed in savanna and forest populations of *S. damnosum s.l.* (48). Differences may also exist between the Guatemalan and northern Venezuelan forms of the parasite but this still needs confirmation (119). Different vector species within the same disease focus also show varying capacities to act as hosts to *O. volvulus*. In Guatemala synchronous development of *O. volvulus* to the L3 stage occurs in *S. ochraceum s.l.* but development in *S. callidum*, *S. haematopotum*, and *S. metallicum s.l.* is asynchronous, with many parasites degenerating or remaining in the L1 stage because of physiological incompatibility rather than overcrowding (19, 69, 74, 131). Differential asynchrony of parasite development was also seen in *S. metallicum s.l.* and *S. exiguum s.l.* in Venezuela (42) and is one of the more important factors that contribute to the higher host capacity of the former vector species. The synchrony of parasite development also varies within species complexes. The Cayapa cytospecies of *S. exiguum* is a highly efficient host to *O. volvulus* in Ecuador with most microfilariae developing to the L3 stage in 5–8 days (122). In Colombia, where a different vector cytospecies probably occurs, the vector is less efficient because of a partially asynchronous cycle (126, 134). In Venezuela the increased asynchrony of the parasite's development cycle (42) means that this cytospecies (as yet unstudied) probably only sporadically transmits the parasite. Similarly, *S. metallicum* in Mexico and Guatemala, where cytospecies A, B, H, I, X occur, is a poor vector compared to northern Venezuela where cytospecies D and E are probably the vectors (22, 121).

Host susceptibility to filarial parasites in relation to vector strain differences has been well studied in mosquitoes (93) but the difficulties experienced in colonizing blackflies and the lack of an adequate laboratory animal to infect with *O. volvulus* has severely limited development of genetical studies (7). However, the use of the cattle filaria *O. lienalis* in experimentally infected natural and surrogate hosts has provided information that can be applied to human onchocerciasis (7, 62). Gene expression in relation to metabolism of *O. lienalis* larvae developing in *S. ornatum s.l.* is a promising new line of research (7).

Temperature has an important influence on filarial development in the vector, with no development occurring in *O. volvulus* below 18°C (94, 96). The optimal time for development in different simuliid vectors also varies with temperature; at 21°C infective larvae appeared after 1 week in *S. neavei s.s.* but failed to develop in *S. damnosum s.l.*, whereas in *S. woodi* of the *S. neavei* group fastest development occurred at 24°C. Temperature variation in relation to altitude was also seen to influence filarial development in *S. ochraceum s.l.* in Guatemala. Here increasing altitude and the consequent drop in temperature reduce transmission because the parasite develops to the L3 stage more slowly, the gonotrophic cycle is slower, and hence biting frequencies are reduced and fly populations build up more slowly (119). Comprehensive studies have been carried out only on *O. volvulus* but studies on the development of *M. ozzardi* in simuliids show that a similar temperature related development cycle occurs (145).

When fully developed the L3 larvae of some filariae congregate in the abdomen only moving toward the head when blood feeding begins while in *O. volvulus* they migrate to the proboscis where their presence may cause the fly to feed (93). The route by which infective larvae leave the vector and the stimuli triggering this process are not clearly understood. Infective larvae have been reported to leave the fly by rupturing the labellum (8), but other studies were unable to confirm this (96). Only 40% of infective flies lose larvae when first feeding and hence most of the remaining 60% remain infective for their next feed in forest *damnosum* and only 80% of the infective larvae leave each fly at the first feed (43).

Host–Vector Contact

For a species to be able to transmit filariae it must survive for a long enough period to pick up the parasite, allow its full development, and then its subsequent transmission to the definitive host. After egg laying the fly returns for a further blood meal and the next gonotrophic cycle commences. The time taken for gonotrophic cycles is therefore important as it is this that determines contact between fly and definitive host. In *S. damnosum s.l.* times between blood meals vary from 3½–6 days (40, 72, 105) and in *S. ochraceum s.l.* and *S. metallicum s.l.* 2–4 days (58). Transmission success of the L3 larva now depends on contact between the definitive host and the infective vector, which is influenced by the distribution and biting habits of the vector.

Geographic distribution of vectors is a major limiting factor on the transmission and dispersal of filarial infections. The presence of anthropophilic cytospecies of *S. damnosum s.l.* over extensive areas in West Africa has resulted in large endemic areas of onchocerciasis, whereas in East Africa the discrete foci of the disease are a direct result of the limited distribution of *S. damnosum s.l.* and members of the *S. neavei* group. In Latin America the disease is entirely focal, either because of the limited distributions of efficient vectors (*S. guianense s.l.*), because more widely spread vectors (*S. ochraceum s.l.*, *S. oyapockense s.l.*, *metallicum s.l.*) are inefficient hosts, or because infected individuals are only just starting to migrate from foci in areas where efficient vectors are widely distributed (*S. exiguum s.l.* in Ecuador) (119, 121). Choice of larval breeding grounds greatly affects vector species distribution, with *S. damnosum s.l.* generally preferring rapids in small to large rivers, *S. neavei s.l.* heavily shaded forest streams inhabited by *Potomonautes* crabs, and *S. albivirgulatum* slow flowing reaches of medium and large rivers (28). In Latin America breeding ground preferences vary from trickles (*S. ochraceum s.l.*) to large tributaries of the Amazon (*S. oyapockense s.l.*) (119).

Dispersal of blackflies from their breeding grounds can have a considerable effect on parasite transmission. In Latin America the flight range of three human onchocerciasis vectors (*S. ochraceum s.l.*, *S. metallicum s.l.*, and *S. callidum*) is never greater than 15 km and is of little epidemiological importance (31). However, in West Africa dispersal of the *S. damnosum* cytospecies affects parasite transmission and therefore control strategies (28, 60, 72, 105). *Simulium sirbanum* and *S. damnosum s.s.* occur principally in the savanna but are also present in the adjacent forest zone and can migrate up to 500 km, *S. squamosum* of the savanna forest mosaic area can migrate up to 150 km. The forest vectors are either nonmigratory (*S. yahense*) or will disperse up to 50 km from their breeding grounds (*S. soubrense* and *S. sanctipauli*). During the dry season many rivers in the savanna dry out completely, breeding ceases and biting populations of the two savanna species decrease or disappear. Reinvasion by *S. sirbanum* and *S. damnosum s.s.* along savanna river courses from adjacent and distant breeding grounds in forests to the south is associated with south westerly monsoon winds at the onset of the wet season. Control by larviciding of *S. damnosum s.s.* and *S. sirbanum* in savanna areas in the Onchocerciasis Control Programme of West Africa was highly effective in controlling local populations but invasion of parous and hence potentially infective flies from adjacent uncontrolled areas allowed transmission to continue, albeit at a lower rate.

Transmission of filariae is more effective if large numbers of vector species bite the definitive host. This is dependent not only on the ability of the species to produce large populations but also on its host preferences. In temperate regions species are either univoltine or multivoltine with several generations per year. All tropical vectors of onchocerciasis are multivoltine with up to 15 annual generations and larval development lasting as little as 4 days in the West African savannas. Larval life can be as long as 3–10 weeks in *S. neavei s.l.* and large population sizes are never reached because larvae live phoretically on crabs and food is available only when crabs move into running water (28). A major constraint in tropical countries is water availability rather than temperature, except at high altitudes where low temperatures extend the length of the life cycle and reduce the number of generations. Larval populations generally build up during the rains and adult biting peaks are seen toward the end of the wet season and the beginning of the dry season in Latin American species that breed in both large rivers (*S. exiguum s.l.*, *S. oyapockense s.l.*) and streams (*S. ochraceum s.l.*, *S. metallicum s.l.*) (119). The converse can be true in parts of West Africa where *S. sirbanum* and *S. damnosum s.s.* disappear from some areas of the savanna as rivers dry up into pools.

Blackflies bite by the day and in the open, although *S. ochraceum s.l.* (31), *S. amazonicum*, and *S. pertinax* (Shelley, unpublished data) will bite man in small numbers in houses and at nightfall when biting populations are high. Transmission of filariae occurs mainly in individuals working in fields or on rivers and the enormous numbers of biting *S. oyapockense s.l.* and *S. amazonicum* seen alongside rivers in the Brazilian Amazon immediately disappear a few meters inside the forest. African vectors never bite in the large numbers seen in Latin America. *Simulium damnosum s.l.* has rates of 30–60 per hour in the savanna and 200 in the forest toward the end of the wet season when populations are highest; rates for the *S. neavei* group are generally much lower (28, 72). In Latin America biting rates of up to 1000 flies per hour have been recorded for *S. ochraceum s.l.* in Guatemala (74) and 6780 per day in Brazil for *S. oyapockense s.l.* (126). One of the main vector species of *M. ozzardi*, *S. amazonicum*, also bites in enormous numbers along southern Amazon tributaries such as the R. Purus (Shelley, personal observation).

Considerable diurnal and seasonal variation is seen in biting and great attention has been paid to this for the African onchocerciasis vector *S. damnosum s.l.* (28, 105). Innate diurnal rhythms are affected by many physical factors, flies preferring to bite in certain temperature ranges with low wind speeds, high humidities, and during morning and afternoon times of increasing and decreasing light intensity. Biting patterns are usually similar from day to day but can be disrupted by climatic changes such as thunderstorms, which depress activity. Individual flies only bite once in their gonotrophic cycle (about 3 days in the tropics) after laying eggs and the age of the fly can influence biting times. Nevertheless variation occurs in conditions preferred by different species. *Simulium callidum* breeding in shaded streams commences biting at a lower temperature, but will avoid direct sunlight during the middle of the day, unlike *S. ochraceum s.l.* (31). In many vector species (*S. damnosum s.l.*, *S. ethiopiense* of the *S. neavei* group), *S. callidum*, *S. guianense s.l.*, *S. oyapockense s.l.* biting peaks generally occur in early morning and late afternoon (28). However, both unimodal and bimodal activity patterns can be seen within the same species, e.g., savanna cytospecies of the *S. damnosum* complex in West Africa (28), *S. exiguum s.l.* and *S. quadrivittatum* in Ecuador (122), and *S. ochraceum s.l.* in Guatemala (104). Seasonal variation in biting occurs in tropical areas with wet and dry seasons and in regions with seasonal temperature extremes (28). Thus in West African savannas breeding and biting by *S. damnosum s.s.* and *S. sirbanum* ceases in the dry season and begins in the wet season as the region is recolonized by flies migrating from the forested areas to the south. In other tropical areas such as the Amazon Basin *S. oyapockense s.l.* bites throughout the year with population peaks occurring at the end of the period when rain is more intense (119). In some areas seasonal biting peaks can vary for different species with the proportions of *S. squamosum*, *S. damnosum s.s.*, and *S. sirbanum* of the *S. damnosum* complex in Togo differing seasonally, largely due to population fluctuations in *S. squamosum* (16). In the hypoendemic lowland areas of the Brazil Amazonia focus three species bite throughout the year with *S. oyapockense s.l.* predominating at the end of the more rainy months, *S. exiguum s.l.* in the least rainy months and *S. guianense s.l.* always in low numbers (Shelley, unpublished). However, seasonal population cycles in vectors have a more pronounced effect on the transmission of some animal filaria species. Transmission of *O. cervipedis* (as *Wehrdikmansia cervipedis*) to Columbian black-tailed deer in California occurs only in spring because the vector *Prosimulium impostor* is univoltine, transmission of *O. tarsicola* in European cervids only during the warmer months of the year (118), while *S. fallisensis* transmission occurs only in the spring and summer months (2). Transmission of cattle *Onchocerca* species is only seasonal in temperate regions during the warmer months when the multivoltine vectors are present (78), while in tropical areas transmission is probably continuous except in savanna areas during the dry season when vector breeding grounds dry up.

Choice of host by the vector is another important factor that affects filarial transmission, being well illustrated by simuliids that transmit human onchocerciasis. In this case the more zoophilic a vector species the greater the chance of infective larvae of *O. volvulus* being passed to an animal rather than human host when feeding. In Guatemala *S. metallicum s.l.* bites man in large numbers and is often the prevalent species in some localities. However, its lower suitability as a host to *O. volvulus* and its pronounced zoophily (31, 104) make it a far less efficient vector than the highly anthropophilic *S. ochraceum s.l.* The zoophilic tendencies of *S. callidum* (31) also reduce its value as a vector to *O. volvulus*. In northern Venezuela the apparent high anthropophily of *S. metallicum s.l.*, together with its suitability as a host to *O. volvulus* relative to the mainly zoophilic *S. exiguum s.l.*, makes it the only significant vector there (42, 76). Differences in degrees of anthropophily may also be seen between sibling species and can determine whether a sibling is a vector. Thirteen of the more than 40 members of the *S. damnosum* complex are vectors, the others being largely zoophilic (26, 28). In Africa high anthropophily is seen in the savanna siblings of the *S. damnosum* complex while some of the forest vectors such as *S. sanctipauli* will significantly bite animals (28, 56, 59). The different vector capacities of *S. exiguum s.l.* in Ecuador, Colombia, and Venezuela is probably partially due to different cytospecies showing different degrees of anthropophily (121). Another feature of the biting habits of vectors such as preferred biting site on the host was dealt with in the "parasite availability in the definitive host" section.

The transmission of filariae has been quantified for *O. volvulus* by relating man–vector contact to vector infectivity rates (41). Biting catches are used to estimate the monthly or annual biting rates of a species and parous females are then dissected for the presence of infective larvae in the head of each fly. From these data the monthly or annual transmission potential, which is the theoretical number of infective larvae carried by flies biting one person during the relevant time period, can be calculated. Annual biting rates and annual transmission potentials (ATPs) can be used for assessing the importance of transmission at different localities and for quantifying the effects of parasite control with ivermectin or vector control through larviciding. In West Africa ATPs of up to 5000 in forest areas where onchocerciasis was hypoendemic were not associated with blindness, whereas in savanna areas blindness occurred when ATPs exceeded 220. The use of ivermectin to kill skin microfilariae of *O. volvulus* and larviciding against *S. damnosum s.s.* and *S. sirbanum* has prevented transmission in many savanna areas (146) with a resultant fall in ATPs. Insufficient data on vector bionomics are available for Latin American species for ATP comparisons to be made.

Effects of Parasitism on Vectors

Parasitism of flies and the consequent reduction in population size can seriously affect filaria transmission. Simuliids are infected by many parasites ranging from unicellular Microsporidians to mermithid nematodes and this subject has been well reviewed (28). Here we are concerned only with the effect that filariae have on adult simuliids and how this affects their transmission. The main effects are on the vector's immunity to infection, and its longevity, dispersal, and fecundity.

Various mechanisms exist to block parasite development in hematophagous insects but encapsulation and melanisation of the parasite do not occur in blackflies (7, 62). Work on filariae and simuliids has focused on the cattle filaria *O. lienalis* in its vectors *S. ornatum s.l.* and *S. erythrocephalum* using inoculation of microfilariae into the vector, as a model for the more important *O. volvulus* parasitic in man. Two mechanisms to combat the filaria probably exist. Once the microfilaria has reached the hemocoel cecropins, lectins and possibly attacins appear in the hemolymph and can reduce the number of microfilariae that will successfully develop to L3 larvae in the thoracic muscles. Immunity to filariae can also be acquired from previous infections since reinfection of the vectors with *O. lienalis* produces a large reduction in development of the second infection and to a lesser extent the development of some larvae from the first infection is affected. The significance of these findings in natural infections needs to be determined especially in areas where filariae of cattle and man use the same simuliid species as a vector.

Simuliids are able to survive quite large filarial loads, 72 normal sized L3 larvae being recorded in single *S. neavei s.s.* in East Africa (96) and 107 *O. lienalis* in *S. ornatum s.l.* (7). However, high microfilarial intakes of *O. volvulus* have been shown to reduce longevity and even result in death of *S. damnosum s.l.* and *S. neavei s.l.*, particularly within 48 hr of the infective blood meal (36–38, 56, 94). Microfilarial intake is often high in Latin American vectors as a result of the concentration effect. The effect that this has on the fly depends on whether the cibarium is armed or not. The damage caused by cibarial teeth to microfilariae in *S. ochraceum s.l.*, *S. oyapockense s.l.*, *S. quadrivittatum*, and *S. haematopotum* (99, 122, 126, 131, 132) considerably reduces the number of live microfilariae that can penetrate the mid gut wall and so protects the fly from death. In *S. guianense s.l.*, *S. metallicum s.l.*, *S. callidum*, and probably *S. exiguum s.l.*, which have unarmed cibaria, no mechanical damage occurs to ingested microfilariae, which can then kill the fly if ingested in large numbers. One or more of the following events may cause mortality: reverse migration of parasites to the anterior tubular mid-gut causing injury and disintegration to the epithelium, extensive damage to stomach epithelial cells releasing gut contents into the hemocoel, damage to the peritrophic membrane causing failure to digest blood, obstruction of the hind gut by parasitized blood, and injury to internal organs of the fly by large numbers of microfilariae thereby affecting the fly's behavior, particularly locomotion and vision (99). Laboratory studies on *S. ochraceum s.l.*, *S. metallicum s.l.*, and *S. callidum* and *S. damnosum s.l.* (20, 36, 37, 74) showed that following high initial mortalities in heavily infected flies short-term survival rates of infected and uninfected flies did not significantly differ unless infected with an incompatible strain of parasite (38). Under natural conditions in Guatemala the longevity of flies is considerably reduced by filarial infection (18, 31, 32, 108) although high natural infectivity and parous rates (58) show that survival of infected females is sufficiently high to allow continuous transmission to occur.

Filarial infection of flies was also believed to reduce flight range of vector species in Guatemala (32) and in *S. ornatum s.l.* infected with *O. lienalis* (24), but more recent work (60) in West Africa has shown that high L3 loads occur in savanna and forest cytospecies of the *S. damnosum* complex that had migrated over considerable distances. Infection with filariae can also reduce fecundity in *S. ornatum s.l.* and *S. lineatum* infected with *O. lienalis* either by inoculation or ingested in a blood meal (63). In nature *S. damnosum s.l.* infected with *O. volvulus* have fewer maturing oocytes than uninfected flies (17).

Effects of Control on Transmission

Direct control of vectors, principally through the use of various larvicides, can eliminate transmission, particularly in large river breeders such as *S. damnosum s.l.* but is less effective for trickle breeding species such as *S. ochraceum s.l.* in Guatemala and Mexico (119). Deforestation, either for agricultural purposes or for control of simuliids, has been effective in reducing populations of *S. neavei* and *S. woodi* of the *S. neavei* group in East Africa because of the drying of some streams and the lack of suitable shading for breeding in others (91). Deforestation in West Africa also affects the breeding grounds of the forest species of the *S. damnosum* complex and the southern encroachment of the savanna increases the distribution of *S. damnosum s.s.* and *S. sirbanum* the vectors of the more pathogenic strain of *O. volvulus* (139). The regular biannual use of ivermectin in Guatemala reduced parasite transmission by *S. ochraceum s.l.* by 80–90% and in Africa the drug had similar effects on transmission by *S. sirbanum*, *S. damnosum s.s.*, *S. yahense*, *S. soubrense*, and *S. sanctipauli* of the *S. damnosum* complex (29). Difficulties in administering ivermectin to infected local Indian tribes and the high vector efficiency of *S. exiguum s.l.* and *S. guianense s.l.* mean that transmission would probably be effectively reduced only through a combination of parasite and vector control campaigns (121). *In vitro* studies and inoculation of ivermectin exposed microfilariae showed that *O. lienalis* is similarly affected by this drug in that varying exposure times to ivermectin directly affect the subsequent development of the parasite in *S. ornatum s.l.* (138).

Conclusions

Considerable progress has been made over the last two decades in understanding the transmission of strains of *O. volvulus* by cytospecies of the *S. damnosum* complex in West Africa. The reliable identification of *Onchocerca* species that infect cattle and that are transmitted by the same vector species as *O. volvulus* is near to being resolved with the development of more sensitive DNA probes, which can already detect parasite strain differences. Speciation within the *S. damnosum* complex in West Africa is now well on the way to being understood, although no single identification method can yet be used that will differentiate all vector cytospecies. The use of these more precise identification methods together with developing work on insect immune systems have allowed further investigation of parasite transmission mechanisms so that the epidemiology and control of onchocerciasis are more fully understood. The success of the microfilaricide ivermectin and larviciding against *S. damnosum s.l.* in West Africa has placed more emphasis on research into human onchocerciasis in Latin America. Advances have recently been made in clarifying speciation in the 12 vector species and in incriminating local vectors. Future work will make use of the techniques developed and experience obtained from West Africa. Research will need to further clarify speciation in vectors with the development of a method for differentiating adult vector cytospecies and their filarial parasites so that transmission studies can be made to obtain a better understanding of the epidemiology of onchocerciasis in the many unique Latin American foci of the disease. Investigations are also needed on the role of simuliids in transmitting animal filariae (particularly in monkeys), which probably occur sympatrically with the human filariae *O. volvulus* and *M. ozzardi*.

References

1. Addison, E.M. 1980. Transmission of *Dirofilaria ursi* Yamaguti, 1941 (Nematoda: Onchocercidae) of black bears (*Ursus americanus*) by blackflies (Simuliidae). *Can. J. Zool.* **58**:1913–1922.
2. Anderson, R.C. 1956. The life cycle and seasonal transmission of *Ornithofilaria fallisensis* Anderson, a parasite of domestic and wild ducks. *Can. J. Zool.* **34**:485–525.
3. Anderson, R.C., and Bain, O. 1976. CIH keys to genera of the order Spirurida, Part 3: Diplotriaenoidea, Aproctoidea and Filarioidea. Commonwealth Agricultural Bureaux, Farnham, Royal Bucks, UK, pp. 59–116.

4. Bain, O. 1968. Morphology of the larval stages of *Onchocerca volvulus* in *Simulium damnosum* and a redescription of the microfilaria. *WHO/ONCHO/68.70*. ii.1–8 (unpublished mimeographed document).
5. Bain, O., and Chabaud, A.G. 1986. Atlas des larves infestantes de filaires. *Trop. Med. Parasitol.* **37**:237–340.
6. Basáñez, M.G., Yarzabal, L., Takaoka, H., Suzuki, H., Noda, S., and Tada, I. 1988. The vectoral role of several blackfly species (Diptera: Simuliidae) in relation to human onchocerciasis in the Sierra Parima and Upper Orinoco regions of Venezuela. *Ann. Trop. Med. Parasitol.* **82**:597–611.
7. Bianco, A.E. 1991. Onchocerciasis. In C.N.L. Macpherson and P.S. Craig, eds., *Parasitic Helminths and Zoonoses in Africa*. Unwin Hyman, London, pp. 138–203.
8. Blacklock, D.B. 1926. The development of *Onchocerca volvulus* in *Simulium damnosum*. *Ann. Trop. Med. Parasitol.* **20**:1–48.
9. Botto, C., Escalante, A., Arango, M., and Yarzabal, L. 1988. Morphological differences between Venezuelan and African microfilariae of *Onchocerca volvulus*. *J. Helminth.* **62**:345–351.
10. Campbell, C.C., Collins, R.C., Huong, A.Y., and Figueroa Marroquin, H. 1980. Quantitative aspects of the infection of *Simulium ochraceum* by *Onchocerca volvulus*: The relation of skin microfilarial density to vector infection. *Tropenmed. Parasitol.* **31**:475–478.
11. Chabaud, A.G. 1954. Sur le cycle évolutif des spirurides et de nématodes ayant una biologie comparable. Valeur systématique des caractères biologiques. *Ann. Parasitol.* **29**:42–88, 206–249, 358–425.
12. Chabaud, A.G. 1965. Superfamille des Filarioidea, [Systematique des Nematodes]. *Traite Zool.* **4**:1102–1151.
13. Chabaud, A.G. 1974. CIH keys to the nematode parasites of vertebrates, No. 1: General introduction, Class Nematoda, Keys to subclasses, orders and superfamilies. Commonwealth Agricultural Bureaux, Farnham, Royal Bucks, UK, pp. 6–17.
14. Chabaud, A.G., and Anderson, R.C. 1959. Nouvel essai de classification des filaires (Superfamille des Filarioidea) II. *Ann. Parasitol.* **34**:64–87.
15. Charalambous, M., Ready, P.D.R., Shelley, A.J., Arzube, M., and Lowry, C.A. 1993. Cytological and isoenzyme analysis of the Bucay and Quevedo cytotypes of the onchocerciasis vector *Simulium exiguum* (Diptera: Simuliidae) in Ecuador. *Mems. Inst. Oswaldo Cruz.* **88**:39–48.
16. Cheke, R.A., and Garms, R. 1983. Reinfestations of the south-western flank of the Onchocerciasis Control Programme area by wind borne vectors. *Phil. Trans. R. Soc. Ser. B* **302**:471–484.
17. Cheke, R.A., Garms, R., and Kerner, M. 1982. The fecundity of *Simulium damnosum* s.l. in northern Togo and infections with *Onchocerca* spp. *Ann. Trop. Med. Parasitol.* **76**:561–568.
18. Collins, R.C. 1979. Onchocerciasis transmission potentials of four species of Guatemalan Simuliidae. *Am. J. Trop. Med. Hyg.* **28**:72–75.
19. Collins, R.C. 1979. Development of *Onchocerca volvulus* in *Simulium ochraceum* and *Simulium metallicum*. *Am. J. Trop. Med. Hyg.* **28**:491–495.
20. Collins, R.C., Campbell, C.C., Wilton, D.P., and Newton, L. 1977. Quantitative aspects of the infection of *Simulium ochraceum by Onchocerca volvulus. Tropenmed. Parasitol.* **28**:235–243.

21. Coluzzi, M., and Trabucchi, R. 1968. Importanza dell'armatura bucco-faringea in *Anopheles* e *Culex* in relazione alle infezioni con *Dirofilaria*. *Parassitologia* **10**:47–59.
22. Conn, J. 1988. A cytological study of the *Simulium metallicum* complex (Diptera: Simuliidae) from Central and South America. In M.W. Service, ed., *Biosystematics of Haematophagous Insects*. Clarendon Press, Oxford, pp. 221–243.
23. Coombs, I., and Crompton, D.W.T. 1991. *A Guide to Human Helminths*. Taylor and Francis, London.
24. Cooter, R.J. 1982. Studies on the flight of blackflies (Diptera: Simuliidae). 1. Flight performance of *Simulium ornatum* Meigen. *Bull. Entomol. Res.* **72**:303–317.
25. Crosskey, R.W. 1981. Geographical distribution of Simuliidae. In M. Laird, ed., *Blackflies: The Future for Biological Methods in Integrated Control*. Academic Press, London, pp. 57–68.
26. Crosskey, R.W. 1987. A taxa summary for the *Simulium damnosum* complex, with special reference to distribution outside the control areas of West Africa. *Ann. Trop. Med. Parasitol.* **81**:181–192.
27. Crosskey, R.W. 1987. An annotated checklist of the world black flies (Diptera: Simuliidae). In K.C. Kim and R.W. Merritt, eds., *Black Flies: Ecology, Population Management, and Annotated World List*. Pennsylvania State University, University Park, pp. 425–520.
28. Crosskey, R.W. 1990. *The Natural History of Blackflies*. John Wiley, Chichester.
29. Cupp, E.C. 1992. Treatment of onchocerciasis with ivermectin in Central America. *Parasitol. Today* **8**:212–214.
30. Cupp, E.W., Bernardo. M.J., Kiszewski, A.E., Collins, R.C., Taylor, H.R., Aziz, M.A., and Greene, B.M. 1986. The effects of ivermectin on transmission of *O. volvulus*. *Science* **231**:740–742.
31. Dalmat, H.T. 1955. The black flies (Diptera, Simuliidae) of Guatemala and their role as vectors of onchocerciasis. *Smithsonian. Misc. Coll.* **125**:vii–425.
32. Dalmat, H.T., and Gibson, C.L. 1952. A study of flight ranges and longevity of black flies (Diptera, Simuliidae) infected with *Onchocerca volvulus*. *Ann. Entomol. Soc. Am.* **45**:605–612.
33. Denke, A.M., and Bain, O. 1978. Données sur le cycle d'*Onchocerca ochengi* chez *Simulium damnosum* s.l. au Togo. *Ann. Parasitol.* **53**:757–760.
34. Dissanayake, S., and Piessens, W.F. 1992. Identification of filarial larvae in vectors by hybridization. *Parasitol. Today* **8**:67–69.
35. Dohnal, J., Blinn, J., Wahl, G., and Schulz-Key, H. 1990. Distribution of microfilariae of *Onchocerca lienalis* and *Onchocerca gutturosa* in the skin of cattle in Germany and their development in *Simulium ornatum* and *Culicoides nubeculosus* following artificial infestation. *Vet. Parasitol.* **36**: 325–332.
36. Duke, B.O.L. 1962. Studies on the factors influencing the transmission of onchocerciasis. I: The survival rate of *Simulium damnosum* under laboratory conditions and the effect upon it of *Onchocerca volvulus*. *Ann. Trop. Med. Parasitol.* **56**:130–135.
37. Duke, B.O.L. 1962. Studies on the factors influencing the transmission of onchocerciasis. II: The intake of *Onchocerca volvulus* microfilariae by

Simulium damnosum and the survival of the parasites in the fly under laboratory conditions. *Ann. Trop. Med. Parasitol.* **56**:255–263.

38. Duke, B.O.L. 1966. *Onchocerca-Simulium* complexes. III: The survival of *Simulium damnosum* after high intakes of microfilariae of incompatible strains of *Onchocerca volvulus*, and the survival of the parasites in the fly. *Ann. Trop. Med. Parasitol.* **60**:495–500.
39. Duke, B.O.L. 1967. Infective filaria larvae, other than *Onchocerca volvulus* in *Simulium damnosum. Ann. Trop. Med. Parasitol.* **61**:200–205.
40. Duke, B.O.L. 1968. Studies on factors influencing the transmission of onchocerciasis. V: The stages of *Onchocerca volvulus* in wild "forest" *Simulium damnosum*, the fate of the parasite in the fly and the age distribution of the biting population. *Ann. Trop. Med. Parasitol.* **62**:107–116.
41. Duke, B.O.L. 1968. Studies on factors influencing the transmission of onchocerciasis. IV: The biting-cycles, infective biting, density and transmission potential of "forest" *Simulium damnosum. Ann. Trop. Med. Parasitol.* **62**:95–106.
42. Duke, B.O.L. 1970. *Onchocerca-Simulium* complexes. VI: Experimental studies on the transmission of Venezuelan and West African strains of *Onchocerca volvulus* by *Simulium metallicum* and *S. exiguum* in Venezuela. *Ann. Trop. Med. Parasitol.* **64**:421–431.
43. Duke, B.O.L. 1973. Studies on factors influencing the transmission of onchocerciasis. VIII: The escape of infective *Onchocerca volvulus* larvae from feeding "forest" *Simulium damnosum. Ann. Trop. Med. Hyg.* **67**: 95–99.
44. Duke, B.O.L. 1990. Human onchocerciasis—an overview of the disease. *Acta Leidensia* **59**:9–24.
45. Duke, B.O.L., and Beesley, W.N. 1958. The vertical distribution of *Simulium damnosum* bites on the human body. *Ann. Trop. Med. Parasitol.* **52**:274–281.
46. Duke, B.O.L., and Lewis, D.J. 1964. Studies on factors influencing the transmission of onchocerciasis. II: Observations on the effect of the peritrophic membrane in limiting the development of *Onchocerca volvulus* microfilariae in *Simulium damnosum. Ann. Trop. Med. Parasitol.* **58**: 83–88.
47. Duke, B.O.L., Lewis, D.J., and Moore, P.J. 1966. *Onchocerca-Simulium* complexes. 1: Transmission of forest and Sudan-savanna strains of *Onchocerca volvulus*, from Cameroon, by *Simulium damnosum* from various West African bioclimatic zones. *Ann. Trop. Med. Parasitol.* **60**:318–336.
48. Duke, B.O.L., Moore, P.J., and de Leon, J.R. 1967. *Onchocerca-Simulium* complexes. V: The intake and subsequent fate of microfilariae of a Guatemalan strain of *Onchocerca volvulus* in forest and Sudan-savanna forms of West African *Simulium damnosum. Ann. Trop. Med. Parasitol.* **61**:332–337.
49. Eichler, D.A. 1973. Studies on *Onchocerca gutturosa* (Neumann, 1910) and its development in *Simulium ornatum* (Meigen, 1818). *J. Helminth.* **47**:73–88.
50. Eichner, M., Renz, A., Wahl, G., and Enyong, P. 1991. Development of *Onchocerca volvulus* microfilariae injected into *Simulium* species from Cameroon. *Med. Vet. Entomol.* **5**:293–297.

51. Erttmann, K.D., Meredith, S.E.O., Greene, B.M., and Unnasch, T.R. 1990. Isolation and characterization of form specific DNA sequences of *O. volvulus*. *Acta Leidensia* **59**:253–260.
52. Flockhart, H.A. 1982. The identification of some *Onchocerca* spp. of cattle by isoenzyme analysis. *Tropenmed. Parasitol.* **33**:51–56.
53. Flockhart, H.A., Cibulskis, R.E., Karam, M., and Albiez, E.J. 1986. *Onchocerca volvulus*: Enzyme polymorphism in relation to the differentiation of forest and savanna strains of this parasite. *Trans. R. Soc. Trop. Med. Hyg.* **80**:285–292.
54. Franz, M., and Renz, A. 1980. Scanning electron microscope studies of infective filaria larvae of Type D and *Onchocerca volvulus*. *Tropenmed. Parasitol.* **31**:31–33.
55. Fuglsang, H., and Anderson, J. 1976. Seasonal variation in the concentration of *Onchocerca volvulus* microfilariae in the skin? *Tropenmed. Parasitol.* **24**:365–369.
56. Garms, R. 1973. Quantitative studies on the transmission of *Onchocerca volvulus* by *Simulium damnosum* in the Bong Range, Liberia. *Tropenmed. Parasitol.* **24**:358–372.
57. Garms, R. 1985. Morphological differentiation from *Onchocerca volvulus* of first stage larvae of an unknown filarial species commonly found in *Simulium sanctipauli* s.l. in Liberia. *Trop. Med. Parasitol.* **36**:255–256.
58. Garms, R., and Ochoa, J.O. 1979. Further studies on the relative importance of Guatemalan blackfly species as vectors of *Onchocerca volvulus*. *Tropenmed. Parasitol.* **30**:120–128.
59. Garms, R., and Voelker, J. 1969. Unknown filarial larvae and zoophily in *Simulium damnosum* in Liberia. *Trans. R. Soc. Trop. Med. Hyg.* **63**:676–677.
60. Garms, R., and Walsh, J.F. 1987. The migration and dispersal of blackflies: *Simulium damnosum* s.l., the main vector of human onchocerciasis. In K.C. Kim and R.W. Merritt, eds., *Black Flies: Ecology, Population Management, and Annotated World List*. Pennsylvania State University, University Park, pp. 201–214.
61. Greene, B. 1992. Ivermectin and immunodiagnosis are filariasis expert's top priorities. *TDR News* **39**:3.
62. Ham, P. 1992. Immunity of haematophagous insect vectors to parasitic infection. *Adv. Dis. Vect. Res.* **9**:101–149.
63. Ham, P., and Banya, A.J. 1984. The effect of experimental *Onchocerca* infections on the fecundity and oviposition of laboratory reared *Simulium* sp. (Diptera, Simuliidae). *Tropenmed. Parasitol.* **35**:61–66.
64. Ham, P.J., and Garms, R. 1985. Development of forest *Onchocerca volvulus* in *Simulium yahense* and *Simulium sanctipauli* following intrathoracic injection and ingestion of microfilariae. *Trop. Med. Parasitol.* **36** (Suppl. 1):25.
65. Ham, P.J., and Garms, R. 1987. Failure of *Onchocerca gutturosa* to develop in *Simulium soubrense* and *Simulium yahense* from Liberia. *Trop. Med. Parasitol.* **38**:135–136.
66. Hashiguchi, Y., Kawabata, M., Ito, S., and Recinos, M.M. 1981. Limited fly load and development of *Onchocerca volvulus* microfilariae in Guatemalan *Simulium ochraceum*. *J. Helminth.* **55**:189–196.

67. Hashiguchi, Y., Kawabata, M., Tanaka, I., Okazawa, T., Flores C.O., and Recinos, M.M. 1981. Seasonal variation in the microfilarial skin density of *Onchocerca volvulus* and in the biting activity of *Simulium* spp. in Guatemala. *Trans. R. Soc. Trop. Med. Hyg.* **75**:839–845.
68. Hawking, F., and Worms, M. 1961. Transmission of filarioid nematodes. *Ann. Rev. Entomol.* **6**:413–432.
69. Ito, S., Tanaka, I., and Ochoa, J.O. 1980. Comparative studies on the affinities of two black flies, *Simulium metallicum* and *S. ochraceum* for the larvae of *Onchocerca volvulus* in Guatemala. *Jpn. J. Sanit. Zool.* **31**:261–270.
70. Katz, S.P., Raccurt, C.P., Lowrie, R.C., Boncy, J., and Leiva, L.M. 1986. *Mansonella ozzardi* in Haiti. IV: Evaluation of antibody reactivity to heterologous antigens. *Am. J. Trop. Med. Hyg.* **35**:303–307.
71. Laurence, B.R. 1966. Intake and migration of the microfilariae of *Onchocerca volvulus* (Leuckart) in *Simulium damnosum* Theobald. *J. Helminth.* **40**:337–342.
72. Le Berre, R. 1966. Contribution a l'étude biologique et ecologique de *Simulium damnosum* Theobald, 1903 (Diptera, Simuliidae). *Mém. O.R.S.T.O.M.* **17**:xv.1–204.
73. Lavoipierre, M.M.J. 1958. Studies on the host-parasite relationships of filarial nematodes and their arthropod hosts. II: The arthropod as a host to the nematode: A brief appraisal of our present knowledge, based on a study of the more important literature from 1878 to 1957. *Ann. Trop. Med. Parasitol.* **52**:326–345.
74. Leon, J.R., and Duke, B.O.L. 1966. Experimental studies on the transmission of Guatemalan and West African strains of *Onchocerca volvulus* by *Simulium ochraceum*, *S. metallicum* and *S. callidum*. *Trans. Soc. Trop. Med. Hyg.* **60**:735–752.
75. Lewis, D.J. 1953. *Simulium damnosum* and its relation to onchocerciasis in the Anglo-Egyptian Sudan. *Bull. Entomol. Res.* **43**:597–644.
76. Lewis, D.J., and Ibañez de Aldecoa, R. 1962. Simuliidae and their relation to human onchocerciasis in northern Venezuela. Bull. W.H.O. **27**: 449–464.
77. Lobos, E.N., and Weiss, N. 1985. Immunological comparison between worm extracts of *Onchocerca volvulus* from savanna and rain forest. *Parasite Immun.* **7**:333–347.
78. Lok, J.B., Cupp, E.W., and Bernardo, M.J. 1983. *Simulium jenningsi* Malloch (Diptera:Simuliidae): A vector of *Onchocerca lienalis* Stiles (Nematoda:Filarioidea) in New York. *Am. J. Vet. Res.* **44**:2355–2358.
79. Mackenzie, C.D., Rivas-Alcala, A.R., Flockhart, H.A., and Nelson, G.S. 1982. Morphological and isoenzyme patterns of an *Onchocerca* sp. found in cattle in Southern Mexico. *Tropenmed. Parasitol.* **33**:181–187.
80. McCall, P.J., Townson, H., and Trees, A.J. 1992. Morphometric differentiation of *Onchocerca volvulus* and *O. ochengi* infective larvae. *Trans. R. Soc. Trop. Med. Hyg.* **86**:63–65.
81. McGreevy, P.B., Bryan, J.H., Oothuman, P., and Kolstrup, N. 1978. The lethal effects of the cibarial and pharyngeal armatures of mosquitoes on microfilariae. *Trans. R. Soc. Trop. Med. Hyg.* **72**:361–368.

82. Meredith, S.E.O., Unnasch, T.R., Karam, M., Piessens, W.F., and Wirth, D.F. 1989. Cloning and characterization of an *Onchocerca volvulus* specific DNA sequence. *Mol. Biochem. Parasitol.* **36**:1–10.
83. Mikhailyuk, A.P. 1967. Study of the causative agent of onchocerciasis in cattle in the forest steppe zone of the Ukraine. *Veterinariya* **11**:62–67.
84. Miller, K.M., Lustigman, S., Brotman, B., Andrus, L., Weiss, N., Hotze, C., and Prince, A.M. 1991. *Onchocerca volvulus*: Characterization of monoclonal antibodies reactive with surface components of third stage larvae. *Trop. Med. Parasitol.* **42**:85–90.
85. Millest, A.L. 1992. Identification of members of the *Simulium ochraceum* species complex in the three onchocerciasis foci in Mexico. *Med. Vet. Entomol.* **6**:23–28.
86. Moraes, M.A.P. 1976. *Mansonella ozzardi* microfilariae in skin snips. *Trans. R. Soc. Trop. Med. Hyg.* **70**:16.
87. Moraes, M.A.P. 1985. Oncocercose no Brasil. In La oncocercosis en America. *CAICET Publ. Cientifica* **3**:15–21.
88. Muller, R. 1975. *Worms and Disease. A Manual of Medical Helminthology.* William Heinemann Medical Books Limited, London.
89. Muller, R. 1979. Identification of *Onchocerca.* In A.E.R. Taylor and R. Muller, eds. *Problems in the Identification of Parasites and Their Vectors*, 17th Symposium of the British Society for Parasitology. Blackwell Scientific Publications, Oxford, pp. 175–206.
90. Muller, R. 1983. Species recognition in human filarioids. In A.R. Stone, H.M. Platt, and L.F. Khalil, eds., *Concepts in Nematode Systematics.* Academic Press, London, pp. 339–349.
91. Muro, A.S., and Raybould, J.N. 1990. Population decline of *Simulium woodi* and reduced onchocerciasis transmission at Amani, Tanzania, in relation to deforestation. *Acta Leidensia* **59**:153–159.
92. Mwaiko, G.L. 1981. The development of *Onchocerca gutturosa* Neuman to infective stage in Simulium vorax Pomeroy. *Tropenmed. Parasitol.* **32**:276–277.
93. Nelson, G.S. 1963. Factors influencing the development and behaviour of filarial nematodes in their arthropodan hosts. In A.E.R. Taylor, ed., *Host Parasite Relationships in Invertebrate Hosts*, Second Symposium of the British Society for Parasitology. Blackwell Scientific Publications, Oxford, pp. 75–119.
94. Nelson, G.S. 1970. Onchocerciasis. *Adv. Parasitol.* **8**:173–224.
95. Nelson, G.S. 1991. Human onchocerciasis: Notes on the history, the parasite and the life cycle. *Ann. Trop. Med. Parasitol.* **85**:83–95.
96. Nelson, G.S., and Pester, F.R.N. 1962. The identification of infective filarial larvae in Simuliidae. *Bull. W. H. O.* **27**:473–481.
97. Omar, M.S. 1978. Histochemical enzyme-staining patterns of *Onchocerca volvulus* microfilariae and their occurrence in different onchocerciasis areas. *Tropenmed. Parasitol.* **29**:462–472.
98. Omar, M.S. 1984. Some current and new techniques for differential identification of filarial parasites. In B.N. Newton and F. Midal, eds., *Tropical Disease Research Series No. 5., New Approaches to the Identification of Parasites and Their Vectors*. Schwabe, and Basel, pp. 361–380.

99. Omar, M.S., and Garms, R. 1975. The fate and migration of microfilariae of a Guatemalan strain of *Onchocerca volvulus* in *S. ochraceum* and *S. metallicum*, and the role of the buccopharyngeal armature in the destruction of microfilariae. *Tropenmed. Parasitol.* **26**:183–190.
100. Omar, M.S., and Garms, R. 1981. Histochemical differentiation of filarial larvae found in *Simulium damnosum* s.l. in West Africa. *Tropenmed, Parasitol* **32**:259–264.
101. Omar, M.S., and Kuhlow, F. 1978. Histochemical differentiation of filarial larvae developing in *Simulium damnosum* in West Africa. *WHO/ONCHO/78* **144**:1–11 (mimeographed document).
102. Omar, M.S., Denke, A.M., and Raybould, J.N. 1979. The development of *Onchocerca ochengi* (Nematoda: Filarioidea) to the infective stage in *Simulium damnosum* s.l. with a note on the histochemical staining of the parasite. *Tropenmed. Parasitol.* **30**:157–162.
103. Omar, M.S., Prost, T.F., and de C., Marshall, T.F. 1982. Histochemical enzyme variation in *Onchocerca volvulus* microfilariae from rain forest and Sudan-savanna areas of the Onchocerciasis Control Programme in West Africa. *Bull. W.H.O.* **60**:933–944.
104. Onofre Ochoa, J. 1982. Studies on the anthropophilic blackfly species in Guatemala, with special reference to the transmission of onchocerciasis in the southeastern endemic area. *Jpn. J. Sanit. Zool.* **33**:129–138.
105. Philippon, B. 1977. Etudé de la transmission d'*Onchocerca volvulus* (Leuckart, 1893) (Nematoda, Onchocercidae) para *Simulium damnosum* Theobald, 1903 (Diptera, Simuliidae) en Afrique tropicale. *Trav. Docum. O.R.S.T.O.M. No.* **63**:1–308.
106. Plaisier, A.P., van Oortmarssen, G.J., Remme, J., and Habbema, J.D.F. 1991. The reproductive lifespan of *Onchocerca volvulus* in West African savanna. *Acta Trop.* **48**:271–284.
107. Pledger, D.J., Samuel, M.W., and Craig, D.A. 1980. Black flies (Diptera, Simuliidae) as possible vectors of legworm (*Onchocerca cervipedis*) in moose of central Alberta. *Proc. N. Am Moose Conf. Workshop* **16**:171–202.
108. Porter, C.H., and Collins, R.C. 1985. The gonotrophic cycle of wild *Simulium ochraceum* and the associated development of *Onchocerca volvulus. Am. J. Trop. Med. Hyg.* **34**:302–309.
109. Post, R.J., and Crampton, J.M. 1988. The taxonomic use of variation in repetitive DNA sequences in the *Simulium damnosum* complex. In M.W. Service, ed., *Biosystematics of Haematophagous Insects*. Clarendon Press, Oxford, pp. 245–256.
110. Post, R.J., Murray, K.A., Flook. P., and Millest, A.L. 1991. Molecular taxonomy in the control of West African onchocerciasis. In G.M. Hewitt, et al., eds., *Molecular Techniques in taxonomy*, NATO ASI Series Vol H57. Springer-Verlag, Berlin pp. 271–281.
111. Procunier, W.S. 1989. Cytological approaches to simuliid biosystematics in relation to the epidemiology and control of human onchocerciasis. *Genome* **32**:559–569.
112. Quillévéré, D., Pendriez, B., Sechan, Y., and Philippon, B. 1977. Etudé du complexe *Simulium damnosum* en Afrique de l'ouest. VII: Etudé de la bioecologie et du pouvoir vecteur des femelles *S. sanctipauli*, *S. soubrense* et *S. yahense* en Côte d'Ivoire. *Cah. O.R.S.T.O.M. Sér. Entomol. Med. Parasitol.* **15**:301–329.

113. Reid, G.D.F., and Lehane, M.J. 1984. Peritrophic membrane formation in three temperate simuliids, *Simulium ornatum*, *S. equinum* and *S. lineatum*, with respect to the migration of onchocercal microfilariae. *Ann. Trop. Med. Parasitol.* **78**:527–539.
114. Richards, A.G., and Richards, P.A. 1977. The peritrophic membrane of insects. *Annu. Rev. Entomol.* **22**:219–240.
115. Schacher, J.F. 1973. Laboratory models in filariasis: A review of filarial life-cycle patterns. *SE. Asian J. Trop. Med. Publ. Hlth.* **4**:337–349.
116. Schibel, J.M., and Renz, A. 1990. Differentiation of non-*Onchocerca volvulus* filarial larvae in anthropophilic *Simulium* spp. in north-Cameroon. *Acta Leidensia* **59**:460.
117. Schulz-Key, H. 1990. Observations on the reproductive biology of *Onchocerca volvulus*. *Acta Leidensia* **59**:27–28.
118. Schulz-Key, H., and Wenk, P. 1981. The transmission of *Onchocerca tarsicola* (Filarioidea: Onchocercidae) by *Odagmia ornata* and *Prosimulium nigripes* (Diptera:Simuliidae). *J. Helminth.* **55**:161–166.
119. Shelley, A.J. 1988. Vector aspects of the epidemiology of onchocerciasis in Latin America. *Annu. Rev. Entomol.* **33**:337–366.
120. Shelley, A.J. 1988. Biosystematics and medical importance of the *Simulium amazonicum* group and the *S. exiguum* complex in Latin America. In M.W. Service, ed., *Biosystematics of Haematophagous Insects*. Clarendon Press, Oxford, pp. 203–220.
121. Shelley, A.J. 1991. Simuliidae and the transmission and control of human onchocerciasis in Latin America. *Cadern. Saude Publ.* **7**:310–327.
122. Shelley, A.J., and Arzube, M. 1985. Studies on the biology of Simuliidae (Diptera) at the Santiago onchocerciasis focus in Ecuador, with special reference to the vectors and disease transmission. *Trans. R. Soc. Trop. Med. Hyg.* **79**:328–338.
123. Shelley, A.J., Luna Dias, A.P.A., and Moraes, M.A.P. 1980. *Simulium* species of the *amazonicum* group as vectors of *Mansonella ozzardi* in the Brazilian Amazon. *Trans. R. Soc. Trop. Med. Hyg.* **74**:784–788.
124. Shelley, A.J., Charalambous, M., and Arzube, M. 1990. *Onchocerca volvulus* development in four *Simulium exiguum* cytospecies in Ecuador. *Bull. Soc. Fr. Parasitol.* **8**:1145.
125. Shelley, A.J., Pinger, R.R., Moraes, M.A.P., and Hayes, J. 1979. Concentration of microfilariae of *Onchocerca volvulus* by *Simulium sanguineum* during feeding: Use in mapping parasite distribution in the skin. *J. Med. Entomol.* **16**:48–54.
126. Shelley, A.J., Luna Dias, A.P.A., Moraes, M.A.P., and Procunier, W.S. 1987. The status of *Simulium oyapockense* and *S. limbatum* as vectors of human onchocerciasis in Brazilian Amazonia. *Med. Vet. Entomol.* **1**:219–234.
127. Sonin, M.D. 1966. Filariata of animals and man and diseases caused by them, Part 1: Aproctoidea. In K.I. Skrjabin, ed., *Essentials of Nematodology*. Izdatel'stvo "Nauka," Moscow, pp. v+365. [Translated from Russian by Israel Program for Scientific Translations.]
128. Spencer Jones, M.E., and Gibson, D.I. 1987. A list of old and recently erected genus-group names not included in the "CIH" keys to nematode parasites of vertebrates and invertebrates. *Syst. Parasitol.* **9**:125–136.

129. Strong, S.P., Sandground, J.H., Bequaert, J.C., and Ochoa, M.M. 1934. *Onchocerciasis with Special Reference to the Central American Form of the Disease.* Harvard University Press, Cambridge, MA.
130. Takaoka, H., Baba, M., and Bain, O. 1989. Natural infections of *Simulium bidentatum* (Diptera:Simuliidae) with larvae of *Onchocerca* spp., in relation to a human zoonotic onchocerciasis in Oita, Japan. *Jpn. J. Trop. Med. Hyg.* **17**:279–284.
131. Takaoka, H., Suzuki, H., Noda, S., Onofre Ochoa, A.J., and Tada, I. 1984. The intake, migration and development of *Onchocerca volvulus* microfilariae in *Simulium haematopotum* in Guatemala. *Jpn. J. Sanit. Zool.* **35**:121–127.
132. Takaoka, H., Suzuki, H., Noda, S., Tada, I., Basáñez, M.G., and Yarzabal, L. 1984. Development of *Onchocerca volvulus* larvae in *Simulium pintoi* in the Amazonas region of Venezuela. *Am. J. Trop. Med. Hyg.* **33**:414–419.
133. Tidwell, M.A., and Tidwell, M.A. 1982. Development of *Mansonella ozzardi* in *Simulium amazonicum*, *S. argentiscutum*, and *Culicoides insinuatus* from Amazonas, Colombia. *Am. J. Trop. Med. Hyg.* **31**:1137–1141.
134. Tidwell, M.A., Tidwell, M.A., and Muñoz de Hoyos, P. 1980. Development of *Mansonella ozzardi* in a black fly species of the *Simulium sanguineum* group from Eastern Vaupes, Colombia. *Am. J. Trop. Med. Hyg.* **29**:1209–1214.
135. Titus, R.G., and Ribeiro, J.M.C. 1990. The role of vector saliva in the transmission of arthropod borne disease. *Parasitol. Today* **6**:157–160.
136. Townson, H., and Meredith, S.E.O. 1979. Identification of the Simuliidae in relation to onchocerciasis. In A.E.R. Taylor and R. Muller, eds., *Problems in the Identification of Parasites and Their Vectors*, 17th Symposium of the British Society for Parasitology. Blackwell Scientific Publications, Oxford, pp. 145–174.
137. Townson, H., Post, R.J., and Phillips, A. 1987. Biochemical approaches to black fly taxonomy. In K.C. Kim and R.W. Merritt, eds., *Black Flies: Ecology, Population Management and Annotated World List.* The Pennsylvania State University, University Park, pp. 24–38.
138. Townson, S., and Tagboto, S.K. 1991. The effects of ivermectin on the viability of *Onchocerca lienalis* microfilariae in vitro and on their subsequent development in the blackfly vector, *Simulium ornatum. Trop. Med. Parasitol.* **42**:31–37.
139. Vajime, C.G., and Gregory, W.G. 1990. Species complexes of vectors and epidemiology. *Acta Leidensia* **59**:235–252.
140. Voelker, J., and Garms, R. 1972. Zur morphologie unbekannter Filarienlarven aus dem Onchocercose-überträger *Simulium damnosum* und aus *S. kenyae* in Liberia und zur Frage der möglichen Endwirte. *Tropenmed. Parasitol.* **23**:285–301.
141. Wahl, G., and Renz, A. 1991. Transmission of *Onchocerca dukei* by *Simulium bovis* in North Cameroon. *Trop. Med. Parasitol.* **42**:368–370.
142. Weinmann, C.J., Anderson, J.R., Longhurst, W.M., and Connolly, G. 1973. Filarial worms of Colombian black-tailed deer in California. 1: Observations in the vertebrate host. *J. Widl. Dis.* **9**:213–220.
143. Wilkinson, P.R. 1949. Some observations on microfilariae in Africans at Jinja. *E. Afr. Med. J.* **26**:344–346.

144. Wilson, M.D., Post, R.J., and Boakye, D.A. 1992. Studies on environmentally-induced colour variation in *Simulium sirbanum* (Diptera: Simuliidae) using a portable rearing system. *Ann. Trop. Med. Parasitol.* **86**:169–174.
145. Wollen, M. 1982. The enigma of *Mansonella ozzardi*. A critical review of the literature. B.Sc. dissertation, University of Aberdeen, 115 pp. Unpublished.
146. World Health Organization. 1985. Ten years of onchocerciasis control in West Africa. Review of the work of the Onchocerciasis Control Programme in the Volta River Basin area from 1974–1984, *OCP/GVA/85.1B*. Geneva: WHO, 113 pp. (unpublished mimeographed document).
147. Yarzabal, L., Botto, C., Arango, M., Raga, L.M., Wong, F., Allan, R., Jaimes, I.L., and Sanchez-Beaujon, R. 1985. Epidemiological aspects of onchocerciasis in the Sierra Parima, Federal Territory of Amazonas, Venezuela. In La oncocercosis en America. *CAICET Publ. Cientifica* **3**:271–281.
148. Yarzabal, L., Petralanda, I., Arango, M., Lobo, L., and Botto, C. 1983. Acid phosphatase patterns in microfilariae of *Onchocerca volvulus* s.l. from the Upper Orinoco Basin, Venezuela. *Tropenmed. Parasitol.* **34**:109–112.

8
Strategies for Controlling African Cassava Mosaic Geminivirus

J.M. Thresh and G.W. Otim-Nape

Introduction

The disease now known as African cassava mosaic was first reported in 1894. It is prevalent in many parts of Africa and causes serious losses. The geminivirus responsible (ACMV) and its whitefly vector (*Bemisia tabaci* Gennadius) have been studied extensively and much attention has been given to possible control measures. This paper considers the various strategies that have been or could be adopted and the opportunities for their use on a large scale.

Four important features must be considered in assessing the need and justification for adopting virus disease control measures:

- The severity of the damage caused, as determined by the virulence of the prevalent strains of virus and the sensitivity to infection of the varieties grown.
- The proportion of plants that becomes infected.
- The stage of growth when infection occurs.
- The economics of the crop in relation to the cost of control measures.

On these criteria there is an overwhelming case for controlling ACMV, with the important proviso that the measures used must be simple, inexpensive and within the limited capacity of the farmers concerned. This conclusion is based on the known prevalence of ACMV in Africa, even though detailed information is lacking on the incidence of infection in the different regions and on the sensitivity of the numerous varieties grown (58). In many important cassava-growing areas virtually all plantings are almost totally infected from the outset due to the use of

J.M. Thresh, Natural Resources Institute, Chatham Maritime, Kent ME4 4TB, United Kingdom.
G.W. Otim-Nape, Namulonge Agricultural and Animal Production Research Institute, P.O. Box 7084, Kampala, Uganda.

infected cuttings. Yields are seriously affected and throughout the entire continent it has been suggested that annual losses may exceed 30 million tones (22). This estimate was based on FAO crop production data and on the assumption that all plants in Africa are infected and sustain an average yield loss of 37%. The estimated loss is similar if recalculated on more plausible assumptions (58).

Effective control measures used on a large scale would increase productivity per unit area and lead to greater production, or release land and labor for other uses and permit longer periods of fallow. This emphasizes the importance of developing effective control strategies that are appropriate for the many peasant farmers involved and also for the relatively few large commercial mechanised farms. Subsistence farmers have scant resources of capital, equipment, or technical expertise and very limited access to extension personnel or other advisers. Moreover, the low value of cassava as a cash crop precludes the use of expensive control measures. These are formidable constraints but are of paramount importance in considering the various options available.

Possible Control Measures

There are three possible approaches to decreasing the losses due to a virus disease:

- Decrease the proportion of plants that become infected.
- Delay infection to such a late stage of crop growth that losses become unimportant.
- Decrease the severity of the damage sustained after infection has occurred.

These objectives can be achieved in diverse ways and the main possibilities for controlling ACMV are considered here.

Sanitation: ACMV-Free Planting Material

Cassava is a vegetatively-propagated crop and a basic approach to control is to use ACMV-free cuttings for all new plantings. The benefits are considerable as healthy cuttings establish more readily and grow faster than infected ones. Yields are also substantially greater, even if plants are infected later by the whitefly vector (22).

The feasibility and effectiveness of the sanitation approach to control are dependent on the rapidity with which ACMV-free material is infected by whiteflies and on the availability of adequate stocks of cuttings at prices farmers can afford. Herein lie the difficulties as only some African farmers have ready access to planting material of improved varieties and even fewer can obtain ACMV-free stocks. Moreover, even if ACMV-free material becomes available in quantity there may be problems due to subsequent infection by vectors. Unless such spread is very slow it would be necessary to introduce ACMV-free cuttings at frequent intervals, or to adopt very stringent selection procedures to ensure that only ACMV-free plants are used to provide cuttings for further plantings. These are serious limitations and add to the difficulty of developing a simple and acceptable set of measures for use by farmers.

PRODUCTION OF ACMV-FREE MATERIAL

There are no technical difficulties in producing basic stocks of ACMV-free cassava. This has been done simply by careful selection among the stocks already available, as in Zanzibar (59), Tanzania (12, 52), Uganda (34; G.W. Otim-Nape, unpublished), Malawi (R.F.N. Sauti, unpublished), Kenya (6) and Côte d'Ivoire (27). It has also been possible to exploit the recovery phenomenon referred to as "reversion". This term is used when ACMV-free plants develop from cuttings derived from infected source plants and seems to occur because of the apparent failure of ACMV to become fully systemic, at least in some of the more resistant cultivars (27, 46).

Heat and meristem-tip therapy have also been used to develop ACMV-free plants of totally infected clones (11, 39, 40). Originally, freedom from ACMV was inferred from the complete absence of symptoms during prolonged periods of observation. Quick, sensitive serological and biochemical tests for ACMV are now available and such tests can also be used to ensure freedom from other viruses, including the one(s) associated with cassava brown streak disease in Tanzania and some other parts of East Africa.

Once ACMV-free foundation plants have been produced they can be multiplied quickly by micropropagation, single-node cuttings, mist propagation, or other standard techniques. Initially this can, if necessary, be done in insect-proof structures to provide a protected favorable year-round environment and to avoid infection by whitefly vectors. However, stocks can be produced in the great quantity required for distribution to farmers only by using large outdoor propagation sites. The plants may then be at risk of contamination and it is important to utilize sites where there is a low probability of infection by incoming whitefly vectors.

There are advantages in using experimental or state farms under official control and sound management and away from the main areas of cassava production. Suitable sites have been identified in both upland and lowland areas of East Africa (6, 52; G.W. Otim-Nape, unpublished) and in the Guinea savannah region of Côte d'Ivoire (25, 26). However, the use of very remote sites can create management and distribution problems, as well as add to the costs of transporting large quantities of bulky material into the main cassava-producing areas.

Rates of Infection

The fate of ACMV-free cuttings when exposed in the field depends on their inherent susceptibility to infection and on the overall "infection pressure" encountered. The concept of "infection pressure" is an important one and indicates the degree to which plants are at risk. This is determined by the magnitude, activity and infectivity of whitefly vector populations, which are in turn influenced by environmental factors and by the size, proximity, distribution and potency of sources of infection, especially cassava plantings located upwind and nearby (21).

The interaction between varietal susceptibility and "infection pressure" is evident from experience at mid-altitudes in Uganda and in the lowland forest zones of Côte d'Ivoire and Nigeria. Susceptible varieties can be almost totally infected within a few months of planting at sites where there is relatively little spread to very resistant ones (20, 28, 32; G.W. Otim-Nape, unpublished). Experimental plantings have also been made in coastal regions of Kenya to evaluate the infection pressure in different climatic zones and to rank varieties on the basis of their resistance so that appropriate ones can be recommended according to the risk of infection (8, 49). There is an urgent need for comparable information on rates of spread in other regions of Africa and on the relative susceptibility of the main varieties being grown in each country. Only then will it be possible to determine the minimum degree of resistance required in each area and the prospects of exploiting ACMV-free planting material.

Use of ACMV-Free Material

The merits of using ACMV-free planting material have long been apparent. They were first discussed by Storey (52), who noted that farmers in the Morogoro Region of what is now Tanzania obtained cuttings from the highland areas where there was a low incidence of ACMV. Official schemes for the release of ACMV-free material were first introduced in Zanzibar (9). A similar approach was adopted later in Uganda, where attempts were made to establish plots of ACMV-free material in each chiefdom of the worst-affected areas (34, 60). The extent to which such practices have been and are being adopted elsewhere in Africa is unclear from the limited information available on the activities of Departments of Agriculture and other organizations.

In several countries the ACMV-free material released has been of improved varieties that have greater resistance to infection or other advantages compared with those grown previously. For example, special arrangements were made in Tanzania to release uninfected material of two ACMV-resistant, high-yielding hybrids (12). Material for distribution was raised at isolated sites or by arrangement with selected farmers and the aim was to displace the local ACMV-infected varieties from entire localities within 2–3 years.

A similar policy was adopted in the Teso District of Uganda (now Soroti and Kumi Districts) following the severe drought of 1943–1944, when there was a big increase in the incidence of ACMV due to the widespread use of infected cuttings (34). Uninfected material of ACMV-resistant varieties was introduced to the area in attempts to displace the infected stocks being grown. The uninfected stocks were raised initially at a "quarantine" site at the Serere Experimental Station about 1.7 km from the nearest infected cassava. Secondary multiplication on prison farms maintained by local government authorities quickly produced sufficient material to supply a complete administrative unit of *ca.* 300 ha. Farmers were advised to remove all existing cassava before replanting and were promised assistance if food shortages occurred before the new crops of cassava could be harvested. This approach was highly successful and was extended later to other areas by adopting a systematic "advancing front" plan, which involved clearing cassava from whole areas before replanting began, to decrease the opportunity for infection to occur. By 1951 Teso District had sufficient planting material to supply all needs and legislation was introduced to enforce the destruction of any remaining infected cassava.

The Uganda scheme operated very effectively for over a decade but it lapsed during the 1970s and is only now being reintroduced (G.W. Otim-Nape, unpublished). A similar scheme is being operated in parts of Malawi, where ACMV has been almost entirely displaced from some northern areas (R.F.N. Sauti, personal communication). Replies to a 1987 questionnaire addressed to staff of national programmes (3) indicate that ACMV-free planting material is also being used in some other parts of Africa, including Liberia, Cameroon, Burundi and Benin. However, few details are available and there is an urgent need for up-to-date assessments of the production and distribution procedures employed, the problems encountered, the results achieved and the response of farmers.

In each locality much will depend on the extent to which cassava is being grown and on the overall infection pressure encountered. There are obvious advantages in a systematic approach, with the emphasis on supplying entire areas with ACMV-free cuttings to avoid carryover of inoculum and to ensure that infected material is replaced expeditiously. However, there are likely to be difficulties in areas where it is necessary to obtain the participation and cooperation of many individuals, each farming small contiguous land holdings. There are also likely to be problems in producing sufficient ACMV-free material unless enough propagation sites are available in low-risk areas or the varieties grown are so resistant to infection that there is little spread by vectors.

Sanitation: Roguing

A possible method of maintaining or even improving the health of cassava plantings is to inspect them regularly to remove all mosaic-affected plants at an early stage of growth and before they have had much opportunity to act as foci for further spread. Such roguing procedures have been recommended repeatedly as a means of controlling ACMV and most recently by Robertson (49) and Guthrie (31) from their experience in Kenya. Roguing could be advantageous in some circumstances, but in others may be ineffective or inappropriate. This is apparent from contrasting experiences in East and West Africa, even though there has been no systematic evaluation of possible roguing procedures and the circumstances in which they are most likely to be effective.

Roguing can be done most easily and efficiently soon after planting, as the cuttings begin to grow and when crops are most readily accessible. Early symptoms of ACMV tend to be particularly conspicuous and are associated with the greatest effects on growth and yield. Moreover, young infected plants are easily removed and can be replaced by ACMV-free cuttings or by other crop plants with little or no loss of stand. Another approach is to plant more cuttings than required, to allow for losses due to roguing.

Roguing later in the growing season is more difficult, especially with straggling, multistemmed varieties that form dense stands. Vacancies are not easily filled so that there is some loss of crop. Moreover, symptoms of ACMV tend to become inconspicuous as plants age and may be masked by the effects of drought, mineral deficiencies, mealybug (*Phenacoccus manihoti* Mat.-Ferr.), green mites (*Mononychellus tanajoa* Bondar and others), or foliar pathogens. Nevertheless, inspection at crop maturity is advocated in India to mark plants affected by Indian cassava mosaic geminivirus so that they can be avoided when collecting cuttings for further plantings (43).

These considerations explain why early roguing is so strongly advocated and why it has been successful in improving the health of planting material at sites in East Africa (7, 12, 34) and also in the Guinea savannah region of Côte d'Ivoire (27), where there was little spread from internal or external sources of infection. Roguing is *not* effective where there is such rapid spread from outside sources that internal foci within crops are relatively unimportant. This is apparent from experience near Abidjan in the lowland forest region of Côte d'Ivoire. There the main spread is between and not within cassava plantings and repeated roguing created numerous gaps and decreased overall yields, but had little effect on the final incidence of infection (17, 20).

Feasibility

Experience with a wide range of tropical and temperate crops is that roguing is an irksome and generally unpopular control measure (57). Farmers are understandably reluctant to remove infected plants that still contribute to yield and it is difficult to ensure that roguing is practised with the efficiency and diligence required to control ACMV. Moreover, individual farmers will not obtain the full benefits of roguing unless the practice is also adopted by their neighbours and preferably throughout whole localities.

This presents a challenging problem to extension workers promoting a sanitation approach to control ACMV and there is no precedent for such a vast undertaking involving so many farmers. The most immediate application of roguing is likely to be in operating official schemes for producing ACMV-free planting material and in large-scale commercial farms or plantations under strong central management. This would provide the example for small farmers to follow, especially if the introduction of roguing is associated with the release of high-yielding ACMV-resistant varieties that require much less roguing than those now grown.

Cultural Practices and Crop Disposition

Planting dates, cultural practices and crop disposition can greatly influence vector populations and virus spread (56). This has been demonstrated in studies on ACMV that may be of considerable practical relevance. However, only limited information is available and additional work is required before farmers can be given definitive advice on the most appropriate cropping practices to adopt. Even then there are likely to be great difficulties in implementation due to the small size of many farms and to the limited availability of suitable sites, which imposes serious constraints on field size, shape and disposition. Moreover, there is the need to produce continuous supplies of cassava for family use, or local sale, or for processing and transport to urban markets. This facilitates the spread of ACMV by leading to successive plantings in close proximity and in overlapping sequence. The problems arising are less acute on large commercial farms or at official establishments used to produce ACMV-free cuttings for distribution to farmers. Much can be done in these circumstances by adopting cropping schedules and spatial arrangements that decrease spread and facilitate control.

Planting Date

Cassava is so easily established from cuttings that farmers usually have considerable latitude in the choice of planting date. The aim should be to facilitate the control of ACMV by not exposing plants to serious risk of infection by whiteflies when they are at a young and vulnerable stage of growth. There are good prospects of increasing yields in this way by decreasing the incidence of ACMV and because plants infected late in growth are not seriously affected.

In coastal districts of Kenya the main spread of ACMV by vectors occurs during the early rains from mid-May to mid-July and there are likely to be advantages in planting later in the year, provided that crop establishment and subsequent growth are not impaired (49). A similar situation exists in the forest areas of Côte d'Ivoire, where virus spread occurs throughout the year but is most rapid from March to July and least from August to November during the latter part of the rainy season (19).

The full implications, practicability and likely benefits of manipulating planting dates to decrease the losses due to ACMV have not been studied. Hence the need for additional information from the various agroecological zones before attempts are made to persuade farmers to change traditional practices, which usually involve planting soon after the onset of the rains. Such studies merit high priority because they could bring great benefits at no great inconvenience or expense. Major shifts in attitude are possible if there is sufficient incentive and justification, as demonstrated in parts of Central America where cassava planting is now delayed to facilitate control of bacterial blight caused by *Xanthomonas campestris p.v. manihotis* Berthet & Bondar: Dye (42).

Crop Spacing

There is evidence from Côte d'Ivoire that the spread of ACMV is influenced by host plant population (20). The incidence of infection expressed as a percentage of the total stand was greatest at the widest spacing adopted and alongside footpaths or around gaps in otherwise continuous stands. These findings are of obvious practical importance and can be utilized immediately by farmers who should aim to establish uniform dense stands.

There is little evidence on the spread of ACMV in plantings of cassava grown with other crops as compared with monocultures (18). This is a serious omission because in many parts of Africa cassava is usually grown in mixed or relay cropping systems (30). These involve many other crop species that may have beneficial effects by improving overall land productivity and by decreasing whitefly populations, whitefly activity and virus spread.

Plot Size and Shape

A recurring feature of experiments over several years in the lowland forest areas of Côte d'Ivoire was that whitefly numbers and incidence of ACMV were greatest in the outermost rows of plantings and especially those oriented across the direction of the prevailing southwesterly wind (21). This suggests that it is advantageous to plant in large compact blocks and orient elongated plots along rather than across the direction of the prevailing wind, to decrease the proportion of plants in the most vulnerable peripheral areas (Fig. 8.2). It may also be appropriate to discard the outermost rows of propagation plots being used to raise ACMV-free cuttings for distribution. An alternative approach may be to plant windbreaks or barrier rows of a suitably tall crop other than cassava to intercept and impede the movement of incoming vectors and so restrict spread of ACMV. However, little attention has been given to this possibility and, in the only experiment reported, there was considerable infection along the windward edge of a plot surrounded by rows of sugarcane 2.5 m high (21).

Crop Disposition

The main spread of ACMV in the lowland forest zone of Côte d'Ivoire is into and not within plantings (20). This emphasizes the importance of crop disposition and the opportunity to facilitate control by selecting sites where there is only limited risk of infection from outside sources. In practice this means using suitably isolated sites, but there is scant information available on the minimum degree of isolation required and on the scope and extent of the gradients from infected plantings of different magnitude and potency. Nevertheless, studies in Côte d'Ivoire have demonstrated spread by wind-borne adult whiteflies over distances of several kilometres (25). The importance of older sources of infection located upwind was also established (21, 25, 26). Thus new plantings tend to be at great risk where sources of ACMV are upwind and nearby, but not when the nearest sources are downwind and remote (Fig. 8.1). Spread is also likely to be decreased by establishing sequential plantings in an upwind direction and not downwind and by orienting plantings so as to restrict the interfaces along which spread is most likely to occur (Fig. 8.2).

The opportunity for utilizing such information is obviously influenced by the overall patterns of wind direction, land use and land ownership. The greatest scope is in areas where cassava is not widely grown and there is considerable separation between plantings and also where cassava

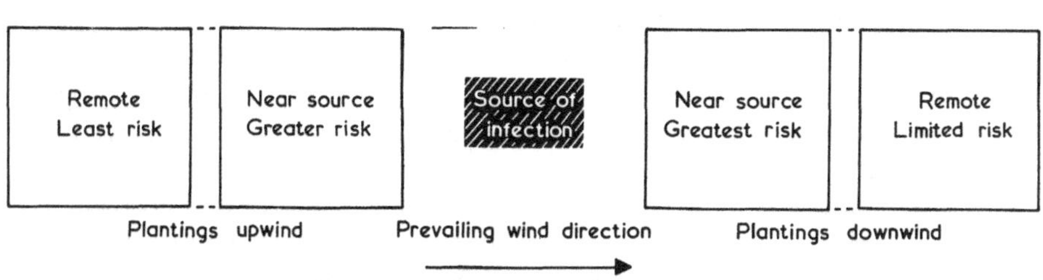

Figure 8.1. Diagrammatic representation of the implications of planting near or far from a major source of infection and in an upwind or downwind direction in relation to the prevailing wind.

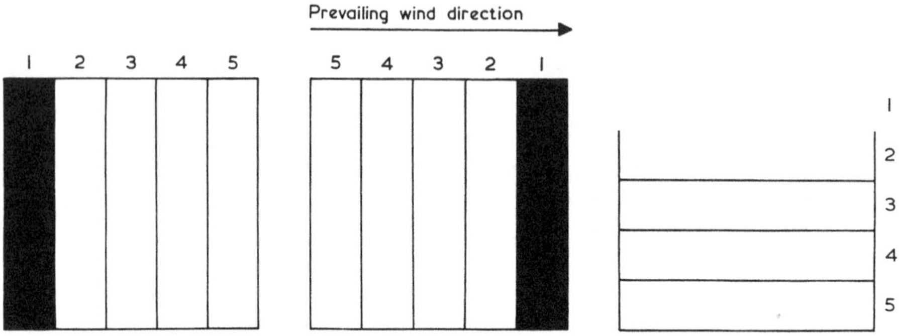

FIGURE 8.2. Diagrammatic representation of the implications of sequential rectangular plantings. Spread from old (1) to successively younger plantings (2–5) is facilitated when they are made in a downwind direction (left) but not when they are made upwind (center) or along the axis of the prevailing wind (right).

is established synchronously in large uniform blocks of similar age. However, this is seldom the situation in the main areas of cassava production in West and Central Africa.

Vector Control by Insecticides or Other Means

INSECTICIDES

A possible means of controlling arthropod-borne viruses is by using pesticides against the vectors. There are several reasons why this approach has seldom been adopted to control ACMV, quite apart from toxicological and environmental considerations. The most cogent is that insecticides are least effective in controlling insect-borne viruses if the main spread is into and not within crops. Thus insecticides are unlikely to kill incoming viruliferous whiteflies before they have had an opportunity to introduce ACMV.

Experience with cotton and several other crops is that whiteflies are not easily controlled by insecticides and they tend to be killed less readily than their natural enemies (14, 16). There can be a rapid resurgence of whitefly populations soon after insecticides have been applied and any benefits of treatment tend to be shortlived. Furthermore, subsistence farmers can seldom afford to purchase or gain access to spray machines and insecticides. Thus there is little immediate prospect that insecticides will be used to control ACMV. The situation could change if more effective insecticides or repellents become available and subsidies are provided to farmers, or if crop production becomes concentrated into larger and more profitable farming units.

Biological Control

There seems to be limited scope for implementing biological control of whitefly vectors on cassava by natural enemies and it is likely that the main effort on these lines will continue to be against the green mites and cassava mealybug. As recently introduced pests from South America they are more amenable to biological control by exotic parasitoids than *B. tabaci*, which has been in Africa for many years.

Vector-Resistant Varieties

Cassava is not a very favorable host plant for *B. tabaci* and adult populations tend to be small and mainly restricted to the young expanding leaves (23). A possible approach to decreasing whitefly populations and spread of ACMV is to breed varieties that are even less susceptible to *B. tabaci* than those now grown. This possibility was considered by Leuschner (41) who emphasized the difficulties likely to be encountered in developing and exploiting such varieties. More recent studies have identified possible whitefly-resistant genotypes and shown that resistance to vector and to ACMV are independent attributes that can be combined to give a very advantageous combination of features (28). Such varieties may have an important role in future ACMV control strategies.

ACMV-Resistant Varieties

The use of resistant or tolerant varieties has obvious advantages in seeking to decrease the losses due to viruses. This was appreciated by early workers on ACMV in East Africa who were influenced by the great success of efforts to breed sugarcane varieties resistant to sugarcane mosaic potyvirus (52). Local and introduced varieties of cassava (*Manihot esculenta* Crantz) and intra-specific crosses between cassava varieties were soon found to be insufficiently resistant to infection. Attention then turned to hybrids between cassava and other species, of which Ceara rubber (*M. glaziovii* Muell.-Arg.) proved to be the most suitable (44). There was no evidence of immunity to infection, but *M. esculenta* × *M. glaziovii* hybrids had considerable resistance. This was also expressed in progenies obtained by back-crossing such hybrids to cassava.

Resistance was assessed at first by exposing batches of clonal plants to infection and recording the number of months before *each* plant developed symptoms. The selection procedures were later simplified and resistance to infection was assessed more rapidly, but less precisely, from the number of months before the *first* plant of a clone developed symptoms. However, it was appreciated that the response of plants to infection was also important and that resistant varieties would be of limited value if they yielded less satisfactorily than susceptible ones after they became infected.

From subsequent experience in Tanzania it became apparent that selections resistant to infection tended to develop inconspicuous symptoms. Moreover, symptoms were sometimes ephemeral and restricted to parts of only one or two shoots (35). Consequently, recording procedures were revised to take account of both the *incidence* of infection and the *intensity* of symptoms. Some of the varieties selected in this way proved suitable for East African conditions and were released to farmers (15). Seeds from promising selections were also sent to Nigeria in the 1950s for use in the National breeding programme (37). Nigerian seedling selection 58308, which originated from this material, has since featured prominently as a parent in resistance breeding at the International Institute of Tropical Agriculture (IITA), Ibadan, Nigeria.

In recent decades the most comprehensive and influential cassava breeding programme in Africa has been at IITA, where resistance to ACMV is one of the main attributes sought (32). Progenies are evaluated using a scale ranging from no symptoms (score 1) to severe leaf mosaic and distortion (score 5). A limitation of this approach is that virus incidence and symptom severity are not clearly distinguished and symptomless plants could be ACMV-free escapes or they could be infected and extremely tolerant. Moreover, a low average score for a progeny or selection could mean that a few plants are infected and show severe symptoms, or that many succumb but are only slightly affected.

Several IITA selections have been categorized as "highly resistant", "resistant" or "moderately resistant" to ACMV and they have been released for use by farmers in Nigeria and for assessment in some other countries. IITA clones and seed stocks have also been widely distributed for use as parents, or for selection in national breeding programmes. TMS 30395 and TMS 30572 are two of the improved IITA varieties that are resistant to infection with ACMV, yet both sustain a serious loss of yield when infected (54, 55). The most resistant of the IITA varieties, notably TMS 30001, are even more difficult to infect and when infected develop symptoms of restricted distribution. ACMV is incompletely systemic in such varieties and tends to be localized at the base of the main stem, as described in East Africa (36, 38).

There is only limited information available on the relative performance of infected and uninfected clones of the varieties released by IITA and designated as in some way virus-resistant. An important feature of these varieties is their ability to grow and yield satisfactorily, even in areas where ACMV is prevalent and causes serious losses in traditional local varieties. Some of the IITA varieties including TMS 4(2)1425 usually develop conspicuous symptoms only during the early stages of growth and then recover. Other varieties including TMS 30572, TMS 30395 and TMS 30001 show a strong tendency to "revert", as many of the cuttings from

ACMV-infected sources grow into uninfected plants. Such varieties never become totally infected, even when grown for years in conditions of intense infection pressure. However, the reversion phenomenon is poorly understood and the overall loss in yield caused by ACMV in mixed stands of infected and uninfected plants of IITA varieties has not been determined. Nor has it been established whether there is any advantage in roguing such varieties and in developing ACMV-free stocks.

Little information is available on the extent to which IITA varieties are being grown in the different countries of Africa. A survey in Nigeria established that IITA varieties, including some resistant to ACMV, account for *ca.* 20% of all plantings in Ondo State (1, 2, 33). The results of a more recent questionnaire (3) indicate that IITA varieties are much less widely grown outside Nigeria, except in Republic of Benin where they are said to account for over 60% of the total area. Uptake has been limited in many other countries and several problems are cited by respondents. They include a shortage of planting material and of information on overall performance and suitability. There are also complaints about the flavor, bitterness, or cyanide content of the roots, the branched growth habit, storage characteristics and lack of adaptation to local conditions. Moreover, only some countries have the resources to establish and multiply the improved varieties that are introduced as ACMV-free plantlets in nutrient culture. There may also be limited facilities to grow and assess the seed material distributed from IITA for local evaluation and selection.

These are serious constraints that will not be easily resolved, although cassava is propagated so easily that a shortage of planting material should not be a problem given a suitable degree of commitment. There is an obvious need to make greater use of the material already available and to develop new varieties with all the main attributes required. The general use of ACMV-resistant varieties is an essential component of any disease control strategy in areas of high infection pressure. It may then become possible to exploit the benefits of ACMV-free planting material and roguing in areas where such measures are otherwise ineffective. Alternatively, it may be possible to develop varieties that are so difficult to infect or in which the reversion characteristic is so marked that roguing or other control measures become unnecessary.

ACMV-Tolerant Varieties

The losses due to many important virus diseases of temperate and tropical crops are greatly decreased by using tolerant varieties that grow and yield satisfactorily when infected. Nevertheless, there is often some reluctance to introduce tolerant varieties because they may act as sources of infection and decrease the effectiveness of control by sanitation or other means (57). Such concerns may be justified, but tolerant varieties have obvious advantages if immune or hypersensitive ones are not available and there is rapid spread of virus into or within crops that is difficult to control by other means (47). This is the situation in those parts of West Africa where ACMV poses intractable problems and where currently there are limited prospects of control by sanitation. In circumstances of such high infection pressure tolerant varieties would have great advantages, especially if tolerance is associated with a low virus content. Such varieties would then be tolremic (*sensu* Buddenhagen (10)) as infected plants are poor sources of virus and spread tends to be slow. There are good prospects of exploiting tolremicity now that ELISA can be used routinely to assess the virus content of ACMV-infected cassava (23).

The original objective of the cassava breeding programmes in East and West Africa was to develop varieties with high levels of resistance to infection by ACMV. The amount of attention that has since been given to tolerance is unclear, although there is evidence that resistance to infection is associated with slight symptom expression. However, as discussed by Bock (7) and in general terms by Russell (50), symptom expression is not necessarily related closely to effects on growth or productivity and individual plants of at least some "resistant" varieties sustain a serious loss of yield when infected (51, 54, 55).

This is a serious defect in conditions of high infection pressure where ACMV spreads rapidly and sanitation is not practised. Hence the need to consider the yield of ACMV-infected and uninfected plants in evaluating breeders' selections. There is certainly scope for developing very tolerant varieties for use as an alternative to sanitation.

Mild Strain Protection

A possible means of alleviating the effects of virus infection on growth and yield is by prior inoculation with a mild strain of the same virus. The ability of mild strains to protect plants from the effects of closely related virulent ones has long been recognized and the phenomenon has been used extensively by virologists to assess the relationships between viruses and virus strains (4). However, there were several reasons for an initial reluctance to adopt this approach to control. First, there was concern that mild strains might mutate to more virulent forms or spread to other and more sensitive crops. It was also feared that mild strains might have synergistic effects in combination with unrelated viruses. Moreover, the deliberate dissemination of mild strains is incompatible with a control strategy based on sanitation.

Despite these reservations mild strains have in recent years been disseminated widely and successfully to control tomato mosaic, passion fruit woodiness, citrus tristeza and papaya ringspot viruses (29). The technique has also been considered as a possible means of controlling cocoa swollen shoot virus in the worst affected areas of Ghana (45, 48). Each of these viruses is so prevalent and so difficult to control by other means that the deliberate dissemination of mild strains poses no additional hazards. This is the situation in the areas worst affected by ACMV. The dissemination of cuttings containing a mild strain has great possibilities in such circumstances, provided that the strain used confers adequate protection against the prevalent virulent ones.

Advantages of this approach would be that established locally adapted varieties could be used and cuttings containing the mild strain could be produced in quantity by propagating from ACMV-infected source plants. No special isolation would be required and an initial release of cuttings would suffice for repeated cycles of crop production. Thus the costs would be less and the logistics would be simpler than for sanitation.

At present any consideration of mild strain protection for cassava is hypothetical. No attention has been given to such an approach and no suitable mild strain has been reported. The precedents from cross-protection studies with beet curly top geminivirus are not encouraging (5). Moreover, Storey and Nichols (53) failed to demonstrate protection between two Tanzanian strains of ACMV, although some of their challenge inoculations were made by grafts. These provide a more stringent test of protection than inoculation by vectors, as would occur in nature. One possible difficulty in exploiting mild strain protection could arise from the apparently erratic distribution of ACMV in infected plants, which may facilitate the establishment of virulent strains. However, the prospects for deploying mild strains will not become clear until detailed studies have been undertaken and such work is long overdue.

Integration of Control Measures

Only limited and intermittent progress has been made in controlling ACMV in many parts of Africa, even though the virus has been studied for more than 50 years and much information is available on possible control measures (58).

One constraint is that cassava is a neglected crop, despite its dietary importance in many countries and its potential for export, or as a source of animal feed or fuel. Only limited resources have been allocated to cassava for research and extension activities and in Africa in recent years there has been a major diversion of effort to combat green mites, cassava mealybug and bacterial blight. In some countries these now cause damage that is more obvious than that due to ACMV, which has relatively insidious effects on growth and yield. Consequently, mosaic tends to be accepted as almost inevitable by many farmers and even by some advisers, researchers and policy makers.

Cassava in Africa has a generally low status as a crop because it is mainly grown for local consumption by subsistence farmers. It does not receive the attention given to export crops or those grown extensively by influential landowners or commercial companies on large mechanized farms. Hence the great difficulties encountered by the extension services and non-governmental organizations in implementing cassava improvement programs. Progress is likely to be slow and inevitably it will be easier to make research findings than to ensure their implementation in the field.

Particular problems will arise in the many countries of Africa where the extension and advisory services are limited by a grave shortage of financial resources, facilities and personnel, and where the need to control ACMV must be considered in relation to the more obvious losses caused by other pests and diseases of cassava and also to the requirements of other important crops. Farmers should, ideally, have access to varieties with resistance not only to ACMV and its vector but also to other pests and diseases. Moreover, the control measures used against ACMV must be fully compatible with the overall crop protection strategy being implemented and also with recommended cropping practices.

Only limited progress has been made in developing an integrated approach to the control of cassava pests and diseases. One strategy is to rely on resistant varieties and routine screening tests against pests and diseases feature prominently in the cassava improvement program at IITA and in many national programs. Much attention has also been given to the biological control of green mites and the cassava mealybug. However, these contrasting approaches to control by resistant varieties and by natural enemies have not been closely coordinated. The role of sanitation, including the use of planting material free of the various pests and pathogens of cassava in the overall context of the cropping systems employed, has also received scant attention. There is obvious cope for developing improved practices and cassava is one of the target crops being considered by the International Task Force set up recently to promote Integrated Crop Protection in the Tropics.

Apart from the problems of mounting an effective crop improvement and integrated pest and disease control program for cassava there are likely to be difficulties due to the very diverse ecological conditions under which the crop is grown in Africa. It is planted over a wide altitudinal and climatic range, as a major or minor crop and in areas having very different growing seasons and patterns of crop production. Consequently, there are likely to be differences in whitefly populations, virus strains and virus spread between regions and control measures or varieties that are appropriate in one situation may be ineffective elsewhere. This is evident from the contrasting experiences in Kenya, Côte d'Ivoire and Uganda, but there is a dearth of epidemiological information from elsewhere. This is a serious constraint in devising appropriate control measures for each region to integrate within an overall crop protection strategy.

There is no immediate prospect of using insecticides, biological control, mild strain protection, tolerant varieties, or genetically engineered forms of resistance and the basic approach to the control of ACMV in all regions involves ACMV-resistant varieties, sanitation and crop deployment.

Unfortunately, an integrated approach along these lines is seldom feasible. Acceptable varieties are not always available with the necessary combination of attributes, or they are released in insufficient quantity, or the minimum degree of resistance required has not been determined. Furthermore, farmers may be unwilling to rogue or they have not been encouraged to do so. They seldom have much latitude in the choice of site because of the limited land available. They also have little or no control over the activities of their neighbors, who may retain severely diseased plantings nearby.

Despite these constraints, the available evidence suggests that ACMV *can* be controlled effectively in large areas of Africa by releasing ACMV-free material of local or improved varieties and by persuading farmers to rogue effectively and to adopt other basic measures of sanitation and crop deployment. Experience suggests that there are no intrinsic difficulties in adopting such an approach, *provided that suitably resistant varieties are used*.

Circumstances in eastern Africa are generally propitious, as the overall infection pressure appears to be low throughout much of the region. Whitefly populations and spread of ACMV seem to be limited by prolonged dry periods and much of the cassava is grown at upland sites above 1000 m. Moreover, the crop is not widely grown compared with the humid lowland forest zones of West and Central Africa and there is usually at least some spatial separation between plantings. This suggests that the main problems likely to be encountered in controlling ACMV in eastern Africa will relate to the logistics of producing and distributing ACMV-free cuttings on the scale required and with the need to change the attitudes and practices of farmers.

Conditions in the lowland forest areas of West Africa are very different as they permit whitefly reproduction and rapid spread of ACMV for much of the year. Cassava is widely grown throughout the region and the infection pressure is usually high. Thus, it is hardly surprising that there appears to be little prospect of control by sanitation using many of the varieties now generally grown. However, control will become feasible once acceptable varieties have been introduced that have adequate resistance to infection. This is apparent from experience with the most resistant of the IITA varieties and from trials with Aipin Valenca in Côte d'Ivoire (13, 28).

The situation in the lowland forest zone will be transformed once such resistant varieties become generally available and they are accepted by farmers, processors and consumers. Meanwhile, the use of ACMV-free planting material, deployment and roguing with existing varieties might be successful on large commercial farms isolated from outside sources of infection. Sanitation is also likely to be effective in the adjoining southern Guinea savannah areas where cassava is not yet widely grown and where the infection pressure is correspondingly low (26). It is unclear whether a similar approach would be successful in the important cassava-producing areas of Zaire, Congo and other countries of Central Africa. ACMV is prevalent in these areas, but whether this is due mainly to the dissemination of infected planting material, or to rapid spread by whiteflies has not been determined. Thus the current scope for using ACMV-free planting material and the way in which it should be deployed is uncertain.

Conclusions

Much could be achieved by utilizing the existing information on ACMV and research has in some respects outstripped the ability to apply the results obtained. However, the control measures now advocated depend on the availability of ACMV-free cuttings and of suitably resistant varieties and also on the willingness of farmers to adopt this material and to rogue as appropriate. This emphasizs the importance of developing additional varieties and the means to introduce them on an adequately large scale. There is also a need for further information on the epidemiology and control of ACMV, and on the prevalence of different strains, especially in the many parts of Africa where there have been no previous studies.

It has still to be determined whether the present approach to control is appropriate in all circumstances. In some of the worst-affected areas the adoption of mild strain protection or ACMV-tolerant varieties may be a more satisfactory alternative than the use of ACMV-free material, virus-resistant varieties and roguing. Whatever the strategies advocated it will be necessary for the extension services and other agencies to introduce greatly improved methods to ensure that the measures advocated are widely adopted.

Acknowledgments. Grateful acknowledgments are due to G.I. Atiri, K.R. Bock, I.W. Buddenhagen, D. Fargette, C. Fauquet, S.K. Hahn, H. Herren, S.N. Lyonga, E. Ntawuyirusha, F. Ogbe, H. Rossel, R.F.N. Sauti, A.A. Seif and G. Thottappilly for their hospitality and helpful discussions on visits to Benin, Burundi, Cameroun, Côte d'Ivoire, Ghana, Kenya, Malawi, Nigeria, Tanzania, Uganda, Zambia, and Zimbabwe funded by the U.K. Overseas Development Administration and the Government of Uganda.

References

1. Akoroda, M.O., Gebremeskel, T., and Oyinlola, A.E. 1989. Impact of IITA cassava varieties in Oyo State, Nigeria, 1976–1985. *Trop. Agric. (Trin.)* **66**:113–120.
2. Akoroda, M.O., Oyinlola, A.E., and Gebremeskel, T. 1987. Plantable stem supply system for IITA cassava varieties in Oyo State Nigeria. *Agric. Syst.* **24**:305–317.
3. Anonymous. 1988. *National Questionnaires: African Cassava Mosaic Disease and Its Control.* Special Report to International Seminar on African Cassava Mosaic Disease and its Control. Yamoussoukro, Côte d'Ivoire, 4–8 May 1987. CTA/FAO/ORSTOM/IITA/CPI.
4. Bawden, F.C. 1950. *Plant Viruses and Virus Diseases*, Chronica Botanica Company, Waltham, Ma.
5. Bennett, C.W. 1971. *The Curly Top Disease of Sugarbeet and Other Plants.* Monograph 7: American Phytopathological Society.
6. Bock, K.R. 1982. Kenya: Avoiding cassava mosaic disease. *Span* **25**:11–13.
7. Bock, K.R. 1983. Epidemiology of cassava mosaic disease in Kenya. In R.T. Plumb and J.M. Thresh, eds., *Plant Virus Epidemiology.* Blackwell Scientific, Oxford, pp. 337–347.
8. Bock, K.R. 1988. Some aspects of the epidemiology of African cassava mosaic virus in coastal districts of Kenya. In C. Fauquet and D. Fargette, eds., *Proceedings International Seminar on African Cassava Mosaic Disease and Its Control.* Yamoussoukro, Côte d'Ivoire, 4–8 May 1987. CTA/FAO/ORSTOM/IITA/IAPC, pp. 125–130.
9. Briant, A.K., and Johns, R. 1940. Cassava investigations in Zanzibar. *East Afr. Agric. J.* **5**:404–412.
10. Buddenhagen, I.W. 1981. Conceptual and practical considerations when breeding for tolerance or resistance. In R.C. Staples and G.H. Toenniessen, eds., *Plant Disease Control: Resistance and Susceptibility.* John Wiley, New York, pp. 221–234.
11. Chant, S.R. 1959. A note on the inactivation of mosaic virus in cassava (*Manihot utilissima* Pohl.) by heat treatment. *Emp. J. Exp. Agric.* **27**:55–58.
12. Childs, A.H.B. 1957. Trials with virus resistant cassavas in Tanga Province, Tanganyika. *East Afr. Agric. J.* **23**:135–137.
13. Colon, L., and Fauquet, C. 1984. *Contribution à l'étude de la résistance variétale du manioc vis à vis de la mosaïque Africaine du manioc.* Rapport de Stage Centre d'Adiopodoume, ORSTOM, Côte d'Ivoire.
14. Dittrich, V., Hassan, S.O., and Ernst, G.H. 1985. Sudanese cotton and the whitefly: A case study of the emergence of a new primary pest. *Crop Prot.* **4**:161–176.
15. Doughty, L.R. 1958. Cassava breeding for resistance to mosaic and brown streak viruses: a review of twenty-one years' work. *Ann. Rep. East Afr. Agric. For. Res. Org. for 1958*, pp. 48–51.
16. Eveleens, K.G. 1983. Cotton-insect control in the Sudan Gezira: Analysis of a crisis. *Crop Prot.* **2**:273–287.
17. Fargette, D. 1987. *Épidémiologie de la mosaïque Africaine du manioc en Côte d'Ivoire.* Éditions de l'Orstom. Collection Etudes et Thèses.

18. Fargette, D., and Fauquet, C. 1988. A preliminary study on the influence of intercropping maize and cassava on the spread of African cassava mosaic virus by whiteflies. *Aspects Appl. Biol.* **17**:195–202.
19. Fargette, D., Fauquet, C., Fishpool, L., and Jeger, M.J. 1994 (in press).
20. Fargette, D., Fauquet, C., Grenier, E., and Thresh, J.M. 1990. The spread of African cassava mosaic virus into and within cassava fields. *J. Phytopathol.* **130**:289–302.
21. Fargette, D., Fauquet, C., and Thouvenel, J.-C. 1985. Field studies on the spread of African cassava mosaic. *Ann. Appl. Biol.* **106**:285–294.
22. Fargette, D., Fauquet, C., and Thouvenel, J.-C. 1988. Yield losses induced by African cassava mosaic virus in relation to the mode and the date of infection. *Trop. Pest Manage.* **34**:89–91.
23. Fargette, D., Thouvenel, J.-C., and Fauquet, C. 1987. Virus content of leaves of cassava infected by African cassava mosaic virus. *Ann. Appl. Biol.* **110**:65–73.
24. Fauquet, C., and Fargette, D., eds. 1988. *Proceedings International Seminar on African Cassava Mosaic Disease and Its Control.* Yamoussoukro, Côte d'Ivoire, 4–8 May, 1987. CTA/FAO/ORSTOM/IITA/IAPC.
25. Fauquet, C., Fargette, D., and Thouvenel, J.-C. 1988. Some aspects of the epidemiology of African cassava mosaic virus in Ivory Coast. *Trop. Pest Manage.* **34**:92–96.
26. Fauquet, C., Fargette, D., and Thouvenel, J.-C., 1988. Epidemiology of African cassava mosaic on a regional scale in the Ivory Coast. In C. Fauquet and D. Fargette, eds., *Proceedings International Seminar on African Cassava Mosaic Disease and Its Control.* Yamoussoukro, Côte d'Ivoire, 4–8 May, 1987. CTA/FAO/ORSTOM/IITA/IAPC, pp. 131–135.
27. Fauquet, C., Fargette, D., and Thouvenel, J.-C. 1988. Selection of healthy cassava plants obtained by reversion in cassava fields. In C. Fauquet and D. Fargette, eds., *Proceedings International Seminar on African Cassava Mosaic Disease and Its Control.* Yamoussoukro, Côte d'Ivoire, 4–8 May, 1987. CTA/FAO/ORSTOM/IITA/IAPC, pp. 146–149.
28. Fauquet, C., Fargette, D., and Thouvenel, J.-C. 1988. The resistance of cassava to African cassava mosaic. In C. Fauquet and D. Fargette, eds., *Proceedings International Seminar on African Cassava Mosaic Disease and Its Control.* Yamoussoukro, Côte d'Ivoire, 4–8 May, 1987. CTA/FAO/ORSTOM/IITA/IAPC, pp. 183–188.
29. Fulton, R. 1986. Practices and precautions in the use of cross protection for plant virus disease control. *Annu. Rev. Phytopathol.* **24**:67–81.
30. Gold, C.S., Altieri, M.A., and Bellotti, C. 1989. Cassava intercropping and pest incidence: A review illustrated with a case study from Colombia. *Trop. Pest Manage.* **35**:339–344.
31. Guthrie, J. 1990. *Controlling African Cassava Mosaic Disease.* C.T.A. Wageningen, Netherlands.
32. Hahn, S.K., Terry, E.R., and Leuschner, K. 1980. Breeding cassava for resistance to cassava mosaic disease. *Euphytica* **29**:673–683.
33. IITA. 1986. Impact of IITA cassava varieties. *Annual Report and Research Highlights: 1985*. International Institute for Tropical Agriculture, Ibadan, Nigeria, pp. 103–104.
34. Jameson, J.D. 1964. Cassava mosaic disease in Uganda. *East Afr. Agric. For. J.* **29**:208–213.

35. Jennings, D.L. 1957. Further studies in breeding cassava for virus resistance. *East Afr. Agric. J.* **22**:213–219.
36. Jennings, D.L. 1960. Observations on virus diseases of cassava in resistant and susceptible varieties. I. Mosaic disease. *Emp. J. Exp. Agric.* **28**: 23–34.
37. Jennings, D.L. 1976. Breeding for resistance to African cassava mosaic disease. In B.L. Nestel, ed., *Report of an Inter-disciplinary Workshop held at Muguga, Kenya, 19–22 February 1976*. IDRC-07le, Ottawa, Canada, pp. 39–44.
38. Jennings, D. 1988. Host-virus relationships of resistant cassava and ACMV: Some implications for breeding and disease control. In C. Fauquet and D. Fargette, eds., *Proceedings International Seminar on African Cassava Mosaic Disease and Its Control*. Yamoussoukro, Côte d'Ivoire, 4–8 May, 1987. CTA/FAO/ORSTOM/IITTA/IAPC, pp. 170–176.
39. Kaiser, W.J., and Louie, R. 1982. Heat therapy of cassava infected with African cassava mosaic disease. *Plant Dis.* **66**:475–477.
40. Kaiser, W.J., and Teemba, L.R. 1979. Use of tissue culture and thermotherapy to free East African cassava cultivars of African cassava mosaic and cassava brown streak diseases. *Plant Dis. Rep.* **63**:780–784.
41. Leuschner, K. 1977. Whiteflies: biology and transmission of African mosaic disease. In T. Brekelbaum, A. Bellotti, and J.C. Lozano, eds., *Proceedings Cassava Protection Workshop*. CIAT, Cali, Colombia, pp. 51–58.
42. Lozano, J.C. 1986. Cassava bacterial blight: a manageable disease. *Plant Dis.* **70**:1089–1093.
43. Malathi, V.G., Thankappan, M., Nair, N.G., Nambisan, B., and Ghosh, S.P. 1988. Cassava mosaic disease in India. In C. Fauquet and D. Fargette, eds., *Proceedings International Seminar on African Cassava Mosaic Disease and Its Control*. Yamoussoukro, Côyte d'Ivoire, 4–8 May, 1987. CTA/FAO/ORSTOM/IITA/IAPC, pp. 189–198.
44. Nichols, R.F.W. 1947. Breeding cassava for virus resistance. *East Afr. Agric. J.* **12**:184–194.
45. Ollennu, L.A.A., Owusu, G.K., and Thresh, J.M. 1989. The control of cocoa swollen shoot disease in Ghana. *Cocoa Growers' Bull.* **42**:25–35.
46. Pacumbaba, R.P. 1985. Virus-free shoots from cassava stem cuttings infected with cassava latent virus. *Plant Dis.* **69**:231–232.
47. Posnette, A.F. 1969. Tolerance of virus infection in crop plants. *Rev. of Appl. Mycol.* **48**:113–118.
48. Posnette, A.F., and Todd, J.McA. 1955. Virus diseases of cacao in West Africa. IX. Strain variation and interference in virus IA. *Ann. Appl. Biol.* **43**:433–453.
49. Robertson, I.A.D. 1987. The whitefly *Bemisia tabaci* (Gennadius) as a vector of African cassava mosaic virus at the Kenya Coast and ways in which the yield losses in cassava *Manihot esculenta* Crant caused by the virus can be reduced. *Insect Sci. Appl.* **8**:797–801.
50. Russell, G.E. 1978. *Plant Breeding for Pest and Disease Resistance*. Butterworth, London.
51. Seif, A.A. 1982. Effect of cassava mosaic virus on yield of cassava. *Plant Dis.* **66**:661–662.
52. Storey, H.H. 1936. Virus diseases of East African plants: VI. A progress report on studies of the disease of cassava. *East Afr. Agric. J.* **2**:34–39.

53. Storey, H.H., and Nichols, R.F.W. 1938. Studies of the mosaic diseases of cassava. *Ann. Appl. Biol.* **25**:790–806.
54. Terry, E.R. 1982. Resistance to African cassava mosaic and productivity of improved cassava cultivars. In E.H. Belen and M. Villanueva, eds., *Proceedings 5th Symposium International Society Tropical Root and Tuber Crops*. PCARRD, Philippines. pp. 415–429.
55. Terry, E.R., and Hahn, S.K. 1980. The effect of cassava mosaic disease on growth and yield of a local and an improved variety of cassava. *Trop. Pest Manage.* **26**:34–37.
56. Thresh, J.M. 1982. Cropping practices and virus spread. *Annu. Rev. Phytopathol.* **20**:193–218.
57. Thresh, J.M. 1988. Eradication as a virus disease control measure. In B.C. Clifford and E. Lester, eds., *Control of Plant Diseases: Costs and Benefits*. Blackwell, Oxford, pp. 155–194.
58. Thresh, J.M., Fishpool, L.D.C., Fargette, D., and Otim-Nape, G.W. 1994. African cassava mosaic virus: an underestimated and unsolved problem, *Trop. Sci.*: In Press.
59. Tidbury, G.E. 1937. A note on the yield of mosaic-diseased cassava, *East Afr. Agric. J.* **3**:119.
60. Tothill, J.D., ed. 1940. *Agriculture in Uganda*. Oxford University Press, London.

9
Practices and Perspective of Control of Papaya Ringspot Virus by Cross Protection

Shyi-Dong Yeh and Dennis Gonsalves

Introduction

Papaya (*Carica papaya* L.) is widely grown in tropical and subtropical areas for its edible fruit and delicate taste. The plant grows fast and the fruit can be harvested 8–10 months after transplanting the tree in the field. It continues producing fruit for 2–3 years under normal conditions. The delicious and nutritious fruit contains a popular protease, papain, which helps to digest food for assimilation. The extensive adaption of this plant and wide acceptance of the fruit offer considerable promise for papaya as a commercial crop for local and export purpose. Like banana, pineapple, and mango, papaya is one of the important cash crops in the tropics and subtropics.

However, a destructive disease caused by papaya ringspot virus (PRV) is a major obstacle to wide-scale planting of this fruit tree. PRV has been reported as a major limiting factor for growing papaya in Hawaii (46, 52), Florida (12, 16), Caribbean countries (2, 46, 92), South America (43, 46), Africa (49), India (9, 91), and the Far East (101). In papaya it causes mottling and distortion of leaves, ringspots on fruit, and water-soaking streaks on stems and petioles. It stunts the plant and drastically reduced the size of the fruit (15).

PRV is a member of *Potyvirus*, with filamentous particles of 780 × 12 nm. It is stylet-borne by aphids and sap-transmissible. PRV has a narrow host range, which includes species of three dicotyledonous families, Caricaceae, Chenopodiaceae, and Cucurbitaceae (83). Most strains of the virus belong to one of two major types, which are serologically very closely related. Type P (papaya infecting) isolates cause

Shyi-Dong Yeh, Department of Plant Pathology, National Chung Hsing University, Taichung, Taiwan, Republic of China.
Dennis Gonsalves, Department of Plant Pathology, New York State Agricultural Experiment Station, Cornell University, Geneva, New York 14456, USA.

important disease in papaya. Type W (watermelon mosaic virus 1) isolates cause important diseases in watermelon and other cucurbits, but do not infect papaya. PRV contains a single-stranded positive sense RNA of about 40 S (20, 106). The virus has a single type of capsid protein (CP) of M_r 36,000 (36K) (38, 82). It induces cylindrical (pinwheel)inclusion (CI) (81) and amorphous inclusion (AI) (63) in the cytoplasm of host cells. The former consists of a protein of 70K (cylindrical inclusion protein, CIP; 105) and the latter consists of a protein of 51K (amorphous inclusion protein, AIP; 21, 22). The nucleotide sequence corresponding to the 3′ end region of the viral genome including the complete coat protein gene was determined (84). Recently, the complete nucleotide sequence of the RNA genome and the genetic organization of PRV has been elucidated (111).

Although tolerant selections of papaya have been described (13, 14), resistance to PRV does not occur within *C. papaya* (13, 14, 16, 101). Some species of *Carica* are resistant to PRV (13, 14, 16, 67). Unfortunately these species are incompatible with *C. papaya* and conventional interspecific hybridization is difficult (67). *In vitro* embryo culture techniques have been successfully employed to rescue hybrid embryos (48, 61, 62), however, backcrossing of the interspecific hybrids with *C. papaya* remained difficult. A diligent roguing program has been practiced successfully in Hawaii to suppress the spread of PRV in certain areas of the state (72). However, roguing is not a permanent solution for an area without geographic isolation, and it is impossible to eradicate the virus sources where the disease has become endemic. Thus, the unavailability of PRV-resistant papaya varieties, the restrictive host range, the difficulty for eradication, and the great loss caused by PRV make cross protection an attractive method of controlling this virus.

Cross protection, first described by McKinney in 1929 (65) with tobacco mosaic virus (TMV), describes the phenomenon in which plants systemically infected with one strain of a virus are protected from infection by a second related strain of the same virus. However, wide-scale adoption of this technique for control of TMV did not occur until Rast (86) produced an almost symptomless mutant (MII-16) from a common tomato strain of TMV using the nitrous acid mutagenic treatment (36). The symptomless mutant has been manufactured commercially (87) and has been applied to a high proportion of glasshouse-grown tomato crops in The Netherlands and the United Kingdom since 1970 (29, 30, 87). Successful wide-scale control of tomato mosaic disease with an attenuated mutant (L11 A), which was isolated from the tomato strain of TMV in plants treated with high temperature, was also reported in Japan (76). Cross protection also has been used on a large scale to control citrus tristeza virus (CTV), a closterovirus, that is important worldwide (79). Naturally occurring mild strains of CTV have been selected and can offer protection in the field (31, 39, 40, 69, 74, 75, 93). In Brazil, the number of protected sweet orange trees exceeded 8 million in 1980, and no breakdown in protection was observed (17).

The key for practical application of cross protection is the availability of a protective virus strain. In the past decade, our laboratories have been searching for suitable mild strains of PRV for use of cross protection as a control measure. Two mild mutants, PRV HA 5-1 and 6-1, were obtained by nitrous acid induction (104). The mutants have been carefully evaluated for their cross protection effectiveness under greenhouse and field conditions (104). The success and the drawbacks of this unique control method are discussed in this article. In accordance with the newly elucidated complete nucleotide sequence of PRV (111), an approach for generation of stable attenuated virus strains by recombinant DNA techniques is described. Current development in genetically engineered cross protection, in which coat protein gene of a virus is integrated and expressed in its host encouraged us to shift our reach on the coat protein induced resistance in papaya. Current achievements of the new approach are also summarized.

Generation of Mild Virus Strains of PRV

Efforts to select naturally occurring mild strains of PRV by field collection or single-lesion isolation from natural virion population were not successful (51, 104). In an attempt to induce production of mild strains, crude sap from PRV-infected squash was treated with nitrous acid and the sap was used to inoculate *Chenopodium quinoa*, a local-lesion host. Two mutants, designated PRV HA 5-1 and PRV HA 6-1, that produced no conspicuous symptoms on papaya in the greenhouse were selected (104). Papaya seedlings inoculated with these mild mutants remained symptomless or showed diffuse mottling with no reduction in plant growth. Under greenhouse condition, protection was observed when PRV HA 5-1 was used to protect papaya against the parental strain (104). The results indicated that the mutant could be used as a protectant for control of PRV.

The mild PRV stains were introduced from Cornell University to Taiwan in 1983. Tests were conducted at Feng-Shan Tropical Horticultural Experiment Station under greenhouse and field conditions to study the potential of cross protection for control of ringspot disease of papaya in Taiwan. Under greenhouse conditions, both mutants caused only mild or symptomless infection on test plants of Chenopodiaceae and Cucurbitaceae families and on the major papaya cultivar of Taiwan, Tainung No. 2 (102). Also, the mild mutants provided a high degree of protection in papaya against the severe effects of the wilting and the mosaic strains of PRV that are prevalent in Taiwan (10). These results indicated that both mutants had great potential for controlling PRV by cross protection in Taiwan.

Preliminary field tests were conducted in PRV endemic areas of southern Taiwan in autumn 1983 (102). When protected papaya plants were planted with unprotected controls randomly or row by row under high disease pressure, protected plants showed severe symptoms 1–2 months later than the controls and the protection was not economically beneficial. However, in a test plot where protected and unprotected plants were established in solid blocks and where the disease pressure inside the test orchard was minimized by roguing once every 10 days up to the flowering stage, protected trees showed 82% higher yield than unprotected trees (102). This resulted in a 111% increase in income because of better fruit quality from protected trees (102). The success in the preliminary field tests brought hope for the use of cross protection to control the devastating PRV in Taiwan.

Large Scale Application in Taiwan

Since papaya is propagated by seeds, an efficient method of inoculating large numbers of seedlings is essential for the practical application of cross protection in the field. Mass inoculation by pressure spray was developed in 1983 (102). Inoculum was propagated in *Cucumis metuliferus* and ground in cold potassium phosphate buffer. The extracts were strained through cheesecloth, then mixed with the abrasive carborundum in a metal tank connected to a spray-gun attached to an air compressor. By this method, a person can inoculate 10,000 seedlings within 2 hr. In Taiwan, individual seedlings are grown singly in small plastic bags and arranged in wooden boxes (100 plants per box) for easy handling during transplanting, so the additional cost of spraying is minimal.

Large-scale field trials in the spring (44,000 protected plants for 22 ha) and autumn of 1984 (200,000 protected plants for 100 ha) were carried out by the support of the Council of Agriculture of the Republic of China. These tests reflected the results of the initial trial of 1983 with more than 100% increase in total yield in the protected orchards (110). The method was warmly welcomed by the growers. In the autumn of 1986, the government transferred the inoculation responsibilities to two private and several government nurseries and expanded the project. The inoculations and seedling distribution went smoothly because of well-organized teamwork, strong support by the central and local government, and growers' great appreciation. Up to 1991, the protected papaya orchards have reached 1722 ha and near three and a half million papaya seedlings have been treated with the mild virus strains (Table 9.1). Use of cross protection for control of PRV has become a routine practice in Taiwan.

TABLE 9.1. Application of cross protection for control of PRV in Taiwan.[a]

Year	Planting time	Hectares
1884	Spring	22
1984	Spring	100
1985	Autumn	250
1986	Autumn	500
1987	Autumn	250
1988	Autumn	250
1989	Autumn	50
1990	Autumn	100
1991	Autumn	200
Total		1722

[a] Only seedlings inoculated under authorized license are recorded. A total of 3,444,000 papaya seedlings were treated (2000 plants/ha).

Application in Hawaii

Cross protection trials in farmer fields were started on the island of Oahu in 1986 where the virus has limited papaya production for a number of years. Results from Taiwan suggested that cross protection broke down under heavy disease pressure and over time, which indicated that the lack of long lasting cross protection might be due to strain differences. Thus, it was very interesting to determine how cross protection would do in Hawaii since the mild strains originated from a severe Hawaiian strain (104).

Extensive field trials clearly showed that cross protection was very effective in Hawaii, as compared to Taiwan and other areas mentioned below. A trial in Waianae on the Island of Oahu showed that cross protection allowed economic papaya production for at least 2 years (64). Several conclusions could be drawn from these trials: first, protection was effective and long lasting since breakdown was not observed even 2.5 years after the trial commenced; second, cross-protected trees showed obvious foliage symptoms and slight but significant reduction in fruit size and tree height; and third, cross protection was a viable approach to papaya production in areas of Hawaii where the virus causes severe damage. Since these trials were done with only one variety of solo papaya, "Line 8," the question still remained as to whether cross protection would be practical with the other more popular varieties that are grown on Oahu, mainly "Waimanalo" and "Sunrise." Subsequent tests on Oahu with these varieties (Ferreira et al., unpublished data) gave similar results as the trials for "Line 8," with the important exception that severity of the ringspots on fruits varied with the varieties. "Sunrise" was especially affected. Ringspots were very obvious and often sunken. As a whole, it appeared that effect of the mild strain was least noticeable on fruit of "Line 8," intermediate on "Waimanalo," and very obvious on "Sunrise." Thus, these trials showed that cross protection was more effective in Hawaii than in Taiwan, and that differences in effectiveness were largely due to differences in strains.

Application in Other Parts of the World

Trials in other parts of the world also showed the influence of viral strains on the effectiveness of cross protection. Starting in 1986, a number of field trials were carried out in Northeast Thailand using the Hawaiian mild strains. Unlike the previous cross protection trials, efforts were aimed at controlling the disease in papaya plants that were grown in "backyards" of farmers. This represented a major challenge in logistics. However, work by Vilai Prasartsee and colleagues at the Northeast Regional Office of Agriculture in Tha Pra clearly showed that the logistics of applying cross protection to "backyards" of farming villages could be worked out, but unfortunately, cross protection experiments using the Hawaiian mild strains were not effective. Comparative greenhouse tests clearly showed that the lack of protection was due to differences in strains. That is, the Thailand strain of PRV broke down the protection afforded by the Hawaiian mild strains.

Limited cross protection trails have been done in Florida and Mexico. A field trial in Florida using a papaya variety that has tolerance to PRV indicated that cross protection may be effective against Florida strains (66). However, trials will need to be done with susceptible varieties to draw firm conclusions. The trial in Mexico indicated that cross protection would not work using the Hawaiian mild strains.

Potential Disadvantages

During the past 8 years of application, we have learned several potential disadvantages of the application. Using this method to obtain maximal economic benefit, factors affecting the effectiveness of cross protection have to be noticed (37).

Effects of mild virus strains on papaya: Under warm temperatures, the papaya plants inoculated with the mild mutants grew well and did not show conspicuous symptoms. However, when the temperature dropped below 20°C and under shady and rainy conditions, chlorotic spots on leaves and ringspots on fruits set at this period. There were no deformations or reduction in fruit size. New leaves did not develop chlorotic spots when the temperature rose above 25°C, and ringspots on fruits also became less apparent when the fruits were ripe. These mild effects of the mutants did not adversely affect plant growth and fruit production. The mild symptoms could be repressed by applying more fertilizers to the protected trees.

Different varieties of papaya reacted slightly different with the mild strains. As described in previous sections, they are mild on the variety of Tainung No. 2 (Taiwan), Kapoho Solo, and Line 8 (Hawaii), moderate on Waimanalo (Hawaii), but rather severe on the Sunrise variety. Therefore, the protective strains were not recommended to be used on the Sunrise papaya.

Inoculation and indexing: The pressure spray is a very efficient method for mass inoculation. The additional cost is only about U.S. $28 for 1 ha (2000 seedlings) (110), whereas the economical benefits were several hundred times the additional cost. To keep the spray-gun at the right distance and the right pressure needs practice and sometimes the infection rate varies from person to person. To increase the chances of infection, seedlings were inoculated twice at an interval of 3–5 days. To monitor the results of the inoculation, 1% of the seedlings treated were randomly selected from each wooden box for ELISA test 3 weeks after inoculation (110). However, this increased the difficulty of handling during transportation because the seedlings grew out of the wooden box after 3 weeks. When seedlings were treated under high temperature, such as 30–35°C in summer, the infection rate was very low, this might be due to the instability of the virus or unfavorable physical conditions of the papaya seedlings. Thus, inoculation under high temperature conditions should be avoided.

Long-term storage and propagation of the inoculum: In greenhouse tests, severe symptoms caused by possible revertants were not seen, indicating that the mild mutants are stable. However, severe symptoms on papaya was occasionally observed when PRV HA 5-1 was recovered from $CaCl_2$ dried tissue of *C. metuliferus* after long-term storage at 4°C. Therefore, precautions must always be taken to index and evaluate the inocula before large-scale application. In addition, propagation of mild strains in *C. metuliferus* for mass inoculation has to be carried out in large quantity and in a well-isolated screenhouse to avoid the possible contamination from nature.

Unfavorable winter conditions: The mild virus strain infected papaya seedlings systemically and persisted throughout the plants' lifetime, so protected plants did not require reinoculation. In Taiwan, protected trees normally began to produce fruit 8–10 months after being transplanted and continued to yield fruit for another 5–6 months. The orchards were then abandoned because of unfavorable winter conditions and high incidence of breakdown in cross protection. Control of PRV in Taiwan by cross protection has not restored papaya plants to perennial status yet, but growers no longer risk losing the whole crop due to PRV. As noted in Hawaii, protected trees produced fruit for 2 years.

Disease pressure: In Taiwan, when the protected papaya orchards were located in an area with high disease pressure, such as adjacent to an old diseased orchard or surrounded with severely infected orchards, the protection may not be economically beneficial because the breakdown might happen before flowering. Also, the secondary spread of the virus in a papaya orchard is much faster than the primary spread in which the virus was introduced by aphids from outside sources. Therefore, even with the protected plants a grower still has to pay attention to the sanitation of the orchards to remove diseased plants as soon as possible and to clean the virus sources surrounding the protected orchard to minimize the challenge pressure. In contrast to Taiwan, the protection in Hawaii lasted longer than 2 years and it seems the disease pressure is not a problem there.

Possible revertants: Under screenhouse condition, papaya inoculated with mild HA 5-1 did not show any severe symptoms even though the plants were kept for 2–3 years, the mild mutant were considered rather stable. However, because the mild strains were originally induced by nitrous acid treatment that resulted in only point mutations (36, 70, 90), the possibility of revertant arisen from natural mutations cannot be excluded. Because of lack of suitable serology and nucleic acid probes to distinguish the possible severe revertant from the mild strains, it is difficult to determine whether the revertants do happen under field conditions or not.

Problems of Strain-Specific Protection

In the greenhouse and field tests, protection by the mild virus strains delayed the expression of severe symptoms only under certain conditions, and superinfection with the severe strain did occur (102, 104). When protection broke down before flowering stage, there was no economical return. There are several possibilities for the breakdown: (1) the challenging virus is introduced into the nonexpanded young leaves around apex, where the concentration of mild virus strains are lower and thus higher incidence of breakdown was noticed (104); (2) severe challenge pressure, such as a protected papaya orchard adjacent to severely infected orchards, usually resulted in high frequencies of breakdown; (3) inoculated papaya seedlings escape infection with the mild strain and could not provide any degree of cross protection; (4) poor management and neglect of removing superinfected plants in the protected orchards, which would build up tremendous challenge pressure inside the orchards and drastically shorten the protection period; and (5) severe strains different from the parental virus of mild mutants exist. Some of these problems can be easily avoided, such as inoculating papaya seedlings carefully to ensure an infection rate of 100%, planting the protected orchard in a cleaner place, and eradicating the severely infected trees in the protected orchard to minimize the challenge pressure.

One of the major concerns was the strain-specific protection by the mild virus strains. The mild strains were induced from a severe strain HA originated from Hawaii (38). Under greenhouse conditions a very high degree of protection (90–100%) against HA was observed (104), but relatively low protection rates (50–60%) were obtained when these mild mutants were used to protect papaya against the severe strains from Taiwan (102). The strain-specific protection was clearly reflected in the application of the mild strains in different parts of the world. In Hawaii, the protection was excellent, even where the protected plants were randomly mixed with the unprotected ones. Moreover, protected orchards planted in solid block produced fruit for 2 years (64). In Taiwan, the protection is economically beneficial, but high frequency of break-down still limited papaya as an annual crop. In Mexico and Thailand, the protection was not economically viable. Thus, the strain-specific protection limits the application of PRV HA 5-1 and HA 6-1 in different areas of the world. Efforts should be made to obtain mild strains from a local severe strain to improve the protection effectiveness.

Search for Mild Strain Derived from Local Severe Strains

To minimize the incomplete protection that resulted from the strain-specific protection problem, a mild strain derived from a local prevalent strain becomes essential. Substantial research has been conducted on nitrous-acid mutagenic treatment (104) of two local severe strains of Taiwan, PRV-Wilt and PRV-Mosaic (10, 51). Through a total of more than 3000 single-lesion isolation from 35 independent experiments from 1985 to 1988 in Taiwan, more than 50 attenuated strains that caused milder symptoms on papaya than the parental strains were obtained (S.D. Yeh, unpublished). However, none of them was proved to be stable or mild enough for practical use. Compared to the PRV HA 5-1 and HA 6-1 which were obtained from 663 single-lesion isolation from only 7 experiments (104), the endeavors on the artificial induction on Taiwan severe strains were enormous. It seems that the chance to obtain a useful attenuated strain varied with different virus strains.

Cucumis metuliferus (Naud.) Mey. Acc. 2459 has been used as the best propagation host for PRV for the purpose of purification and other basic studies (80, 106, 109). PRV HA 5-1 and HA 6-1 were found to be able to overcome the resistance governed by a single dominant gene *Wmv* in *C. metuliferus* line PI 292190, in which resistance was defined by immunity to the parent severe strain PRV HA and susceptibility to the mild strains as manifested by systemic mosaic and necrosis symptoms (80, 108).

Moreover, the systemic and necrosis symptoms caused by HA 6-1 were found in all F_1 test plants of Acc. 2459 × PI292190 and perfectly followed the segregation of the *Wmv* gene in the F_2 population, indicating that the symptoms are direct interactions between the mild virus strain and the *Wmv* gene (108). The possibility of using the resistant line of *C. metuliferus* to screen useful attenuated strains of PRV for cross protection was investigated. Virus in crude sap from susceptible *C. metuliferus* Acc. 2459 infected with PRV TM, a local strain prevalent in Taiwan (10), was treated with nitrous acid, enrich-propagated in plants of line Acc. 2459, and then transferred to plants of the resistant line PI 292190 (108). Virus isolates that induced systemic symptoms on line PI 292190 were selected and followed by repeated serial dilutions on the same line. The isolates were then transferred to papaya to determine their pathogenicity. Among 20 isolates selected, six induced mild symptoms on papaya and one, PRV TM-1, caused infection without conspicuous symptoms. Under greenhouse conditions, papaya inoculated with PRV TM-1 were completely protected against PRV TM when challenge inoculation were at 23, 45, or 90 days after preimmunization (108). Unfortunately, the mild strain caused severe infection under field conditions and was not applicable for cross protection (S.D. Yeh, unpublished). Nevertheless, the results indicated that the resistant *C. metuliferus* can be used as a selective host for screening attenuated virus stains of PRV from artificial induction. Although the mechanism involved in this unique system needs to be further studied, the system provides a good alternative for the selection of attenuated strains in addition to the single-lesion isolation method.

Cross Protection Integrated with Tolerant Varieties

Although resistant *C. papaya* species was not available, selections with tolerance to PRV have been reported (13, 14, 67). A tolerant variety introduced from Florida (kindly donated to R.J. Chiu by the late R.A. Conover) has been released in the field in Taiwan. When this variety was protected with HA 5-1, fruit yield increased significantly, but poorer fruit quality and dioecious sexuality made it less acceptable than other commercial varieties. In 1988, a new tolerant variety Tainung No. 5 bred by Feng-Shan Tropical Horticultural Experiment Station was released. It was derived from a Florida variety and a Costa Rica variety. The new variety possesses high degree of tolerance to the virus and with hermaphroditic sexuality. However, the fruit quality still needs to be improved.

Because of the dramatic increase in yield, the planting acreage of Tainung No. 5 has been increasing. Due to the tolerance, the growers usually were reluctant to remove the diseased plants that still produced fruits. This created a tremendous reservoir for PRV and built up an extremely high inoculating pressure for other commercial varieties, thus significantly shortening the effectiveness of cross protection by mild strains. To solve these side effects of using the tolerant variety, the integration of cross protection with the tolerant variety is strongly recommended.

Construction of Attenuated Strains by Recombinant DNA Technique

The mutations induced by nitrous acid which causes deamination of the bases are not predictable (90). Selection of attenuated PRV strains after nitrous acid treatment is tedious and laborious, and sometimes depends on luck. The advance in molecular plant virology encouraged us to shift our efforts for generating mild mutants by recombinant DNA techniques.

Infectious RNA transcript generated from full-length cDNA clones of plant and animal RNA viruses have been reported (3, 19, 41, 68, 95, 99). This technique has become important for studying the molecular biology of RNA viruses. Site-directed mutagenesis as well as other manipulation of the DNA template have facilitated the study of plant virus gene functions, such as the internal promoter involved for coat protein synthesis (32) and a region in the CP gene of potyvirus that determines aphid transmissibility (6). A strain of TMV that produces mild symptoms on tobacco plants has been molecularly cloned to identify the region of the genome responsible for symptom attenuation (45). Exchange of restriction fragments between the infectious mild strain cDNA and an infectious U1 strain cDNA indicated that the determinants involved in symptom attenuation reside in the open reading frame encoding the 126/183-kDa proteins of TMV (45).

Although the potyviral genomes are longer than most of the other plant virus groups, three infectious transcripts generated from different potyviruses have been reported, including tobacco vein mottling virus (25), plum pox virus (88), and zucchini yellow mosaic virus (34). The severe strain of PRV HA and the mild strains HA 5-1 and HA 6-1 provide an excellent system for analyzing the gene functions of PRV in relation to pathogenicity of the virus. Our efforts has shifted to construct the infectious transcripts of the severe and the mild PRV strains. Through appropriate recombination at the cDNA level, the mutation correlated to attenuation of the virus can be pinpointed.

The nucleotide sequence of the 3′-terminal region of PRV HA 5-1, including most of the NIb gene, complete CP gene, and the 3′ untranslated region, has been determined (71, 84). The comparison with the nucleotide sequence of the parental severe strain HA showed that the two strains shared a 99.4% identity in their 3′-terminal 2235 nucleotides, and there were five and two amino acid residues different at NIb gene and CP gene, respectively (100). The complete nucleotide sequence of PRV HA has recently been elucidated (111). The virus generates at least eight proteins and its genetic organization is similar to other reported potyviruses (4, 24, 47, 60, 89, 98) except that the first protein processed from the N-terminus of the polyprotein has an M_r of 63K, 18–34K larger than those of the other potyviruses (22, 85, 106, 107, 111). These recent advances in the studies of PRV RNA provide a solid base that may enable us to construct infectious transcripts both from HA and HA 5-1. The understanding of the essential changes responsible for the mildness of HA 5-1 can further extend our work to construct a nonpathogenic mutant by modifying the virus at the cDNA level. The mild strain derived from the recombinant DNA techniques by deleting the essential domain for the pathogenicity can avoid the potential disadvantages of the mild strains derived by nitrous acid induction. This approach should produce more stable and predictable mild strains that can be used for control of the virus by cross protection.

Innovative Approach on Genetically Engineered Cross Protection

A number of models have been proposed to explain the mechanism of cross protection (35, 78). The most acceptable one proposed that capsid protein (CP) produced by the first virus infection prevents the uncoating or encapsidates the RNA of the challenging strain, thereby interfering with its replication (23). Based on this hypothesis the CP gene of tobacco mosaic virus has been introduced into tobacco by Ti vector (1, 8). Plants regenerated from transformed cells expressed TMV mRNA and CP gene and symptom development was delayed when they were challenged with TMV. This genetically engineered resistance also has been demonstrated in several other plant viruses, such as alfalfa mosaic virus (59, 94, 96), cucumovirus (18), Ilarvirus (97), tobravirus (96), potexvirus (42), and potyvirus (50). The genetic characteristic of these transgenic plants is that the virus CP gene becomes the nuclear trait of the plants and they become resistant to infection by the corresponding viruses. The phenomenon of CP-induced protection is very similar to the conventional cross protection. A true genetically engineered cross protection was demonstrated by Yamaya et al. (103) where a full length cDNA to the RNA genome of an attenuated strain of TMV was integrated into tobacco by Ti vector, expressed in the cells, and replicated in the transgenic plants as the original virus. This transgenic plants provided a much higher degree of protection as compared to the transgenic plants expressing CP alone (103).

The innovative approach of genetically engineered cross protection can avoid many of the potentially disadvantages of conventional cross protection as a control measure, such as the cost of inoculation, possible revertants, dissemination of the mild virus to other crops, and adverse effects of the attenuated strains on the host. Our efforts has shifted to clone, transfer, and express the capsid protein gene of PRV in transgenic papaya that may then become resistant to severe infection of the virus. A report by Nelson et al. (73) indicated that a transgenic tomato containing U1-TMV CP are more resistant to U1-TMV than to other severe strains such as PV230 and a strain of tomato mosaic virus (strain L). Therefore, the local severe strain should be chosen as the working virus to avoid the possible problem generated by strain-specific protection.

Since the nucleotide sequence of PRV CP has been elucidated (84, 111), to clone and construct an expressible PRV CP gene in plants become workable. In addition to the construction of the target PRV gene and the availability of Ti-vector (7, 44), the regeneration of papaya plantlets from the callus or protoplast level by tissue culture techniques also plays an important role for successful papaya transformation. Regeneration from callus from different explants such as shoot tips (112), seedling petiole segments (5), ovules (54, 55, 56, 57), cotyledon (58) root tips (11), and immature embryo of papaya (26) has been reported. Thus, an efficient system to regenerate transgenic papaya can be achieved by using or improving the available tissue culture techniques.

The PRV-CP gene of the mild strain (PRV HA 5-1) was engineered and subsequently transformed in tobacco and papaya. Ling et al. (53) reported that transgenic tobacco expressing PRV CP provided significant delay in symptom development and the symptoms were attenuated when challenge inoculated with tobacco etch, potato virus Y, and pepper mottle viruses. However, PRV was not used as a challenge virus since it does not infect tobacco. With papaya, Pang and Sanford (77) transferred nopaline synthase gene in papaya cells by an oncopositive Ti vector. Their transformed calli were tumor cells and did not regenerate into normal plants, however.

Recently, Fitch et al. (27) successfully incorporated the PRV CP gene of PRV HA 5-1 into papaya via microprojectile bombardment and obtained plants that expressed the CP gene. Moreover, they obtained R_0 transgenic plants that were resistant to infection by mechanical inoculation with the severe "homologous" PRV HA strain (28). One line (55-1) showed virtual immunity to infection by PRV HA in that none of the inoculated subcloned R_0 plants became infected and virus was not recovered from inoculated plants. R_0 plants of this line have been established in a field trial in Hawaii and initial results are very promising (R. Manshardt et al., unpublished data).

The development of transgenic papaya has allowed us to compare the effect of strain specificity in classical and engineered cross protection because the CP gene used in the transformation of papaya is of the mild strain, PRV HA 5-1. R_1 plants from line 55-1 were recently tested for resistance against 13 isolates of PRV from Hawaii, Mexico, Bahama Islands, Florida, Australia, Brazil, China, Ecuador, Guam, Jamaica, and Thailand (D. Gonsalves et al., unpublished data). The results showed a strain specificity that was similar to that observed with classical cross protection. For example, complete resistance was observed with plants inoculated with severe PRV HA while no resistance was observed for plants inoculated with the Thailand strain. In summary, no resistance was observed against PRV isolates from Australia, Brazil, China, Ecuador, Guam, Jamaica, and Thailand, while intermediate resistance was observed for strains from Mexico, Bahama Islands, and Florida. In the latter category, the resistance was not impressive since 31–70% of the inoculated plants became infected.

Thus transgenic plants expressing the coat protein gene of PRV HA 5-1 provide a very promising way for controlling PRV in Hawaii. However, our observations predict that transgenic papaya expressing the coat protein gene of PRV HA 5-1 will not be effective in many regions outside of Hawaii. These observations are similar to our results with classical cross protection using PRV HA 5-1.

Concluding Remarks

Although cross protection is a general phenomenon with plant viruses, not all plant diseases caused by viruses can be controlled by a protective mild strain. Fulton (33) pointed out that careful attention to the selection of the best protective isolates of a virus and to their introduction into the crop to be protected is essential. The mild strains of the PRV have the general criteria as valuable protective strains: 1) do not cause severe damage to the protected papaya plants, 2) are stable for a long period, 3) protect plants against the effects of severe strains, 4) are suitable for infecting large numbers of plants by pressure spray, 5) do not affect other crops in the vicinity of the crop protected, and 6) have no synergistic reactions with other viruses. Cross protection in Taiwan has helped papaya growers to produce a fruitful crop. Results in Hawaii also showed that cross protection allows profitable production of papaya. However, during the past 8 years of application we have learned several drawbacks of this control measure. These include the adverse effects of mild strains on papaya under cool and rainy conditions, the additional cost of inoculating and indexing the seedlings, propagation and preservation of inoculum, possible occurrence of severe revertants, breakdown under severe disease pressure, and strain-specific protection. Most of these problems can be solved by different approaches. Application of cross protection integrated with cultural practices and the tolerant variety where applicable is recommended.

The most concern for breakdown seems to be the strain-specific protection. A mild strain derived from a local prevalent strain should be considered as the best protective strain. Since the traditional single-lesion isolation after nitrous acid mutagenic treatment is tedious and not a reliable or reproducible approach, our current efforts emphasize understanding the regions correlated to the attenuation of HA 5-1 by recombination at the cDNA level. The information obtained thus can be used to modify PRV at the cDNA level. A mild strain constructed by recombinant DNA techniques would be more stable and predictable and can minimize the potential advantages mentioned above.

The development of transgenic papaya that are resistant to PRV provides a promising approach for controlling this disease. However, initial results indicate that, like classical cross protection, resistance by engineered cross protection is strain specific. Nevertheless, genetically engineered cross protection avoids the potential disadvantages of classical cross protection. Data from field experiments will soon determine whether this approach will bring a revolution for papaya production in areas where the disease is severe.

References

1. Abel, P.P., Nelson, R.S., De, B., Hoffmann, N., Rogers, S.G., Fraley, R.T., and Beachy, R.N. 1986. Delay of disease development in transgenic plant that express the tobacco mosaic virus coat protein gene. *Science* **232**:738–743.
2. Adsuar, J. 1946. Studies on virus disease of papaya (*Carica papaya*) in Puerto Rico, I. Transmission of papaya mosaic. *Puerto Rico Agric. Exp. Stn. Tech. Pap.* 1.
3. Ahlquist, P., French, R., Janda, M., and Loesch-Fries, L.S. 1984. Muticomponent RNA plant virus infection derived from cloned viral cDNA. *Proc. Natl. Acad. Sci. U.S.A.* **81**:7066–7070.
4. Allison, R.F., Johnston, R.E., and Dougherty, W.G. 1986. The nucleotide sequence of the coding region of tobacco etch virus genomic RNA: Evidence for the synthesis of a single polyprotein. *Virology* **154**:9–20.
5. Arora, I.K., and Singh, R.N. 1978. Growth hormones and *in vitro* callus formation of papaya, *Sci. Hortic.* **8**:357–361.
6. Atreya, C.D., Raccah, B., and Pirone, T.P. 1990. A point mutation in the coat protein abolishes aphid transmissibility of a potyvirus. *Virology* **178**:161–165.
7. Bevan, M. 1984. Binary *Agrobacterium* vectors for plant transformation. *Nucl. Acids Res.* **12**:8711–8721.
8. Bevan, M.W., Mason, S.E., and Goelet, P. 1985. Expression of tobacco mosaic virus coat protein by cauliflower mosaic virus promoter in plants transformed by *Agrobacterium. EMBO J.* **4**:1921–1926.
9. Capoor, S.P., and Varma, P.M. 1948. A mosaic disease of *Carica papaya* L. in the Bombay province. *Current Sci.* **17**:265–266.
10. Chang, C.A. 1979. Isolation and comparison of two isolates of papaya ringspot virus in Taiwan. *J. Agric. Res. China* **28**:207–216.

11. Chen, M.H., Wang, P.J., and Maeda, E. 1987. Somatic embryogenesis and plant regeneration in *Carica papaya* L. tissue culture derived from root explants. *Plant Cell Rep.* **6**:348–351.
12. Conover, R.A. 1964. Distortion ringspot, a severe disease of papaya in Florida. *Proc. Fla. State Hortic. Soc.* **77**:440–444.
13. Conover, R.A. 1976. A program for development of papayas tolerant to the distortion ringspot virus. *Proc. Fla. State Hortic. Soc.* **89**:229–231.
14. Conover, R.A., and Litz, R.E. 1978. Progress in breeding papayas with tolerance to papaya ringspot virus. *Proc. Fla. State Hortic. Soc.* **91**:182–184.
15. Cook, A.A. 1972. Virus diseases of papaya. *Fla. Agr. Exp. Sta. Bull.* 750 (Tech.), 19pp.
16. Cook, A.A., and Zettler, F.W. 1970. Susceptibility of papaya cultivars to papaya ringspot and papaya mosaic virus. *Plant Dis. Rep.* **54**:893–895.
17. Cost, A.S., and Muller, G.W. 1980. Tristeza control by cross protection: A U.S.-Brazil cooperative success. *Plant Dis.* **64**:538–541.
18. Cuozzo, M., O'cnnell, K.M., Kaniewski, W., Fang, R.X., Chua, N.H., and Tumer, N. 1988. Viral protection in transgenic tobacco plants expressing the cucumber mosaic virus coat protein or its antisense RNA. *Biotechnology* **6**:549–557.
19. Dawson, W.O., Beck, D.L., Knorr, D.A., and Grantham, G.L. 1986. cDNA cloning of the complete genome of tobacco mosaic virus and production of infectious transcript. *Proc. Natl. Acad. Sci. U.S.A.* **83**:1832–1836.
20. De La Rosa, M., and Lastra, R. 1983. Purification and partial characterization of papaya ringspot virus. *Phytopathol. Z.* **106**:329–336.
21. De Mejia, M.V.G., Hiebert, E., and Purcifull, D.E. 1985. Isolation and partial characterization of the amorphous cytoplasmic inclusions associated with infections caused by two potyviruses. *Virology* **142**:24–33.
22. De Mejia, M.V.G., Hiebert, E., Purcifull, D.E., Thornbury, D.W., and Pirone, T.P. 1985. Identification of potyviral amorphous inclusion protein as a nonstructural, virus-specific protein related to helper component. *Virology* **142**:34–43.
23. De Zoeten, G.A., and Fulton, R.W. 1975. Understanding generates possibilities. *Phytopathology* **65**:221–222.
24. Domier, L.L., Franklin, K.M., Shahabuddin, M., Hellmann, G.M., Lomonossoff, G.P., Shaw, J.G., and Rhoads, R.D. 1986. The nucleotide sequence of tobacco vein mottling virus RNA. *Nucl. Acids Res.* **13**:5417–5430.
25. Domier, L.L., Franklin, K.M., Hunt, A.G., Rhoads, R.E., and Shaw, J.G. 1989. Infectious *in vitro* transcript from cloned cDNA of a potyvirus, tobacco vein mottling virus. *Proc. Natl. Acad. Sci. U.S.A.* **86**:3509–3513.
26. Fitch, M.M.M., and Manshardt, R.M. 1990. Somatic embryogenesis and plant regeneration from immature zygotic embryos of papaya (*Carica papaya L.*). *Plant Cell Rep.* **9**:320–324.
27. Fitch, M.M.M., Manshardt, R.M., Gonsalves, D., Slightom, J.L., and Sanford, J.C. 1990. Stable transformation of papaya via microprojectile bombardment. *Plant Cell Rep.* **9**:189–194.
28. Fitch, M.M.M., Manshardt, R.M., Gonsalves, D., Slightom, J.L., and Sanford, J.C. 1992. Virus resistant papaya plants derived from tissues

bombarded with the coat protein gene of papaya ringspot virus. *Bio/ Technology* **10**:1466–1472.
29. Fletcher, J.T. 1978. The use of avirulent virus strains to protect plants against the effects of virulent strains. *Ann. Appl. Biol.* **89**:110–114.
30. Fletcher, J.T., and Rowe, J.M. 1975. Observations and experiments on the use of an avirulent mutant strain of tobacco mosaic virus as a means of controlling tomato mosaic. *Ann. Appl. Biol.* **81**:171–179.
31. Frazer, L.R., Long, K., and Cox, J. 1968. Stem pitting of grapefruit-field protection by the use of mild virus strains. *Proceeding, 4th Conference of the International Organization of Citrus Virologists*, pp. 27–31.
32. French, R., and Ahlquist, P. 1988. Characterization and engineering of sequences controlling *in vivo* synthesis of brome mosaic virus subgenomic RNA. *J. Virol.* **52**:2411–2420.
33. Fulton, R.W. 1986. Practices and precautions in the use of cross protection for plant virus disease control. *Annu. Rev. Phytopathol.* **24**:67–81.
34. Gal-On, A., Antignus, Y., Rosner, A., and Raccah, B. 1991. Infectious *in vitro* RNA transcripts derived form cloned cDNA of the cucurbit potyvirus, zucchini yellow mosaic virus. *J. Gen. Virol.* **72**:2639–2643.
35. Gibbs, A. 1969. Plant virus classification. *Adv. Virus Res.* **14**:263–328.
36. Gierer, A., and Mundry, K.W. 1958. Production of mutants of tobacco mosaic virus by chemical alteration of its ribonucleic acid. *Nature (London)* **182**:1457–1458.
37. Gonsalves, D., and Garnsey, S.M. 1989. Cross protection techniques for control of virus diseases in the tropics. *Plant Dis.* **73**:592–597.
38. Gonsalves, D., and Ishii, M. 1980. Purification and serology of papaya ringspot virus. *Phytopathology* **70**:1028–1032.
39. Grant, T.J., and Costa, A.S. 1951. A mild strain of the tristeza virus of citrus. *Phytopathology* **41**:114–122.
40. Grant, T.J., and Higgins, R.P. 1957. Occurrence of mixture of tristeza virus strains in citrus. *Phytopathology* **47**:272–276.
41. Heaton, L.A., Carrington, J.C., and Morris, T.J. 1989. Turnip crinkle virus infection from RNA synthesized *in vitro*. *Virology* **170**:214–218.
42. Hemenway, C., Fang, R.X., Kaniewski, W., Chua, N.H., and Tumer, N.E. 1988. Analysis of the mechanism of protection in transgenic plants expressing potato virus X protein or its antisense RNA. *EMBO J.* **7**:1273–1280.
43. Herold, F., and Weibel, J. 1962. Electron microscopic demonstration of papaya ringspot virus. *Virology* **18**:307–311.
44. Hoekema, A., Hirsch, P.R., Hooykaas, P.J.J., and Schilperoort, R.A. 1983. A binary plant vector strategy based on separation of *Vir* and T-region of the *Agrobacterium tumefaciens* Ti-plasmid. *Nature (London)* **303**:179–180.
45. Holt, C.A., Hodgson, R.A., Coker, F.A., Beachy, R.N., and Nelson, R.S. 1990. Characterization of the masked strain of tobacco mosaic virus: Identification of the region responsible for symptom attenuation by analysis of an infectious cDNA clone. *Mol. Plant-Microbe Interact.* **3**:417–423.
46. Jensen, D.D. 1949. Papaya ringspot virus and its insect vector relationships. *Phytopathology* **39**:212–220.

47. Johansen, E., Rasmussen, O.F., Heide, M., and Borkhardt, B. 1991. The complete nucleotide sequence of pea seed-borne mosaic virus RNA. *J. Gen. Virol.* **72**:2625–2632.
48. Khuspe, S.S., Hendre, R.R., Mascarenhas, A.F., Jagannathan, V., Thombre, M.V., and Joshi, A.B. 1980. Utilization of tissue culture to isolate interspecific hybrids in *Carica* L. In P.S. Rao, M.R. Heble, and M.S. Chadha, eds., *Plant Tissue Culture, Genetic Manipulation and Somatic Hybridization of Plant Cells*. Bhabha Atomic Research Center, Bombay, India, pp. 198–205.
49. Lana, A.F. 1980. Transmission and properties of virus isolated from *Carica papaya* in Nigeria. *J. Hortic. Sci.* **55**:191–197.
50. Lawson, C., Kaniewski, W., Haley, L., Rozman, R., Newell, C., Sanders, P., and Tumer, N.E. 1990. Engineering resistance to mixed virus infection in a commercial potato cultivar: resistance to potato virus X and potato virus Y in transgenic russet burbank. *Biotechnology* **8**:127–134.
51. Lin, C.C. 1980. Strains of papaya ringspot virus and their cross protection. Ph.D. thesis, National Taiwan University, Taipei.
52. Linder, R.C., Jensen, D.D., and Ikeda, W. 1945. Ringspot: new papaya plunderer. *Hawaii Farm Home* **8**:10–14.
53. Ling, K., Namba, S., Gonsalves, C., Slightom, J.L., and Gonsalves, D. 1991. Protection against detrimental effects of potyvirus infection in transgenic tobacco plants expressing the papaya ringspot virus coat protein gene. *Biotechnology* **9**:752–748.
54. Litz, R.E. 1984. Papaya. In D.A. Evans, W.R. Sharp, P.V. Ammirato, and Y. Yanada, eds., *Handbook of Plant Cell Culture*, Vol. 2. New York, Macmillan, pp. 349–368.
55. Litz, R.E., and Conover, R.A. 1981. Effect of sex type, season, and other factors on *in vitro* establishment and culture of *Carica papaya* L. explants. *J. Am. Soc. Hortic. Sci.* **106**:792–794.
56. Litz, R.E., and Conover, R.A. 1982. *In vitro* somatic embryogenesis and plant regeneration from *Carica papaya* L. ovular callus. *Plant Sci. Lett.* **26**:153–158.
57. Litz, R.E., and Conover, R.A. 1983. High-frequency somatic embryogenesis from *Carica* suspension culture. *Ann Bot.* **51**:683–686.
58. Litz, R.E., O'Hair, S.K., and Conover, R.A. 1983. *In vitro* growth of *Carica papaya* L. cotyledons. *Sci. Hort.* **19**:287–293.
59. Loesch-Fries, L.S., Merlo, D., Zinnen, T., Burhop, L., Hill, K., Krahn, K., Jarvis, N., Nelson, S., and Halk, E. 1987. Expression of alfalfa mosaic virus RNA 4 in transgenic plants confers virus resistance. *EMBO J.* **6**:1845–1851.
60. Maiss, E., Timpe, U., Brisske, A., Jelkmann, W., Casper, R., Himmler, G., Mattanovich, D., and Katinger, H.W.D. 1989. The complete nucleotide sequence of plum pox virus RNA. *J. Gen. Virol.* **70**:513–524.
61. Manshardt, R.M., and Wenslaff, T.F. 1989. Zygotic polyembryony in interspecific hybrids of *Carica papaya* and *C. cauliflora*. *J. Am Soc. Hort. Sci.* **114**:684–689.
62. Manshardt, R.M., and Wenslaff, T.F. 1989. Interspecific hybridization of papaya with other *Carica* species. *J. Am. Soc. Hort. Sci.* **114**:689–694.
63. Martelli, G.P., and Russo, M. 1976. Unusual cytoplasmic inclusions induced by watermelon mosaic virus. *Virology* **72**:352–362.

64. Mau, R.F.L., Gonsalves, D., and Bautista, R. 1990. Use of cross-protection to control papaya ringspot virus at Waianae. *Proc. 25th Annual Papaya Industry Association Conference*, 1989, pp. 77–84.
65. McKinney, H.H. 1929. Mosaic diseases in the Canary Islands, West Africa, and Gibraltar. *J. Agric. Res.* **39**:557–578.
66. McMillan, Jr., R.T., and Gonsalves, D. 1987. Effectiveness of cross-protection by a mild mutant of papaya ringspot virus for control of ringspot disease of papaya in Florida. *Proc. Fla. State Hort. Soc.* **100**:294–296.
67. Mekako, H.U., and Nakasone, H.Y. 1975. Interspecific hybridization among six *Carica* species. *J. Am. Soc. Hortic. Sci.* **100**:237–242.
68. Meshi, T., Ishikawa, M., Motoyoshi, F., Semba, K., and Okada, Y. 1986. *In vitro* transcription of infectious RNAs from full-length cDNAs of tobacco mosaic virus. *Proc. Natl. Acad. Sci. U.S.A.* **83**:5043–5047.
69. Muller, G.W., and Costa, A.S. 1977. Tristeza control in Brazil by preimmunization with mild strains. *Proc. Int. Soc. Citric.* **3**:868–872.
70. Mundry, K.W. 1959. The effect of nitrous acid on tobacco mosaic virus: Mutation not selection. *Virology* **9**:722–726.
71. Nagel, J., and Hiebert, E. 1985. Complementary DNA cloning and expressing of the papaya ringspot potyvirus sequences encoding capsid protein and a nuclear inclusion-like protein in *Escherichia coli. Virology* **143**:435–441.
72. Namba, R., and Higa, S.Y. 1977. Retention of the inoculativity of the papaya mosaic virus by the green peach aphid. *Proc. Hawaii Entomol. Soc.* **22**:491–494.
73. Nelson, R.S., McCormick, S.M., Delannay, X., Dube, P., Layton, J., Anderson, E.J., Kaniewska, M., Proksch, R.K., Horsch, R.B., Rogers, S.G., Fraley, R.T., and Beachy, R.N. 1988. Virus tolerance, plant growth, and field performance of transgenic tomato plants expressing coat protein from tobacco mosaic virus. *Biotechnology* **6**:403–409.
74. Olson, E.O. 1956. Mild and severe strains of the tristeza virus in Texas citrus. *Phytopathology* **46**:336–341.
75. Olson, E.O. 1958. Responses of lime and sour orange seedlings and four scion rootstock combinations to infection by strains of the tristeza virus. *Phytopathology* **48**:454–459.
76. Oshima, N. 1975. The control of tomato mosaic disease with attenuated virus of tomato strain of TMV. *Rev. Plant Prot. Res.* **8**:126–135.
77. Pang, S.Z., and Sanford, J.C. 1988. Agrobacterium-mediated gene transfer in papaya. *J. Am. Soc. Hort. Sci.* **113**:287–291.
78. Ponz, F., and Bruening, G. 1986. Mechanism of resistance to plant viruses. *Annu. Rev. Phytopathol.* **24**:355–381.
79. Price, W.C. 1970. Citrus tristeza virus. CMI/AAB Description of Plant Viruses, No. 33.
80. Provvidenti, R., and Gonsalves, D. 1982. Resistance to papaya ringspot virus in *Cucumis metuliferus* and its relationship to watermelon mosaic virus 1. *J. Hered.* **73**:239–240.
81. Purcifull, D.E., and Edwardson, J.R. 1967. Watermelon mosaic virus: tubular inclusion in pumpkin leaves and aggregates in leaf extracts. *Virology* **32**:393–401.
82. Purcifull, D.E., and Hiebert, E. 1979. Serological distinction of watermelon mosaic virus isolates. *Phytopathology* **69**:112–116.

83. Purcifull, D.E., Edwardson, J., Hiebert, E., and Gonsalves, D. 1984. Papaya ringspot virus. CMI/AAB Descriptions of Plant Viruses, No. 292.
84. Quemada, H., L'Hostis, B., Gonsalves, D., Reardon, I.M., Heinrikson, R., Hiebert, E.L., Sieu, L.C., and Slightom, J.L. 1990. The nucleotide sequences of the 3′-terminal regions of papaya ringspot virus strains W and P. *J. Gen. Virol.* **71**:203–210.
85. Quiot-Douine, L., Purcifull, D.E., and de Mejia, M.V.G. 1986. Serological relationships and *in vitro* translation of an antigenically distinct strain of papaya ringspot virus. *Phytopathology* **76**:346–351.
86. Rast, A.T.B. 1972. M II-16, an artificial symptomless mutant of tobacco mosaic for seedling inoculation of tomato crops. *Neth. J. Plant Pathol.* **78**:110–112.
87. Rast, A.T.B. 1975. Variability of tobacco mosaic virus in relation to control tomato mosaic in greenhouse tomato crops by resistance breeding and cross protection. *Agr. Res. Rept.* **834**:1–76. Inst. Phytopathol. Res. Wageningen, The Netherlands.
88. Riechmann, J.L., Lain, S., and Garcia, J.A. 1990. Infectious *in vitro* transcripts from plum pox potyvirus cDNA clone. *Virology* **177**:710–716.
89. Robaglia, C., Durand-Tardif, M., Tronchet, M., Boudazin, G., Astier-Manifacier, S., and Casse-Delbart, F. 1989. Nucleotide sequence of potato virus Y (N strain) genomic RNA. *J. Gen. Virol.* **70**:935–947.
90. Siegel, A. 1965. Artificial production of mutants of tobacco mosaic virus. *Adv. Virus Res.* **11**:25–60.
91. Singh, A.B. 1969. A new virus disease of *Carica papaya* in India. *Plant Dis. Rep.* **53**:267–269.
92. Story, G.E., and Halliwell, R.S. 1969. Identification of distortion ringspot virus disease of papaya in the Dominican Republic. *Plant Dis. Rep.* **53**:757–760.
93. Stubbs, L.L. 1964. Transmission and protective inoculation studies with viruses of the citrus tristeza complex. *Aust. J. Agr. Res.* **15**:752–770.
94. Tumer, N.E., O'Connell, K.M., Nelson, R.S., Sanders, P.R., Beachy, R.N., Fraley, R.T., and Shah, D.M. 1987. Expression of alfalfa mosaic virus coat protein gene confers cross-protection in transgenic tobacco and tomato plants. *EMBO J.* **6**:1181–1188.
95. van der Werf, S., Bradley, J., Wimmer, E., Studier, F.W., and Dunn, J.J. 1986. Synthesis of infectious poliovirus RNA by purified T7 RNA polymerase. *Proc. Natl. Acad. Sci. U.S.A.* **83**:2330–2334.
96. van Dun, C.M.P., Bol, J.F., and van Vloten-Doting, L. 1987. Expression of alfalfa mosaic virus and tobacco rattle virus protein genes in transgenic tobacco plants. *Virology* **159**:299–305.
97. van Dun, C.M.P., Overduin, B., van Vloten-Doting, L., and Bol, J.F. 1988. Transgenic tobacco streak virus or mutated alfalfa mosaic virus coat protein does not cross-protect against alfalfa mosaic virus infection. *Virology* **164**:383–389.
98. Verchot, J., Koonin, E.V., and Carrington, J.C. 1991. The 35-kDa protein from the N-terminus of the potyviral polyprotein functions as a third virus-encoded proteinase. *Virology* **185**:527–535.
99. Vos, P., Jaegle, M., Wellink, J., Verver, J., Eggen, R., van Kammen, A., and Goldbach, R. 1988. Infectious RNA transcripts derived from full-length

DNA copies of the genomic RNAs of cowpea mosaic virus. *Virology* **165**:33–41.

100. Wang, C.H., and Yeh, S.-D. 1992. Nucleotide sequence comparison of the 3'terminal region of severe, mild, and non-papaya infecting strains of papaya ringspot virus. *Arch. Virol.* **127**:345–354.
101. Wang, H.L., Wang, C.C., Chiu, R.J., and Sun, M.H. 1978. Preliminary study on papaya ringspot virus in Taiwan. *Plant Prot. Bull.* (Taiwan) **20**:133–140.
102. Wang, H.L., Yeh, S.-D., Chiu, R.-J., and Gonsalves, D. 1987. Effectiveness of cross-protection by mild mutants of papaya ringspot virus for control of ringspot disease of papaya in Taiwan. *Plant Dis.* **71**:491–497.
103. Yamaya, J., Yoshioka, M., Meshi, T., Okada, Y., and Ohno, T. 1988. Expression of tobacco mosaic virus RNA in transgenic plants. *Mol. Gen. Genet.* **211**:520–525.
104. Yeh, S.-D., and Gonsalves, D. 1984. Evaluation of induced mutants of papaya ringspot virus for control by cross protection. *Phytopathology* **74**:1086–1091.
105. Yeh, S.-D., and Gonsalves, D. 1984. Purification and immunological analyses of cylindrical-inclusion protein induced by papaya ringspot virus and watermelon mosaic virus 1. *Phytopathology* **74**:1273–1278.
106. Yeh, S.-D., and Gonsalves, D. 1985. Translation of papaya ringspot virus RNA *in vitro*: Detection of a possible polyprotein that is processed for capsid protein, cylindrical-inclusion protein, and amorphous-inclusion protein. *Virology* **143**:260–271.
107. Yeh, S.-D., and Bih, F.Y. 1989. Comparative studies on *in vitro* translation of a severe strain and a mild strain of papaya ringspot virus. *Plant Prot. Bull.* (Taiwan, R.O.C.) **31**:276–289.
108. Yeh, S.-D., and Cheng, Y.-H. 1989. Use of resistant *Cucumis metuliferus* for selection of nitrous-acid induced attenuated strains of papaya ringspot virus. *Phytopathology* **79**:1257–1261.
109. Yeh, S.-D., Gonsalves, D., and Provvidenti, R. 1984. Comparative studies on host range and serology of papaya ringspot virus and watermelon mosaic virus 1. *Phytopathology* **74**:1081–1085.
110. Yeh, S.-D., Gonsalves, D., Wang, H.-W., Namba, R., and Chiu, R.-J. 1988. Control of papaya ringspot virus by cross protection. *Plant Dis.* **72**:375–380.
111. Yeh, S.-D., Jan, F.-J., Chiang, C.-H., Doong, T.-J., Chen, M.-C., Chung, P.-H., and Bau, H.-J. 1992. Complete nucleotide sequence and genetic organization of papaya ringspot virus RNA. *J. Gen. Virol.* **73**:2531–2541.
112. Yie, S.T., and Liaw, S.I. 1977. Plant regeneration from shoot tips and callus of papaya. *In Vitro* **13**:564–567.

10
Tomato Yellow Leaf Curl Virus, a Whitefly-Borne Geminivirus of Tomatoes

S. Cohen and Y. Antignus

Introduction

The first evidence of economic damage to vegetable crops caused by *Bemisia tabaci* Gennadius in Israel was recorded in 1931 (8), and since 1935 it has been a permanent pest, mainly in the Jordan Valley. The fluctuations in population size in different years were always directly correlated with the damage due to tomato yellow leaf curl virus (TYLCV). A TYLCV-like disease was first reported in Israel in 1939–1940 and was associated with outbreaks of *B. tabaci*. The disease involved leaf curling, short internodes, leathery leaves, flower drop, and general dwarfing in comparison with healthy plants (8). Twenty years later, in 1959, when there was a heavy outbreak of *B. tabaci* in the Jordan and Bet She'an Valleys, the entire tomato crop was completely destroyed by a disease with the following syndrome: "Severe stunting of growth, the shoots are erect, and the leaflets are markedly smaller and misshapen. Those leaflets that appear soon after infection are cupped down and inwards, while subsequently developing leaves are strikingly chlorotic and show an upward curling of the leaflet margins. When young plants are attacked they lose vigor and hardly produce any marketable fruits" (25). The viral nature of the disease and its spread by *B. tabaci* were confirmed this time. The virus geminate shape was first observed in 1980 in electron microscope thin sections (82), but only in 1988 was TYLCV purified (32). From 1959 to date, TYLCV has spread to every region in Israel and

[1] Contribution from the Agricultural Research Organization, The Volcani Center, Bet Dagan, Israel. No. 3636-E, 1992 series.

S. Cohen, Virus Laboratory, Agricultural Research Organization, The Volcani Center, P.O. Box 6, Bet Dagan 50250, Israel.
Y. Antignus, Virus Laboratory, Agricultural Research Organization, The Volcani Center, P.O. Box 6, Bet Dagan 50250, Israel.

become the major limiting factor in tomato production during both the summer and the winter.

TYLCV-like viruses were reported from different countries around the Mediterranean basin, Africa, India, and the Far East (3, 16, 31, 36, 37, 54, 67, 70). However this review is focused on the type strain of TYLCV (25). The great economic significance of the disease stimulated studies of the virus, its interaction with the vector, and control of both the virus and vector. These topics will be discussed in detail in the following sections.

Host Range

Host range studies were carried out under greenhouse conditions (26) using viruliferous *B. tabaci* for inoculation and virus-free whiteflies for recovery tests. Plants from six botanical families were found susceptible to TYLCV: Asclepiadaceae, Compositae, Leguminosae, Malvaceae, Solanaceae, and Umbelliferae.

The following plants were susceptible to TYLCV and evinced clear symptoms of disease: *Datura stramonium* (L.), *Hyoscyamus desertorum* (Asch.) Eig, *Lycopersicon esculentum* (L.) Mill.—more than 50 cultivars were tested and all reacted with typical symptoms, *Nicotiana benthamiana*, and *Nicotiana glutinosa*.

The following plants were susceptible symptomless carriers of the virus: *Chaerogphyllum* sp., *Cynanchum acutum* L., *Lens esculenta* Moench, *Malva nicaensis* All., *Malva parviflora* L., *Nicotiana tabacum* L., "Samsun" and "Yellow Special Flue cured Virginia," *Phaseolus vulgaris* L. Bulgarit, and *Sonchus oleraceus* L. Of the species tested, only *C. acutum* and *M. parviflora* were found to be naturally affected.

Plants apparently immune to TYLCV are Amaranthaceae—*Amaranthus retroflexus* L.; Apocynaceae—*Nerium oleander* L.; Asclepiadaceae—*Calotropis procera* (Willd.) R.Br.; Capparidaceae—*Capparis aegyptia* Lam.; Caprifoliaceae—*Lonicera* sp.; Chenopodiaceae—*Atriplex* sp., *Beta vulgaris* L., *Chenopodium amaranticolor* Coste & Reyn.; Compositae—*Ageratum houstonianum* Mill., *Xanthium strumarium* L.; Convolvulaceae—*Ipomea batatas* (L.) Lam. "21" and "Gokoku"; Cucurbitaceae—*Cucumis sativus* L. "Beit Alpha," *Ecbalium elaterium* (L.) Rich, *Momordica balsamina* L.; Euphorbiaceae—*Euphorbia cybirensis* Boiss., *Euphorbia paralias* L., *Ricinus communis* L.; Leguminosae—*Arachis hypogaea* L., *Medicago sativa* L. "Peruvian," *Pisum sativum* L. var. *arvense* (L.) Poir, "Dunn," *Prosopis farcata* (Banks et Sol.) Eig, *Trifolium alexandrinum* L. "Faheli"; Malvaceae—*Althaea* sp., *Gossypium hirsutum* L. Acala 4-42, *Hibiscus rosa—Sinensis* L., *Lavatera cretica* L., *Sida rhombifolia* L.; Plumbaginaceae—*Plumbago capensis* Thunb., *Polygonum equisetiforme* s. efs.; Polygonaceae—*Polygonum equisetiforme* S. ets.; Resedaceae—*Ochradenus baccatus*

Del.; Solanaceae—*Lycium* sp., *Nicotiana repanda* Willd., *N. rustica* L., *N. tabacum* L. "White Burley," *Physalis floridana* Rydb., *Solanum incanum* L., *S. villosum* (L.) Lam., *Withania somnifera* (L.) Dun.; Tamaricaceae—*Tamarix* sp.; Zygophyllaceae—*Tribulus* sp.

Cytopathology of TYLCV

Two main kinds of abnormalities have been observed in whitefly-transmitted geminiviruses, one consisting of ordered or disordered aggregates of virus-like particles (39, 62) and the other comprising rings of fibrilar material (62, 63).

Electron microscope thin sections from TYLCV-infected tomato tissues sampled 30 days after inoculation showed some major changes that occurred in the nuclei of phloem parenchyma cells. Altered nuclei were easily recognized by their uniformly granular texture, reduction of the chromatin to a few small peripheral clumps, and apparent absence of nucleoli. Some nuclei contained intensely electron opaque ring-shaped structures varying from 0.4 to 3 µm. These rings appeared to be made of an amorphous fibrillar matrix. The periphery of the rings was lined with rounded electron-dense bodies resembling virus particles. Similar virus-like particles were plentiful in the nucleoplasm of all modified nuclei. All these cytopathic effects were confined to the nuclei of infected cells and none was observed in the cytoplasm of infected cells (82). The ultrastructural effects of TYLCV resemble those of other whitefly-transmitted geminiviruses (78, 82).

Virus Purification

The difficulties encountered in purifying TYLCV inhibited the characterization of this virus for many years. However partially purified preparations of the virus were obtained by extraction of the infected tissue with 0.1 M sodium phosphate pH 7.0, 2.5 mM EDTA, 10 mM sodium sulfite, 0.1% (v/v) 2-mercaptoethanol, and 1% (v/v) Triton X-100 (Fig. 10.1). This was followed by a chloroform clarification step and separation on 10–40% sucrose gradient (32). Virus yields obtained by this procedure were approximately 1.2 mg/kg of infected tissue. *D. stramonium* plants were superior to tomato plants as a source for virus purification. Improved yields were obtained when *N. benthamiana* infected with a TYLCV-like geminivirus served as a source plant for virus purification. In this case tissue extraction was carried out in the presence of 0.5 M phosphate buffer, overnight incubation in 0.1% driselase, and a final separation on $CsSO_4$ gradient (66). The reported yields for this procedure were 5–10 mg/kg of starting material.

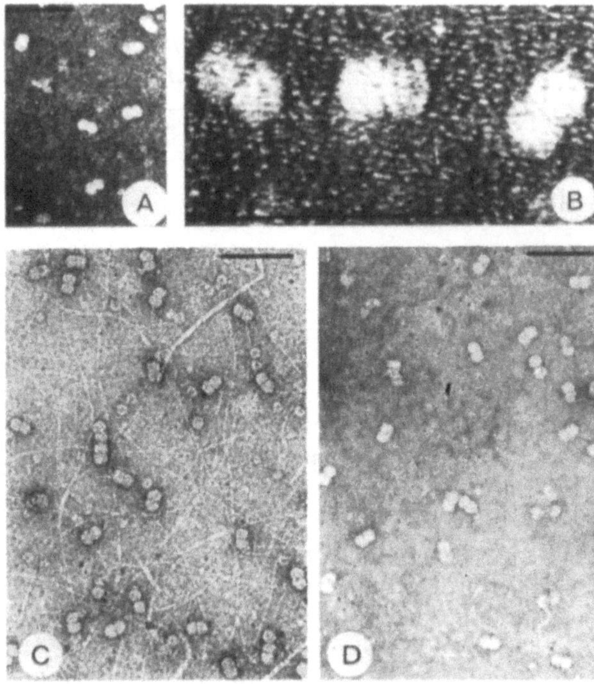

FIGURE 10.1. Electron micrograph of tomato yellow leaf curl virus (TYLCV). Partially purified particles isolated from *Datura stramonium* (A, B, C) and particles isolated from tomato (D). Staining was done by 2% uranyl acetate. Bar represents 100 nm. (See Czosnek et al. (32).)

Serological Relationships

Strong serological relationships were observed among all the whitefly-transmitted geminiviruses, in agreement with the highly conserved sequences found in their coat protein genes (49). These findings are compatible with the hypothesis that structural features of the particle surface are required for virus transmission by their common vector, *B. tabaci* (14, 79, 92).

A polyclonal antiserum against TYLCV was prepared (7) but was found to be of low titer and to react nonspecifically with healthy plant material in ELISA. Similar results were obtained with an antiserum against a TYLCV-like virus in Italy (66).

Based on these results it is suggested that the virus is of low immunogeneity. However, the Israeli antiserum reacted positively and specifically with the viral coat protein when used for immunostaining of western blots.

This procedure was used to study the serological relationships of the virus with other whitefly-borne geminiviruses. Samples of a partially purified preparation of TYLCV were separated on polyacrylamide gels, blotted onto nitrocellulose membranes, and subjected to immunostaining with the antisera against TYLCV, squash leaf curl virus (SLCV), bean golden mosaic virus (BGMV), and african cassava mosaic virus (ACMV). A positive reaction was obtained with all the tested antisera (Y. Antignus, unpublished) (Fig. 10.2).

ELISA tests with monoclonal antibodies have shown that TYLCV appears to share 8 of 17 epitopes with the type strain of ACMV (West African form) and only 2 of 9 epitopes with the Indian cassava mosaic virus (ICMV) (B.D. Harrison, personal communications).

The Molecular Biology of TYLCV

Geminiviruses are small DNA-containing plant viruses characterized by their unique capsid morphology, which consists of double icosahedral virions. The viral coat protein encapsidates a circular single-stranded DNA genome approximately 2.8 kb in size.

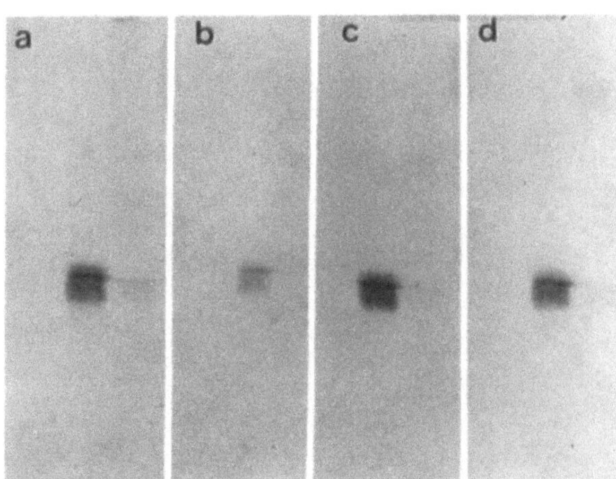

FIGURE 10.2. Serological relationships of tomato yellow leaf curl virus (TYLCV) with different whitefly-borne geminiviruses. Partially purified preparations of TYLCV virions were boiled in the presence of SDS and separated on SDS 10% polyacrylamide gel. The protein band was electroblotted onto nitrocellulose membrane and immunostained with antisera against the following viruses: African cassava mosaic virus (ACMV) (lane a), squash leaf curl virus (SLCV) (lane b), TYLCV (lane c), and BGMV (lane d).

On the basis of their host range and genome organization geminiviruses are classified into two subgroups: (1) All monocot-infecting members have a genome comprising a single genomic component of about 2.7 kb, and are transmitted by leafhoppers. (2) Dicotyledon-infecting members with a bipartite genome comprising similar-sized (2.8 kb) DNAs (designated A and B). The members of this subgroup are whitefly-borne viruses, which are transmitted exclusively by the whitefly *B. tabaci*. The molecular biology of geminiviruses has been studied and reviewed intensively over the last few years (49, 52, 65, 86, 87).

TYLCV Particle Components

Analysis of the viral coat protein by SDS-polyacrylamide gel electrophoresis indicated that it consists of a single protein subunit with an estimated molecular weight of 33,000 (7) (Fig. 10.3). The viral nucleic acid was obtained from partially purified virions using proteinase K and phenol extraction. It was found resistant to RNase A but sensitive to DNase I and S1 nuclease, indicating its single-stranded nature. It appeared

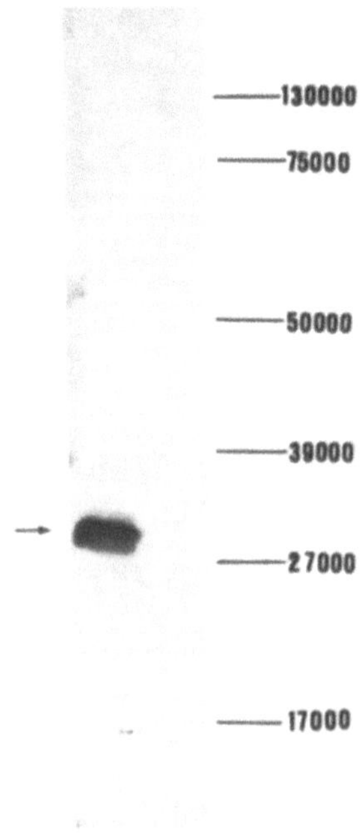

FIGURE 10.3. Estimation of the molecular weight of tomato yellow leaf curl virus (TYLCV) coat protein subunit. Partially purified preparations of TYLCV virions were boiled in the presence of SDS and separated on SDS 10% polyacrylamide gel. The protein band was electroblotted onto nitrocellulose membrane and immunostained with TYLCV specific antiserum.

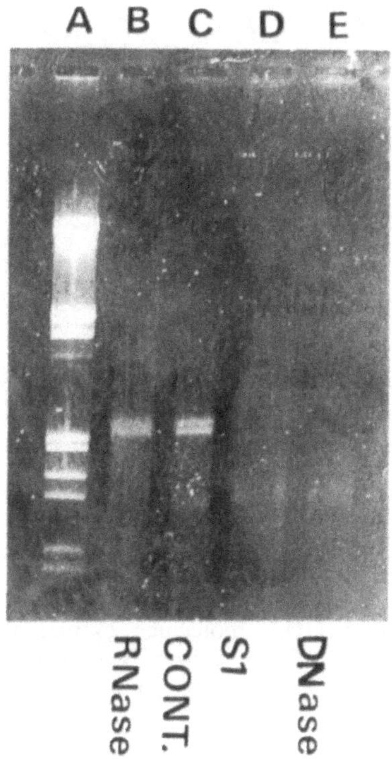

FIGURE 10.4. Evidence for the single-stranded nature of tomato yellow leaf curl virus (TYLCV) genomic DNA. Molecular weight size marker (A). RNase treatment (B), Control (C), S_1 treatment (D), DNase I treatment (E).

on agarose gels as two closely migrating bands, where the one with the higher mobility represents linear ssDNA molecules and the one with the lower mobility represents the circular form of this DNA (Fig. 10.4). Electron microscope spreads of virion DNA demonstrated the presence of circular and linear molecules. Most of the circular molecules were 850 nm in size, while a few were 400 nm (32) (Fig. 10.5). A similar phenomenon was reported for tomato golden mosaic virus (TGMV) (47) and ACMV (88, 90). It is suggested that the small molecules representing subgenomic DNA comprise a family of closely related DNA B molecules of approximately half the size of the full-length component in which both genes have been disrupted. This defective DNA was shown to interfere with virus proliferation manifested as a delay in symptom development and symptom amelioration (90).

TYLCV DNA Forms in Infected Tissues

Single-stranded viral DNA as well as its double-stranded form could be detected in TYLCV-infected tomatoes (33, 81). The ssDNA molecules

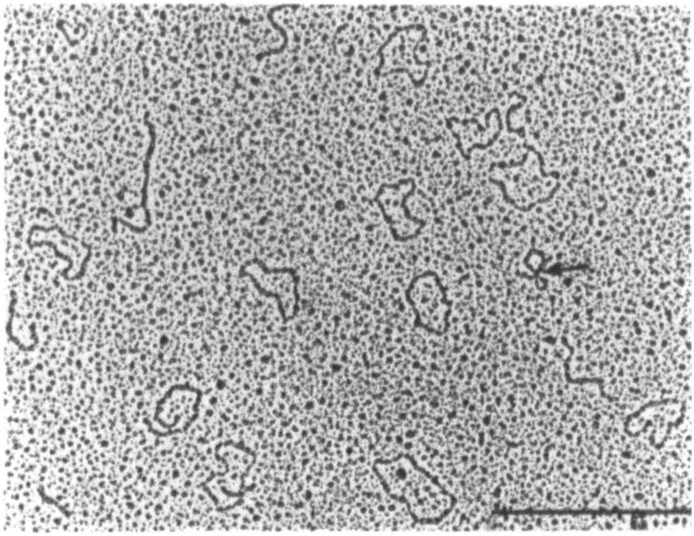

FIGURE 10.5. Electron micrograph of spread of tomato yellow leaf curl virus (TYLCV) single-stranded DNA molecules extracted from partially purified virions. Linear and ciruclar molecules are shown. The arrow indicates a half size molecule. Bar represents 1 μm. (See Czosnek et al. (32).)

were of two sizes, the 2.8-kb full-length DNA and molecules of approximately half this size representing probably defective forms of the viral genome, as was found in different geminiviruses (90). Circular supercoiled and relaxed dsDNA could also be detected in some of the tested plants, depending on their stage of infection. Similar viral double-stranded molecules have been described in plants infected by ACMV and TGMV, and it was suggested that they represent the replicative intermediates of the virus, which exists in a free state in infected cells (46, 90). It is assumed that TYLCV, like other geminiviruses, replicates via a rolling circle mechanism (84, 91). It was found that the replicative form (RF) DNA of TYLCV is copurified with the gemini particles by our routine purification procedure (Y. Antignus, unpublished). A similar phenomenon was described for abutilon mosaic virus (AbMV) where it was suggested that these copurified DNA molecules represent a minichromosome composed of viral DNA and host histones, a complex that might be associated with the viral replication mechanism (1).

Accumulation and Spread of TYLCV in Infected Tissues

Although symptom appearance in infected plants is usually observed 15 days postinoculation, viral DNA could be detected by a riboprobe as early as 3 days postinoculation. The viral DNA reached a peak 12–15 days postinoculation (81). The highest concentrations of TYLCV-DNA were found in rapidly growing tissues (shoot apex, young leaves, roots, and stems), and the lowest in the older leaves and cotyledons. Young leaves and apices were found to be the best sites for the successful virus inoculation. In these tissues the viral DNA replicated and was further transported to the roots, then to the shoot apex and to the neighboring leaves and flowers. Inoculation through the oldest leaves was inefficient (9). The association of replication ability of geminiviruses and cell division has already been shown (94, 96) and also confirmed for TYLCV in a protoplast system. Successful transfection of tobacco protoplasts with TYLCV virions was obtained only when protoplasts were grown on a medium that supported their division activity (Y. Antignus and A. Gera, unpublished).

The accumulation of TYLCV-DNA in tolerant genotypes of tomatoes was found to be 15–40% lower than its accumulation levels in susceptible genotypes. The viral DNA level in these lines was positively correlated to the level of tolerance as reflected by symptom severity. The mechanism of tolerance in the tested tomato lines is still unknown; however, the lower accumulation levels found in the tolerant lines may be due to a reduction in the capacity of the virus to replicate or to spread in the infected tissue of these specific genotypes (81).

TYLCV Genome Organization

A replicative form of TYLCV isolated from field-infected tomatoes was cloned and sequenced (72). Nucleotide sequence analysis of these clones showed that it comprises 2787 nucleotides, encoding six open reading frames, two on the virion strand and four on the complementary strand. The genome organization of this TYLCV isolate resembles DNA A of ACMV (72), a whitefly-transmitted geminivirus with which it shares the highest sequence homology (73%) (89, 93). No evidence was found in this case for the presence of a TYLCV B component, which is typical for all the former described whitefly-borne geminiviruses (35). When dimeric copies of the cloned TYLCV-DNA were introduced into tomato plants by agroinfection, severe typical disease symptoms were observed. It was also shown that the progeny virus produced in the agroinfected plants was transmitted successfully by the vector, *B. tabaci* (72). The genome of TYLCV was also cloned by a different strategy, where the purified virion single-stranded DNA served as a template for the synthesis of its double-stranded form (5). In this case the original culture of TYLCV served as a source for virus purification (25).

Restriction analysis of the *in vitro*-synthesized viral dsDNA indicated that DNA fragments of an overall size of 5.6 kbp were obtained, twice the virus suggested genome unit size. These results suggest that TYLCV virion DNA consists of two populations reflecting some sequence differences. Hybridization tests with a cloned DNA representing these DNA populations demonstrated a high sequence homology. A head-to-tail dimer construct of a 2.8-kbp DNA representing one of the populations was introduced into TYLCV host plants via agroinoculation resulting in systemic viral infection. However, at least in some hosts symptoms were different from those observed in infections with the wild-type virus. A very mild or symptomless infection was observed when tolerant tomato genotypes (TY-20, 7170, 7171) were agroinfected with the cloned DNA. These symptoms were very mild compared with those induced in these tomato lines by the wild-type virus. In the highly susceptible variety Marmand agroinfection with the cloned DNA starts as a normal TYLCV infection but later on the plants partially recover showing milder symptoms and regeneration of the ability for fruit setting (Y. Antignus and S. Cohen, in preparation). Moreover, some preliminary results indicate that infection with the cloned DNA can cross-protect efficiently susceptible tomato varieties from the severe wild-type infection (Y. Antignus and S. Cohen, in preparation). Reports have been published on a systemic infection of whitefly-transmitted viruses following agroinoculation in the absence of DNA B. Agroinfection with DNA A of TYLCV from Thailand resulted in attenuated symptoms compared with symptoms originating from infection with the two genomic components (80). Agroinoculation with component A of ACMV resulted in spread of the viral DNA and virus assembly, but plants remained asymptomatic (64).

Two working hypotheses can be put forward to explain the dimorphism found for the genomic DNA of TYLCV (type virus) and the unusual disease phenology resulting from the agroinfection with a single genomic component:

- TYLCV, like other whitefly-transmitted viruses, has a genome with two genomic components with different viral genes. A normal disease phenology is obtained only in the presence of both components.
- The original culture of TYLCV is practically a mixture of two strains of a geminivirus with a single genome component, differing in parts of their sequence as well as in the symptoms they induce in their common host plants.

TYLCV-Like Geminiviruses

The dramatic rise in the populations of *Bemisia tabaci* worldwide during the last few years, and the following virus epidemics, led to increased research aimed at characterizing whitefly-transmitted geminiviruses. Nonmechanically transmissible whitefly-borne geminiviruses causing significant damage in tomatoes were described from Senegal and Tunisia (17, 37). Ultrastructural observations in tissues infected with these viruses have shown cytopathological changes in the phloem cells typical to those described for other geminiviruses.

An Egyptian isolate of TYLCV was found to have 96% sequence homology with the ALI ORF of the Israeli isolate (71). In 1988 and 1989, severe outbreaks of a TYLCV-like disease occurred for the first time in the northern part of the Mediterranean basin, in Sicily and Sardinia, causing heavy losses in greenhouse grown tomatoes (31, 40). The Sardinian isolate (TYLCV-S) was cloned and sequenced (60) and found to have the same genome organization as the Israeli strain (72); however, there was only 77% nucleotide identity between these two viruses. The degree of DNA diversity between the Israeli strains TYLCV-S and TYLCV-T (23 and 25%, respectively) is significantly higher than the expected similarities among strains of the same virus (60).

A disease syndrome similar to that caused by TYLCV was described in tomatoes in Thailand. The geminivirus isolated from the diseased plants (TYLCV-T) was cloned and sequenced. In contrast to the Israeli and the Sardinian isolates, this virus was found to have two distinct genomic components as is the case of all other reported whitefly-transmitted geminiviruses (80).

A leaf curl disease of tobacco has been reported from Japan (74). The causal agent of this disease is a whitefly-transmitted geminivirus, tobacco leaf curl virus (TLCV), which induces TYLCV-like symptoms in tomatoes. Using polyclonal antibodies it was found that TLCV and the Thai isolate of TYLCV are serologically identical (16).

A devastating disease of tomatoes was reported from many parts of India (83). It is caused by Indian tomato leaf curl (ITmLCV), which is transmitted persistently by *B. tabaci* (15). The genome organization of this virus is not yet known, but it reacted with only 1 of 17 monoclonal antibodies specific for different epitopes in the particle protein of ACMV and 5 or 6 of 10 monoclonal antibodies to the particle protein of Indian cassava mosaic virus (ICMV) (70). These results indicate a distant serological relationship with the Israeli TYLCV-type culture, which seems to be more closely related to ACMV (West African form) than to ICMV (see Serological Relationships section of this review). From the above mentioned data it is clear that practically, the name TYLCV was adopted to describe different viruses collected at different geographic regions of the world. Most of these viruses are probably too distant to be considered strains of the original TYLCV-type virus (25). The nomenclature of geminiviruses inducing a similar disease syndrome in tomatoes should be based on serological and biological criteria, which were already described in the past (48). Application of these criteria in addition to sequencing data and the use of monoclonal antibody technology may prevent the confusion introduced by the present nomenclature system. A similar confusing situation was described in the case of BGMV-PR and BGMV-BZ, which may not be strains of the same virus but two distinct viruses (53).

The Phylogeny of TYLCV

The coat protein and the replication associated protein of 16 sequenced geminiviruses were used to construct a phylogenetic tree. This tree has two main branches, which correspond to the host specificities of the viruses for either monocots or dicots. The branch containing the viruses that infect dicots divided into two subbranches: one includes whitefly-borne viruses of the New World [AbMV, beet curly top virus (BCTV) BGMV-PR, BGMV-BZ, and TGMV]; those of the other branch [ACMV, Mung bean yellow mosaic virus (MYMV) and TYLCV] were collected in the Old World. A very similar tree was obtained when the coat proteins of these viruses were analyzed (53).

It was suggested that geminiviruses with bipartite genomes had evolved from ancestors with monopartite genomes (52, 61). It is not clear yet whether TYLCV is a monopartite virus or has a genome consisting of two homologous DNA populations. Either way, it may represent a phyllogenetic link between the monopartite and the bipartite geminiviruses.

Virus–Vector Relationships Biological Interaction with the Vector

1. Transmission efficiency: Females were more efficient (43 of 133—32%) in transmitting TYLCV. This transmission efficiency was 6-fold higher than that of males (5 of 104—5%). Transmission rates of 32, 83, 84, 86, and 100% were obtained by 1, 3, 5, 10, and 15 inoculative females, respectively (30). About twice the number of females (42%) acquired TYLCV from young fully developed leaves of infected *D. stramonium*, in comparison with only 23% that acquired the virus from the third leaf from the top (18).

2. Acquisition and inoculation thresholds: Between 15 and 30 min. However, at least 4 hr is required to obtain high infection rates (30).

3. Latent period: Between 21 and 24 hr (30).

4. Persistence: Following 48 hr of acquisition feeding, 2 of 39 females retained the virus for 20 days. Most of the females did not transmit TYLCV later than 10 days after the end of the acquisition feeding period. Shorter feedings resulted in shorter retention of the virus. Thus, maximum retention of 13 days was measured after a 6-hr acquisition period (18). The virus is not transmitted to the progenies but is acquired by the larval stages and 28% of the enclosed adults were inoculative (30).

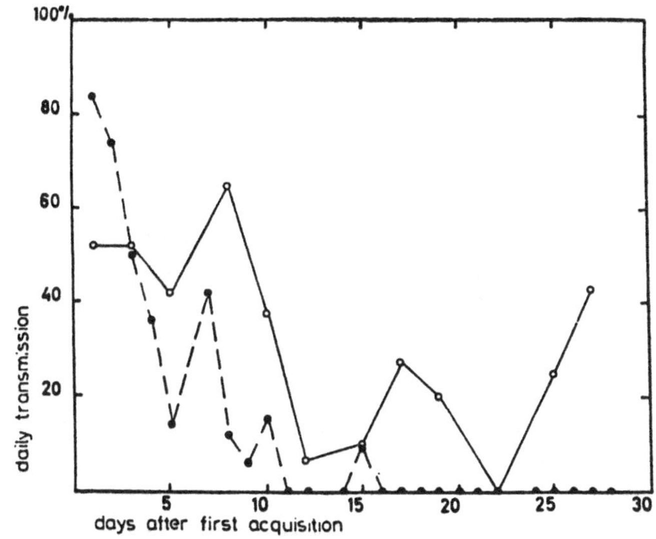

FIGURE 10.6. The periodic acquisition phenomenon. Daily transmission percentage of tomato yellow leaf curl virus (TYLCV) by *Bemisia tabaci*. Insect group fed alternately on TYLCV-infected source and healthy test plants (——). Control group following a single 48-hr acquisition feeding (– –). (See Cohen and Harpaz (25).)

5. *The "periodic acquisition" phenomenon*: The interaction of TYLCV with its whitefly vector *B. tabaci* was studied by Cohen and Harpaz (25). It was shown that whiteflies carrying TYLCV progressively lose infectivity and 10 days after completion of acquisition feeding most insects do not transmit the virus. Progressive loss of inoculativity of the vector cannot be prevented by repeated or prolonged acquisition on an infected plant. Inoculativity can be restored only when virus transmission has ceased (Fig. 10.6). This unique phenomenon in plant–virus–insect relationships was termed "periodic acquisition" (25) and was observed also with tomato yellow mosaic virus (TYMV) in India (95).

The periodic acquisition phenomenon indicates more than a passive role for certain geminiviruses in their whitefly vectors. A factor designated periodic acquisition-related factor (PARF)—apparently related to the above mentioned phenomenon—was found in homogenates of insects that were fed for 48 hr on an infected source. This factor inhibited acquisition transmission and retention of TYLCV by nonviruliferous females who were fed on PARF immediately or 2 days before being transferred to source plants for 24 hr of acquisition access. PARF was not found in homogenates of insects that were approaching the end of the inoculativity cycle, i.e., 7 days after acquisition access (19, 20).

PARF was detected in the whitefly bodies only 48 hr after the beginning of the acquisition feeding, indicating that an incubation time is needed for the production of detectable amounts of the factor in the insect's body.

Further purification of PARF was achieved by ammonium sulfate fractionation (by increasing the concentration from 40 to 70%) yielding an anti-TMV fraction. The occurrence of this antiviral fraction in the insect homogenates was associated with PARF and therefore was designated PARFa (27). However, when fed through membranes to whiteflies, no anti-TYLCV activity was detected. Injections of PARFa into whiteflies prior to, or after, TYLCV acquisition reduced the whiteflies inoculativity. Moreover, the formation of PARFa in whiteflies could be provoked by membrane feeding of PARF.

Therefore, it was suggested that PARF and PARFa are the products of a chain of reactions taking place following the entry of TYLCV into the whitefly body (68, 69).

PARFa was sensitive to temperatures above 35°C, resistant to pH in the range of 2.2–9.0, resistant to RNase, but lost its activity after Pronase digestion. These characteristics may indicate that the active antiviral component of PARFa is a protein.

Based on the activity and nature of the above-mentioned antiviral factors, an analogy was suggested between these factors and the interferon system described in virus-infected vertebrates (20). In view of the accumulating information on the interaction of geminiviruses with their whitefly vector, the following mechanisms were suggested to explain the periodic acquisition phenomenon:

a. An active mechanism based on the production of an antiviral factor is triggered by the virus. This factor interacts with particles that are introduced into the hemocoel following subsequent acquisitions, thus inhibiting—for some unknown reason(s)—their mobilization into the salivary gland. The whiteflies inoculativity is restored only when the antiviral activity comes to an end.
b. The bond between virus particles and the membrane's putative receptors is an irreversible one. As a consequence, these receptors are removed from the salivary gland membrane during the release of the virus particle into the saliva. The time required for the regeneration of these specific receptors site reflects the period during which the insect cannot restore its inoculativity.
c. Inactive virus particles or degradation products of the viral coat protein block the salivary gland surface thus inhibiting its ability to absorb a fresh influx of viral particles. The insect inoculativity is restored only after the removal of these blocking elements by the insect's metabolism.

6. *Interactions at the molecular level*: TYLCV-specific probes were used to detect the viral DNA in the whitefly vector (6, 34). TYLCV DNA was detected in *B. tabaci* as early as 1 hr after the beginning of an acquisition access. The accumulation level of the viral nucleic acid in the vector and nonvector whitefly *Trialeurodes vaporariorum* was correlated with the length of the acquisition access at least up to 96 hr. The retention time for the viral nucleic acid in the vector was at least 23 days. Time course of the retention of the viral nucleic acid for 23 days after acquisition access indicated a statistically nonsignificant increase in the accumulation of the viral DNA. Moreover, no viral replicative forms were detected in extracts of viruliferous whiteflies, which apparently indicates that the virus does not multiply in its whitefly vector (Y. Antignus and S. Cohen, in preparation). This is in agreement with previous findings that geminiviruses are able to pass through the molt from larva to adult but not through the egg to progeny insects (49).

The viral nucleic acid was found in high rates in the nonvector whitefly *T. vaporariorum* as well as in the aphid *Myzus persicae*, indicating a nonspecific acquisition of the virus by these insects. Interestingly, different retention patterns were found for the viral coat protein. Immunostained western blots of insect homogenates served to follow the retention time of the viral coat protein in viruliferous whiteflies. Preliminary results indicate that the viral coat protein was not detectable in the bodies of viruliferous *B. tabaci* as soon as 8 days after an 48 hr acquisition access (Y. Antignus and S. Cohen, unpublished). It was also possible to show the presence of TYLCV coat protein in *T. vaporariorum*. A similar retention time for the viral coat antigen was found in studies of the interaction of SLCV with *B. tabaci* (24). It was shown that the virus antigen was still detectable 6 days after removal of the insects from the source, but at one-third of its initial level (24). The above-described results support the suggestion that TYLCV, like other geminiviruses, has a nonpropagative nature.

The differences in retention time of the viral nucleic acid and the coat protein in the insect may be explained by the impossibility of detecting the viral protein due to its interaction with receptors on the membrane of the accessory salivary glands and the disintegration of the coat protein of unbound particles (24). However, more information on the route of the virus in its insect vector is required to explain the relatively high levels of the viral nucleic acid present in the insect body at the time when its inoculativity approaches zero.

7. *An hypothetic transmission mechanism*: Based on similarities in the mode of transmission with aphid-borne luteoviruses, it is suggested that the hind gut is the site where the virus particles pass from the gut lumen to the hemocoel. From the hemocoel, the virus particles enter cells of the accessory salivary glands, and then the saliva (24, 41–45). According to

this model, the virus transport in the insect body is in the form of coated vesicles which are able to interact with the insect cell membrane (41).

The association of the whitefly salivary system with transmission efficiency of geminiviruses was shown recently in studies of transmission efficiency of different whitefly biotypes. The whitefly biotype IV-90 is characterized by its ability to induce silver phytoxemia in squash plants by its saliva secretion (23). This biotype accumulates significantly higher amounts of SLCV than biotype IV-81, but its transmission is markedly less efficient. This difference in transmission rates might be due to differences in the physiology of the biotypes affecting the turnover of the virus in the insect body, or to the direct effect of the saliva component on the virus particles. Indications for such effects were demonstrated recently (J.E. Duffus, S. Cohen, and H.Y. Liu, unpublished).

The capsids of all the whitefly-transmitted geminiviruses have one or more antigen epitopes in common (79) and it has been suggested that these may be the determinants of vector specificity, which would indicate that the coat protein plays a predominant role in virus transmission, as was found for some luteoviruses (41–45). It was shown that mutants of ACMV that do not produce coat protein were not acquired by the vector whitefly (14). The role of the geminivirus coat protein in determining insect–vector specificity was demonstrated recently by the construction of a chimeric virus in which the coat protein gene of whitefly-transmitted ACMV was replaced by that of the leafhopper-transmitted BCTV (14). The resulting chimeric virus was thus transformed to a leafhopper transmitted type.

Based on the above described data it is proposed that the transmission of TYLCV as well as that of other geminiviruses consists of two phases. First, acquisition involves passage of the virus through the gut wall into the hemocoel. The second phase involves the passage of the virus from the hemocoel into the salivary gland. It is also suggested that the viral coat protein is responsible for virus–vector specificity and that this specificity resides at the hemocoel–salivary gland barrier. Using the membrane feeding procedure it was shown that *B. tabaci* can acquire and transmit purified intact virions of TYLCV but failed to do so after acquisition feeding on (1) virions with a coat protein impaired by UV irradiation, (2) naked virion ssDNA, or (3) TYLCV cloned dsDNA (6). From these results, an additional function may be attributed to the virus coat protein in the transmission mechanism. It is suggested that epitopes on the virus coat protein interact with the insect gut wall membrane to enable virus mobilization into the hemocoel. Once these epitopes are removed, the viral nucleic acid is unable to leave the gut lumen.

The nonspecific acquisition of TYLCV by different insects indicates that virus–vector specificity among geminiviruses lies in the hemocoel–salivary gland barrier rather than in the gut–hemocoel barrier.

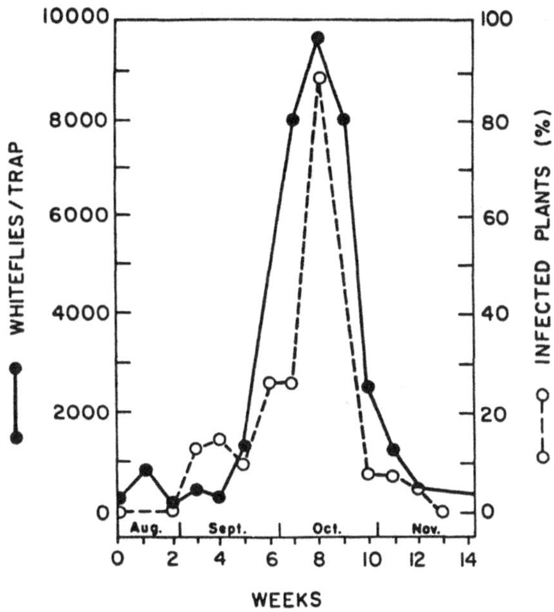

FIGURE 10.7. The association between number of *Bemisia tabaci* whiteflies trapped and incidence of tomato yellow leaf curl (TYLCV) virus on plants in the Jordan Valley, Israel. (See Cohen et al. (26).)

Epidemiology

In the Jordan Valley, where TYLCV was first observed, the spread of the virus is significantly correlated with the population size of *B. tabaci* (26) (Fig. 10.7). The whitefly population usually reaches its peak between the first week of September and mid-October. During this period, a maximum of 5.4% of the total of 6500 whiteflies collected weekly in the fields, transmitted TYLCV (26). *Cynanchum acutum* (Fig. 10.8), which is a common wild weed growing on the banks of the Jordan River, has been found to be the only perennial natural host in the region, and more than 50% of 65 sampled plants were found infected. Whiteflies marked with fluorescent dust while feeding naturally on *C. acutum* along the Jordan River were subsequently trapped within the main tomato production area 7 km away.

Based on these data, an epidemiological cycle of TYLCV in the Jordan Valley was proposed. During December–February only the subterranean parts of *C. acutum* survive. The plants start to grow again only in spring reaching full vegetation in August–September, when flowering starts. From then on, the vegetation declines. During spring and early summer the numbers of migrating *B. tabaci* are very small. Only in June are the first viruliferous whiteflies observed, and their number increases sharply with the increase in the total number of migrating whiteflies. The timing of the whitefly peak population coincides with the transplanting season of tomatoes in this region. The preferred time for flights of *B. tabaci* is during the morning hours (28). At that time the prevailing wind during the summer months is from east to west. It is suggested that the winds carry viruliferous whiteflies from the infected *C. acutum* source into the tomato-growing area. Once TYLCV has been introduced into tomatoes, a secondary spread takes place. *M. parviflora* serves also as a natural reservoir for TYLCV spread. The *M. parviflora* population is very low during the summer and therefore cannot serve as a host bridge for TYLCV between seasons (Fig. 10.9). It is important to note that in plots

FIGURE 10.8. Infected common hosts of tomato yellow leaf curl virus (TYLCV). *Cynancum acutum* (a), *Datura stramonium* (b), *Lycopersicon esculentum* (c).

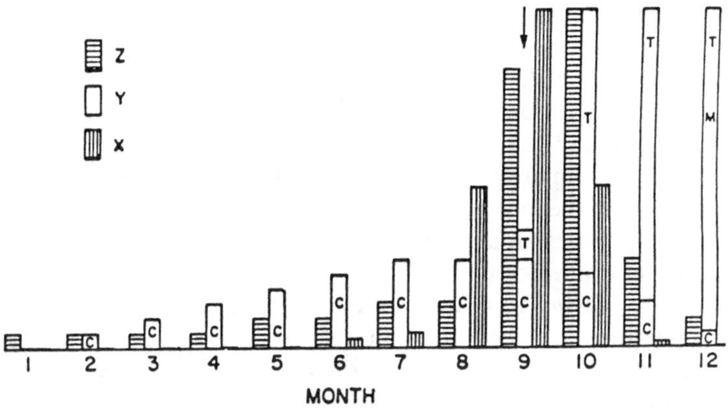

FIGURE 10.9. Schematic diagram of the epidemiology of tomato yellow leaf curl virus (TYLCV). Z, Flying *Bemisia tabaci*; Y, hosts and proportion infected; C, *Cynancum acutum*; T, tomatoes (*Lycopersicon esculentum*); M, *Malva parviflora*; X, infective *B. tabaci* population; arrow, beginning of tomato transplanting. (See Cohen et al. (26).)

surrounded by windbreaks, both the whitefly population and TYLCV incidence were high compared within open fields.

Control

1. Eradication of source plants: Interference in the epidemiological cycle of TYLCV can be achieved by eradicating *C. acutum* in June–July before the beginning of the *B. tabaci* migration, and thereby reduce the primary spread of TYLCV in the region. However, in the Arava region south of the Dead Sea, no infected *C. acutum* plants were found, and still TYLCV incidence in this region may reach 100% at the end of the summer. A thorough survey among the wild and cultivated flora of the region revealed no natural host of the virus that can serve as a bridge between seasons. Based on the trapping of whiteflies in the mid-season at distances of at least 10 miles from plant sources, we assume that the virus is being introduced yearly into the region by long distance wind carried viruliferous whiteflies from the east or the northern part of the Jordan Valley.

In Cyprus (54), in a relatively isolated region, the spread of TYLCV was successfully controlled by eradication of primary inoculum sources in overwintered tomato crops in the spring before emergence of adult whiteflies. In three consecutive years when this measure was undertaken, primary virus spread to spring plantings was almost completely prevented, while further secondary spread to summer plantings was kept below 5%, compared with 40–50% in previous years.

2. Chemical Control: As in the case of other persistent viruses, the spread of TYLCV can be partially controlled by killing the vector through insecticide sprays (12, 29, 85).

Laboratory and microplot techniques have been developed in Israel (12) to screen for the proper insecticide. These screening techniques are based on the practical minimum inoculation time (4 hr) that is required for viruliferous whiteflies to infect an healthy plant successfully (30).

Control of *B. tabaci* to a level that will result in a significant reduction in virus infection has been difficult (85). Whiteflies feed on the under surface of leaves, and the waxy covering of immature insect stages provides protection from insecticides. The development of resistance (38) and the loss of natural predators and parasites after repeated application of insecticides, contribute to the control problems. At the end of summer and during autumn Israeli farmers spray daily. Disease incidence in treated fields may reach 25–50%, but when severe whitefly outbreaks occur 100% infection is very common despite the sprayings.

3. Cultural methods: *Yellow mulches*: Mulching of tomato and cucumber fields, with sawdust, straw, or yellow polyethylene sheets markedly reduced *B. tabaci* populations and consequently the spread of

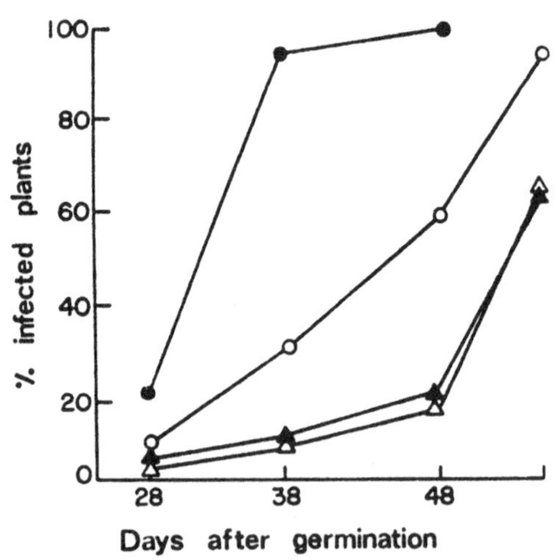

FIGURE 10.10. The effect of different control means on the spread of tomato yellow leaf curl virus (TYLCV). (○) Straw mulch plus sprays of azinphos-methyl starting 10 days after germination; (△) mulching with yellow polyethylene sheets plus sprays of azinphos-methyl starting 10 days after germination; (▲) mulching with yellow polyethylene sheets plus sprays of azinphos-methyl starting 20 days after germination; (●) control. (See Cohen and Melamed-Madjar (28).)

the whitefly-borne viruses TYLCV and cucumber vein yellowing virus (CVYV) (8, 22, 28, 29, 59, 73). Whiteflies are attracted to the yellow color of the mulches and are killed later by the reflected heat (29). The control effect of mulches lasts for a relatively short period of 20–30 days. Fading of the mulch color and changes in the ratio of canopy to mulch are probably the reasons for the reduction in the control effect (Fig. 10.10).

Bait crops: It was shown that by planting alternate rows of tomatoes and cucumbers (planted 30 days before tomato transplantation) the spread of TYLCV in the tomatoes was significantly delayed during the first 2 months (4).

The controlling mechanism of the cucumber bait plants is probably due to the combined effects of (1) The huge difference between the canopy volume of cucumbers and tomatoes. The cucumber leaves cover the soil surface at the time of tomato transplanting (2). The cucumber is a much better host for the whiteflies and once they land on this host they do not tend to leave it as long as the plants are fresh.

Tests in a flight chamber showed no preferences of adult whiteflies in landing on tomato or cucumber leaves, but once they land, cucumbers are prefered over tomatoes (26).

Physical barriers: In general, any materials that are fine enough to block the insects' infiltration but not too fine to provide the plants with light and adequate ventilation can be used for this purpose. Among the materials that are being used are perforated polyethylene, polypropylene sheets (Agryl), and fine mesh nets (50 mesh was found in Israel to be very effective). However, during mid-summer, ventilation and/or cooling systems should be operated in the protected greenhouses to avoid overheating of the plants (13, 22, 55).

4. Breeding for resistance: This is probably the best approach to control virus diseases. Two types of resistance have been described:

a. Very high resistance (close to immunity) was found among the *L. chilense* LA 1969 accession (97). None of the 58 plants tested in the field and the 10 plants inoculated under controlled conditions showed symptoms, and only in two plants was TYLCV-DNA detected using the molecular hybridization technique. The possibility of infection escape ("false resistance") as a result of repellency of the vector from the tested plants was rejected by the authors, since they observed the whiteflies feeding on the *L. chilense* leaves.

b. Different degrees of tolerance—no symptoms or mild symptoms and a detectable amount of virus were found among *L. cheesmanii*, *L. hirsutum*, *L. peruvianum*, and *L. pimpinellifolium* (30, 50, 51, 58, 75, 76, 97). These results published by various workers are not comparable due to the lack of standardization in environmental conditions, inoculum pressure, and virus isolate. Only in two cases (75, 76) were the studies performed under controlled conditions and with a defined isolate of

TYLCV. Thus, in the *L. pimpinellifolium* group incomplete dominance of single gene was described in accession LA121 by Pilowsky and Cohen (75), but Hassan et al. (51) found that the resistance derived from LA121 or LA373 is controlled quantitatively by partially recessive gene action. Kasrawi (57) has described monogenic dominantly controlled resistance in the *L. pimpinellifolium* Hirsute—INRA and LA1478, and a polygenic recessive pattern in *L. cheesmanii*. Dominant resistance in *L. hirsutum* was described by Hassan et al. (51).

The resistance sources described so far were all of wild tomatoes. In our experience, determination of the resistance level in the wild tomato types can be complicated. In most cases the vigorous growth of the plants may mask symptom expression despite being infected. Only when the resistance is being introduced into the background of a cultivated tomato can the true level of the resistance be determined. Indeed, this was the case when the resistance that originated in *L. pimpinellifolium* LA121 was found to be unsatisfactory once it was introduced into *L. esculentum* cultivated tomatoes (76).

TY-20 is the first commercial cultivated tomato hybrid with tolerance to TYLCV (77) (Fig. 10.11). When infected with TYLCV, the leaves of

FIGURE 10.11. Commercial tomato cultivars tolerant to tomato yellow leaf curl virus (TYLCV): TY-20 (a), TY70 (b, d), TY-71 (c), susceptible cultivar (e), fruits of the tolerant cultivars (f).

TY-20 plants exhibit a mild interveinal chlorosis. In mature plants leaflets usually become slightly cupped, but the plants give an acceptable yield in spite of the infection. This tolerance originated from *L. peruvianum* (L.) Mill. PI126935 (M. Pilowsky and S. Cohen, unpublished).

To study the inheritance of this tolerance type, crosses were made between susceptible cultivated tomato and a tolerant line (M-60), which is an F6 inbred in the third backcross generation (BC_3F_6) from the cross between susceptible tomato lines and the TYLCV-tolerant accession PI126935 of *L. peruvianum*. Tolerance was found to be controlled by five recessive genes. Whitefly transmission tests indicated that the tolerant hybrid TY-20 was as good a source of TYLCV as was a susceptible hybrid (76). However, when the plants were infected at the first true leaf stage, their TYLCV-DNA titer (81) was markedly lower in the TY-20 hybrid and the M-60 tolerant line than in susceptible varieties. These findings may indicate that the tolerance mechanism operates through the reduction of the multiplication and/or spread of the virus in the plants (81).

Seedlings of TY-20 and some control susceptible cultivars have been inoculated in the nursery at the first leaf stage (the most severe conditions for the plants) and then transplanted in the field. The susceptible cultivars did not recover, remained dwarf, and gave no yield, while TY-20 plants yielded (about 12 tons/acre of marketable fruits) 45% of the maximum potential of healthy plants of this variety. However, when TY-20 seedlings were protected in the nursery for a month and then were exposed in the field to nature infection the yield was 22 tons/acre compared to 1.6 tons/acre in the susceptible variety despite 100% disease incidence. These results were repeatedly confirmed in different regions of Israel and under different climatic conditions (2, 77). It should be noted that to obtain optimal results, the tolerant varieties should be protected during the first month after transplanting. The best protection can be achieved by a combination of insecticide sprays and yellow mulch. These treatments are essential, especially in some of the hottest region of Israel where the whitefly populations are very high, thus causing direct damage to the plants, which are sometimes killed by the feeding of large numbers of adults and larvae.

The destructive potential of *B. tabaci* as a pest has led M. Pilowsky and M. Berlinger in Israel to start a breeding program aimed at incorporating resistance to *B. tabaci* into tomatoes. In the screening tests the rate of TYLCV transmission and honeydew excretion have been used as a means to evaluate resistance. Partial resistance was found among *L. hirsutum* LA771418, LA1777 and *L. hirsutum* f. *glabratum* LA407, PI34418, and high resistance in one *L. pennellii* LA716 accession (10, 11). These accessions are susceptible to TYLCV, but a sticky excretion from the leaf hairs prevents the whiteflies from feeding and colonizing, thus blocking both virus transmission and direct damage to the plants.

Concluding Remarks

The great economic importance of TYLCV and other whitefly-borne TYLCV-like diseases throughout the world served to catalyze intensive research efforts to find practical solutions for the problems caused by these devastating pests. Epidemiological and behavioral studies of the whitefly vector *B. tabaci* led to the development of cultural means to reduce disease damage. However, the limited efficiency of these methods makes chemical control more attractive for the farmers. The increase in public awareness of the damage caused by pesticides to the environment, and the development of resistance against insecticides in the insect population, stimulated a search for other and better options. Thus, conventional breeding programs were initiated and some commercial tolerant varieties, which offer a better solution, are already on the market.

Despite the impressive progress in our understanding of the molecular biology of geminiviruses and their interaction with the vector there is still much to be learned. A better understanding of the mechanisms involved in virus acquisition and transmission and in the interaction of the virus with its host plants is the key factor in the development of innovative control means in the future.

References

1. Abouzid, A.M., Frischmuth, T., and Jeske, H. 1988. A putative replicative form of the abutilon mosaic virus (gemini group) in a chromatin-like structure. *Mol. Gen. Gen.* **212**:252–258.
2. Adler, U., Omar, S., Kern, J., and Ben-Nur, Z. 1989. Autumn tomatoes in Bet Shean Valley. *Hassadeh* **70**:230–232.
3. Al-Hitty, A.A., and Sharik, H.L. 1987. Studies on host-plant preference of *Bemisia tabaci* (Genn.) on some crops and effects of using host on the trap spread of tomato yellow leaf curl virus to tomato in the plastic houses. *Arab J. Plant Prot.* **5**:19–23.
4. Al-Musa, A. 1982. Incidence, economic importance and control of tomato yellow leaf curl in Jordan. *Plant Dis.* **66**:561–563.
5. Antignus, Y., and Cohen, S. 1991. Agroinfection with a single DNA component of the whitefly-borne tomato yellow leaf curl geminivirus. *Proceedings of the Third International Congress of Plant Molecular Biology* (Tucson, Arizona), Abstr. #1161.
6. Antignus, Y., Adler, O., and Cohen, S. 1991. The use of a molecular probe to study the interaction of TYLCV with its vector *B. tabaci*. *Proceedings of the Third International Congress of Plant Molecular Biology* (Tucson, Arizona), Abstr. #1160.
7. Antignus, Y., Cohen, S., Czosnek, H., Ber, R., Navot, N., and Zamir, D. 1988. Isolation and characterization of TYLCV. Abstracts *5th International Congress of Plant Pathology* (Kyoto, Japan).

8. Avidov (Klein) Z. 1944. *Bemisia tabaci in Israel.* Agricultural Research Station, Rehovot. Hassadeh, Tel Aviv.
9. Ber, R., Navot, N., Zamir, D., Antignus, Y., Cohen, S., and Czosnek, H. 1990. Infection of tomato by tomato yellow leaf curl virus: susceptibility to infection, symptom development, and accumulation of viral DNA. *Arch. Virol.* **112**:169–180.
10. Berlinger, M.J., and Dahan, R. 1988. Importance of plant resistance in the control of whiteflies and whitefly-borne viruses in tomato and the development of screening methods. *International Symposium on Integrated Management Practices* (Taiwan, March 21–26, 1988), pp. 240–247.
11. Berlinger, M.J., Dahan, R., Berlinger, O.C., and Mordechi, S. 1989. Honeydew excretion as a possible tool to screen tomato resistance to virus transmission by *Bemisia tabaci. Proceedings of the Working Group "Breeding for Resistance to Insects and Mites"* (Marcelin, Switzerland, 4–6 September 1989), pp. 121–131.
12. Berlinger, M.J., Dahan, R., and Mordechi, S. 1986. Prevention of TYLCV by controlling its vector, the *Bemisia tabaci. Hassadeh* **66**:686–689.
13. Berlinger, M.J., Mordechi, S., Liper, A., Piper, A., Katz, J., and Levav, N. 1991. The use of nets to prevent the penetration of *Bemisia tabaci* into greenhouses. *Hassadeh* **71**:1579–1583.
14. Briddon, R., Pinner, M.S., Stanely, J., and Markham, P.G. 1990. Geminivirus coat protein gene replacement alters insect specificity. *Virology* **177**:85–94.
15. Butter, N.S., and Ratual, H.S. 1977. The virus-vector relationship of tomato leaf curl virus (TLCV) and its vector, *Bemisia tabaci* Gennadius (Hemiptera: Aleurodidae). *Phytoparasitica* **5**:173–186.
16. Chiemsombat, P., Murayama, A., and Ikegami, M. 1991. Tomato yellow leaf curl virus in Thailand and tobacco leaf curl in Japan are serologically identical. *Ann. Phytopathol. Soc. Jpn.* **57**:595–597.
17. Cherif, C., and Russo, M. 1983. Cytological evidence of the association of a geminivirus with the tomato yellow leaf curl disease in Tunisia. *Phytopathol. Z.* **108**:221–225.
18. Cohen, S. 1965. Investigations on plant virus transmitted by the tobacco whitefly (*Bemisia tabaci* Gennadius) in Israel. Ph.D. thesis, Faculty of Agriculture, Hebrew University of Jerusalem, Rehovot, Israel.
19. Cohen, S. 1967. The occurrence in the body of *Bemisia tabaci* of a factor apparently related to the phenomenon of "periodic acquisition" of tomato yellow leaf curl virus. *Virology* **31**:180–183.
20. Cohen, S. 1969. *In vivo* effects in whiteflies of a possible antiviral factor. *Virology* **37**:448–454.
21. Cohen, S. 1982. Control of whitefly vectors of viruses by colour mulches. In K.F. Harris and K. Maramorosch, eds., *Pathogens, Vectors and Plant Diseases, Approaches to Control.* Academic Press, New York, pp. 45–56.
22. Cohen, S., and Berlinger, M.J. 1986. Transmission and cultural control of whitefly-borne viruses. *Agric. Ecosyst. Environ.* **17**:89–97.
23. Cohen, S., Duffus, J.E., and Liu, H.Y. 1992. A new *Bemisia tabaci* biotype in the southwestern United States and its role in silverleaf of squash and transmission of lettuce infectious yellows virus. *Phytopathology* **82**:86–90.

24. Cohen, S., Duffus, J.E., and Liu, H.Y. 1989. Acquisition, interference, and retention of cucurbit leaf curl viruses. *Phytopathology* **79**:109–113.
25. Cohen, S., and Harpaz, I. 1964. Periodic, rather than continual acquisition of a new tomato virus by its vector, the tobacco whitefly (*Bemisia tabaci* Gennadius). *Entomol. Exp. Appl.* **7**:155–166.
26. Cohen, S., Kern, J., Harpaz, I., and Ben-Joseph, R. 1988. Epidemiological studies of the tomato yellow leaf curl virus (TYLCV) in the Jordan Valley, Israel. *Phytoparasitica* **16**:259–270.
27. Cohen, S., and Marco, S. 1970. Periodic occurrence of an anti-TMV factor in the body of whiteflies carrying the tomato yellow leaf curl virus (TYLCV). *Virology* **40**:363–368.
28. Cohen, S., and Melamed-Madjar, V. 1978. Prevention by soil mulching of the spread of tomato yellow leaf curl virus transmitted by *Bemisia tabaci* (Gennadius) (Homoptera: Aleyrodidae) in Israel. *Bull. Entomol. Res.* **68**:465–470.
29. Cohen, S., Melamed-Madjar, V., and Hameiri, J. 1974. Prevention of the spread of tomato yellow leaf curl virus transmitted by *Bemisia tabaci* (Gennadius) (Homoptera, Aleyrodidae) in Israel. *Bull. Entomol. Res.* **64**:193–197.
30. Cohen, S., and Nitzany, F.E. 1966. Transmission and host range of the tomato yellow leaf curl virus. *Phytopathology* **56**: 1127–1131.
31. Credi, R., Betti, L., and Canova, A. 1989. Association of a geminivirus with a severe disease of tomato in Sicily. *Phytopathol. Medit.* **28**:2232–226.
32. Czosnek, H., Ber, R., Antignus, Y., Cohen, S., Navot, N., and Zamir, D. 1988. Isolation of tomato yellow leaf curl virus, a geminivirus. *Phytopathology* **78**:508–512.
33. Czosnek, H., Ber, R., Navot, N., Antignus, Y., Cohen, S., and Zamir, D. 1989. Tomato yellow leaf curl virus DNA forms in the viral capsid, in infected plants and in the insect vector. *J. Phytopathol.* **125**:47–54.
34. Czosnek, H., Ber, R., Navot, N., Zamir, D., Antignus, Y., and Cohen, S. 1988. Detection of tomato yellow leaf curl virus in lysates of plants and insects by hybridization with a viral DNA probe. *Plant Dis.* **72**:949–951.
35. Davies, W., and Stanely, J. 1989. Geminivirus genes and vectors. *Trend Genet.* **5**:77–81.
36. Defranc, Q., D'hondt, M., and Russo, M. 1985. Tomato yellow leaf curl in Senegal. *Phytopathol. Z.* **112**:153–160.
37. D'hondt, M.D., and Russo, M. 1984. Tomato yellow leaf curl in Senegal. *Phytopathol. Z.* **122**:153–160.
38. Dittrich, S., and Ernst, G.H. 1990. Chemical control and insecticide resistance of whiteflies. In D. Gerlig, ed., *Whiteflies—Their Bionomics, Pest Status and Management*. Intercept, UK, pp. 263–280.
39. Francki, R.I.B., Hatta, T., Grylls, N.E., and Grievel, C.J. 1979. The particle morphology and some other properties of chloris striate mosaic virus. *Ann. Appl. Biol.* **91**:51–59.
40. Gallitelli, D., Luisoni, E., Martelli, G.P., Cacciagli, P., Milne, R.C., Accoto, G.P., and Antignus, Y. 1991. L'accartocciamento fogliare giallo del pomodoro in Sardegna. *Inform. Fitopatol.* **41**:42–46.
41. Gildow, F.E. 1982. Coated vesicle transport of Luteoviruses through salivary glands of *Myzus persicae*. *Phytopathology* **72**:1289–1296.

42. Gildow, F.E. 1985. Transcellular transport of barley yellow dwarf virus into the haemocoel of the aphid vector, *Rhopalisiphum padi*. *Phytopathology* **75**:292–297.
43. Gildow, F.E., and Rochow, W.F. 1980. Role of accessory salivary glands in aphid transmission of barley yellow dwarf virus. *Virology* **104**:97–108.
44. Gildow, F.E. and Rochow, W.F. 1980. Transmission interference between two isolates of barley yellow dwarf virus in *Macrosiphum avenae*. *Phytopathology* **70**:122–126.
45. Gildow, F., and Rochow, W.F. 1980. Importance of capsid integrity for interference between two isolates of barley yellow dwarf virus in an aphid. *Phytopathology* **70**:1013–1015.
46. Hamilton, W.D.O, Bisaro, D.M., and Buck, K.W. 1982. Identification of novel DNA forms in tomato golden mosaic virus infected tissue. Evidence for a two component viral genome. *Nucl. Acid Res.* **10**:4901–4912.
47. Hamilton, W.D.O., Bisaro, D.M., and Buck, K.W. 1983. Demonstration of the bipartite nature of the genome of a single stranded DNA plant virus by infection with the cloned DNA components. *Nucl. Acid Res.* **11**:7387–7396.
48. Hamilton, R.I., Edwardson, J.R., Francki, R.I.B., Hsu, H.T., Hull, R., Koenig, R., and Milne, R.G. 1981. Guidelines for identification and characterization of plant viruses. *J. Gen. Virol.* **54**:223–241.
49. Harrison, B.D. 1985. Advances in geminivirus research. *Annu. Rev. Phytopathol.* **23**:55–82.
50. Hassan, A.A., Mazyad, H.M., Moustafa, S.E., and Nakhla, M.K. 1982. Assessment of tomato yellow leaf virus resistance in the genus *Lycopersicon*. *Egypt. J. Hortic.* **9**:13–116.
51. Hassan, A.A., Mazyad, H.M., Moustafa, S.E., Nassar, S.H., Sims, W.L., and Nakhla, M.K. 1984. Genetics and heritability of tomato yellow leaf curl virus tolerance derived from *L. pimpinellifolium*. *Eur. Assoc. Res. Plant Breed. Tomato Working Group*, Wageningen, The Netherlands.
52. Howarth, A.J. 1986. Geminiviruses, The plant viruses with single stranded DNA genomes. *Genet. Eng.* **8**:5–99.
53. Howarth, A.J., and Vandemarak, G.J. 1989. Phyllogeny of geminiviruses. *J. Gen. Virol.* **70**:2717–2727.
54. Ioannou, N. 1987. Cultural management of tomato yellow leaf curl disease in Cyprus. *Plant Pathol.* **36**:367–373.
55. Ioannou, N., and Hadjinicolis, A. 1991. Epidemiology and control of tomato yellow leaf virus in Cyprus. *Proceedings of the Seminar of EEC control DGXII-TS2-A-055F (CD) Partners*, 4–7 September 1991, Montfavet-Avignon, France, pp. 3–5.
56. Ioannou, N., Kyriakou, A., and Hadjinicolis, A. 1987. *Host Range and Natural Reservoirs of Tomato Yellow Leaf Curl Virus*. Tech. Bull. 85, Agric. Res. Inst., Nicosia, Cyprus.
57. Kasrawi, M.A. 1989. Inheritance of resistance to tomato yellow leaf curl virus (TYLCV) in *Lycopersicon pimpinellifolium*. *Plant Dis.* **73**:435–437.
58. Kasrawi, M.A., Suwwan, M.A., and Mansour, A. 1988. Sources of resistance to tomato yellow leaf curl virus in *Lycopersicon* species. *Euphytica* **37**:61–64.
59. Kern, J., Decco, Z., Cohen, S., and Ben-Joseph, R. 1991. Reduction of damage from TYLCV and tobacco whitefly in tomatoes by yellow plastic mulch. *Hassadeh* **71**:864–868.

60. Kheyr-Pour, A., Bendhmane, M., Matzeit, V., Accoto, G.P., Crespi, S., and Gronenborn, B. 1992. Tomato yellow leaf curl virus from Sardinia is a whitefly transmitted monopartite geminivirus. *Nucl. Acid Res.* **24**:6763–6769.
61. Kikuno, R., Toh, H., Hayashida, H., and Miyata, T. 1984. Sequence similarity between putative gene products of geminiviral DNAs. *Nature (London)* **308**:562.
62. Kim, K.S., Shock, T.L., and Goodman, R.M. 1978. Infection of *Phaseolus vulgaris* by bean golden mosaic virus: Ultrastructural aspects. *Virology* **89**:22–23.
63. Kim, K.S., and Flores, E.M. 1979. Nuclear changes associated with euphorbia mosaic virus transmitted by the whitefly. *Phytopathology* **69**:980–984.
64. Klinkenberg, F.A., and Stanely, J. 1990. Encapsidation and spread of African cassava mosaic virus DNA A in the absence of DNA B when agroinoculated to *Nicotiana benthamiana. J. Gen. Virol.* **71**:1409–1412.
65. Lazarowitz, S.G. 1987. The molecular characterization of geminiviruses. *Plant Mol. Biol. Rep.* **4**:177–192.
66. Luisoni, E., Milne, R.G., and Vecchiati, M. 1992. Purification and serology of tomato yellow leafcurl geminivirus. *Recent Advances in Vegetable Virus Research. 7th Conference ISHS, Vegetable Virus Working Group* (Athens, Greece, 12–16 July 1992), pp. 56–58.
67. Makkouk, K.M., and Laterrot, H. 1983. Epidemiology and control of tomato yellow leaf curl virus. In R.T. Plumb and J.M. Thresh, eds., *Plant Virus Epidemiology*. Blackwell Scientific Publications, Oxford, pp. 315–321.
68. Marco, S., Cohen, S., Harpaz, I., and Birk, Y. 1972. *In vivo* suppression of plant virus transmissibility by an anti-TMV factor occurring in an inoculative vector's body. *Virology* **17**:761–766.
69. Marco, S., Cohen, S., Harpaz, I., and Birk, Y. 1975. An anti-TMV factor in the tobacco whitefly after acquisition of tomato yellow leaf curl virus. *J. Insect Physiol.* **21**:1821–1826.
70. Muniyappa, V., Swanson, M.M., Duncan, G.H., and Harrison, B.D. 1991. Particle purification properties and epitope variability of Indian tomato leaf curl geminivirus. *Ann. Appl. Biol.* **118**:595–604.
71. Nakhla, M.K., Rojas, M.R., McLaughlin, W., Wyman, J., and Maxwell, D.P. 1992. Molecular characterization of tomato yellow leaf curl virus from Egypt. *Plant Dis.* **76**:538.
72. Navot, N., Picherski, E., Zeidan, M., Zamir, D., and Czosnek, H. 1991. Tomato yellow leaf curl virus: A whitefly-transmitted geminivirus with a single genomic component. *Virology* **185**:151–161.
73. Nitzany, F.E., Geisenberg, H., and Koch, B. 1964. Tests for the protection of cucumbers from a whitefly-borne virus. *Phytopathology* **54**:1059–1061.
74. Osaki, T., and Inoyue, T. 1981. Tobacco leaf curl virus. CMI/AAB Descriptions of plant Viruses No. 232.
75. Pilowsky, M., and Cohen, S. 1974. Inheritance of resistance to tomato yellow leaf curl virus in tomatoes. *Phytopathology* **64**:632–635.
76. Pilowsky, M., and Cohen, S. 1990. Tolerance to tomato yellow leaf curl virus derived from *Lycopersicon peruvianum. Plant Dis.* **74**:248–250.

77. Pilowsky, M., Cohen, S., Ben-Joseph, R., Shlomo, A., Chen, L., Nahon, S., and Krikun, J. 1989. TY20 A tomato cultivar tolerant to tomato yellow leaf curl virus. *Hassadeh* **69**:1212–1215.
78. Roberts, I.M. 1989. Indian cassava mosaic virus: Ultrastructure of infected cells. *J. Gen. Virol.* **70**:2729–2739.
79. Roberts, I.M., Robinson, D.J., and Harrison, B.D. 1984. Serological relationships and genome homologies among geminiviruses. *J. Gen. Virol.* **65**:1723–1730.
80. Rochester, D.E., Kositratana, W., and Beachy, R. 1990. Systemic movement and symptom production following agroinoculation with a single DNA of tomato yellow leaf curl geminivirus (Thailand). *Virology* **178**:520–526.
81. Rom, M., Antignus, Y., Gidoni, D., Pilowsky, M., and Cohen, S. 1992. Comparative study of tomato yellow leaf curl (TYLCV) DNA accumulation in tolerant and susceptible tomato lines. *Plant Dis.* **77**:253–257.
82. Russo, M., Cohen, S., and Martelli, G.P., 1980. Virus-like particles in tomato plants affected by the yellow leaf curl disease. *J. Gen. Virol.* **49**:209–213.
83. Sastry, K.S., and Singh, S.J. 1973. Assessment of losses in tomato by tomato leaf curl virus. *Indian J. Mycol. Plant Pathol.* **3**:50–54.
84. Saunders, K., Lucy, A., and Stanely, J. 1991. DNA forms of the geminivirus African cassava mosaic virus consistent with a rolling circle mechanism of replication. *Nucl. Acid Res.* **19**:2325–2330.
85. Sharaf, N. 1986. Chemical control of *Bemisia tabaci*. *Agric. Ecosyst. Environ.* **17**:111–127.
86. Stanely, J. 1985. The molecular biology of geminiviruses. *Adv. Virus Res.* **30**:139–173.
87. Stanely, J., and Davies, J.W. 1985. Structure and function of the DNA genome of geminiviruses. In *Molecular Plant Virology*, II. CRC Press, Boca Ratan, FL, pp. 192–214.
88. Stanely, J., Frischmuth, T., and Ellwood, S. 1990. Defective viral DNA ameliorates symptoms of geminivirus infection in transgenic plants. *Proc. Natl. Acad. Sci. U.S.A.* **87**:6291–6295.
89. Stanely, J., and Gay, M.R. 1983. Nucleotide sequence of cassava latent virus DNA. *Nature (London)* **301**:260–262.
90. Stanely, J., and Townsend, R. 1985. Characterization of DNA forms associated with cassava latent virus infection. *Nucl. Acid Res.* **13**:2189–2206.
91. Stenger, D.C., Revington, G.N., Stevenson, M.C., and Bisaro, D.M. 1991. Replicational release of geminivirus genomes from tandemly repeated copies: Evidence for rolling-circle replication of a plant viral DNA. *Proc. Natl. Acad. Sci. U.S.A.* **88**:8029–8033.
92. Thomas, J.E., Massalski, P.R., and Harrison, B.D. 1986. Production of monoclonal antibodies to African cassava mosaic virus and differences in their reactivities with other whitefly-transmitted geminiviruses. *J. Gen. Virol.* **67**:2739–2748.
93. Townsend, R., Stanely, J., Curson, S.J., and Short, M.N. 1985. Major polyadenylated transcripts of cassava latent virus and location of the gene encoding coat protein. *EMBO J.* **4**:33–37.

94. Townsend, R., Watts, J., and Stanely, J. 1986. Synthesis of viral DNA forms in *Nicotiana plumbaginifolia* protoplasts inoculated with cassava latent virus (CLV); evidence for the independent replication of one component of the CLV genome. *Nucl. Acid Res.* **14**:1253–1265.
95. Verma, H.N., Srivastava, K.M., and Mathur, A.K. 1975. A whitefly-transmitted yellow mosaic virus disease of tomato from India. *Plant Dis. Rep.* **59**:494–498.
96. Woolston, C.J., Reynolds, H.V., Stacey, N.J., and Mullineaux, P.M. 1989. Replication of wheat dwarf virus DNA in protoplasts and analysis of coat protein mutants in protoplasts and plants. *Nucl. Acid Res.* **17**:6029–6041.
97. Zakay, Y., Navot, N., Zeidan, M., Kedar, N., Rabinowitch, H., Czosnek, H., and Zamir, D. 1991. Screening *Lycopersicon* accessions for resistance to tomato yellow leaf curl virus: Presence of viral DNA and symptom development. *Plant Dis.* **75**:279–281.

11
Propagative Transmission of Plant and Animal Viruses by Insects: Factors Affecting Vector Specificity and Competence

El Desouky Ammar

Introduction

Until recently, viruses were generally divided on the basis of their host range into bacterial, plant, vertebrate, and invertebrate viruses (82). However, it was clear that strict division of viruses according to host range is not always possible, since some families (e.g., Reoviridae, Rhabdoviridae, and more recently Bunyaviridae and Picornaviridae) included members that infect invertebrates, vertebrates, and/or plants (3, 83, 103). Furthermore, with complete nucleotide sequences available for many viruses, long stretches of amino acid sequence similarity in replicase proteins for certain groups of viruses of both plants and animals suggested that some "superfamilies," which cross the host range lines, may have descended from a common ancestor (135). Thus, the concept of "unity of virology as a whole" became gradually established (82a), and in many cases the dividing line between "animal" and "plant" viruses is no longer valid.

With the above considerations in mind, it is expected that researchers working on animal or plant viruses have much to learn from each other, since certain areas of research seem to be more advanced in animal than in plant virology, and vice versa. To illustrate this point in the field of virus transmission by insects, I will give the following examples. In the early days of studying virus–vector relationships, only two categories of arthropod transmission were recognized for both plant and animal viruses. First, *mechanical transmission* in which the vector is merely a carrier of the virus, usually (it was assumed) through contaminated mouth parts, from an infection source to a new host; and second, *biological transmission* in which the arthropod plays a necessary biological role in

El Desouky Ammar, Department of Economic Entomology, Faculty of Agriculture, Cairo University, Giza, Egypt. Present address: Department of Plant Pathology, University of Kentucky, Lexington, Kentucky, 40546, USA.

the transmission and maintenance of the pathogen (25, 65). Although the concept of mechanical transmission still is being held by animal arbovirologists (53, 147), this concept is no longer used to describe *nonpersistent transmission* of some plant viruses previously assumed to be transmitted "mechanically" through contaminated aphid stylets (57a, 57b, 79, 95, 108). Several studies on potyviruses and caulimoviruses indicated that they require a "helper factor or component" (a virus-encoded protein in the host plant) for transmission by aphids to occur (108). Furthermore, potyviruses apparently are retained not only in the stylets but also in the foregut of aphids (20). Thus, what seemed to be a simple mechanical process, in aphid transmission of potyviruses and some other nonpersistent plant viruses, is now known to be a more complex biological one. On the other hand, with the biologically transmitted "propagative" viruses, much more is known about the intrinsic factors that affect vector competence of mosquito-borne (animal) viruses, compared to those of aphid-, leafhopper-, or planthopper-borne (plant) viruses.

Since this chapter deals mainly with viruses that propagate in their insect vectors, I have used examples of both animal and plant viruses to illustrate the concepts of transmission barriers and other factors affecting vector specificity and vector competence. However, due to the large body of literature concerning animal viruses only a few examples of these, in each case, will be discussed. Pertinent reviews dealing with transmission of animal arboviruses include those by De Folliart et al. (31), Hardy (53), Hardy et al. (55), Leake (76), Stollar (133), and Turell (147). Reviews on the transmission of propagative plant viruses include those by Ammar (7), Harris (57b, 58, 58a), Nault (94), Nault and Ammar (95), Sinha (128), and Sylvester (138). A list of the propagative viruses frequently mentioned in this chapter, their abbreviations, taxonomic grouping, host range, and vectors is presented in Table 11.1. Virus taxonomy follows Matthews (83), Nuss and Dall (100), and Strauss and Strauss (135), while virus names will follow the I.C.T.V. system as adopted by Matthews (83).

Transmission Characteristics of Propagative Insect-Borne Viruses

Propagative viruses are those that multiply in their vectors. Such multiplication can be demonstrated by transmission studies in which virus is passed serially from one insect to another, either by needle injection or through transovarial transmission, until the dilution attained in the final inoculative (infective) insects exceeds the dilution endpoint of the initial inoculum (16, 21, 28, 42, 53, 57b, 95, 129). Transmission electron microscopic studies also can give circumstantial evidence for virus multiplication in the vector, by observation of aggregated virus particles or sites of virus

TABLE 11.1. Propagative arthropod-borne viruses frequently mentioned in this chapter, their abbreviations, taxonomic groups, host range and vectors.[a]

Family/group and virus name	Abbreviation	Hosts/vectors
Bunyaviridae		
La Crosse	LACV	Vertebrates, arthropods/
Rift Valley Fever	RVFV	mosquitoes, ticks
Tomato spotted wilt	TSWV	Plants/thrips
Flaviviridae		
Dengue	DeV	Vertebrates/
Yellow fever	YFV	mosquitoes, ticks
Marafivirus		
Maize rayado fino	MRFV	Plants/leafhoppers
Oat blue dwarf	OBDV	
Reoviridae		
Fijivirus		
Fiji disease	FDV	Plants/planthoppers
Maize rough dwarf	MRDV	
Phytoreovirus		
Rice dwarf	RDV	Plants/leafhoppers
Rice gall dwarf	RGDV	
Wound tumor	WTV	
Rhabdoviridae		
Plant rhabdoviruses		
Potato yellow dwarf	PYDV	Plants/leafhoppers
Rice transitory yellowing	RTYV	
Wheat striate mosaic	WSMV	
Barley yellow striate mosaic	BYSMV	Plants/planthoppers
Iranian maize mosaic	IMMV	
Maize mosaic	MMV	
Sowthistle yellow vein	SYVV	Plants/aphids
Strawberry crinkle	SCV	
Vesiculovirus		
Vesicular stomatitis	VSV	Vertebrates, arthropods/ mosquitoes, simuliid flies
Tenuivirus		
European wheat striate mosaic	EWSMV	Plants/planthoppers
Maize stripe	MStV	
Rice hoja blanca	RHBV	
Rice stripe	RStV	
Togaviridae (alphaviruses)		
Eastern equine encephalomyelitis	EEEV	Vertebrates/mosquitoes
Western equine encephalomyelitis	WEEV	

[a] Information compiled from Matthews (83), Nault and Ammar (95), Nuss and Dall (100), Strauss and Strauss (135), and Turell (147). This list is not inclusive of all known viruses in any category.

assembly in vector cells and tissues (10, 53, 121). However, more direct evidence is obtained by quantitative serology, using enzyme-linked immunosorbent assay (ELISA) or other methods, that show an increase in virus titer in the vector after its removal from the virus source (36, 49, 97). Propagative viruses require longer mean latent periods in their vectors (days to weeks) than do *circulative-nonpropagative* viruses (hours to days). Vectors of both virus groups, however, usually remain inoculative for life, and some propagative viruses are vertically transmitted through the females to subsequent generations (30, 128, 129).

The above definitions and terminologies have been used so far mainly with insect-borne plant viruses, but on the whole they are also applicable to arthropod-borne animal viruses (arboviruses). However, the circulative-nonpropagative category of transmission, where some plant viruses pass through the gut to the hemolymph and salivary glands without any detectable multiplication in their aphid, leafhopper, or whitefly vectors (46, 57b, 95, 138), apparently has not been reported with mosquito-borne arboviruses (53). Also, the term "latent or incubation period" used by plant virologists is usually referred to as the "extrinsic incubation period" by arbovirologists (53, 120, 147).

At least 49 plant viruses belonging to five taxonomic groups are propagative in their insect vectors: 3 marafiviruses transmitted by cicadellid leafhoppers, 5 tenuiviruses by delphacid planthoppers, 13 plant reoviruses by either leafhoppers or planthoppers, 27 plant rhabdoviruses by leafhoppers, planthoppers, or aphids, and more recently, one bunyavirus transmitted by thrips (94, 95, 149). Several of these viruses cause serious or potentially significant diseases to economic crops, including rice, maize, sugarcane, wheat, potato, and tomato in many parts of the world (28, 60, 138, 145).

The arboviruses include about 500 members belonging mainly to five families: Rhabdoviridae, Reoviridae, Togaviridae (alphaviruses), Flaviviridae and Bunyaviridae. More than 200 arboviruses are transmitted by mosquitoes; among them are some of the most extensively studied agents of human disease, including yellow fever, dengue, and Japanese and St. Louis encephalitides (31). Other arboviruses are transmitted by ticks, sandflies, and biting midges. The transmission cycles of arboviruses alternate between a variety of vertebrate hosts including man and his domestic animals, birds, and some lower vertebrates, and blood sucking arthropod vectors (76). Most of the arboviruses are transmitted in a propagative manner by their vectors, although mechanical transmission, presumably via contaminated mosquito mouthparts, has been demonstrated experimentally for some viruses, e.g., yellow fever and vesicular stomatitis viruses (53, 147). However, Stollar (133) defined the arboviruses as those that, in order to be maintained in nature, (a) must be alternately transmitted between a vertebrate host and an arthropod host, and (b) must replicate well in both hosts.

Sequence of Events in a Propagative Transmission Cycle

Several studies using virus-infectivity, immunofluorescent, or immunoperoxidase assays, in addition to light and electron microscopy, have revealed the sequence of events in the propagative transmission cycle for several arthropod-borne plant or vertebrate viruses. Hardy (53) summarized the route of infection for arboviruses in mosquitoes, that can be slightly modified to fit all propagative (and circulative) plant or vertebrate viruses in their insect vectors, as follows. Before an uninfected arthropod can transmit any propagative virus horizontally, that is orally, the following steps must occur: (a) ingestion of a viremic meal (vertebrate blood or plant sap), (b) infection of the cells of the midgut, (c) release or escape of virus from these cells into the hemocoel, (d) infection of the salivary glands with or without secondary multiplication of virus in other cells and tissues, (e) release of virus into the saliva, and (f) feeding on a susceptible host (plant or vertebrate animal) and inoculation with "viremic" saliva.

With animal viruses, this sequence of events has been demonstrated for several arboviruses in their mosquito vectors using fluorescent antibody staining of dissected organs, virus titration in cell cultures of whole mosquitoes, dissected organs, and hemolymph, or transmission electron microscopy of infected mosquitoes at different intervals following oral acquisition of the studied virus (53, 55, 116a, 120). In general, flaviviruses and bunyaviruses require 1–7 days for infection in gut epithelium to be detected, and 4–14 days for virus to disseminate into the salivary glands, after which oral transmission of virus becomes possible (120). Studies with alphaviruses, however, show a trend toward more rapid virus replication and dissemination in mosquitoes. For example, ultrastructural studies indicated that eastern equine encephalomyelitis *Alphavirus* (EEEV) infected and replicated in the salivary glands and was released into salivary material within 55–69 hr following oral acquisition of virus from a blood meal (119).

With plant viruses, a similar sequence of events has been demonstrated for several viruses in leafhoppers, planthoppers, and aphids, starting with studies by Sinha (125) on wound tumor *Phytoreovirus* (WTV, Reoviridae) in its leafhopper vector *Agallia constricta*, and by Sinha and Chiykowski (130) on wheat striate mosaic rhabdovirus (WSMV) in the leafhopper *Endria inimica*. Using infectivity bioassays, WSMV was recovered from the alimentary canal of this vector on day 2, and from the salivary glands on day 4, following a 1-day acquisition access period on diseased plants. More recently, the sequential multiplication of maize stripe *Tenuivirus* (MStV) in the planthopper vector *Peregrinus maidis* has been followed by quantitative serology (97). Mean absorbance ELISA values for MStV increased significantly from days 2 to 23 postacquisition. On day 7, higher proportions of midguts than ovaries were infected, whereas no virus was detected in the salivary glands 7–9 days postacquisition. On days 16 and 23, however, most tested specimens of these three organs were infected.

Intrinsic Factors Affecting Vector Specificity, Efficiency, and Competence

The terms "vector specificity," "vector efficiency," and "vector comptence" are often used to describe the comparative ability of certain species, biotypes, or lines of vector insects to transmit a certain virus or a virus strain. The terms "vector competence" and "vector susceptibility" have been used mainly by arbovirologists, to denote the ability of an insect to transmit a virus (that is competence) or just be infected with a virus, regardless of transmission (that is susceptibility) (53, 147). Interspecific as well as intraspecific variations in vector competence have been reported for several plant and animal viruses. Also differences correlated with the age, sex or wing-form (morphs) of vector aphids, leafhoppers, planthoppers, and mosquitoes have been found with several viruses (53, 95, 138). In the final analysis, however, these intrinsic factors that influence vector competence must be related to internal physiological factors that govern infection of a vector and its ability to transmit the virus, if given the opportunity, in addition to innate behavioral traits such as host preference, feeding behavior, and probing activity (14, 53). Apart from feeding and probing behavior, which recently has been reviewed for homopteran vectors of plant viruses by Backus (14), propagative transmission of plant and animal viruses is probably influenced by several physiological and anatomical barriers in the vectors as will be illustrated below. In addition to these intrinsic factors, virus transmission and epidemiological patterns are influenced by numerous extrinsic or environmental factors such as temperature and virus–host–vector interactions (28, 31), as will be elucidated later in this chapter.

Following is a more detailed analysis of the transmission cycle in propagative viruses. Four groups of barriers to this cycle will be discussed: (1) midgut infection barrier, (2) dissemination barriers, (3) salivary gland escape barrier, and (4) transovarial and venereal transmission barriers.

Midgut Infection Barrier

The insect midgut or mesenteron consists mainly of a single layer of epithelial cells, with extensive microvilli on the lumen side, and a porous basal lamina on the hemocoel side. Initiation of virus infection in the mosquito requires that the female ingest a sufficient concentration of virus to overcome a "threshold of infection." Midgut infection thresholds have been demonstrated for several arboviruses with various mosquito species or strains (53). However, with plant viruses the term "acquisition threshold" usually refers to the minimum time required for exposure to diseased plants, after which an insect vector can become virus-infected (viruliferous). This "acquisition threshold" possibly involves both the time necessary for the insect stylets to reach the plant tissue from which the virus can be acquired (e.g., mesophyll or phloem), as well as a "threshold titer" of virus that must be ingested before infection of the vector can occur. For instance, acquisition thresholds of less than 1 min for WSMV and 5–15 min for rice transitory yellowing virus (RTYV) (both rhabdoviruses) are thought to reflect the more general distribution of WSMV in mesophyll and phloem cells, and the restricted distribution of RTYV in phloem and vascular parenchyma of their gramineous host plants (27, 77). The feeding behavior of several homopterans has been studied in some detail (14), but the "microevents" taking place during virus acquisition or inoculation, from or into host plants, remain largely unknown, except for some studies on the short probes resulting in transmission of nonpersistent viruses by aphids (79), and the feeding behavior resulting in semipersistent virus transmission by leafhoppers (155c). Several studies, however, indicate that longer acquisition or inoculation access periods are correlated with higher efficiency of transmission for many propagative plant viruses (95). For example, the percentage of leafhoppers (*Nephotettix nigropictus*) transmitting rice gall dwarf *Phytoreovirus* (RGDV) increased from 12 to 96% with longer acquisition access periods (from 3 hr to 12 days) on diseased plants, although the latent period of RGDV in the vector did not differ significantly (13.6–14.6 days) throughout this wide range of acquisition periods (66).

In mosquitoes, after initation of infection, some arboviruses multiply rapidly in the midgut and reach peak titers within 1–3 days, as indicated for the western equine encephalomyelitis *Alphavirus* (WEEV). Houk et al. (61) observed significant increases in titers of WEEV in dissected midguts of *Culex tarsalis* and *Aedes dorsalis* within 8 to 12 hr after ingestion of 10^4 to 10^6 pfu (plaque-forming units) of virus. Peak virus titers of 10^5 to 10^6 pfu per midgut occurred at 24 to 36 hr after oral acqusition of virus. Several studies using immunofluorescent antibody technique with some arboviruses suggest that cell-to-cell spread of virus can occur in the mosquito midgut, probably via adsorption and penetration through basolateral plasma membranes (53). The existence of a midgut infection (MI) barrier to an arbovirus in mosquitoes was first demonstrated in 1935 by Merrill and TenBroeck (86). They demonstrated that *A. aegypti* females, which were completely resistant to oral infection with EEEV, could be induced to transmit this virus if it was inoculated directly into the hemocoel, or if the abdomen was punctured immediately after ingestion of a blood meal. This phenomenon had been demonstrated earlier by Storey (134) in *Cicadulina* leafhoppers for transmission of maize streak *Geminivirus*, which is now considered circulative-nonpropagative in its vectors (95). More recently, Hardy et al. (56) reported differences in virus titer, following ingestion or intrathoracic inoculation of WEEV, between four species of mosquitoes differing in vector competence. They also reported that two strains of *C. tarsalis*, selected for susceptibility or resistance to WEEV, differed in susceptibility of the midgut to infection rather than general resistance to viral replication (54). Genetic susceptibility to viral infection within an insect species appears to be virus specific, and even strains of the same virus may differ greatly in their ability to infect the midgut of a given strain or biotype of the vector (147).

With propagative plant viruses, genetic variability in the transmission efficiency of homopteran vectors of several viruses has been documented (44, 95, 127), but in most cases it has not been specifically attributed to midgut infection or other barriers. However, a gut barrier has been implicitly demonstrated in leafhopper transmission of WTV. Efficiency of virus transmission in WTV and some other propagative plant viruses usually decreases with the age of the vector following oral acquisition of virus from plants. Data from abdominal puncture and fluorescent antibody techniques on WTV leafhopper vectors suggest that both the susceptibility of the midgut epithelial cells to infection and gut permeability to this virus decrease with increasing vector age (124). Additionally, in a study by Falk and Tsai (36), using ELISA with maize mosaic rhabdovirus (MMV), a much higher proportion of planthoppers (*Peregrinus maidis*) tested were positive for MMV following injection (85%), compared to those that acquired MMV orally (42%). A more striking example of a possible gut barrier in planthoppers comes from a recent study on Iranian maize mosaic rhabdovirus (IMMV), which is related serologically to some, but not other, MMV isolates and is transmitted naturally by *Ribautodelphax notabilis* (67). Experimentally, however, IMMV has been transmitted by *P. maidis* with very low efficiency (0.4–1.6%) when acquired orally, and much higher efficiency (64%) when acquired by injection into the hemocoel (E.D. Ammar, R. Gomez-Louengo, and D.T. Gordon, unpublished). Also, the latent period in the vector, for several propagative plant viruses, is much shorter when virus is acquired by injection than when it is acquired orally. For example, with maize rayado fino *Marafivirus* (MRFV), latent periods for injected and orally acquired virus in the leafhopper vector *Dalbulus maidis* were 1–3 and 7–31 days,respectively, with increased transmission rate when virus was acquired by injection (96). Similarly, with MMV in its planthopper vector *P. maidis*, the average time between virus injection and detection by ELISA was only 4 days compared to 12.3 days after oral acquisition (36). These results suggest the involvement of a midgut (infection or escape) barrier for these viruses in their vectors.

Various hypotheses, reviewed by Hardy (53) and Hardy et al. (55), have been proposed to explain the MI barrier to arbovirus infection in mosquitoes. These include diversion of the blood meal into diverticula, inactivation of virus by digestive enzymes in the lumen of the midgut, occlusion of virus by the peritrophic membrane, absence or reduced number of cellular receptor sites of virus attachment on the microvillar membrane, and/or abortive replication of virus in midgut epithelial cells. However, current evidence indicates that the resistance of *Culex pipiens* females to oral infection with the alphavirus WEEV appears to be related to failure of this virus to bind to the microvillar membrane. This involves the attachment, penetration, or uncoating steps of the virus multiplication cycle, rather than inability of virus to replicate in the midgut epithelial cells (62). Using electron microscopy, Houk et al. (61) found that WEEV particles apparently entered the epithelial cells via fusion of the virion envelope with the microvillar membranes rather than endocytosis. Maturation of virions occurred either by budding of nucleocapsids through the basal plasma membrane, or in more susceptible mosquitoes in association with endoplasmic reticulum membranes. Experiments using ^{3}H- or ^{32}P-labeled WEEV with isolated brush border fragments suggest that microvillar membranes of the resistant species *C. pipens* lack specific receptor sites for WEEV, which are available in those of the susceptible *C. tarsalis* (12, 53). Virion surface proteins apparently are important in receptor recognition; the interaction of La Crosse virus (LACV, Bunyaviridae) with the midgut of *A. triseriatus* is influenced by the (G1) envelope glycoprotein of the virus (136). Also, Fraser et al. (41) proposed that the receptor-recognition interaction that initiates reovirus infection in host cells is mediated by the Sigma 1 protein located at the vertices of the icosahedral virion. Thus, in addition to the vector genetics, the role of viral genetics is important in vector specificity. For example, small plaque variants and a rodent strain of WEEV are less infective for, and less transmissible by, *C. tarsalis* than are large plaque variants and a mosquito strain of WEEV (J.L. Hardy, personal communication). These data indicated that (a) transmission barriers in *C. tarsalis* are recognized by some strains of WEEV but not others, and (b) only strains of virus that can overcome the transmission barriers will be transmitted in nature.

Interesting results, apparently linked to the MI barrier in mosquitoes, have been reported by Ramasamy et al. (112). Females of *A. aegypti* showed a significant reduction in susceptibility to infection with Ross River *Alphavirus* (Togaviridae) and Murray Valley encephalitis virus (Flaviviridae) when they were fed on a blood–virus mixture containing rabbit antibodies to mosquito midgut components. Presence of the antibodies, however, did not affect virus titers in infected mosquitoes or transmission of virus from these to vertebrates. The authors suggested that reduction in susceptibility of mosquitoes may be due to the presence of antibodies directed against components of the midgut epithelium that serve as receptors to the virus. Earlier, Feinsod et al. (39) demonstrated that antivector sera effectively neutralized *A. aegypti*-propagated Sindbis *Alphavirus*, which suggests the incorporation of vector antigens on the viral envelope. Based on these and other studies, Ramasamy et al. (112)

suggested that it may be possible to reduce the susceptibility of vector mosquitoes to infection with some viruses by immunizing vertebrates using the appropriate mosquito antigen.

Dissemination Barriers

Until recently, dissemination of viruses from the mosquito alimentary canal has been assumed to occur from the posterior midgut because this is the destination of ingested blood (55, 116). However, Romoser et al. (116) provided evidence that Rift Valley fever virus (RVFV, Bunyaviridae) can disseminate into the hemocoel *C. pipiens* from cells located at the foregut/midgut junction, and suggested that this may be an important route of dissemination from the mosquito gut. In a recent immunocytochemical study of RVFV in this vector (116a), virus antigen was detected in all major midgut regions. Among specimens with disseminated infections (infection beyond the midgut epithelium) RVFV antigen was detected in most tissues, including those of the nervous and endocrine systems. A viral dissemination index (DI) based on several tissues was determined for each orally infected specimen and used to estimate the extent of viral dissemination in the hemocoel. Plotting DI values as a function of time after ingestion of an infectious blood meal indicated that dissemination was sporadic, and that once virus escapes from the midgut, its spread to other tissues is rapid.

There seems to be no satisfactory explanation of how viruses are transported through the basal lamina between the midgut and the hemolymph. In mosquitoes, the pore size of this basal lamina is only ca. 10 nm, whereas particles of most arboviruses are much larger (53). Furthermore, although in several cases an almost systemic infection of the vector occurs, it is not clear whether secondary multiplication of virus in other tissues is necessary for the infection of the salivary glands. Weaver (154) observed WEEV virions in fat body cells of *C. tarsalis* within a few hours after ingestion of a viremic blood meal, and before observing virus in midgut epithelial cells. Hardy et al. (55), however, questioned the validity of similar observations that suggest an "intercellular" or a "leaky midgut" route for virus dissemination into the hemocoel.

For many years, the midgut infection barrier was believed to be the principal determinant of the ability of a species to transmit a virus. However, more recently, Hardy et al. (55) demonstrated both a midgut escape (ME) and a salivary gland infection (SGI) barrier for WEEV in *C. tarsalis*. In competent females (38%), virus titer was high in all cell tissues examined. In incompetent females exhibiting an ME barrier (24%), virus titer was low and restricted to the midgut. In incompetent females with SGI barrier (38%), virus titer was normal in midgut, reduced in hemolymph and remnant tissues, and undetectable in the salivary glands. While the ME barrier occurred in *C. tarsalis* only when low doses of

WEEV were ingested, ME barrier was the principal determinant of the ability of *C. pipiens* to transmit RVFV (147). In this mosquito species, there were no significant differences in viral titers of midgut samples among the nondisseminated infected (virus limited to alimentary canal), disseminated infected nontransmitting, and transmitting groups of mosquitoes (38). Thus, multiplication of virus in the midgut cells apparently does not explain the ME barrier in this case.

A mechanism to explain the barriers to arbovirus dissemination in susceptible mosquitoes has yet to be elucidated. However, Hardy (53) suggested that the genetic trait associated with "modulation" of arbovirus replication in mosquitoes may be involved. That the mosquito can control arbovirus infections is suggested by the observation that the expression of dissemination barriers is dose dependent. If the female mosquito is not initially overwhelmed with a high dose of virus, the infection apparently is brought under control at various phases of tissue infection (53). Also, arbovirus mutants or gene products could limit virus multiplication and dissemination in mosquitoes. Various types of *Alphavirus* mutants that specifically interfere with superinfection of the cells with virus have been recovered from persistently infected mosquito cell culture. In *A. albopectus* cells, infected with Sindbis *Alphavirus*, apparently each virion that enters the cell is compartmentalized into a cytoplasmic "virus factory" where progeny virions are produced and then released from the cell by membrane fusion. Thus, continued infection of the cell depends on reinfection by released virus (23). If a similar phenomenon exists in mosquito salivary glands, a decrease in viral titers in the hemolymph below the salivary gland infection threshold level would interrupt infection of the salivary gland cells. However, whether similar mechanisms occur in intact mosquitoes remains to be investigated (53). Further discussion of the apparent modulation of plant and animal viruses in their vectors will be dealt with later in this chapter.

The involvement of a dissemination (gut escape) barrier has been demonstrated recently by Ullman et al. (149, 150) for tomato spotted wilt *Tospovirus* (TSWV, Bunyaviridae) in its vector thrips *Frankliniella occidentalis*. For a long time, TSWV was considered circulative-nonpropagative in its vectors (118). Several lines of evidence, however, including ELISA, immunocytochemical electron microscopy, and specific cDNA probes, strongly suggested that TSWV replicates in its vector (149, 150a, 155b). TSWV is transmitted only when the thrips acquire it by feeding during the larval, but not the adult, stage. Recent ultrastructural studies showed virions lining and apparently fusing to the apical plasmalemma of the midgut microvilli of adult thrips following their acquisition feeds on TSWV-infected plants (149, 150). Furthermore, endocytosis of these virions by the plasmalemma and occurrence of virions in the cytoplasm of the midgut epithelial cells were also observed. However, TSWV, apparently was confined to the digestive tract and midgut cells in adult thrips, whereas in thrips that acquired virus at the larval stage virions were also observed in the hemocoel and later in the salivary glands (150, 155b). Additionally, in cells of thrips that acquired virus as larvae, virions and viroplasms were labeled with TSWV antibodies, and fibrous inclusions were labeled with antibodies specific to the 52.4K (TSWV-encoded) nonstructural protein, which indicates replication of TSWV in these cells (150a, 155b). The above results suggest a midgut escape barrier to TSWV in adults but not in larvae of vector thrips. Investigations on defective TSWV isolates that have lost insect transmissibility showed that viral membrane glycoproteins are altered or absent, which indicates the potential importance of these structural proteins in vector specificity of TSWV (106, 149). Among animal-infecting bunyaviruses membrane glycoproteins play a major role in receptor-mediated endocytosis and cell fusion critical to infection pathway in vertebrates and arthropod vectors (33).

With Fiji disease *Fijivirus* (FDV, Reoviridae), the first, second, and possibly third nymphal instars, but apparently not adults, of the planthopper vector *Perkinsiella saccharicida* can acquire the virus orally from infected plants (32). A very low proportion (*ca.* 15%) of the insects were infected with FDV as detected by ELISA and other serological methods, even though they were reared on infected sugarcane for at least three generations (4 months). Moreover, fewer than half of the infected planthoppers (6%) transmitted FDV following a 7-day inoculation access period. Francki et al. (40) concluded that *P. saccharicida* is an inefficient FDV vector, and suggested that the feeding behavior of the vector may not be compatible with the distribution of FDV in plant tissue (for acquisition) and/or introduction of virus to susceptible plant tissue (for inoculation). Equally plausible, however, is the occurrence of midgut (infection or escape) or other dissemination barriers to FDV in this vector.

Replication of the rhabdovirus MMV in its planthopper vector *P. maidis* appears to be dose-dependent. When virus concentration of inocula, injected into the hemocoel of the vector, ranged between 0.25 and 25 μg/ml, the differences in percentage of ELISA-positive insects/sample, minimum time between injection and first serological detection of MMV, and the average absorbance value (i.e., antigen concentration) for ELISA-positive individuals over time were found to be dose-dependent (36). Similarly, with the tenuivirus MStV acquired orally by the same vector (*P. maidis*), virus concentration as determined by ELISA was substantially higher following an acquisition access period of 7 days, compared to that of 1 day (E.D. Ammar, R.E. Gingery, and L.V. Madden, unpublished). Interestingly, the vertebrate-infecting vesicular stomatitis *Vesiculovirus* (VSV, Rhabdoviridae), propagatively transmitted by simuliid and probably psychodid flies (30a, 155a), has been successfully propagated in the planthopper *P. maidis* (75). This is the first demonstration of a common host for multiplication of both plant- and vertebrate-infecting viruses. However, the thrips-transmitted TSWV apparently does not multiply in mosquito vectors of animal bunyaviruses, although it was detected by ELISA at the original level at least for 1 week following inoculation of mosquitoes (151).

The ability of the planthopper *Sogatodes oryzicola* to support replication of rice hoja blanca *Tenuivirus* (RHBV) appears to be genetically controlled (156). Progenies of nonvector parents from lineages including at least one vector were allowed to acquire RHBV and then were assayed by ELISA for postacquisition increase in virus titer. The progeny segregated in a manner consistent with a single recessive gene controlling planthopper ability to support virus replication. The viruliferous (ELISA-positive) progeny could transmit RHBV after a normal incubation period. There was no evidence of sex linkage for determination of the ability to support RHBV replication; however, a strong maternal influence on progeny transmission ability was detected (156).

Previous ultrastructural and immunological studies on WTV indicated that virus accumulations were high in various organs of an efficient vector *Agallia constricta*, but low (with no virus detected in the salivary glands) in an inefficient vector *Agaliopsis novella* (50). Although cell lines from these two vector species were equally susceptible to WTV, susceptibility of cells from the nonvector *Aceratogallia sanguinolenta* was about 0.064 times that of vector cells (63). In several studies on plant viruses, the latent period is usually shorter in efficient than in inefficient vectors, which probably indicates faster replication and/or transport of virus in the tissues of efficient vectors. Direct evidence of the correlation between lower virus titer and longer latent period has been obtained with MStV in the planthopper vector *P. maidis* (8). Three isolates of MStV, obtained from Florida (USA), Costa Rica, and Africa were acquired and transmitted orally by this vector with efficiencies of 0–18, 18–71 and 60–93%, respectively. ELISA tests indicated a significantly lower titer of the Florida isolate in infective hoppers, compared to that of the other two isolates. Also, the latent period of the Florida isolate was significantly longer than that in other isolates. The above studies suggest that, with these propagative plant viruses in their vectors, ME, SGI, or other dissemination barriers are involved.

The inefficiency with which the aphid *Macrosiphum euphorbiae* transmits sowthistle yellow vein rhabdovirus (SYVV) apparently is not due to the inability of the virus to invade, survive, or multiply in this species, nor does it appear to be related to the concentration of virus in the hemolymph (138). *M. euphorbiae* was able to acquire SYVV from infected plants as readily as did the highly efficient vector *Hyperomyzus lactucae*, and the amount of virus recovered from the former species increased to a concentration, at least temporarily, comparable to that in

H. lactucae. However, the rare transmission of SYVV by *M. euphorbiae* and the failure to detect virus particles in its salivary glands (17, 18) suggest that an SGI barrier may be responsible for the inefficiency of this species to transmit SYVV.

Studies using insect cell monolayers indicate that the phytoreovirus WTV and potato yellow dwarf rhabdovirus (PYDV) can multiply in cell lines from nonvector leafhopper species (64), which is consistent with the possible involvement of ME, SGI, or other dissemination barriers in nonvector insects. That transmission is also dependent on the virus or virus strains as well as on the vector has been demonstrated with strains of PYDV. Adam and Hsu (2) reported that differences in the G (glycosylated) protein might be related to the selective transmission of PYDV strains by their leafhopper vectors. The G protein, which protrudes from the virion envelope, functions in the attachment of rhabdoviruses to host recognition sites on the plasma membrane during the early stages of infection (68); this was apparently the case also with PYDV in insect cell cultures (43). Furthermore, several studies on the replication of WTV indicate that host–cell recognition sites may be involved in vector specificity of this reovirus. Loss of vector transmissibility of WTV, as well as its replication in vector cell monolayers, was found to be associated with the deletion of segments 2 and 5 of the 12 double-stranded RNA segments present in the wild type, leafhopper transmissible, isolates (99, 113). Nuss (99) suggested that the gene products of segments 2 and 5 are required for multiplication of WTV in the vector, but not in the plants. The products of these segments comprise the outer capsid (shell) of the virus, which indicates that these proteins may be involved in the recognition of vector cells, in virus penetration into them, or both (1). Since removal of the outer protein coat by protease treatment apparently caused no loss of infectivity to vector cell monolayers, it was suggested that the products of segments 2 and 5 might perform multiple functions in the replication cycle of WTV in its vector (99).

Salivary Gland Escape Barrier

The existence of a salivary gland escape (SGE) barrier has been demonstrated for some arboviruses in their mosquito vectors. In *A. hendersoni*, the SGE barrier was a major contributor to vector incompetence of the bunyavirus LACV, since 65% of one infected population had salivary gland infections (by infectivity assay) and only 5% of the infected females transmitted this virus (51). An SGE barrier was also implicated in transmission of the same virus in another mosquito, *Aedes brelandi* (105). Hardy (53) speculated on several possible mechanisms that might explain the SGE barrier to transmission of arboviruses by mosquitoes. One mechanism might be that too little virus is being produced in the salivary glands or secreted during the feeding process to infect vertebrate hosts; another mechanism is that the virus titers are modulated in mosquito salivary gland cells to low or undetectable levels. This may explain why transmission rates of Chikungunya *Alphavirus* by *A. aegypti* decreased from 78% at 21 days to 12% at 49 days postinfection (85). However, a decrease in the ability of mosquitoes with infected salivary glands to transmit some alphaviruses may simply be due to cytopathology as shown for Semliki Forest *Alphavirus* in *A. aegypti* (87).

In the salivary glands of non- or low-transmitting mosquitoes, Japanese encephalitis virus (Flaviviridae) infections were frequently confined to cells in the lateral lobes, whereas in efficient transmitters cells of the median lobe became infected (142). Similarly, with the bunyavirus LACV, an SGE barrier in *A. hendersoni* apparently does not involve lack of virus replication in the salivary glands. By 21 days postinfection with this virus, 100% of *A. triseriatus* and 70% of *A. hendersoni* had infected salivary glands, and the geometric mean titer of these in *A. hendersoni* was 10 times higher than in those of *A. triseriatus*. However, when tested for transmission 22 days postinfection, only 9% of *A. hendersoni* with disseminated infections transmitted virus, compared to 71% of *A. triseriatus*. Furthermore, females of *A. hendersoni* that had been infected by injection, transovarially transmitted virus to 25% of their progeny (104). It is possible that compatibility of virus with the salivary secretions, which contain several enzymes (114, 131), may be responsible for inefficiency of oral transmission of some viruses by their vectors. Inhibitory effects of the salivary secretions of aphids on some (plant) aphid-nontransmissible viruses, including tobacco mosaic, potato X, and turnip mosaic viruses, have been reviewed by Nishi (98). On the other hand, it has been reported that tick saliva (or salivary gland extract) potentiates the transmission of Thogoto virus to uninfected ticks (68a); enhancement of virus transmission was observed only when salivary gland extracts were derived from ticks that were competent vectors of this virus. However, there seems to be no reports on the effects of salivary secretions of vectors and nonvectors on plant and animal viruses propagatively transmitted by insects.

With several propagative plant viruses, it has been demonstrated that transmission by infected insects is usually intermittent (95). With the rhabdovirus RTYV in *Nephotettix* leafhoppers (26), more escapes in daily transmission occurred at lower temperature (15–20°C) and also later in the retention period. Similarly, the pattern of transmission of the tenuivirus MStV by the planthopper *P. maidis* was intermittent and infrequent except for one insect that transmitted virus to 18 of 22 test plants over a 7-week test period (97). Using infectivity assays with rice gall dwarf *Phytoreovirus* (RGDV), Omura et al. (102) showed that virus titer in whole insect vectors (*Nephotettix*) remained high up to 40 days postacquisition, although transmission efficiency decreased with advancing age of infected leafhoppers. Additionally, several studies with other plant viruses have shown that virus can be detected serologically in some vector individuals that were unable to transmit the virus. For example, only 10–24% of *Dalbulus* leafhoppers exposed to MRFV-diseased plants transmitted virus although about 80% were positive for this virus in ELISA tests (44). Similar results were obtained with MStV and maize rough dwarf *Fijivirus* in their planthopper vectors (30, 97). The above results suggest the occurrence of SGI and/or SGE barriers for these viruses in their vectors, probably correlated with age of the vector in some cases. More direct evidence of an SGE barrier for a plant virus comes from a study by Nault and Gordon (97) on MStV in its planthopper vector *P. maidis*. Of 31 planthoppers with salivary glands that were positive for MStV by ELISA, 24 insects failed to transmit virus to plants. Similarly, with FDV, only 6% of the planthopper vector *P. saccharicida* that had been reared on infected plants transmitted the virus, although 15% were positive for FDV by ELISA or immunoelectron microscopy (40). In addition to suggesting SGE or SGI barriers in these vectors, these results indicate that such serological techniques may be convenient and useful for monitoring vector populations and forecasting virus epidemics (10), but it must not be assumed that all viruliferous (infected) insects can transmit the virus to susceptible plants.

Sinha (127) reported that WTV probably does not multiply equally well in all lobes of the salivary glands in the leafhopper vector since virus antigens were found mostly restricted to the anterior lobes. Ultrastructural studies on MMV in various organs of the vector *P. maidis* (10) (Figs. 11.1–11.6) revealed differences in budding sites that might explain how this rhabdovirus is produced and transported within the salivary glands. In plant cells and in most vector tissues examined, including nerve cells, midgut epithelium, epidermis, fat tissue, and acinar cells of the "accessory" salivary gland, MMV particles were found to bud mainly through nuclear membranes, and to accumulate in perinuclear spaces (Figs. 11.3–11.8). However in secretory cells of the "principal" salivary glands, MMV particles bud mainly through the plasma membrane and accumulate in intercellular and extracellular spaces (Fig. 11.1). These spaces are apparently connected with the extracellular vacuoles and canaliculi that lead to the salivary ductules and ducts (6). Thus, Ammar and Nault (10) suggested that this intercellular route allows efficient discharge of virus into the saliva. An intercellular route was also suggested for the corn stunt spiroplasma (Mollicute) within the salivary gland of its vector leafhopper, *Dalbulus maidis* (81). This could be one way by which some viruses and mollicutes overcome an SGE barrier in their vectors.

Membrane system preference between host organs and tissues, sometimes dependent on the stage of infection, is also known in vertebrate-infecting rhabdoviruses, e.g., rabies virus (RV). In the fox brain, RV particles bud mainly through endoplasmic reticulum, whereas in the salivary glands they bud primarily through plasma membranes of mucous cells facing the salivary secretion space; this delivers much infectious virus into the saliva for bite transmission and is considered essential for survival of virus in nature (89). With another rhabdovirus, VSV, in all tissues examined from a suspected sandfly vector (*Lutzomyia shannoni*), virus matured primarily by budding from plasma membranes. However, in the midgut, budding occurred exclusively from the basolateral plasma membrane, while maturation in salivary gland cells involved apical budding (155a). Ultrastructural studies on the alphavirus EEEV in vector mosquitoes indicated that this virus infected and replicated in the salivary glands and was released into the salivary matrix within 55–69 hr postinfection. Naked nucleocapsids were abundant below the plasma membrane of the apical cavities within salivary gland acinar cells. Some nucleocapsids appeared to be budding through the plasma membrane surrounding apical cavities and in the process maturing into infective virions (120).

The inability of some circulative-nonpropagative plant viruses to pass through the "accessory" salivary gland of their vectors has been implicated in vector specificity. This has been reported with pea enation mosaic virus (PEMV) and barley yellow dwarf *Luteovirus* (BYDV) in aphids (46, 57), and suggested for squash leaf curl *Geminivirus* in the nonvector whitefly *Trialeurodes abutiloneus* (29). Virions of an aphid-transmissible strain of PEMV, the first isometric, aphid-borne virus to be localized in the salivary glands of its vector, were able to penetrate the basal lamina of the accessory glands into plasma-membrane cisternae of gland cells, whereas virions of a nonaphid-transmissible variant could not (57). These findings and other related reports prompted Harris (57a, 57b) to postulate that the virus-vector specificity phenomena associated with aphid transmission of small isometric viruses can be mediated by the accessory glands and, furthermore, that reciprocity between recognition sites on virus coat protein and accessory gland membranes is required for passage of virions from the hemocoel through the salivary system to plants.

The MAV isolate of BYDV, which is specifically transmitted by *Sitobion avenae*, was found embedded in the accessory salivary gland basal lamina, in cytoplasmic coated vesicles, and in the lumen of secretory canals. Virions of the RPV isolate, transmitted specifically by *Rhopalosiphum padi* but not by *S. avenae*, were also observed in *S. avenae* in the dasal lamina, but not in sites beyond the basal plasmalemma (46). Furthermore, *R. padi*, which acquires and transmits RPV but not MAV, occasionally transmitted both isolates from doubly infected plants. Serological tests indicated that, in these plants, the MAV

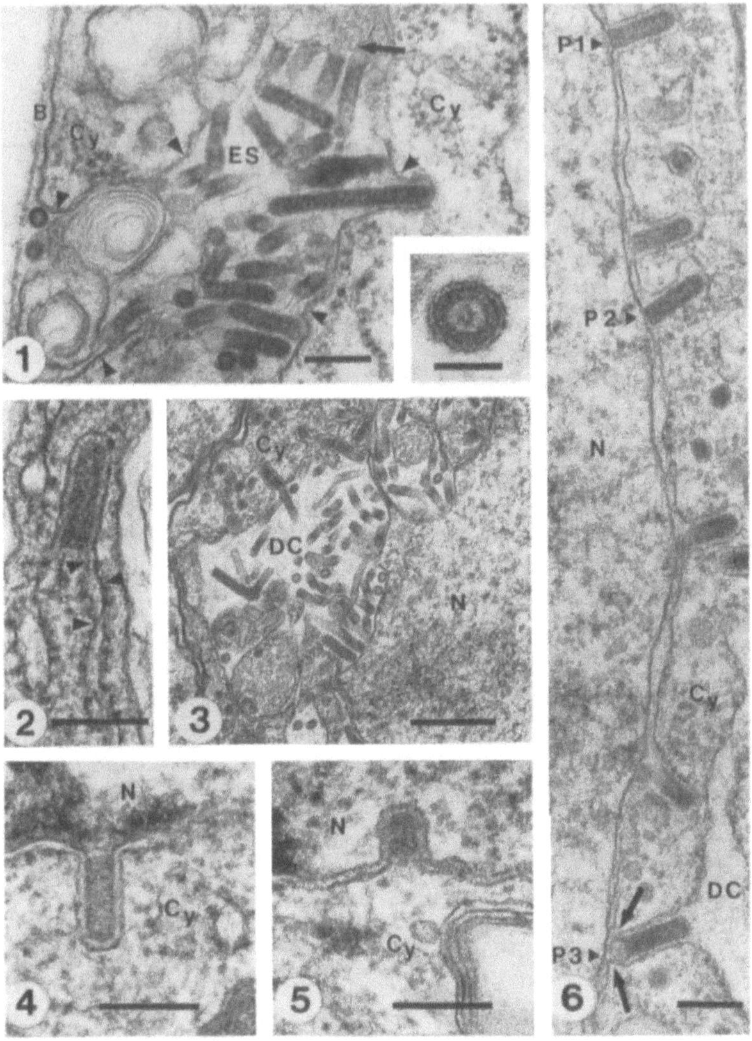

FIGURES 11.1–11.6. Assembly and accumulation sites of rhabdovirus particles of maize mosaic virus (MMV) in the planthopper vector *Peregrinus maidis*. Figure 11.1. Particles accumulated in an extracellular space (ES) between basal lamina (B) and plasma membrane (arrowheads) in a salivary gland secretory cell; some particles appear to bud from obliquely sectioned plasma membrane (arrow); inset shows cross section of one particle; Cy, cytoplasm. Figure 11.2. An MMV particle attached to membranes of the endoplasmic reticulum (arrowheads). Figure 11.3. Particles accumulated within perinuclear dilated cisternae (DC) in the cytoplasm (Cy) of a brain cell; N, nucleus. Figures 11.4 and 11.5. Particles of intermediate length apparently budding through nuclear membranes. Figure 11.6. Particles (P1–P3) apparently budding through the inner nuclear membrane of a brain cell; DC, dilated cisternae connected with outer nuclear membrane at arrows. Bars: 200 nm, except for inset in Figure 11.1 (50 nm) and for Figure 11.3 (500 nm). (From Ammar and Nault (10).)

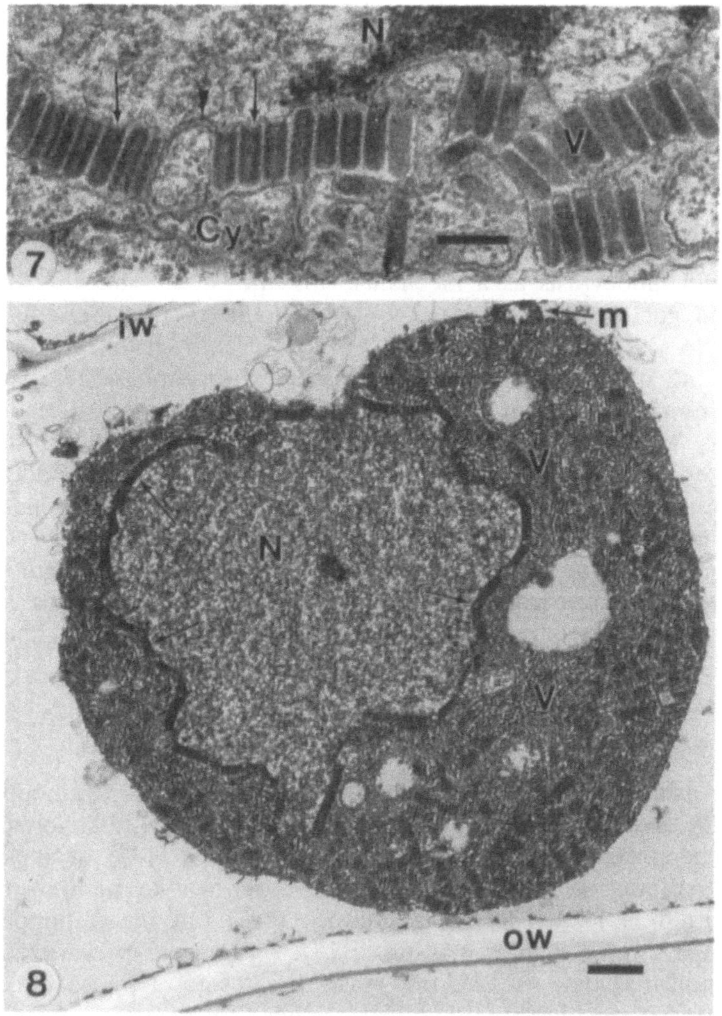

FIGURES 11.7 and 11.8. Particles of Iranian maize mosaic rhabdovirus in mesophyll (Fig. 11.7) and epidermal (Fig. 11.8) maize leaf cells; particles (V and arrows) apparently are budding through nuclear membranes (arrowhead in Fig. 11.7), and accumulated in dilated perinuclear cisternae, eventually occupying or replacing most of the cell cytoplasm (Fig. 11.8, compare with the much smaller aggregates of particles of a related rhabdovirus in cells of the insect vector in Figs. 11.1–11.6); iw and ow, inner and outer walls of the epidermal cell, respectively; m, mitochondrion; N, nucleus. Bars: (11.7) 200 nm; (11.8) 1 μm.

genome was encapsidated in the RPV coat protein during virion assembly (115). Based on these studies, Gildow (46) suggested that vector-specific transmission of luteoviruses is based on an interaction between the virus capsid protein and receptor sites in the accessory salivary gland cell membrane. Gildow et al. (48) later reported some results suggesting that the basal lamina surrounding the accessory salivary gland may act as a selective barrier determining access of some luteoviruses to the cell membrane. Additionally, recent ultrastructural studies implicated receptor-mediated endocytosis as a mechanism of luteovirus acquisition through the hind gut of aphid vectors (46a). With the circulatively transmitted geminiviruses, vector specificity, even at the family level, appears to be dependent on the virus coat protein. For example, beet curly top *Geminivirus* (BCTV) is transmitted only by leafhoppers (Cicadellidae) whereas African cassava mosaic *Geminivirus* (ACMV) is transmitted only by whiteflies (Aleyrodidae) (83). Yet, in a recent study by Briddon et al. (22), a chimeric virus, composed of ACMV genomic components and BCTV coat protein, was transmitted by the leafhopper *Circulifer tenellus*, the regular vector of BCTV. However, apart from the gut barrier suggested for maize streak virus in its leafhopper vectors 60 years ago by Storey (134), the barriers responsible for vector specificity in geminiviruses have not been elucidated.

Transovarial and Venereal Transmission Barriers

Transovarial transmission of viruses in leafhoppers and planthoppers has been documented for many years (122, 128), but only relatively recently has it been convincingly demonstrated in aphids (137) or mosquitoes (153). At present, the number of viruses known to be transovarially transmitted is about 30 in mosquitoes (147), 14 in leafhoppers and planthoppers (95), and 3 in aphids (128). In mosquitoes, various studies have indicated that vertical transmission rates for females infected by intrathoracic inoculation are similar to those for females infected orally (147). Several studies indicated that the efficiency of transovarial transmission appears to be related to the virus taxonomic classification. Thus, most of the bunyaviruses tested are efficiently transmitted transovarially (to more than 20% of the progeny). Most flaviviruses, however, tend to have low rates of transovarial transmission (less than 1%) (147).

Similarly, with propagative plant viruses, the occurrence and efficiency of transovarial transmission vary greatly between virus groups (95, 128, 138). Most tenuiviruses (4/5) are transovarially transmitted at a high rate (50–90%) in their planthopper vectors. All three known phytoreoviruses are transovarially transmitted with efficiencies ranging between 1.8 and 100% in their leafhopper vectors, whereas only a few of the studied Fijiviruses (3/10) are transmitted with lower efficiency (0.2–17%) in their planthopper vectors. Some plant rhabdoviruses are transovarially transmitted in their vector leafhoppers (2/7), planthoppers (2/11), or aphids (2/3), usually with very low efficiency (<1–4%). None of the three studied marafiviruses are transovarially transmitted in their leafhopper vectors. However, within each virus group different rates of transovarial transmission have been reported for different viruses, and even for the same virus in different vectors, e.g., 85% for rice dwarf *Phytoreovirus* (RDV) in *Nephotettix cincticeps*, only 2–10% for WTV (another phytoreovirus) in *A. novella* and up to 80% for WTV in *A. constricta* (128). Rice stripe *Tenuivirus* (RStV) was reported to pass through the eggs of a single infective female of *Leodelphax striatellus* for 40 generations in a classic experiment that lasted 6 years; 95% of the progeny insects of the fortieth generation were inoculative with RStV. Insects that received virus transovarially remained inoculative throughout their nymphal stage and for 2–3 weeks of adult life, but some of them lost their inoculativity later (122). The efficiency of transovarial transmission of WTV in *A. constricta* appears to be under genetic control, and it was reported that those races of leafhoppers with high rate of transovarial transmission also were efficient in oral transmission of WTV to plants (128). Similarly, with MStV, efficiency of transovarial transmission in three isolates obtained from Florida (USA), Costa Rica, and Africa was apparently correlated with that of oral transmission of these isolates in the planthopper vector *P. maidis* (8). With another tenuivirus, (RHBV), active female planthopper vectors transmitted virus transovarially to their progenies, regardless of the male parent and progeny genotype. In progeny receiving RHBV maternally, however, virus titers as determined by ELISA were lower and more variable in insects with an ELISA-negative male parent than in insects with two ELISA-positive parents (156).

Venereal transmission has been reported for some arboviruses in mosquitoes (147). Transovarially infected *A. triseriatus* males transmitted LACV venereally to females, which in turn were able to transmit the virus both horizontally to vertebrates and vertically to their progeny (144). While venereal transmission rates were low if mating took place prior to the female's first blood meal, 49% of the fed females became infected venereally if they had obtained a blood meal prior to copulation (143). Males of *A. albopictus* experimentally infected with Dengue virus (DeV, Flaviviridae) transmitted virus sexually to females of the same species. Such transmission was enhanced if the females had taken a blood meal 2–7 days prior to mating. Male *A. albopectus* also transmitted DeV vertically to the F1 progeny via inseminated females. Infected progeny were found among those derived from eggs laid 73 hr after mating but not among those derived from eggs laid prior to that time. This suggests that virus probably was not transmitted directly to ova but, rather, underwent prior replication in the female genital tract. In the same study, female *A.*

albopectus experimentally infected with DeV did not transmit their infection sexually to males (117). No venereal transmission has been reported for propagative plant viruses in their homopteran vectors (95, 128).

For a virus to be transmitted transovarially, a virus must cross certain barriers in the vector in addition to those encountered in oral transmission. Following successful infection of the hemolymph, the virus must cross the ovarian and the ovariole sheaths and the follicular epithelium, before it can have access to the developing oocytes. If infection of the oocyte does not occur early in oogenesis, presumably the developing chorion would provide another substantial barrier (147). For venereal transmission, comparable barriers, in addition to the ability of the virus to survive in the seminal fluid before, during, and after mating, are probably encountered.

Ultrastructural and fluorescent antibody studies on the alphavirus EEEV in various tissues of its vector mosquito indicated the multiplication and accumulation of this virus in several tissues that did not include the ovarioles; virus isolated from mosquito eggs was considered transovum (outside the egg) rather than transovarial (inside the eggs) (120). With Ross River *Alphavirus*, viral antigens detected in sections of eggs from the ovary of one vector mosquito (*Aedes vigilax*) were reported to provide further evidence of transovarial transmission of this virus in its vector (150b). Some studies in mosquito and homopteran vectors, however, indicated that the detection of virus in the ovaries of a vector is not sufficient evidence that it can transmit that virus transovarially. In the planthopper *P. maidis*, rhabdovirus particles of MMV, which is not known to be transmitted transovarially or venereally, were found by electron microscopy in the follicular cells of female ovaries, and in the epithelium of the ejaculatory ducts in males, as well as in most other organs examined (10). In this study, MMV was detected only in one out of five ovaries examined. Using ELISA with MStV (97), which is transmitted transovarially by the same vector (*P. maidis*) to a large proportion of the progeny, all the tested ovaries (8/8), oviducts (6/6), and bursae copulatrix (6/6) contained MStV antigens. ELISA detected MStV also in single eggs (3/9).

With European wheat striate mosaic *Tenuivirus* (EWSMV), if females of the planthopper vector *Javesella pellucida* acquired the virus orally as adults, little or no transovarial transmission occurred, but if the mother acquired virus during the nymphal stage, most of the progeny inherited the virus and were able to transmit immediately after hatching (127). In a study on daily transovarial transmission rates of the phytoreovirus RGDV in *Nephotettix* leafhoppers, fewer transmissions occurred near the end of the oviposition period (26). Also, with RStV, the rates of transovarial passage decreased with the age of females especially at high temperatures (111).

The above studies suggest the presence of some transovarial transmission barriers, which have not been studied in detail, for several plant and animal viruses. In fact, the mechanism of transovarial transmission remains unstudied in most cases. A notable exception is an ultratructural study by Nasu (93) on RDV in its leafhopper vector *Nephotettix cincticeps*. Apparently, RDV particles were attached selectively to the surface of L-symbiotes, present in the cytoplasm of the mycetocytes, which migrate to the adjacent oocytes. However, other mechanisms for infection of the oocytes in this and other virus–vector systems, e.g., that the virus may enter the oocytes directly through cytoplasmic membranes, are feasible but so far uninvestigated.

Extrinsic Factors Affecting Vector Competence

Environmental Factors

Temperature seems to be the main environmental factor that modifies the rate of virus acquisition and inoculation, in addition to the length of the latent period and the retention time (of inoculativity) in vectors of plant or animal viruses (53, 83, 138). The effect of elevated temperature on the spread of WTV in the leafhopper vector *A. constricta* was studied by Sinha (126). Groups of nymphs were given a 1-day acquisition access on a virus source, then incubated at 27°C for 3 days before being divided into two groups: one group was held at this temperature and another held at 36°C. WTV antigen was detected in the filter chamber of these leafhoppers 3 days postacquisition, but later the higher temperature apparently prevented the spread of virus from the midgut to the hemolymph and salivary glands. Also, temperature had a marked effect on the rate of transovarial transmission of RStV in its planthopper vector *L. striatellus*. Percentage of viruliferous females that passed virus to a high proportion (more than 90%) of their progeny was 82.6% at 17.5°C, compared to 12.5% at 32.5°C (111). In this study, higher temperature also reduced the longevity and fertility of vector planthoppers. Under field conditions, temperature, humidity, rain, wind, and other climatic conditions are known to affect, or be correlated with, outbreaks of arthropod-borne viruses and other pathogens (28, 53, 83).

Several studies have consistently shown that the latent (or extrinsic incubation) period of plant or animal viruses in their vectors is inversely correlated with the temperature at which the vectors are held (95, 147). With arboviruses, environmental temperature appears to affect both infection and transmission rates differently, depending on the virus and vector combinations. For example, reduced infection rates with yellow fever virus (YFV, Flaviviridae) in *Haemagogus* mosquitoes were associated with low temperature (20°C); in *C. tarsalis* reduced infection rates with WEEV were associated with high temperature (32°C), whereas no effect of temperature was found on infection rates of another alphavirus, EEEV, in *A. triseriatus* (147). Additionally, with WEEV in *C. tarsalis* although more rapid transmission occurred at higher temperature, mosquitoes held at 25°C were less efficient vectors than those held at 18°C (74). Thus, environmental temperature may play an important role in vector competence, but few generalizations can be made other than that virus transmission normally occurs faster following acquisition at higher temperatures (147).

Nutritional deprivation of mosquito larvae often produces adult females that are smaller but more competent vectors of an arbovirus (53). In a recent study, Grimstad and Walker (52) reared *A. triseriatus* either as nutritionally deprived or well-fed larvae, and then assessed the vector competence of resulting females for LACV. Small (deprived) females transmitted virus at the rates of 15–90% compared to 0–42% for large (well-fed) females. All females had infected midguts, particularly when mosquitoes fed on a high dose of LACV, but disseminated infection of small females was 100% compared to 69% in large females. Ultrastructural studies revealed that the basal lamina of the midgut in small females had 3–6 laminae (0.14 μm thick), whereas that in large females had 9–16 laminae (0.24 μm thick). These morphological differences were interpreted to indicate that the midgut escape barrier, which accounted for the differences in vector competence between small and large females, may be, in part, a physical barrier modified by nutritional deprivation in larval instars.

The progressive reduction in transmission efficiency of rice white leaf virus (= RHBV) by its planthopper vector *Sogatodes oriziola* was studied under greenhouse conditions in Columbia (44a). An initial mean transmission capacity of 64.5% was linearly reduced to 8.5% by the 8th generation. A similar reduction in transmission efficiency of another tenuivirus, MStV, by the planthopper vector *P. maidis* over several generations of rearing under greenhouse conditions in the USA was observed (E.D. Ammar and L.R. Nault, unpublished). The effects of carbon dioxide anesthesia on the ability of the planthopper *L. striatellus* to acquire and transmit maize rough dwarf *Fijivirus* (MRDV) and barley yellow striate mosaic rhabdovirus (BYSMV) was studied by Caciagli (24). Anesthesia reduced the acquisition of MRDV by about 45% but did not affect acquisition of BYSMV. Anesthesia also caused interruption of transmission for both MRDV and BYSMV in some insects, possibly due to impaired ability to find the phloem. Transmission of MRDV was resumed 3 days after anesthesia, but none of the insects that stopped transmitting BYSMV resumed transmission. The survival of insects that ceased to transmit BYSMV was significantly lower than that of hoppers that continued to transmit. These results suggested possible interaction of CO_2 with BYSMV in vector planthoppers: CO_2 is known to cause paralysis in *Drosophila melanogaster* infected with some rhabdoviruses (23a), and in four species of *Aedes* infected with California encephalitis virus (147a).

Coinfection with Other Pathogens

Reassortment of LACV and Tahyna bunyaviruses in the mosquito vector, *A. triseriatus*, has been studied by Chandler et al. (25a) using a molecular hybridization technique to analyse progeny viruses. High frequency reassortment occurred in this mosquito; all of the expected genotypes resulting from a cross of the two viruses were obtained, but possible effects of this reassortment on transmission efficiency of either virus were not reported. In several other studies, interference with arbovirus infection, multiplication, or transmission has been observed when mosquitoes are coinfected with some, but not other, combinations of arboviruses (53, 147). Generally, homologous interference seems to occur more frequently than heterologous interference, and infection with one virus has to be established in the vector before it can interfere with the second virus (53). However, Hardy (53) questioned the epidemiological significance of these laboratory findings, indicating that dual infection of mosquitoes with two arboviruses in nature may be a rare event, given the usually low arbovirus infection rates in field mosquito populations. He suggested that interference of arbovirus infections in mosquitoes infected with "endogenous" viruses may be more important epidemiologically if a significant proportion of a vector population is infected with such viruses, although apparently no data on such interactions are available. As to dual infection by viruses and other exogenous nonviral pathogens in mosquitoes, a recent study by Paulson et al. (105a) indicated that coinfection of *A. hendersoni* with *Plasmodium gallinaceum* and LACV dramatically increased virus transmission (72% versus 8%), whereas LACV transmission by coinfected *A. triseriatus* was not significantly affected. Additionally, Turell et al. (148) showed that *Aedes taeniorhynchus*, which fed on a gerbil with both an RVF viremia and a *Brugia malayi* microfilaremia, had a 4-fold higher viral dissemination rate, and more than 6-fold higher viral transmission rate, than did mosquitoes that fed on a gerbil infected with RVF virus alone. Moreover, the microfilarie enabled the virus to disseminate into the hemocoel more rapidly than would have occurred normally. Because some microfilariae puncture the mosquito midgut shortly after ingestion, it has been hypothesized that concurrent ingestion of an arbovirus and microfilariae may enhance arbovirus dissemination and transmission in this way (147).

Since a single plant species may be host to several related or unrelated arthropod-borne pathogens, it is expected that some plants in the field might be infected with more than one virus and/or mollicute that are transmissible by the same vector (15, 139). Strains of the circulative-nonpropagative beet curly top *Geminivirus* can be poly-acquired and simultaneously transmitted by the leafhopper *Circulifer tenellus* (19). Apparently, these strains remain independent of each other in the vector, while in the plant host there is a low level of interaction, but no cross-protection. On the other hand, a study by Cohen and Duffus (29) on two other geminiviruses, circulatively transmitted by the whitefly *Bemisia tabaci*, revealed that prior acquisition of one virus (squash leaf curl) by the vector reduced transmission efficiency of the second virus (melon leaf curl). For propagatively transmitted plant viruses, very few combinations have been studied for possible interaction in their common vectors. Simultaneous transmission by the leafhopper *A. constricta* of the phytoreovirus WTV and the rhabdovirus PYDV apparently can occur without any evidence of interaction (90). When two unrelated aphid-borne rhabdoviruses, viz. sowthistle yellow vein virus (SYVV) and strawberry crinkle virus (SCV), were simultaneously or sequentially inoculated into *Chaetosiphon jacobi*, a vector of SCV but not SYVV, both viruses apparently multiplied, but vector host–plant specificity precluded adequate testing for transmission of SYVV by this aphid species (140). Interactions of the rhabdovirus MMV and the tenuivirus MStV in their common planthopper vector *P. maidis* and in maize plants were studied using bioassay and ELISA tests (9). Access of planthoppers to MMV-infected plants within 0–14 days before or after access to MStV-infected plants significantly reduced and delayed transmission of MStV by *P. maidis*. In contrast, access to MStV-infected plants usually had no effect on the acquisition and transmission of MMV. Plants previously infected with either virus were partially protected from infection by the other. These results suggest that the mechanisms of interference between these two viruses may be different in plants and in the insect vector. In both plants and vectors, however, MMV apparently interfered with multiplication of MStV as demonstrated by ELISA tests (9).

Studies on the effects of dual acquisition of aster yellows mycoplasma-like organism (AY-MLO) and oat blue dwarf *Marafivirus* (OBDV) suggested mutual interference between these two pathogens in their leafhopper vector *Macrosteles quadralineatus* (= *fascifrons*) (15). Interference occurred when leafhoppers sequentially acquired both pathogens from plants, or when AY-MLO- or OBDV-inoculative leafhoppers were injected with extracts of the other pathogen. Reduced OBDV titers were detected by ELISA in salivary glands and bodies of insects that acquired OBDV as the challenging agent (13). Interference between certain viruses and MLO in vertebrate tissue cultures has been explained as a result of competition for replicative sites or substrates (123). It is possible that a similar explanation may be true for interference in the vector, between certain propagative viruses and/or mollicutes, since both apparently have a similar route in their insect vectors (81, 95, 110).

Insect vectors of propagative plant viruses may be infected also by their own "endogenous" viruses that include the following examples: two leafhopper reoviruses, leafhopper-A virus (LAV), found in *Cicadulina bimaculata* (101), and *Graminella nigrifrons* virus (GNV) (9a); two planthopper reoviruses, *Peregrinus maidis* virus (PMV) (35) and *Nilaparvata lugens* virus (NLV) (98a); two aphid picorna-like viruses, aphid lethal paralysis virus (ALPV) (59) and *Ropalosiphum padi* virus (RhPV) (47); and one planthopper picorna-like virus found in *L. striatellus* and two other species (52a). At least LAV, NLV, and ALPV are transmitted transovarially, while LAV, RhPV, and probably GNV are transmitted horizontally through the insect's host plants in which, apparently, these viruses do not multiply. LAV and PMV do not seem to have significant deleterious effects on their insect hosts (35, 101), whereas ALPV significantly reduced the longevity and fecundity of two aphid species (75a). RhPV-infection of *R. padi* or *Schizaphis graminum* had no effect on transmission efficiency or vector specificity of the circulatively transmitted luteovirus BYDV in these aphids (47). However, particularly with the transovarially transmitted endogenous viruses, it may be difficult to have good controls for testing the effects of dual infection with other propagative (vectored) viruses, since a large proportion of the insect population might be naturally infected with the endogenous virus (9a, 52a). Furthermore, when investigating virus–vector relationships, one must be aware that insect vectors may naturally harbor different pathogenic, latent, or symbiotic agents (Figs. 11.9–11.12) (6, 11). Thus, immunocytochemical or other serological methods of demonstrable specificity may be necessary to differentiate between these endogenous agents and the vectored pathogens investigated, particularly if they were morphologically similar.

Pathogenicity and "Modulation" of Viruses in Their Insect Vectors

Most of the earlier reports on pathogenic effects of some plant "viruses" on their leafhopper or planthopper vectors now either have been contradicted or turned out to involve "mollicutes" previously assumed to be viruses, e.g., Western X-disease mycoplasma-like agent and corn stunt spiroplasma (110). With the latter pathogen, recent work by Nault and associates on the leafhopper genus *Dalbulus* indicated that in well-adapted, coevolved vector species, mollicutes not only are nonpathogenic, but they may benefit their vectors. In contrast, poorly adapted *Dalbulus* species are frequently inefficient vectors and their life span and fecundity are greatly reduced compared to noninfected leafhoppers (110). With regard to "true" plant viruses, earlier reports that the tenuivirus RStV reduces the longevity and fecundity of its planthopper vector *L. striatellus*

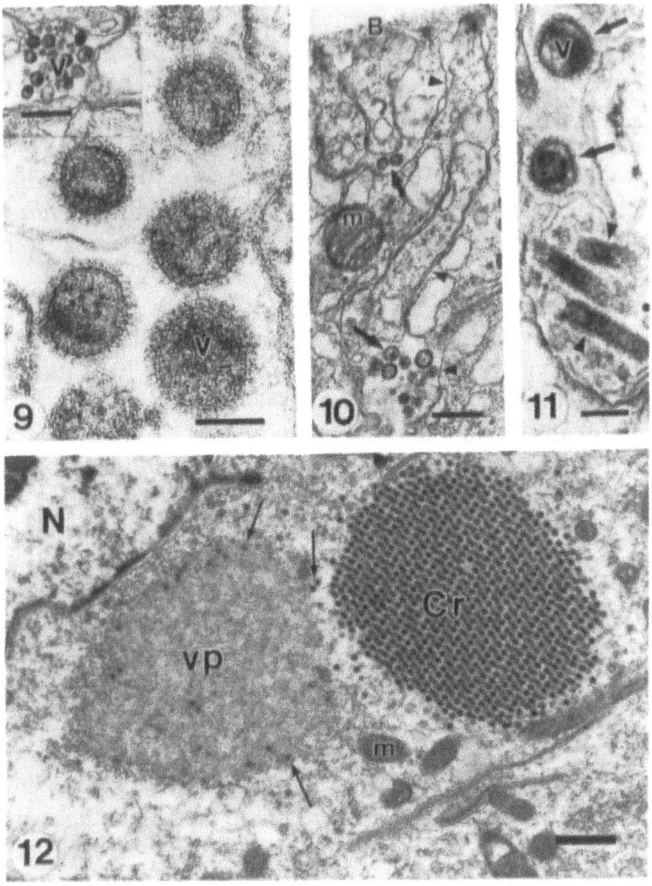

FIGURES 11.9–11.11. Paramyxovirus-like particles (v and arrows) and rhabdovirus particles of maize mosaic virus (MMV) (arrowheads in Fig. 11.11) in salivary gland cells of the planthopper *P. maidis*, vector of MMV; B, basal lamina; m, mitochondrion. Bars: (11.9 and 11.11) 100 nm, (11.9, inset, and 11.10) 500 nm. (From Ammar et al. (11).) FIGURE 11.12. Reovirus-like particles in a muscle cell of an apparently healthy leafhopper, *Graminella nigrifrons*, vector of maize rayado fino *Marafivirus* and maize chlorotic dwarf virus (9a); reovirus-like particles are found in a crystalline array (Cr) or scattered singly (arrows) in the periphery of a viroplasmic region (vp) in the cytoplasm; m, mitochondrion; N, nucleus. Bar: 500 nm.

(92) were refuted later by Kisimoto (71) and Raga et al. (111) who concluded that RStV had no deleterious effects on this efficient vector. Another tenuivirus, EWSMV, was previously reported to reduce the fecundity and increase embryonic abnormalities in its vector *J. pellucida* (152). However, Kisimoto and Watson (72) suggested that these embryonic abnormalities may have been due to inbreeding of the vector in the

laboratory. Later, the effects of EWSMV on this efficient vector were studied by Ammar (4, 5), taking into consideration the effects of inbreeding, sex, wing-form, and feeding on diseased plants by the vector. When EWSMV was acquired orally from diseased plants by *J. pellucida*, no deleterious effects were observed on adult longevity, fecundity, or embryonic or nymphal development; but when EWSMV was acquired transovarially for one or two generations, nymphal mortality was increased by 13–17% and adult longevity was reduced by 14%. However, deleterious effects on *J. pellucida* of inbreeding for two generations were much greater than those of EWSMV (5). With a third tenuivirus, MStV, no pathogenic effects on longevity or fecundity of the vector *P. maidis* have been found (146, E.D. Ammar and L.R. Nault, unpublished). The capsid protein of MStV was detected at a much lower level in infected, inoculative planthoppers compared to that in infected maize, whereas the major noncapsid protein, also encoded by MStV, was detected only in infected plants but not in vectors (37). If MStV replicates differently in plant and vector tissues, this might explain its lack of adverse effects on the vector compared to its effects on maize.

From the above mentioned studies, it seems that at least with tenuiviruses and their planthopper vectors, as in the case with mollicutes and leafhoppers, in well-adapted vectors, genotypes naturally have been selected that minimize or eliminate pathogenicity of plant pathogens to their vectors (95). On the other hand, with the phytoreovirus RDV, adverse effects on the longevity and fecundity of two of its leafhopper vectors, *Inazuma dorsalis* and *Nephotettix cincticeps*, have been reported (91, 122). Additionally, when the rhabdovirus SCV was injected into its aphid vectors, *C. jacobi* and *C. fragaefolii*, longevity and fecundity of both vectors were reduced (141). These two species, however, are inefficient vectors of SCV when this virus is acquired orally.

Generally, it is also believed that animal arboviruses and their vectors have evolved to a degree where most of these viruses no longer cause mortality in their natural vectors (53). Nevertheless, some arboviruses are reported to induce adverse effects on their vectors, including reduced fecundity, increased developmental time, and reduced efficiency in obtaining a blood meal (147). Turell (147), however, argued that the latter effect can be, in fact, to the virus' advantage, since it may enhance the transmission capacity of the infected vector by inducing it to have repeated feedings on potential virus hosts. Cytopathologic lesions in the midgut of *Culiseta melanura*, vector of EEEV, 2–5 days after oral infection with this virus have been reported by Weaver et al. (155). The authors suggested that disruption of the midgut basal lamina could result in bypassing of dissemination barriers allowing rapid transmission to occur. They also suggested that luminal sloughing of heavily infected midgut epithelial cells may serve to modulate virus infection, thus reducing its pathological effect on mosquito vectors.

The question of why or how propagative insect-borne viruses usually have deleterious effects on their plant or vertebrate hosts while having little or no adverse effects on their vectors has been investigated at the cellular level in a few cases. Ultrastructural studies on the rhabdovirus MMV revealed much smaller aggregates of virus particles in any of the tissues examined from the planthopper vector *P. maidis* (Figs. 11.1–11.6) (10), whereas massive and often crystalline aggregates of particles of MMV or its rhabdovirus relative IMMV occurred in most cell types of infected maize leaf tissues (Figs. 11.7–11.8) (84). Additionally, a higher proportion of budding MMV particles with intermediate lengths was observed in insect compared to plant tissues. Thus, Ammar and Nault (10) suggested that these features may indicate a slower rate of MMV multiplication or assembly in insect compared to plant cells, which can explain the lack of effects of MMV on the vector's longevity (E.D. Ammar and L.R. Nault, unpublished). Similarly, with the phytoreovirus RGDV in diseased rice plants, phloem parenchyma cells were severely affected and the cells appeared fully occupied by virus particles and viroplasm, whereas in viruliferous leafhoppers (*N. nigropectus*) virus infection never occupied large areas of infected cells (70). Thus, Kimura and Omura (70) suggested that insect vectors might have some kind of regulatory mechanism that reduces or slows down virus multiplication in the vector compared to its multiplication in plant cells. Furthermore, these authors tentatively suggested that phagocytic-like vesicles, which they found only in RGDV-infected leafhoppers, may provide such a mechanism.

With another phytoreovirus, WTV, despite the cytological changes that occur in infected leafhopper vectors, these vectors apparently show no "behavioral disorders" or decreased fecundity or longevity (80). Nuss and Dall (100) reported that persistent productive WTV infections of cell cultures from the vector cause no apparent changes in cell growth rate or protein synthesis capabilities. The mechanisms responsible for non-cytopathic persistent WTV infections of vector insects are unresolved, but in cultured leafhopper cells, at least, they might be associated with alterations in translational capabilities of viral mRNA. Peterson and Nuss (107) reported that WTV-specific polypeptide synthesis and viral genome RNA accumulation increased to a maximum level during the first 5 days following inoculation and then decreased as infected cells were passaged. In contrast, viral-specific mRNAs were present at approximately the same level in the acute phase and in the early stage of the persistent phase of infection. Transcripts isolated from persistently infected cells were inefficiently translated *in vitro*. These results indicated that the level of viral polypeptide synthesis associated with the persistent phase of WTV infection in vector cells is related to a change in the translational activity of viral transcripts. Similarly, with the rhabdovirus PYDV, which induces marked symptoms in infected plants, no cytopathic effects were observed in an infected vector cell line. However, when the inoculum contained high concentration of PYDV, some cytopathic effects leading to cell death were observed (63).

In susceptible vertebrate cells arbovirus infection is short term and cytocidal, whereas in cultured mosquito vector cells the infection usually is long term and noncytopathic. However, there is no evidence that the replication strategy of the same virus differs in major respects between the two host systems (73). When mosquito tissues or cell cultures are infected with an arbovirus, the virus initially replicates to a high titer, but within a few days the viral titer decreases significantly. The tissue or cell culture is then persistently infected and can continue to release low levels of infectious virus for the life of the mosquito or culture (55). This quantitative limitation of the amount of infectious virus produced by an infected mosquito cell has been called "modulation of viral titer" and has been suggested as a mechanism to explain the cellular resistance of susceptible invertebrate cells to the potential cytopathic effects of arboviruses (88). Although the mechanism involved in the modulation of viral titers in mosquito cells is unknown, Murphy (88) suggested either host-derived factors, e.g., interferon, or virus-derived factors, e.g., defective interfering (DI) particles or temperature-sensitive (TS) mutants. The production of interferon-like substances in *A. albopictus* infected with Sindbis virus, and of DI particles and TS mutants of several arboviruses in mosquito cells, has been reported (34, 132). Also, Lee and Schloemer (78) reported that *A. albopictus* cells infected with a flavivirus (Banzi virus) produced a virus-specific antiviral factor, identical to the viral matrix polypeptide. They postulated that this substance might act by regulating viral RNA synthesis. Persistent alphavirus infection in cultured mosquito cells is accompanied by reduction in viral RNA and protein production, decreased production of infectious virus, a decline in the number of infected cells, and increased production of DI particles (23). Hardy et al. (55) suggested that the vector, not the virus, appears to control the production of the "modulating factors" involved, citing several lines of indirect evidence for this hypothesis with regard to mosquito-borne arboviruses. As indicated earlier, the concept of modulation of viral titers may provide an explanation for the dissemination barriers found in some vectors that are infected with, but cannot transmit, certain viruses.

Conclusions and Prospects

In 1980, Sylvester (138) stated that "Vector specificity remains an intractable problem. Yet to understand specificity may be to understand the entire chain of events that define a vector-virus relationship, and specificity as presently conceived undoubtedly has more than a single cause." The first part of this statement is still true today, but the last part seems now an understatement, in view of our increasing awareness of how complex the problem of vector specificity really is. In this chapter, I have summarized some of the current knowledge on factors affecting vector specificity and vector competence for insect transmission of plant and animal viruses. During the past decade, the concept of transmission barriers has been much more developed with regard to animal arboviruses, paticularly those transmitted by mosquitoes, than with plant viruses transmitted by leafhoppers, planthoppers, or aphids. However, with both plant and animal viruses, several problems related to vector specificity and competence remain to be elucidated. For example, the postulated receptor sites (53, 61), either in the midgut or other tissues of the vector, and the interaction of viral coat or attachment proteins with these sites (1, 41, 100, 136), need further investigation. Also, the question of virus "modulation" in vector tissues (53, 55), and in this respect, whether replication strategies and translation products of viruses differ in their insect vectors versus their vertebrate or plant hosts (10, 37, 100), provide significant and challenging opportunities for further research. Better understanding of the factors underlying "modulation" of viruses in their vectors, e.g., antiviral agents or defective interfering particles (34, 132), may provide important clues to an understanding of resistance to virus infections, in addition to possible mechanisms of vector specificity. Furthermore, the microevents that take place during virus acquisition or inoculation feeding by insect vectors, on plant or vertebrate hosts, need to be examined in much greater detail. For example, the feeding and probing behavior (14), and the possible effects of salivary secretions (98, 114, 131) of vector and nonvector species on transmission or non-transmission of viruses, may explain vector specificity at least in certain virus-vector combinations. Also, possible effects of interactions or "reassortment" of different viruses in the same host or vector on transmission efficiency of plant and animal viruses (9, 25a) deserve further investigation.

Several virological, entomological, and other techniques that may help answer some of the above questions include electronic monitoring of feeding behavior of virus vectors (14, 69), insect cell monolayers (70, 102), and molecular biological techniques used to study virus genomes and their functions (83, 100). Additionally, more sensitive and more specific detection methods for viruses and virus products in plant, insect, and vertebrate tissues are now available, employing various serological assays, immunocytochemical light and electron microscopy (7, 116a, 150a, 150b, 155b), and other immunolabeling techniques and nucleic acid hybridization (25a, 83, 109). With the increasing sensitivity of these detection methods, some of the plant viruses now thought to be circulative-nonpropagative may prove to be inefficiently propagated or "modulated" in their vectors to lower levels than can be detected at present. A recent example, is the plant-infecting bunyavirus, TSWV, which was considered nonpropagative in its thrips vectors for a long time, but is now thought to propagate in its vector (149, 150). Furthermore, detecting new modes of virus transmission by insects may still be possible; an "ingestion–defecation" model of insect transmission, in which the salivary glands are not implicated, has been recently proposed by Gibb and Randles (45, 45a) to explain persistent (but nonpropagative) transmission of velvet tobacco mottle virus by its mirid vector. Finally, it is my belief that better communication and cooperation between animal and plant virus–vector researchers, and between these and specialists from other desciplines (e.g., physiologists, biochemists, and molecular biologists) will greatly enhance the prospects to further our understanding of the complex, but intriguing and potentially very significant, problem of vector specificity for both plant and animal viruses.

Acknowledgments. This chapter was prepared while the author was on sabbatical leave from Cairo University, at the Ohio State University (OSU) and later the University of Kentucky (UK). For critical reading of earlier versions of the manuscript, I am grateful to L.R. Nault (OSU), T.P. Pirone, R. Hunt, U. Järlfors (UK), and J.L. Hardy (University of California, Berkeley). Thanks are also due to S. Karger AG, Basel and Academic Press for permissions to reproduce Figs. 11.1–11.6 and 11.9–11.11, respectively.

References

1. Adam, G. 1984. Plant virus studies in insect vector cell cultures. In M.A. Mayo and K.A. Harrap, eds., *Vectors in Virus Biology*. Academic Press, New York, pp. 37–62.
2. Adam, G., and Hsu, H.T. 1984. Comparison of structural proteins from two potato yellow dwarf viruses. *J. Gen. Virol.* **65**:991–994.
3. Adams, J.R., and Bonami, J.R., eds. 1991. *Atlas of Invertebrate Viruses*. CRC Press, Boca Raton, FL.
4. Ammar, E.D. 1975. Effect of European wheat striate mosaic, acquired by feeding on diseased plants, on the biology of its planthopper vector *Javesella pellucida. Ann. Appl. Biol.* **79**:195–202.
5. Ammar, E.D. 1975. Effect of European wheat striate mosaic, acquired transovarially, on the biology of its planthopper vector *Javesella pellucida. Ann. Appl. Biol.* **79**:203–213.
6. Ammar, E.D. 1985. Internal morphology and ultrastructure of leafhoppers and planthoppers. In L.R. Nault and J.G. Rodriguez, eds., *The Leafhoppers and Planthoppers*. John Wiley, New York, pp. 127–162.
7. Ammar, E.D. 1991. Mechanisms of plant virus transmission by homopteran insects. In K. Mendgen and D.-E. Lesemann, eds., *Electron Microscopy of Plant Pathogens*. Springer-Verlag, Berlin, pp. 133–146.

8. Ammar, E.D., Gingery, R.E., and Madden, L.V. 1990. Geographical isolates of maize stripe virus differing in efficiency of transmission by, and titer in, the planthopper *Peregrinus maidis*. *Phytopathology* **80**:1022 (Abstr.).
9. Ammar, E.D., Gingery, R.E., and Nault, L.R. 1987. Interactions between maize mosaic and maize stripe viruses in their insect vector, *Peregrinus maidis*, and in maize. *Phytopathology* **77**:1051–1056.
9a. Ammar, E.D., Gingery, R.E., and Nault, L.R. 1993. Cytopathology and isolation of reoviruslike particles from the leafhopper *Graminella nigrifrons* (Homoptera, Cicadellidae). *J. Invertebr. Pathol.* (in press).
10. Ammar, E.D., and Nault, L.R. 1985. Assembly and accumulation sites of maize mosaic virus in its planthopper vector. *Intervirology* **24**:33–41.
11. Ammar, E.D., Nault, L.R., Styer, W.E., and Saif, Y.M. 1987. Staphylococcus, paramyxovirus-like, rickettsia-like and other structures in *Peregrinus maidis* (Homoptera, Delphacidae). *J. Invert. Pathol.* **49**:209–217.
12. Arcus, Y.M., Houk, E.J., and Hardy, J.L. 1983. Comparative in vitro binding of an arbovirus to midgut microvillar membranes from susceptible and refractory *Culex* mosquitoes. *Fed. Proc.* **42**:2141 (Abstr.).
13. Atcham, T., and Banttari, E.E. 1986. Histopathology, serology and transmission assay of AY and OBDV causing interference in dually infected aster leafhoppers. *Phytopathology* **76**:1062 (Abstr.).
14. Backus, E.A. 1988. Sensory mechanisms and behaviours which mediate hemipteran plant-feeding: A taxonomic overview. *J. Insect Physiol.* **34**:151–165.
15. Banttari, E.E. 1988. The occurrence and interactions of plant viruses and mollicutes in plant and insect vectors. In K. Maramorosch and S.P.J. Raychaudhuri, eds., *Mycoplasma Diseases of Crops, Basic and Applied Aspects*. Springer-Verlag, New York, pp. 193–208.
16. Banttari, E.E., and Zeyen, R.J. 1976. Multiplication of oat blue dwarf virus in the aster leafhopper. *Phytopathology* **66**:896–900.
17. Behncken, G.M. 1971. Relationship of sowthistle yellow vein virus to an inefficienct aphid vector, *Macrosiphum euphorbiae* (Thomas). Ph.D. Thesis, University of California, Berkeley, CA.
18. Behncken, G.M. 1973. Evidence of multiplication of sowthistle yellow vein virus in an inefficient aphid vector, *Macrosiphum euphorbiae*. *Virology* **53**:405–412.
19. Bennett, C.W. 1967. Apparent absence of cross-protection between strains of the curly top virus in the beet leafhopper, *Circulifer tenellus*. *Phytopathology* **57**:207–209.
20. Berger, P.H., and Pirone, T.P. 1986. The effect of helper component on the uptake and localization of potyviruses in *Myzus persicae*. *Virology* **153**:256–261.
21. Black, L.M., and Brakke, M.K. 1952. Multiplication of wound tumor virus in an insect vector. *Phytopathology* **42**:269–273.
22. Briddon, R.W., Pinner, M.S., Stanley, J., and Markham, P.G. 1990. Geminivirus coat protein gene replacement alters insect specificity. *Virology* **177**:85–94.
23. Brown, D.T., and Condreay, L.D. 1986. Replication of alphaviruses in mosquito cells. In S. Schlesinger and M.J. Schlesinger, eds., *The Togaviridae and Flaviviridae*. Plenum Press, New York, pp. 171–207.

24. Caciagli, P. 1991. Effect of anaesthesia with carbon dioxide on vectoring ability and survival of the planthopper *Leodelphax striatellus. Ann. Appl. Biol.* **119**:257–264.
25. Chamberlain, R.W., and Sudia, W.D. 1961. Mechanism of transmission of viruses by mosquitoes. *Annu. Rev. Entomol.* **6**:371–390.
25a. Chandler, L.J., Hogge, G., Endres, M., Jacoby, D.R., Nathanson, N., and Beaty, B.J. 1991. Reassortment of La Crosse and Tahyna bunyaviruses in *Aees triseriatus* mosquitoes. *Virus Res.* **20**:181–191.
26. Chen, C.-C., and Chiu, R.-J. 1980. Factors affecting transmission of rice transitory yellowing virus by green leafhoppers. *Plant Prot. Bull.* (Taiwan) **22**:297–306.
27. Chen, M.J., and Shikata, E. 1971. Morphology and intracellular localization of rice transitory yellowing virus. *Virology* **46**:786–796.
28. Chiykowski, L.N. 1981. Epidemiology of diseases caused by leafhopper-borne pathogens. In K. Maramorosch and K.F. Harris, eds., *Plant Diseases and Vectors, Ecology and Epidemiology*. Academic Press, New York, pp. 106–159.
29. Cohen, S., Duffus, J.E., and Liu, H.Y. 1989. Acquisition, interference, and retention of cucurbit leaf curl viruses in whiteflies. *Phytopathology* **79**:109–113.
30. Conti, M. 1985. Transmission of plant viruses by leafhoppers and planthoppers. In L.R. Nault and J.G. Rodriguez, eds., *The Leafhoppers and Planthoppers*. John Wiley, New York, pp. 289–307.
30a. Cupp, E.W., Mare, C.J., Cupp, M.S., and Ramberg, F.B. 1992. Biological transmission of vesicular stomatitis (New Jersey) by *Simulium vittatum* (Diptera: Simuliidae). *J. Med. Entomol.* **29**:137–140.
31. DeFolliart, G.R., Grimstad, P.R., and Watts, D.M. 1987. Advances in mosquito-borne arbovirus/vector research. *Annu. Rev. Entomol.* **32**:479–505.
32. Egan, B.T., Ryan, C.C., and Francki, R.I.B. 1989. Fiji disease. In C. Ricaud, B.T. Egan, A.G. Gillaspie, and C.G. Hughes, eds., *Diseases of Sugarcane. Major Diseases*. Elsevier, Amsterdam, pp. 263–287.
33. Elliot, R.M. 1990. Molecular biology of the bunyaviridae. *J. Gen. Virol.* **71**:501–522.
34. Enzmann, P.-J. 1973. Induction of an interferon-like substance in persistently infected *Aedes albopictus* cells. *Arch. Gesamte Virusforsch.* **41**:382–389.
35. Falk, B.W., Kim, K.S., and Tsai, J.H. 1988. Electron microscopic and physicochemical analysis of a reo-like virus of the planthopper *Peregrinus maidis. Intervirology* **29**:195–206.
36. Falk, B.W., and Tsai, J.H. 1985. Serological detection and evidence for multiplication of maize mosaic virus in the planthopper, *Peregrinus maidis. Phytopathology* **75**:852–855.
37. Falk, B.W., Tsai, J.H., and Lommel, S.A. 1987. Differences in the levels of detection for the maize stripe virus capsid and major non-capsid proteins in plant and insect hosts. *J. Gen. Virol.* **68**:1801–1811.
38. Faran, M.E., Romoser, W.S., and Routier, R.G. 1988. The distribution of Rift Valley fever virus in the mosquito *Culex pipiens* as revealed by viral titration of diseased organs and tissues. *Am. J. Tropic. Med. Hyg.* **39**:206–213.

39. Feinsod, F.M., Spielman, A., and Waner, J.L. 1975. Neutralization of sindbis virus by antisera to antigens of vector mosquitoes. *Am. J. Trop. Med. Hyg.* **24**:533–536.
40. Francki, R.I.B., Ryan, C.C., Hatta, T., Rohozinski, J., and Grivell, C.J. 1986. Serological detection of Fiji disease virus antigens in the planthopper *Perkinsiella saccharricida* and its inefficient ability to transmit the virus. *Plant Pathol.* **35**:324–328.
41. Fraser, R.D.B., Furlong, D.B., Trus, B.L., Nibert, M.L., Fields, B.N., and Steven, A.C. 1990. Molecular structure of the cell-attachment protein of reovirus: Correlation of computer-processed electron micrographs with sequence-based predictions. *J. Virol.* **64**:2990–3000.
42. Fukushi, T. 1969. Relationships between propagative rice viruses and their vectors. In K. Maramorosch, ed., *Viruses, Vectors and Vegetation.* John Wiley, New York, pp. 279–301.
43. Gaedigk, K., Adam, G., and Mundry, K.W. 1986. The spike protein of potato yellow dwarf virus and its functional role in the infection of inscet vector cells. *J. Gen. Virol.* **67**:2763–2773.
44. Gamez, R., and Leon, P. 1988. Maize rayado fino and related viruses. In R. Koenig, ed., *The Plant Viruses*, Vol. 3. Plenum, New York, pp. 213–233.
44a. Gaviria, M.E., Martinez, C., and Gonzalez, R. 1989. Reduction of the vector capacity of *Sogatodes oryzicola* (Hom.: Delphacidae) for the rice (*Oryza sativa*) white leaf virus. *Turrialba* **38**:300–305.
45. Gibb, K.S., and Randles, J.W. 1990. Distribution of velvet tobacco mottle virus in mirid vector and its relationship to transmissiblity. *Ann. Appl. Biol.* **116**:513–521.
45a. Gibb, K.S., and Randles, J.W. 1991. Transmission of velvet tobacco mottle virus and related viruses by the mirid *Cyrtopeltis nicotianae*. In K.F. Harris, ed., *Adv. Dis. Vector Res.* **7**:1–17.
46. Gildow, F.E. 1987. Virus-membrane interactions involved in circulative transmission of luteoviruses by aphids. *Curr. Topics Vector Res.* **4**:93–120.
46a. Gildow, F.E. 1993. Evidence for receptor-mediated endocytosis regulating luteovirus acquisition by aphids. *Phytopathology* **83**:270–277.
47. Gildow, F.E., and D'Arcy, C.J. 1988. Barley and oats as reservoirs for an aphid virus and influence on barley yellow dwarf virus transmission. *Phytopathology* **78**:811–816.
48. Gildow, F.E., and Gray, S.M. 1992. Basal lamina as a selective barrier associated with vector-specific transmission of luteoviruses by aphids. *Phytopathology* **82**:1103 (Abstr.).
49. Gingery, R.E., Gordon, D.T., and Nault, L.R. 1982. Purification and properties of maize rayado fino virus from the United States. *Phytopathology* **72**:1313–1318.
50. Granados, R.R., Hirumi, H., and Maramorosch, K. 1967. Electron microscopic evidence for wound tumor virus accumulation in various organs of an inefficient leafhopper vector, *Agalliopsis novella*. *J. Invert. Pathol.* **9**:147–159.
51. Grimstad, P.R., Paulson, S.L., and Craig, G.B., Jr. 1985. Vector competence of *Aedes hendersoni* (Diptera; Culicidae) for La Cross virus and evidence of a salivary-gland escape barrier. *J. Med. Entomol.* **22**:447–453.

52. Grimstad, P.R., and Walker, E.D. 1991. *Aedes triseriatus* (Diptera: Culicidae) and La Crosse virus. IV. Nutritional deprivation of larvae affects the adult barriers to infection and transmission. *J. Med. Entomol.* **28**:378–386.

52a. Guy, P.L., Toriyama, S., and Fuji, S. 1992. Occurrence of picorna-like virus in planthopper species and its transmission in *Laodelphax striatellus*. *J. Invertebr. Pathol.* **59**:161–164.

53. Hardy, J.L. 1988. Susceptibility and resistance of vector mosquitoes. In T.P. Monath, ed., *The Arboviruses: Epidemiology and Ecology*, Vol. 1. CRC Press, Boca Raton, FL, pp. 87–126.

54. Hardy, J.L., Apperson, G., Asman, S.M., and Reeves, W.C. 1978. Selection of a strain of *Culex tarsalis* highly resistant to infection following ingestion of western equine encephalomyelitis virus. *Am. J. Trop. Med. Hyg.* **27**:313–321.

55. Hardy, J.L., Houk, E.J., Kramer, L.D., and Reeves, W.C. 1983. Intrinsic factors affecting vector competence of mosquitoes for arboviruses. *Annu. Rev. Entomol.* **28**:229–262.

56. Hardy, J.L., Reeves, W.C., Bruen, J.P., and Presser, S.B. 1979. Vector competence of *Culex tarsalis* and other mosquito species for western equine encephalomyelitis virus. In E. Kurstak, ed., *Arctic and Tropical Arboviruses*. Academic Press, New York, pp. 157–171.

57. Harris, K.F. 1975. Fate of pea enation mosaic virus in PEMV-injected pea aphids. *Virology* **65**:148–162.

57a. Harris, K.F. 1977. An ingestion–egestion hypothesis of noncirculative virus transmission. In K.F. Harris and K. Maramorosch, eds., *Aphids as Virus Vectors*. Academic Press, New York, pp. 165–220.

57b. Harris, K.F. 1979. Leafhoppers and aphids as biological vectors: vector-virus relationships. In K. Maramorosch and K.F. Harris, eds., *Leafhopper Vectors and Plant Disease Agents*. Academic Press, New York, pp. 217–308.

58. Harris, K.F. 1981. Arthropod and nematode vectors of plant viruses. *Annu. Rev. Phytopathol.* **19**:391–426.

58a. Harris, K.F. 1990. Aphid transmission of plant viruses. In C.L. Mandahar, ed., *Plant Viruses Vol. II, Pathology*. CRC Press, Boca Raton, pp. 177–204.

59. Hatfil, S.J., Williamson, C., Kirby, R., and Von-Wechmar, M.B. 1990. Identification and localization of aphid lethal paralysis virus particles in thin tissue sections of the *Rhopalosiphum padi* aphid by *in situ* nucleic acid hybridization. *J. Invert. Pathol.* **55**:265–271.

60. Hibino, H. 1989. Insect-borne viruses of rice. *Adv. Dis. Vector Res.* **6**:209–241.

61. Houk, E.J., Kramer, L.D., Hardy, J.L., and Chiles, R.E. 1985. Western equine encephalomyelitis virus: *in vivo* infection and morphogenesis in mosquito mesenteronal epithelial cells. *Virus Res.* **2**:123–138.

62. Houk, E.J., Kramer, L.D., Hardy, J.L., and Presser, S.B. 1986. An intraspecific mosquito model for the mesenteronal infection barrier to western equine encephalomyelitis virus (*Culex tarsalis* and *Culex pipiens*). *Am. J. Trop. Med. Hyg.* **35**:632–641.

63. Hsu, H.T. 1978. Cell fusion induced by a plant virus. *Virology* 84:9–18.

64. Hsu, H.T., McBeath, J.H., and Black, L.M. 1977. The comparative susceptibilities of cultured vector and nonvector leafhopper cells to three plant viruses. *Virology* **81**:257–262.

65. Huff, C.G. 1931. A proposed classification of disease transmission by arthropods. *Science* **74**:456–457.
66. Inoue, H., and Omura, T. 1982. Transmission of rice gall dwarf virus by the green rice leafhopper. *Plant Dis.* **66**:57–59.
67. Izadpanah, K. 1989. Purification and serology of the Iranian maize mosaic rhabdovirus. *J. Phytopathol.* **126**:43–50.
68. Jackson, A.O., Francki, R.I.B., and Zuidema, D. 1987. Biology, structure and replication of plant rhabdoviruses. In R.R. Wagner, ed., *The Rhabdoviruses*. Plenum, New York, pp. 427–507.
68a. Jones, L.D., Hodgson, E., Williams, T., Higgs, S., and Nattall, P.A. 1992. Saliva activated transmission (SAT) of Thogoto virus: relationship with vector potential of different haematophagous arthropods. *Med. Veter. Entomol.* **6**:261–265.
69. Kimmins, F.M. 1989. Electrical penetration graphs from *Nilaparvata lugens* on resistant and susceptible rice varieties. *Entomol. Exp. Appl.* **50**:69–79.
70. Kimura, I., and Omura, T. 1988. Leafhopper cell cultures as a means for phytoreovirus research. *Adv. Dis. Vector Res.* **5**:111–135.
71. Kisimoto, R. 1973. Leafhoppers and planthoppers. In A.J. Gibbs, ed., *Viruses and Invertebrates*. Elsevier, New York, pp. 137–135.
72. Kisimoto, R., and Watson, M.A. 1965. Abnormal development of embryos induced by inbreeding in *Delphacodes pellucida* Fab. and *Delphacodes dubia* Kirschbaum (Araepidae, Homoptera), vectors of European wheat striate mosaic virus. *J. Invert. Pathol.* **7**:297–305.
73. Koblet, H. 1989. The "merry-go-round": Alphaviruses between vertebrate and invertebrate cells. *Adv. Virus. Res.* **38**:343–398.
74. Kramer, L.D., Hardy, J.L., and Presser, S.B. 1983. Effect of temperature of extrinsic incubation on the vector competence of *Culex tarsalis* for western equine encephalomyelitis virus. *Am. J. Trop. Med. Hyg.* **32**:1130–1139.
75. Lastra, J.R., and Esparza, J. 1976. Multiplication of vesicular stomatitis virus in the leafhopper *Peregrinus maidis* (Ashm.), a vector of a plant rhabdovirus. *J. Gen. Virol.* **32**:139–142.
75a. Laubscher, J.M., and Von-Wechmar, M.B. 1992. Influence of aphid lethal paralysis virus and *Rhopalosiphum padi* virus on aphid biology at different temperatures. *J. Invertebr. Pathol.* **60**:134–140.
76. Leake, C.J. 1984. Transovarial transmission of arboviruses by mosquitoes. In M.A. Mayo and K.A. Harrap, eds., *Vectors in Virus Biology*. Academic Press, London, pp. 63–91.
77. Lee, P.E. 1967. Morphology of wheat striate mosaic virus and its localization in infected cells. *Virology* **33**:84–94.
78. Lee, C.-H., and Schloemer, R.H. 1981. Mosquito cells infected with Banzi virus secrete an antiviral activity which is of viral origin. *Virology* **110**:402–410.
79. Lopez-Abella, D., Bradley, R.H.E., and Harris, K.F. 1988. Correlation between stylet paths made during superficial probing and the ability of aphids to transmit nonpersistent viruses. *Adv. Dis. Vector Res.* **5**:251–285.
80. Maramorosch, K. 1975. Infection of arthropod vectors by plant pathogens. In K. Maramorosch and R.E. Shope, eds., *Invertebrate Immunity*. Academic Press, New York, pp. 49–53.

81. Markham, P.G. 1983. Spiroplasmas in leafhoppers, a review. *Yale J. Biol. Med.* **56**:741–751.
82. Matthews, R.E.F. 1982. Classification and nomenclature of viruses. *Intervirology* **17**:1–199.
82a. Matthews, R.E.F. 1985. Viral taxonomy for the nonvirologist. *Ann. Rev. Microbiol.* **39**:451–474.
83. Matthews, R.E.F. 1991. *Plant Virology*, 3rd ed. Academic Press, New York.
84. McDaniel, L.L., Ammar, E.D., and Gordon, D.T. 1985. Assembly, morphology, and accumulation of a Hawaiian isolate of maize mosaic virus in maize. *Phytopathology* **75**:1167–1172.
85. McIntosch, B.M., and Jupp, P.G. 1970. Attempts to transmit chikungunaya virus with six species of mosquito. *J. Med. Entomol.* **7**:615–618.
86. Merrill, M.H., and TenBroeck, C. 1935. The transmission of equine encephalomyelitis virus by *Aedes aegypti. J. Exp. Med.* **62**:687–695.
87. Mims, C.A., Day, M.F., and Marshall, I.D. 1966. Cytopathic effects of Semliki Forest virus in the mosquito *Aedes aegypti. Am. J. Trop. Med. Hyg.* **15**:775–784.
88. Murphy, F.A. 1975. Cellular resistance to arbovirus infection. *Ann. N.Y. Acad. Sci.* **266**:197–203.
89. Murphy, F., and Harrison, A. 1980. Electron microscopy of the rhabdoviruses on animals. In D.H. Bishop, ed., *Rhabdoviruses*, Vol. 1. CRC Press, Boca Raton, FL, pp. 65–107.
90. Nagaraji, A.N., and Black, L.M. 1962. Hereditary variation in the ability of a leafhopper to transmit two unrelated plant viruses. *Virology* **16**:152–162.
91. Nakasuji, F., and Kiritani, K. 1970. III. Effects of rice dwarf virus upon its vector, *Nephotettix cincticeps* Uhler (Hemiptera: Deltocephalidae) and its significance for changes in relative abundance of infected individuals among vector populations. *Appl. Entomol. Zool.* **5**:1–12.
92. Nasu, S. 1963. Studies on some leafhoppers and planthoppers which transmit virus diseases of rice in Japan. *Bull. Kyushu Agric. Expt. Sta.* **8**:153–349.
93. Nasu, S. 1965. Electron microscopic studies on transovarial passage of rice dwarf virus. *Jpn. J. Appl. Entomol. Zool.* **9**:225–237.
94. Nault, L.R. 1992. Transmission biology, vector specificity and evolution of planthopper transmitted plant viruses. In R.F. Denno and T.J. Perfect, eds., *Planthoppers. Their Ecology, Genetics and Management.* Chapman and Hall, New York.
95. Nault, L.R., and Ammar, E.D. 1989. Leafhopper and planthopper transmission of plant viruses. *Annu. Rev. Entomol.* **34**:503–529.
96. Nault, L.R., Gingery, R.E., and Gordon, D.T. 1980. Leafhopper transmission and host range of maize rayado fino virus. *Phytopathology* **70**:709–712.
97. Nault, L.R., and Gordon, D.T. 1988. Multiplication of maize stripe virus in *Peregrinus maidis. Phytopathology* **78**:991–995.
98. Nishi, Y. 1969. Inhibition of viruses by vector saliva. In K. Maramorosch, ed., *Viruses, Vectors, and Vegetation.* Interscience, New York, pp. 579–591.

98a. Noda, H., Ishikawa, K., Hibino, H., and Omura, T. 1991. A reovirus in the brown planthopper, *Nilaparvata lugens*. *J. Gen. Virol.* **72**:2425–2430.
99. Nuss, D.L. 1984. Molecular biology of wound tumor virus. *Adv. Virus Res.* **29**:57–90.
100. Nuss, D.L., and Dall, D.J. 1990. Structural and functional properties of plant reovirus genomes. *Adv. Virus Res.* **38**:249–306.
101. Ofori, F.A., and Francki, R.I.B. 1985. Transmission of leafhopper A virus, vertically through eggs and horizontally through maize in which it does not multiply. *Virology* **144**:152–157.
102. Omura, T., Kimura, I., Tsuchizaki, T., and Saito, Y. 1988. Infection by rice gall dwarf virus of cultured monolayers of leafhopper cells. *J. Gen. Virol.* **69**:429–432.
103. Palmer, E.L., and Martin, M.L. 1982. *An Atlas of Mammalian Viruses*. CRC Press, Boca Raton, FL.
104. Paulson, S.L., and Grimstad, P.R. 1989. Replication and dissemination of La Crosse virus in the competent vector *Aedes triseriatus* and the incompetent vector *Aedes hendersoni* and evidence for transovarial transmission by *A. hendersoni* (Diptera; Culicidae). *J. Med. Entomol.* **26**:602–609.
105. Paulson, S.L., Grimstad, P.R., and Craig, G.B., Jr. 1989. Midgut and salivary gland barriers to La Crosse virus disseminated in mosquitoes of the *Aedes triseriatus* group. *Med. Vet. Entomol.* **3**:113–123.
105a. Paulson, S.L., Poirier, S.J., Grimstad, P.R., and Craig, G.B., Jr. 1992. Vector competence of *Aedes hendersoni* (Diptera: Culicidae) for La Crosse virus: lack of impaired function in virus-infected salivary glands and enhanced virus transmission by sporozoite-infected mosquitoes. *J. Med. Entomol.* **29**:483–488.
106. Peters, D., de Avila, A.C., Kitajima, E.W., Resende, R. de O., de Haan, P., and Goldbach, R.W. 1991. An overview of tomato spotted wilt virus. In H. Hsu and R.H. Lawson, eds., *Virus-Thrips-Plant Interaction of Tomato Spotted Wilt Virus. Proceedings of a USDA Workshop*, USDA-ARS-87, pp. 1–14.
107. Peterson, A.J., and Nuss, D. 1986. Regulation of expression of the wound tumor virus genome in persistently infected vector cells is related to changes in the translational activity of viral transcripts. *J. Virol.* **59**:195–202.
108. Pirone, T.P., and Thornbury, D.W. 1984. The involvement of a helper component in nonpersistent transmission of plant viruses by aphids. *Microbiol. Sci.* **1**:191–193.
109. Plumb, R.T. 1989. Detecting plant viruses in their vectors. *Adv. Dis. Vector Res.* **6**:191–208.
110. Purcell, A.H., and Nault, L.R. 1991. Interactions among plant pathogenic prokaryotes, plants, and insect vectors. In P. Barbosa, V.A. Krischik, and C.G. Jones, eds., *Microbial Mediation of Plant-Herbivore Interactions*. John Wiley, New York, pp. 383–405.
111. Raga, I.N., Ito, K., Matsui, M., and Okada, M. 1988. Effects of temperature on adult longevity, fertility, and rate of transovarial passage of rice stripe virus in the small brown planthopper, *Laodelphax striatellus* Fallen (Homoptera: Delphacidae). *Appl. Entomol. Zool.* **23**:67–75.

112. Ramasamy, M.S., Sands, M., Kay, B.H., Fanning, I.D., Lawrence, G.W., and Ramasamy, R. 1990. Anti-mosquito antibodies reduce the susceptibility of *Aedes aegypti* to arbovirus infection. *Med. Vet. Entomol.* **4**:49–55.
113. Reddy, D.V.R., and Black, L.M. 1974. Deletion mutations of the genome segments of wound tumor virus. *Virology* **61**:458–473.
114. Ribeiro, J.M.C. 1987. Role of saliva in blood-feeding by arthropods. *Annu. Rev. Entomol.* **32**:463–478.
115. Rochow, W.F., and Miller, I. 1975. Use of aphids injected with virus-specific antiserum for study of plant viruses that circulate in vectors. *Virology* **63**:282–286.
116. Romoser, W.S., Faran, M.E., and Bailey, C.L. 1987. Newly recognized route of arbovirus dissemination from the mosquito (Diptera: Culicidae) midgut. *J. Med. Entomol.* **24**:431–432.
116a. Romoser, W.S., Faran, M.E., Baily, C.L., and Lerdthusnee, K. 1992. An immunocytochemical study of the distribution of Rift Valley fever virus in the mosquito *Culex pipiens*. *Am. J. Trop. Med. Hyg.* **46**:489–501.
117. Rosen, L. 1987. Sexual transmission of dengue viruses by *Aedes albopictus*. *Am. J. Trop. Med. Hyg.* **37**:398–402.
118. Sakimura, K. 1962. The present status of thrips-borne viruses. In K. Maramorosch, ed., *Biological Transmission of Disease Agents*. Academic Press, New York, pp. 33–40.
119. Scott, T.W., and Burrage, T.G. 1984. Rapid infection of salivary glands in *Culiseta melanura* with eastern equine encephalitis virus: An electron microscopic study. *Am. J. Trop. Med. Hyg.* **33**:961–964.
120. Scott, T.W., and Weaver, S.C. 1989. Eastern equine encephalomyelitis virus: Epidemiology and evolution. *Adv. Virus Res.* **37**:227–328.
121. Shikata, E. 1979. Cytopathological changes in leafhopper vectors of plant viruses. In K. Maramorosch and K.F. Harris, eds., *Leafhopper Vectors and Plant Disease Agents*. Academic Press, New York, pp. 309–325.
122. Shinkai, A. 1962. Studies on insect transmission of rice diseases in Japan. *Bull. Natl. Agric. Sci. (Jpn.) Ser. C* **14**:1–112.
123. Singer, S.H., Barile, M.F., and Kirschstein, R.I. 1973. Mixed mycoplasmavirus infections in cell cultures. *Ann. N.Y. Acd. Sci.* **225**:304–310.
124. Sinha, R.C. 1963. Effect of age of vector and of abdomen puncture on virus transmission. *Phytopathology* **53**:1170–1173.
125. Sinha, R.C. 1965. Sequential infection and distribution of wound tumor virus in the internal organs of a vector after ingestion of virus. *Virology* **26**:673–686.
126. Sinha, R.C. 1967. Response of wound tumor virus infection in insects to vector age and temperature. *Virology* **31**:746–748.
127. Sinha, R.C. 1973. Viruses and leafhoppers. In A.J. Gibbs, ed., *Viruses and Invertebrates*. Elsevier, New York, pp. 493–511.
128. Sinha, R.C. 1981. Vertical transmission of plant pathogens. In J.J. Mckelvey, Jr., B.F. Eldridge, and K. Maramorosch, eds., *Vectors of Disease Agents*, Praeger, New York, pp. 119–121.
129. Sinha, R.C., and Chiykowski, L.N. 1967. Multiplication of wheat striate mosaic virus in its leafhopper vector *Endria inimica*. *Virology* **32**:402–405.

130. Sinha, R.C., and Chiykowski, L.N. 1969. Synthesis, distribution and some multiplication sites of wheat striate mosaic virus in a leafhopper vector. *Virology* **38**:679–684.
131. Sogawa, K. 1982. The rice brown planthopper: Feeding physiology and host plant interactions. *Annu. Rev. Entomol.* **27**:49–73.
132. Stollar, V. 1980. Togaviruses in cultured arthropod cells. In R.W. Schlesinger, ed., *The Togaviruses: Biology, Structure, Replication*. Academic Press, New York, pp. 583–621.
133. Stollar, V. 1987. Approaches to the study of vector specificity for arboviruses—Model systems using cultured mosquito cells. *Adv. Virus Res.* **33**:327–365.
134. Storey, H.H. 1933. Investigations of the mechanism of transmission of plant viruses by insect vectors. I. *Proc. R. Soc. London Ser. B* **113**:463–485.
135. Strauss, J.H., and Strauss, E.G. 1988. Evolution of RNA viruses. *Annu. Rev. Microbiol.* **42**:657–683.
136. Sundin, D.R., Beaty, B.J., Nathanson, N., and Gonzalez-Scarano, F. 1987. A G_1 glycoprotein epitope of La Crosse virus; A determinant of infection of *Aedes triseriatus. Science* **235**:591–593.
137. Sylvester, E.S. 1969. Evidence of transovarial passage of the sowthistle yellow vein virus in the aphid *Hyperomyzus lactucae. Virology* **38**:440–446.
138. Sylvester, E.S. 1980. Circulative and propagative virus transmission by aphids. *Annu. Rev. Entomol.* **25**:257–286.
139. Sylvester, E.S. 1985. Multiple acquisition of viruses and vector-dependent prokaryotes: Consequences on transmission. *Annu. Rev. Entomol.* **30**:71–88.
140. Sylvester, E.S., and Richardson, J. 1981. Inoculation of the aphids *Hypermyzus lactucae* and *Chaetosiphon jacobi* with isolates of sowthistle yellow vein virus and strawberry crinkle virus. *Phytopathology* **71**:598–602.
141. Sylvester, E.S., and Richardson, J. 1990. Comparison of vector-virus relationships of strawberry crickle plant rhabdovirus in two aphids (*Chaetosiphon fragaefolii* and *C. jacobi*) infected by injection. *Hilgardia* **58**:1–25.
142. Takahashi, M. 1982. Differential transmission efficiency for Japanese encephalitis virus among colonized strains of *Culex tritaeniorhynchus*. Jpn. J. Sanit. Zool. **33**:325–330.
143. Thompson, W.H. 1979. Higher venereal infection and transmission rates with La Crosse virus in *Aedes triseriatus* engorged before mating. *Am. J. Trop. Med. Hyg.* **28**:890–896.
144. Thompson, W.H., and Beaty, B.J. 1978. Venereal transmission of La Crosse virus from male to female *Aedes triseriatus. Am. J. Trop. Med. Hyg.* **27**:187–196.
145. Tsai, J.H., and Falk, B.W. 1988. Tropical maize pathogens and their associated insect vectors. *Adv. Dis. Vector Res.* **5**:178–201.
146. Tsai, J.H., and Zitter, T.A. 1982. Characteristics of maize stripe virus transmission by corn delphacid. *J. Econ. Entomol.* **75**:397–400.
147. Turell, M.J. 1988. Horizontal and vertical transmission of viruses by insect and tick vectors. In T.P. Monath, ed., *The Arboviruses: Epidemiology and Ecology*, Vol. 1. CRC Press, Boca Raton, FL, pp. 127–152.
147a. Turell, M.J., Hardy, J.L., and Reeves, W.C. 1982. Sensitivity of carbon dioxide in mosquitoes infected with California serogroup arboviruses. *Am. J. Trop. Med. Hyg.* **31**:389–394.

148. Turell, M.J., Rossignol, P.A., Spielman, A., Rossi, C.A., and Baily, C.L. 1984. Enhanced arboviral transmission by mosquitoes that concurrently ingested microfilariae. *Science* **225**:1039–1041.
149. Ullman, D.E., Cho, J.J., Mau, R.F.L., Hunter, W.B., Westcot, D.M., and Custer, D.M. 1992. Thrips–tomato spotted wilt virus interactions: morphological, behavioral and cellular components influencing thrips transmission. *Adv. Dis. Vector Res.* **9**:195–240.
150. Ullman, D.E., Cho, J.J., Mau, R.F.L., Westcot, D.M., and Custer, D.M. 1992. A midgut barrier to tomato spotted wilt virus acquisition by adult western flower thrips. *Phytopathology* **82**:1333–1342.
150a. Ullman, D.E., German, T.L., Sherwood, J.L., Westcot, D.M., and Cantone, F.A. 1993. *Tospovirus* replication in insect vector cells: immunocytochemical evidence that the nonstructural protein encoded by the S RNA of tomato spotted wilt tospovirus is present in thrips vector cells. *Phytopathology* **83**:456–463.
150b. Vale, T.G., Dowling, M.L., and Cloonan, M.J. 1992. Infection and multiplication of Ross River virus in the mosquito vector *Aedes vigilax* (Skuse). *Austral. J. Zool.* **40**:35–41.
151. Wang, M., Mitchell, C.J., Hu, J.S., Gonsalves, D., and Calisher, C.H. 1992. Determination of whether tomato spotted wilt virus replicates in *Toxorhynchites amboinensis* mosquitoes and relatedness of the virus to phleboviruses. *Intervirology* **33**:32–40.
152. Watson, M.A., and Sinha, R.C. 1959. Studies on the transmission of European wheat striate mosaic virus by *Delphacodes pellucida* Fabricius, *Virology* **8**:139–163.
153. Watts, D.M., Pantuwatana, S., DeFoliart, G.R., Yuill, T.M., and Thompson, W.H. 1973. Transovarial transmission of La Crosse virus (California encephalitis group) in the mosquito *Aedes triseriatus*. *Science* **182**:1140–1141.
153a. Wayadande, A.C., and Nault, L.R. 1993. Leafhopper probing behavior associated with maize chlorotic dwarf virus transmission in maize. *Phytopathology* **83**:522–526.
154. Weaver, S.C. 1986. Electron microscopic analysis of infection patterns for Venezuelan equine encephalomyelitis virus in the vector mosquito, *Culex (Melanoconion) taeniopus*. *Am. J. Trop. Med. Hyg.* **35**:624–631.
155. Weaver, S.C., Scott, T.W., Lorenz, L.H., Lerdthusnee, K., and Rosomer, W.S. 1988. Togavirus-associated pathogenic changes in the midgut of a natural mosquito vector. *J. Virol.* **62**:2083–2090.
155a. Weaver, S.C., Tesh, R.B., and Guzman, H. 1992. Ultrastructural aspects of replication of the New Jersey serotype of vesicular stomatitis virus in a suspected sand fly vector, *Lutzomyia shannoni* (Diptera: Psychodidae). *Am. J. Trop. Med. Hyg.* **46**:201–210.
155b. Wijkamp, I., van Lent, J., Kormelink, R., Goldbach, R., and Peters, D. 1993. Multiplication of tomato spotted wilt virus in its insect vector, *Frankliniella occidentalis*. *J. Gen. Virol.* **74**:341–349.
156. Zeigler, R.S., and Morales, F.J. 1990. Genetic determination of replication of rice hoja blanca virus within its planthopper vector, *Sogatodes oryzicola*. *Phytopathology* **80**:559–566.

12
Vectors of Plant Parasites of the Genus *Phytomonas* (Protozoa, Zoomastigophorea, Kinetoplastida)

Erney P. Camargo and Franklin G. Wallace

Introduction

The occurrence of flagellated protozoa as internal parasites in plants was first reported by Lafont in 1909 who found them in the latex of *Euphorbia pilulifera* in Mauritius (78). From the start, hemipterans were suspected of transmitting these plant parasites. However, as described in this chapter, only in a few instances the vector role of phytophagous bugs has been experimentally evidenced.

Many reports followed Lafont's discovery of plant trypanosomes (10, 16, 26, 45, 50, 52, 64, 65, 67, 70). Most of the early publications dealt with plants in the families Euphorbiaceae and Asclepiadaceae and it was thought that flagellates were in laticiferous plants only but in the 1930s they began to be found in the sieve tubes of other plants. Currently, plants of the following families are known as hosts of *Phytomonas*: Euphorbiaceae, Asclepiadaceae, Apocynaceae, Cecropiaceae, Compositae, Moraceae, Urticaceae, Sapotaceae, Palmae, and Rubiaceae (2, 11, 14, 16, 24, 28, 29, 35, 38, 40, 42, 44, 46, 48, 65, 76, 77, 79, 81, 88, 105, 112, 123, 124, 127, 128, 159).

Flagellates have also been found in fruits of the families Anacardaceae (cashew), Oxalidaceae (carambola), Passifloraceae (passion fruit), Punicaceae (pomegranate), Rosaceae (peach), Rutaceae (bergamot) and several species of Solanaceae including tomato and red pepper (23, 37, 60, 71, 75), and in seeds of *Phaseolus vulgaris* (brown bean) and *Glycine max* (soybean) (72, 73).

In most plants *Phytomonas* do not cause documented disease but there are three cases of economically important infections.

Erney P. Camargo, Institute of Biomedical Sciences, Department of Parasitology, University of São Paulo, São Paulo 05508, Brazil.
Franklin G. Wallace, 2603 Cohansey St. Paul, Minnesota, 55113, USA.

Phytomonas proliferate in the sieve tubes of various palm species: *Elaeis guineensis* (oil palm, dendê), *Cocos nucifera* (coconut palm), *Attalea funifera*, *Bentickia nicobarica*, *Carvota mites*, *Maximiliana maripa*, and *Roystones regia* (11, 134). In coconut and oil palms obstruction of sieve tubes by infecting flagellates leads to browning and falling of leaves and inflorescence, rotting of the stem and eventually death (11, 108, 133, 134). The phytoflagelloses of palms are prevalent in many South and Central American countries (16) where they are known by the names of "marchitez sorpressiva" (oil palms) and "hartrot" (coconut palms) (11, 25, 29, 89, 109).

In cassava, *Manihot esculenta*, the proliferation of flagellates in the laticiferous tubes leads to root atrophy, watering of the latex, and death (146). The disease, known by the regional name of "chochamento das raizes" (empty roots), occurred as an epidemic in 1985 in the state of Espirito Santo in Brazil, destroying 50% of the plants on some farms (16).

The phytoflagellosis of coffee, "phloem necrosis," was described by Stahel in Suriname in 1931 (135) causing a severe epidemic, which recurred in the 1940s. Infected trees die within 3 to 12 months (135, 148, 149). The disease has also been reported on a limited scale in Guyana, Brazil, and Colombia and affects the species *Coffea abeocutae*, *C. arabica*, *C. excelsa*, *C. liberica*, and *C. stenophyla* (16).

Family Trypanosomatidae

The order Kinetoplastida comprises animal-like flagellates (Class Zoomastigophorea) with one or two flagella and a single mitochondrion, which contains a conspicuous DNA-containing body, the kinetoplast, located near the flagellar base. The family Trypanosomatidae is defined as those kinetoplastids with a single flagellum. All are parasitic. Four genera are monoxenic parasites of insects and a few other invertebrates. These are *Blastocrithidia*, *Crithidia*, *Leptomonas*, and *Herpetomonas*. They are informally referred to as the lower trypanosomatids and are defined as follows. *Leptomonas* and *Herpetomonas* have the flagellum, whose base is near the anterior end, emerging from the flagellar pocket at the tip of the cell (promastigote). In addition to the promastigote form, the genus *Herpetomonas* exhibits forms in which the flagellar base is in the posterior part of the cell and the flagellum extends through the cell to emerge anteriorly (opisthomastigote). In *Blastocrithidia* (epimastigote) the flagellum emerges on the side of the cell and extends anteriorly along an undulating membrane to become the anterior free flagellum. *Crithidia* are characterized by choanomastigote forms, which are much like promastigotes but are short, blunt, and have a funnel-like opening to the flagellar pocket from which the flagellum emerges. *Crithidia* and *Blastocrithidia* are relatively easy to identify thanks to their peculiar

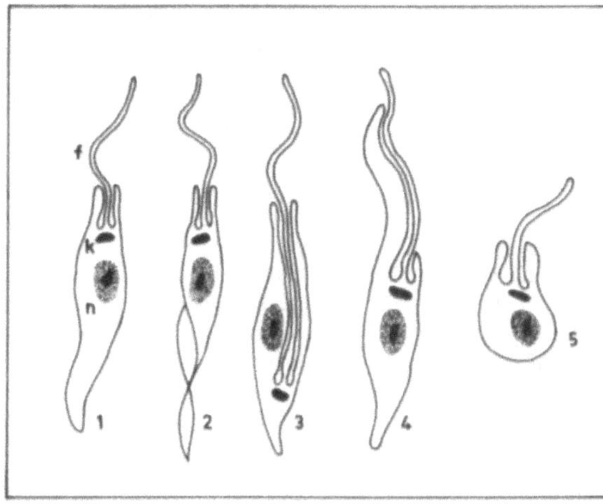

FIGURE 12.1. Developmental forms of lower trypanosomatids: 1, 2, promastigotes; 3, opisthomastigote; 4, epimastigote; 5, choanomastigote.

morphology. The distinction between *Leptomonas* and *Herpetomonas* is revealed only by the erratic occurrence of opisthomastigotes in the latter (Fig. 12.1).

Phytomonas often occur in plants as very large promastigotes with twisted bodies (Fig. 12.1). In insects they also occur as promastigotes but without characteristics that distinguish them from the other genera with promastigote stages. Although they have two hosts they are often grouped with the lower trypanosomatids above.

The genera *Trypanosoma*, *Leishmania*, and *Endotrypanum* are dixenous parasites with a vertebrate and an invertebrate host. They constitute the higher trypanosomatids.

Difficulties in morphological distinction between *Leptomonas*, *Herpetomonas*, and *Phytomonas* led to a search for alternative criteria (155). Enzymes of the arginine–ornithine metabolism helped to separate the genus *Leptomonas* from *Herpetomonas* and *Phytomonas* (15, 18, 36, 156). Monoclonal antibodies specific for *Phytomonas* were used to recognize members of this genus among insect isolates (129, 140) and restriction enzyme analysis of ribosomal genes and synthetic oligonucleotide rDNA probes have been used with the same objective (17). Thus it is possible with a reasonable degree of confidence to identify to genus isolates from insects.

Genus *Phytomonas*

Lafont (78) named the flagellates found in the latex of *Euphorbia pilulifera*, *Leptomonas davidi*. In the same year, 1909, Donovan (31)

TABLE 12.1. Hemiptera associated with infected plants.

Parasite	Plant	Insect	Family	Reference
P. elmassiani	*Asclepias* spp.	*Oncopeltus fasciculatus*	Lygaeidae	(68, 95)
P. elmassiani		*O. sandarachatus*	Lygaeidae	(97)
P. elmassiani		*Lygaeus kalmii*	Lygaeidae	(97)
P. elmassiani		*O. cingulifer*	Lygaeidae	(8, 145)
P. elmassiani		*O. unifasciatellus*	Lygaeidae	(8, 145)
P. elmassiani	*Pergularia extensa*	*O. famelicus*	Lygaeidae	(150)
P. davidi	*Euphorbia pilulifera*	*Nysius euphorbiae*	Lygaeidae	(80)
P. davidi	*E. hypericifolia*	*Nysius euphorbiae*	Lygaeidae	(80)
P. davidi	*E. hirta*	*Dieuches humilis*	Lygaeidae	(14)
P. davidi	*E. pinea*	*Dicranocephalus agilis*	Coreidae	(27, 52)
P. davidi	*E. peplus*	*Dicranocephalus agilis*	Coreidae	(27, 52)
P. davidi	*E. pilulifera*	*Pachybrachius bilobata scutellatus*	Coreidae	(99)
P. davidi	*E. hypericifolia*	*P. bilobata scutellatus*	Coreidae	(99)
P. davidi	*E. hirta*	*P. bilobata*	Coreidae	(74)
P. davidi	*E. hyssopifolia*	*P. bilobata*	Coreidae	(74)
Phytomonas sp.	*Cecropia palmata*	*Edessa loxdali*	Pentatomidae	(74)
P. staheli	*Elaeis guineensis*	*Lincus lethifer*	Pentatomidae	(22, 111)
P. staheli	*Elaeis guineensis*	*L. lobuliger*	Pentatomidae	(121)
P. staheli	*Cocos nucifera*	*L. croupius*	Pentatomidae	(87)
P. serpens	*Lycopersicon esculenta*	*Nezara viridula*	Pentatomidae	(60)
P. serpens	*Lycopersicon esculenta*	*Phthia picta*	Coreidae	(71)

proposed the name *Phytomonas* as generic name for the same organism but this terminology was not accepted for many years and plant flagellates were often referred to the genera *Leptomonas* or *Herpetomonas* until after 1950. Twelve species are now recognized as belonging to the genus *Phytomonas*: *P. davidi* in *Euphorbia pilulifera* (78), *P. elmassiani* in the asclepiad *Araujia* (101), *P. bordasi* in the asclepiad *Morreira odorata* (53), *P. ficuum* in *Ficus edulis* (34), *P. ganorae* in *Ficus hochstetteri* (49), *P. bancrofti* in *Ficus* sp. (69), *P. leptovasorum* in coffee (135), *P. tirucalli* in *Euphorbia tirucalli* (120), *P. serpens* (71) in tomato, *P. tortuosa* in *Ficus* sp. (126), and *P. staheli* in palms (98).

Until 1982 cultures of *Phytomonas* were not available but since then at least 30 isolates from various plants have been cultured (6–7, 27, 71, 75, 77, 100, 146).

Like the trypanosomatids of animals, *Phytomonas* are transmitted by insects. But while animal trypanosomatids are carried by insects of nine families in three orders plus leeches and mites, *Phytomonas* vectors so far described are limited to three families of the Hemiptera: Lygaeidae, Coreidae, and Pentatomidae (Table 12.1).

Vectors of plant flagellates are relatively large insects, often present in large numbers, brightly colored, and often have flagellates in their digestive organs. However, the finding of flagellates in phytophagous bugs cannot be taken as evidence that the bugs are vectors of plant parasites because of the many trypanosomatids that occur in the same groups of insects.

Trypanosomatids in Hemiptera

The phytophagous Hemiptera harbor many monoxenous trypanosomatidae (114, 152, 154). Table 12.2 updates the list of phytophagous hemiptera found to harbor trypanosomatids. In the world literature about 100 different species of flagellates (not all named) have been recorded from 100 species of plant-feeding Hemiptera.

When one considers that there are at least 23,000 known species in the order Hemiptera it is evident that only a small fraction of existing infections is yet known.

Of the insects summarized above, 21 were from the family Lygaeidae, 34 from the Coreidae, 9 from the Pyrrhocoridae, 29 from the Pentatomidae, 6 from the Miridae, and 1 from the Corimelaenidae. Of the protozoa, the large majority is represented by promastigotes, with but a few epimastigotes and choanomastigotes. The latter two morphological types may be identified as *Blastocrithidia* and *Crithidia*, respectively, but the promastigotes might be either *Leptomonas* or *Phytomonas* and eventually *Herpetomonas*.

A recent survey using biochemical and immunological methods to distinguish *Phytomonas* is summarized in a later section.

Hemiptera as Vectors

The Hemiptera are of minor importance in the transmission of other plant diseases. It is believed that their deleterious action is due mainly to mechanical damage and effect of the saliva. Only a few infectious diseases are transmitted by these insects (20).

The Homoptera, on the other hand, transmit a multitude of plant diseases including nearly all the viruses. The Homoptera are virtually free from monoxenous trypanosomatid infection. We know of only one record (107). McGhee and Cosgrove (93) suggested that the reason for the virtual absence of lower trypanosomatids from the Homoptera is their manner of oviposition. Among the phytophagous Hemiptera (and some insectivorous forms) the eggs are laid in masses and there is evidence that

TABLE 12.2. Occurrence of trypanosomatids in phytophagous bugs.

Insect	Trypanosomatid	Reference
Lygaeidae		
Dieuches humilis	*Leptomonas* sp.[a]	(14)
Lygaeus amicta (= *Serinetha amicta*)	*B. cleti*	(66)
	Leptomonas sp.	(125)
L. belfragei	*Leptomonas* sp.	(107)
L. familiaris	*L. familiaris*	(158)
L. fraterna (= *Serinetha fraterna*)	*L. serinethae*	(125)
	Leptomonas sp.	(125)
L. hospes	*L. inhospes*	(31)
	L. indica	(119)
L. kalmii	*L. lygaeorum*	(106)
	C. oncopelti	(106)
L. kalmii angustomarginalis	*Leptomonas* sp.	(107)
L. militaris	*L. lygaei*	(4, 110)
	L. bakeri	(119)
L. pandurus	*B. familiaris*	(142)
L. saxatilis	*Leptomonas* sp.	(40)
	L. familiaris saxatilis	(117)
Nysius euphorbiae	*L. davidi*	(80)
Oncopeltus cingulifer	*Leptomonas* sp.	(106)
Oncopeltus famelicus	*Leptomonas* sp.	(125)
	Leptomonas sp.	(150)
	B. familiaris	(150)
O. fasciatus	*C. acidophili*	(96)
	L. oncopelti	(9, 94, 106)
	H. elmassiani	(68)
Oncopeltus varicolor	Unnamed	(129)
Oncopeltus sp.	*Leptomonas* sp.	(106)
Orthaea bilobata (= *Pachybrachius bilobata*)	*B. ortheae*	(144)
O. vincta (= *Pachybrachius vincta*)	*B. ortheae*	(144)
Oxycarenus hyalinipennis	*L. pierantoni*	(3)
O. lavaterae	*L. pierantoni*	(3)
	B. oxycareni	(39)
Pyrrhocoridae		
Cenaeus carnifex	*B. familiaris*	(57)
	Leptomonas sp.	(57)
Dysdercus nigrofasciatus	*Leptomonas* sp.	(122, 153)
Dysdercus spp.	Unnamed	(129)
D. superstitiosus	*Leptomonas* sp.	(12)
D. suturellus	*L. seymouri*	(153)
Euryophthalmus convivus	*B. euryophthalmi*	(91)
E. davisi	*C. harmosa*	(96)
	Blastocrithida sp.	(92)
	Crithidia sp.	(92)
Euryophthalmus sp.	Unnamed	(129)
Odontopus nigricornis	*L. pyrrhocoris*	(147)
Pyrrhocoris apterus	*Leptomonas* sp.	(82)
	L. pyrrhocoris	(43, 55, 56, 83, 141, 157)

TABLE 12.2. *Continued.*

Insect	Trypanosomatid	Reference
Coreidae		
Anisosceslis foliacea	Unnamed	(129)
Anoplocnemis sp.	*Blastocrithidia* sp.	(131)
Athaumastus sp.	Unnamed	(129)
Camptopus lateralis	*L. tortum*	(118)
Catorhintha selector	*Leptomonas* sp.	(107)
Chariesterus cuspidatus	*Leptomonas* sp.	(136)
Chelenidea vittiger quadripustula	*Leptomonas* sp.	(107)
Cletus bisbipunctatus	*Leptomonas* sp.	(125)
Cletus lituripennis	*Leptomonas* sp.	(125)
C. ochraceus	*L. capsularis*	(59)
C. varius	*B. cleti*	(66)
Coreus marginatus	*B. raabei rostrata*	(113)
Coricinus elongator	Unnamed	(129)
Corizus sp.	Unnamed	(129)
Diactor bilineatus	Unnamed	(129)
Fabrictilis gonagra	Unnamed	(129)
Holymenia clavigera	Unnamed	(129)
Holymenia histrio	Unnamed	(129)
Hypselonotus fulvus	Unnamed	(129)
Leptocoris trivittatus	*B. leptocoridis*	(90)
Leptoglossus membranaceus	*Leptomonas* sp.	(122)
L. phyllopus	*L. leptoglossi*	(62)
	Leptomonas sp.	(107)
Megalotomus sp.	Unnamed	(129)
Mesocerus marginatus	*B. raabei*	(84)
Mirperus jaculus	*L. mirperi*	(125)
Phthia picta	Unnamed	(129)
Sphictyrtus chryseis	Unnamed	(129)
Stalinifera cincta	Unnamed	(129)
Stenocephalus agilis	*Leptomonas* sp.	(52)
Veneza chilensis	Unnamed	(129)
Veneza ingens	Unnamed	(129)
Veneza sp.	Unnamed	(129)
Veneza stigma	Unnamed	(129)
Veneza zonata	Unnamed	(129)
Corimelaenidae		
Corimelaena sp.	*C. corimelaena*	(85)
Pentatomidae		
Acanthocephala femorata	*Blastocrithidia* sp.	(62)
	C. acanthocephali	(62)
Acrosternum impicticorne	Unnamed	(129)
Arvelius albopunctatus	Unnamed	(129)
Aspongopus viduatus	*L. aspongopi*	(1)
Brochymena parva	*Leptomonas* sp.	(107)
B. quadripustula	*Leptomonas* sp.	(107)
Carbula jipensis	*Leptomonas* sp.	(122)
Coenus delius	*Leptomonas* sp.	(61)
Edessa leucogramma	Unnamed	(129)

(*continued*)

TABLE 12.2. *Continued*

Insect	Trypanosomatid	Reference
Erythesina fullo	*Leptomonas* sp.	(19)
	Leptomonas sp.	(151)
	B. erythesinarum	(104)
Eurydema oleracea	flagellate	(44)
E. ornatum (= *Pentatoma ornata*)	*Blastocrithidia* sp.	(44)
E. ornatum pectorale	*Blastocrithidia* sp.	(44)
E. ventralis	*B. vagoi*	(143)
Euschistus conspersus	*Crithidia* sp.	(86)
	C. euchisti	(85)
Euchistus sp.	Unnamed	(129)
E. servus	*B. euschisti*	(62)
	C. epedana	(96)
Graphosoma lineatum var. *italicum*	*Blastocrithidia* sp.	(40)
Holopterna alata	*B. sandoni*	(58)
Loxa sp.	Unnamed	(129)
Nezara viridula	*L. serpens*	(60)
Pentatoma junipera	*Blastocrithidia* sp.	(40)
Peribalus limbolarius	*Leptomonas* sp.	(61)
Piezodorus guildini	Unnamed	(129)
Podisus maculiventris	*Leptomonas* sp.	(61)
Psacasta cerinthi (= *Pentatoma sacastia*?)	*Blastocrithidia* sp.	(40)
Solubea pugnax	*Leptomonas* sp.	(107)
Vulsirea sp.	Unnamed	(129)
Miridae		
Calocoris sexgutatus	*Proteomonas inconstans*	(116)
Lygocoris lucorum	*B. miridarum*	(115)
Adelphocoris quadripunctatus	*B. miridarum*	(115)
Deraeocoris ruber	*B. miridarum*	(115)
Notostira elongata	*B. miridarum*	(115)
Stenodema calcaratum	*B. miridarum*	(115)

[a] Most authors identified to genus exclusively on morphological basis. It is possible that some *Leptomonas* spp. actually are *Phytomonas* spp.

cysts of trypanosomatids survive in the cement binding the eggs together. Homoptera lay their eggs separately and there is no binding cement so this means of transmission does not exist. Coprophagy, however, cannot be discarded as an important means of trypanosomatid dispersion.

Mouthparts and Salivary Glands

The mouthparts of Hemiptera and Homoptera all have the same general plan. The paired stylets are the maxillae and the mandibles. The maxillae are fastened together by two interlocking grooves and ridges along their length. Between them is a relatively large tubular opening, the food channel, and a much smaller salivary tube. The mandibles are slender and armed with backward pointing barbs at the tip. The mandibles and maxillae are enclosed in the fleshy, jointed labium, which is rolled into a tube whose sides come together dorsally. During feeding the labium folds back in the middle while the tip still encloses the stylets. The stylets extend beyond the tip of the labium while the middle of the labium folds away from the stylets. The mandibles touch and probe the surface of the plant while sense organs detect a suitable spot to pierce. The mandibles pierce the surface and then move alternately back and forth, moving into the tissues and bending to find a route to a specific structure, for example, seed, phloem, xylem, or cell. Most studies on the act of piercing relate to Homoptera, which are remarkably precise in selecting target structures.

In *Oncopeltus*, a typical lygaeid, the salivary gland, which fills most of the thorax, consists of three lobes that connect to a common duct. There is an anterior lobe and a deeply bifurcated posterior lobe. The third lobe, which extends over the other two, was called dorsal by Holmes (68) and lateral by Miles (103). We will follow Miles' terminology. The accessory gland is a long tubular gland looped under the salivary gland.

According to Miles the sheath material is formed by two components that come from the anterior and the lateral lobes plus an oxidase from the accessory gland. This sheath material hardens to form the stylet sheath, which surrounds the hole made by the stylets and a collar around the point of insertion. The stylet sheath remains after the mouthparts are withdrawn, constituting a record of the route and termination of the stylet path.

Evidence Linking Hemiptera to Transmission of *Phytomonas* of Latex Plants

Lafont (80) found the lygaeid, *Nysius euphorbiae*, on the plants harboring *P. davidi* and he caused experimental transmission to seedling plants with naturally infected insects. He allowed naturally infected insects to feed on an uninfected branch of an infected plant and also on an uninfected plant and the plants became infected as demonstrated by microscopic examination. In the same year Bouet and Roubaud (14) found flagellates in *Euphorbia hirta* and in the lygaeid, *Dieuches humilis* on the same plants. The authors experimentally transmitted the infection from naturally infected insects to wild plants that were microscopically free from infection. França in Portugal (51–52) reported extensive studies on the transmission of *P. davidi* in *Euphorbia segetalis* and *E. peplus* by the coreid *Dicranocephalus agilis*. In earlier years he had failed to find any putative vector on infected plants although he found signs of insect piercing. In 1919 he found that the insects were active only in the evening and then proceeded to trace the development in the insect. He collected insects in a locality where no infected plants were found, allowed them to feed on infected *Euphorbia* and traced the infection in the insect. The essentials of his conclusions have been confirmed except that he described cysts in the flagellate, which we now know to belong to a monoxenous parasite of the insect.

Strong (136) found flagellates in the coreid *Chariesterus cuspidatus* on *E. hypericifolia*, *E. pilulifera*, and *E. callitrichoides* (= *Chamaesyce prostata*). The insects were seen only in the evening. Wild-caught bugs that were from a locality with uninfected plants were allowed to feed on infected plants and the bugs became infected.

While these early experiments leave little doubt as to the identity of the insect vectors they were deficient in controls and in numbers of plants and insects in the experiments.

Holmes (68, 70) discovered the first *Phytomonas* in North America (*P. elmassiani* in *Asclepias syriaca*, the milkweed) and studied the infection and its transmission in detail. He used *A. curassavica*, a tropical American species, as experimental host as it is well adapted to the greenhouse and may live there for years. He infected greenhouse seedling *A. curassavica* with naturally infected *Oncopeltus fasciatus*. He also transmitted infection to laboratory-reared bugs from infected plants. Although he performed transmissions under controlled conditions, he described cysts, which we now know belong to another flagellate, *L. oncopelti*. He noted that the bugs feed mostly on flowers and fruit and raised the question of infection of seeds but found no need for that hypothesis to account for overwintering.

Vickerman (150) studied *P. elmassiani* in the asclepiad *Pergularia extensa* and the insect *Oncopeltus famelicus* in East Africa. He described the development in the insect and recognized the presence of a species of *Blastocrithidia* and a *Leptomonas* in the same insect. No experimental infections were made.

McGhee and Hanson (95) used many of the methods and materials of Holmes and demonstrated the transmission of *P. elmassiani* to *A. curassavica* by *O. fasciatus*, using laboratory-reared plants and insects. They demonstrated that there is another parasite in the insect, *L. oncopelti*, that cannot be distinguished in many promastigote stages from *P. elmassiani* but produces cysts, attached to the flagella and then freed. Hanson et al. (63) using laboratory-reared plants and insects found that *P. elmassiani*, originating in *A. syriaca*, would infect other members of the Asclepiadaceae namely *Stephanotis floribunda*, *Cynanchum scoparum*, and *C. palustra*.

The above work on parasites of Asclepias was confirmed by Ayala et al. (8) using laboratory-reared *A. curassavica*, and *A. fruticosa*, with *O. cingulifer* and *O. unifasciatellus*. This was again confirmed by Urueta (145) using *A. curassavica* and *O. unifasciatellus*.

Studies on transmission of the Euphorbiaceae were renewed by McGhee and Postell (99) who infected laboratory-reared seedling *Chamaesyce hypericifolia* and *E. pilulifera* (= *hirta*) with naturally infected *Pachybrachius bilobata scutellatus* (Lygaeidae). Kastelein (74) infected *E. hirta* and *E. hyssopifolia* plants from an uninfected place, and also clean seedlings with naturally infected *P. bilobata* in Suriname.

Dollet et al. (27) demonstrated the transmission of *P. davidi* from naturally infected *Dicranocephalus agilis* to laboratory-reared seedling *Euphorbia pinea*.

Cecropia palmata is a small tree, family Cecropiaceae, which has laticifers in the bark. *Phytomonas* was found in the latex. The pentatomid, *Edessa loxdali*, is found feeding on these trees and a large proportion of them are infected with flagellates in the salivary glands in Suriname. Kastelein (74) put 4- to 10-week old seedlings of *Cecropia* grown in a greenhouse in an insect-proof cage and liberated naturally infected *Edessa loxdali*, from a population in which 51% harbored flagellates in the salivary glands in the cage. In three separate experiments all seedlings exposed to the insects became infected while those in a control cage were free from infection. Although *Cecropia* are common around the edges of coconut plantations where the disease, hartrot, occurs and had been suspected of being the source of palm infections, *E. loxdali* were never found feeding on coconut trees. One coconut palm exposed to infected *E. loxdali* in a cage remained free from infection.

Habits of Vectors of Latex *Phytomonas*

Many of the phytophagous Hemiptera feed on seeds. They find seeds inside the fruit with their stylets, inject saliva, and then suck out the partially digested seed contents. Immature seeds are used by many Coreidae, Pyrrhocoridae, Cydnidae, Lygaeidae, and occasionally Pentatomidae, Scutelleridae, and Miridae. Mature seeds are taken by different insects of families Pyrrhocoridae and Lygaeidae. The Lygaeidae are so predominantly seed eaters that they have been termed seed bugs (137). Some have raptorial fore legs but they are used to hold seeds rather than animal prey. Slansky and Panizzi (132) refer to seed-eating Hemiptera as seed suckers.

All of the families represented among vectors of *Phytomonas* include seed suckers. There is very little information on the feeding habits of most known vectors. *Dicranocephalus agilis*, vector of *P. davidi* in *Euphorbia* in Europe is said to feed on fruits of *Euphorbia*. *Pachybrachius* also feeds on fruit and seeds (138–139).

Latex or milky juice is a viscous white liquid that occurs in some plants. It is secreted in vacuoles in certain specialized elongate, branched cells, the latex cells. In the Asclepiadaceae, Euphorbiaceae, Apocynaceae, and Urticaceae, the cells, which may be many centimeters in length, are unfused or separate from one another. In the Compositae and some other families the cells are fused to form a continuous network. The functions of latex are not known with certainty but there is considerable evidence that it protects the plant from phytophagous animals (33).

Although all of the workers on transmission of latex flagellates state unequivocally that the vector insects suck latex we have found no explicit published evidence to that effect. In the literature on insect nutrition we have found no mention of latex.

The large milkweed bug, *Oncopeltus fasciatus*, is the subject of many nutritional studies and may be taken as typical of vectors of *Phytomonas* of laticiferous plants. Seeds are essential to the development of *O. fasciatus* and are sufficient as the insects can be reared indefinitely on dried seeds of the milkweed and water. Sap of green plants is not essential but it may be used as source of water. It is hard to believe that the tiny first stage nymphs can pierce the seed pods and reach developing seeds of the milkweed but we are assured by Blakely (13) that they do. They find the line of attachment of seeds where they are closest to the surface of the seed pod and reach them with their stylets. Nymphs will not complete their development on young plants without seed pods. The seasonal cycle of the insect is coordinated with that of the host plants, eggs being laid only when the plants are fruiting.

Some tropical species of Asclepias are less definitely seasonal with populations being found at various stages of the life cycle at the same time. And a tropical species of milkweed bug (*O. cingulifer*) can complete its development on nonfruiting plants (13, 21).

We have found only one study of the stylet path and feeding site of *Oncopeltus* on the milkweed plant. Miles (102) pictured a section through the stylet path of *O. fasciatus* and stated that the stylets penetrated the phloem. It is possible, however, that the stylets of Hemiptera are much less precise in their action than those of Homoptera and that the Hemiptera ingest a mixture of phloem, xylem, cell sap, and latex (when present).

The question remains, do *Oncopeltus* suck latex by choice and if so why? The answer may be that they take latex for the contained toxins (glycosides or cardenolides) that presumably protect them from predators. Cardenolides are present in seeds and in latex, which the insects may use when on nonfruiting plants. Like many other phytophagous insects, they take enough of the toxin to carry a sublethal dose for a predator, which survives the ingestion of an insect, chastened and wiser. Accompanying this toxin is bright warning (aposematic) coloration that protects the insect from its enemies but at the same time betrays its presence to the investigator (32, 130). According to Scudder and Duffey (130) most genera of Lygaeidae include species with cardenolides. *Oncopeltus* and *Lygaeus* concentrate cardenolides in dorsolateral glands where the concentration is higher than in the hemolymph.

Discovery of Transmission of Phloem-Infecting *Phytomonas*

Desmier de Chenon (22) in Ecuador first identified the vector of palm *Phytomonas*. Among the several genera of Hemiptera found on palms, attention was directed to *Lincus* (Pentatomidae), which was most closely associated with diseased oil palms. Trees were dissected and all bugs collected. *Lincus* were found on all recently infected trees and on some healthy trees. Healthy trees on which some bugs had been found became sick in 2.5–3.5 months. On trees with old infections no bugs were found, presumably because phloem does not flow.

Field type experiments were then conducted. *Lincus lethifer* Dolling (30) from a focus of infection were put on 5 trees (200–500 each) in an area where no disease was found. Two of the five trees showed disease in 95 to 101 days. In another experiment a single tree had 500 bugs from an infection focus put on it. In 125 days the adjacent tree showed disease and in 180 days the experimental tree was diseased. In 1985 Perthuis, Desmier de Chenon and Merland (111) reported further experiments. One tree was enclosed in a cage in an area where no disease occurred (there was a small focus 450 m from the cage). Surrounding trees for 100 m were treated with endrin once a month. Bugs collected in the focus of infection were identified to genus but not species. A total of 64 adults and 182 nymphs were introduced in six occasions in last 6 months of 1984. In February 1985 symptoms appeared in the tree. Adults were identified as *L. lethifer*. Nymphs were not identified.

Lincus are dull colored pentatomids that hide in crevices between bases of palm fronds and are crepuscular in habits. Apparently they do not fly but migrate from tree to tree on the ground. Eggs are laid in groups of 16–18. They hatch in 7–9 days and pass through 5 nymphal stages within 2 months. Adults live about a month.

Rezende et al. (121) studied the transmission of *Phytomonas* by *Lincus lobuliger* in Bahia Brazil. They made two cages and planted a small oil palm and a small coconut palm, 1.5 years old in each. In the experimental cage 12 *Lincus* (8 adults and 4 nymphs) from diseased coconut palms were introduced once a month for 6 months. In the control cage no bugs were introduced. In 5 months the experimental oil palm showed symptoms and *Phytomonas* were found in sap from roots and bulb. This indicates that marchitez and hartrot are the same and should be called *Phytomonas* wilt.

Transmission of hartrot was studied in French Guiana and Suriname. Louise et al. (87) planted 16 coconut palms in each of 2 cages. Insects were collected from foci of infection, from both well and sick trees, and released, all on one tree at one release. In a year 2873 nymphs and 832 adults were released. The vector was not identified to species but three species, *L. croupius*, *L. apollo*, and *L. dentiger*, were included in the experiment. The first appearance of disease came 4 months after the release of bugs on the tree in question. Other trees showed disease with incubation periods of 6–8 months. In 16 months 11 trees were dead in the experimental cage while there was no disease in the control cage. *Lincus croupius* (the major candidate) lives in crevices between bases of fronds in the milieu or debris that accumulates. This disappears in the older crevices but accumulates to favorable amounts only in 4–5 years. Bugs do not move around much as determined by observation of marked individuals. In Suriname *Lincus* were transferred from diseased to healthy coconut palms and disease resulted. The salivary glands of 24% of the bugs from diseased trees had flagellates. Two species, *L. vandoesburgi* and *L. lamelliger*, occurred but were not distinguished during the experiments noted above (5).

Discovery of Transmission of Fruit-Infecting *Phytomonas*

In South Africa, Gibbs (60) found flagellates in adults of the Pentatomid *Nezara viridula* and proceeded to infect tomatoes with these flagellates that he named *Leptomonas serpens*. The transmission of *Phytomonas serpens* was later studied in detail by Jankevicius et al. (71), who were able to infect clean, lab-raised nymphs of the coreid *Phthia picta* with flagellates from infected tomatoes and then infect clean, lab-grown tomatoes with the infected bugs. Infected *Phthia* collected in the wild also infected clean tomatoes. Cultures were derived either from bugs or tomatoes and were used to infect clean, lab-raised fruits and bugs (Fig. 12.2).

Experimental infection of fruits of Solanaceae (tomatoes, red pepper, and fruits of *Solanum erianthum* and *S. viarum*) was achieved by Kastelein and Camargo (75) through the biting of infected pentatomids, *Arvelius albopunctatus*. Infection of this bug from infected tomatoes was also accomplished.

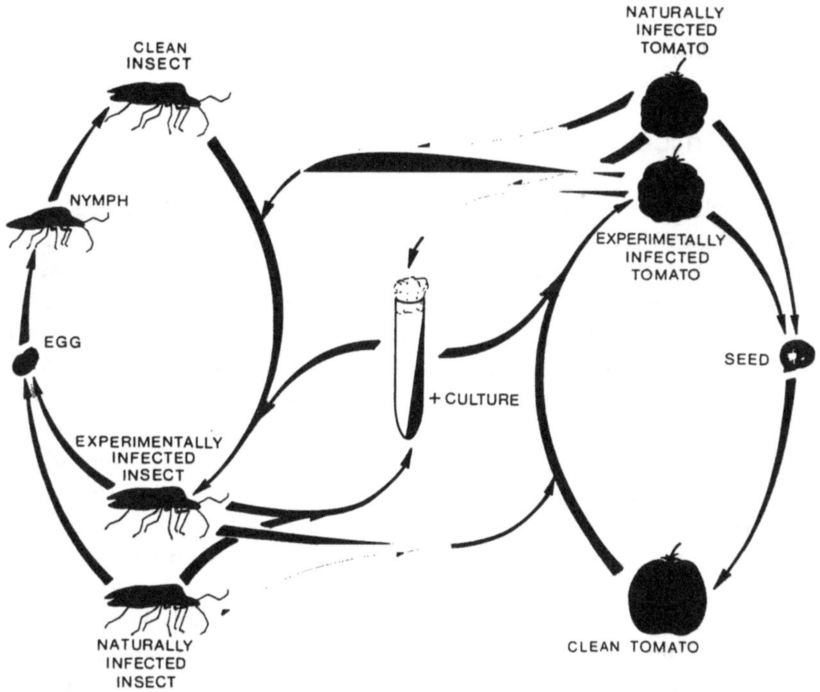

FIGURE 12.2. Experiments involving the tomato/bug/culture cycle of *Phytomonas serpens*.

Discovery of *Phytomonas* in Insects

It is impossible to know with absolute certainty which species belong to the genera *Leptomonas*, *Herpetomonas*, or *Phytomonas* among those listed in Table 12.2 because they all occur as promastigotes in the insect. So, no record before 1989 can be identified as *Phytomonas*.

To clarify the taxonomic status of flagellates occurring in bugs, Sbravate et al. (129) made an extensive survey of insects of the families Coreidae, Lygaeidae, Pentatomidae, and Pyrrhocoridae. Of 53 different species examined, the authors found 28, or 53%, infected with trypanosomatids. Among the 372 individual insects, 133 were positive: a prevalence of 36%. The authors then cultured 52 isolates and tested the cultures for the presence of certain enzymes of the ornithine-arginine metabolism and for antigenicity to monoclonal antibodies (MoAB) raised against *Phytomonas* spp. Cultures not possessing arginase and reacting with anti-*Phytomonas* MoAB were considered to be *Phytomonas*. Sixteen cultures from five species of insects were identified as *Phytomonas* by enzyme and MoAB testing and later confirmed as such by ribosomal DNA restriction analysis (17). Species found harboring *Phytomonas* were

Fabrictilis gonagra, *Holymenia histrio*, *Hypselonotus* sp., *Veneza* sp., and *Veneza zonata*, (family Coreidae).

Cycle in Insect

The life cycle of *Phytomonas* has been described in fived studies. França (52) described *P. davidi* in *Euphorbia peplus* and the insect *Stenocephalus agilis*. Holmes (70) and McGhee and Hanson (95) described *P. elmassiani* in the milkweed *Asclepias* and the coreid *Oncopeltus fasciatus*. Vickerman (150) described *P. elmassiani* in *Pergularia extensa* (Asclepiadaceae) and *Oncopeltus famelicus*. Freymuller et al. (54) and Jankevicius et al. (71) described the development of *P. serpens* in the coreid bug, *Phthia picta*.

These descriptions agree with each other in most respects, the principal difference being that McGhee and Hanson separated the *Phytomonas* from *Leptomonas oncopelti* in the same insect while some other authors attributed stages of *Leptomonas* (mainly the cysts) to *Phytomonas*.

The following description is taken from McGhee and Hanson's account. In latex, *P. elmassiani* averages 13.5 × 2.5 μm. Dividing forms are seen and there are a few amastigotes. After 4 days in the insect, forms similar to those in the plants, except that none is dividing, are found in the pylorus. By the sixth day there is some elongation. In 10 days giant forms are found in the pylorus. At this time similar stages first appear in the hemocoel. After 12 days parasites are found in the salivary glands. These are giant forms. After 20 days giant and small forms are both found in the salivary glands. After 26 days only small forms appear in the salivary glands. Holmes says that in *Oncopeltus* the parasites are found mainly in the lateral lobe (called dorsal by Holmes). Here they are in masses forming a layer lining the lumen. Smaller numbers are found in the anterior lobe while none are seen in the posterior lobe, no matter how heavily infected the lateral lobe is. Giant forms are up to 80 μm long.

In the crop and midgut of newly infected bugs the two parasites, *Phytomonas* and *Leptomonas*, are indistinguishable. In the pylorus, cysts are attached to the flagella of *L. oncopelti*. *P. elmassiani* migrates to the hemocoel where the promastigotes increase in length to attain 80 μm. These giant forms are found in the salivary glands along with small elongate aflagellate forms approximately the size of those in the latex. The route of migration into the hemocoel and into the salivary glands was not observed. The small metacyclic stages are injected into the plant host.

Development of *P. serpens* in the bug *P. picta* also was followed by light and electron microscopy by Freymuller et al. (54). Flagellates colonized the digestive tract of the insect, gained their hemocoel, and invaded their salivary glands where they could be found inside the cells and in the intercellular space (Fig. 12.3).

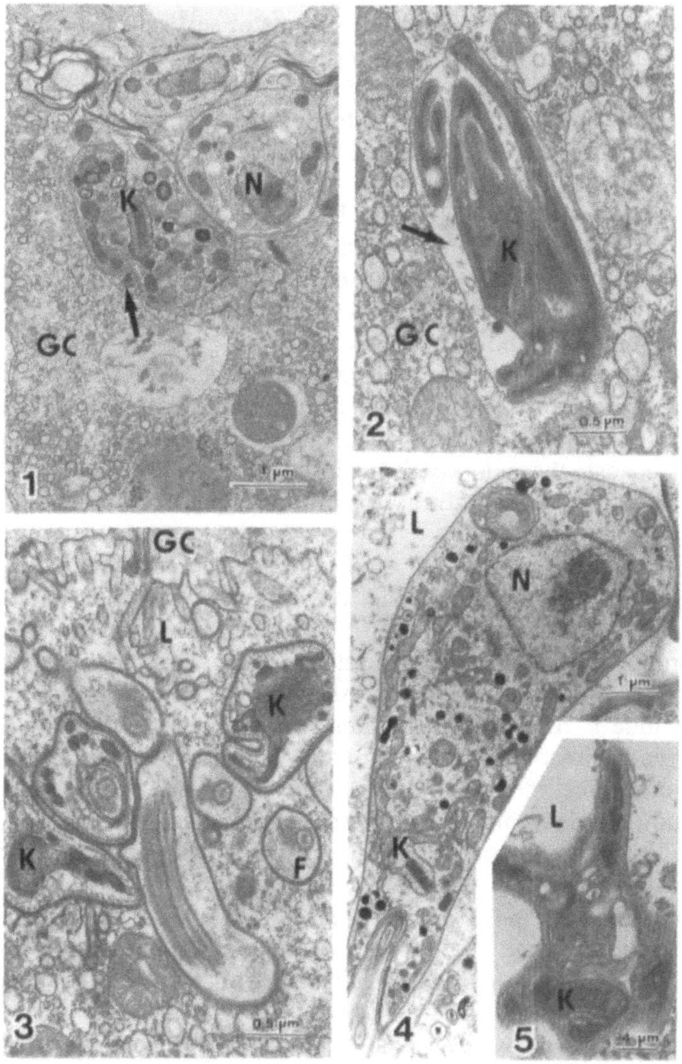

FIGURE 12.3. Intra- and extracellular forms of *Phytomonas serpens* in the salivary glands of *Phthia picta*, its coreid vector. GC, gland cell; L, salivary gland lumen; K, kinetoplast; N, nucleus; F, flagellum. (Originals from Dr. Edna Freymuller.)

Opportunities for Future Work

Vectors of more laticiferous *Phytomonas* should be identified using laboratory (greenhouse-reared) plants and colonies of flagellate-free insects. While there is satisfactory demonstration of the identities of vectors of *Phytomonas* of some common Euphorbias and Asclepias there is no information on vectors for latex plants of several other families. We should not now be satisfied with less than controlled experiments with lab-reared insects and lab-reared plants.

The habits of vectors of laticiferous *Phytomonas* require further studies. Nutritional studies of seed-sucking lygaeids and coreids focus entirely on seeds. Do they suck latex by choice? Stomach contents of insects taken from latex plants and study of stylet paths in sectioned plants are required to demonstrate whether transmission of latex flagellates is due to latex sucking or is incidental to phloem feeding. Occasional flagellates have been seen in phloem and xylem of plants whose latex was infected.

Because of the importance and ubiquity of seed sucking in vectors of *Phytomonas* seeds and fruit of their hosts should be searched assiduously for flagellates. Holmes said that seeds and fruit of milkweed could not be infected because there is no connection between the laticifers and the fruit. But he apparently did not take into account the seed-eating habit of the vectors.

The investigation of possible seed infection in laticiferous plants could well begin with examination of the saliva injected by an infected bug into a seed. The contents of the lateral lobe (described by Holmes) are so heavily infected that flagellates should be evident in the seed if that lobe is the source of the saliva. On the other hand the watery secretion of the posterior lobe is free from flagellates and if that is the material injected into the seed, infection of the seed would be most unlikely.

The biochemical identification of *Phytomonas* isolated in cultures from insects should be exploited in the direction of identifying the species with those isolated from plants, using molecular and immunological methods to identify insect stages with stages from plants.

Transmission of palm *Phytomonas* offers many opportunities for further research. Identity of vectors should be restudied using lab-reared insects. The trypanosomatids should be found and their development traced in the insects and in the plants.

Reservoir hosts of palm flagellates are yet to be found. Are they native palms in South American forests? The suspicion that weeds and cover crops in coconut plantations are sources of infection should be settled by accurate identification and comparison by molecular means of isolates from the various plants and from insect vectors.

References

1. Aders, W.M. 1909. *Herpetomonas aspongopi. Parasitology* **2**:202–207.
2. Alva, Z.B. 1956. Flagelados en el latex de la *Jatropha macrantha (Huanarpo hembra). Arch. Per. Patol. Clin.* **10**:1–12.
3. Antonucci, A. 1941. Sopra un nuovo *Leptomonas* degli emitteri Legeidi (*Leptomonas pierantonii* n.sp.). *Arch. Zool. Ital.* **29**:63–75.

4. Archibald, R.G. 1911. *Herpetomonas lygaei* considered from a protozoological and medical aspect. *Fourth Report Wellcome Tropical Research Laboratories, Karthoum*, Bailliére, Tindall, and Cox, London, pp. 179–184.
5. Asgarali, J., and Ramkalup, P. 1985. Study of *Lincus* sp. (Pentatomidae) as a possible vector of hartrot in coconut. *Surin. Landb.* **33**:56–61.
6. Attias, M., Roitman, I., Camargo, E.P., Dollet, M., and Souza, W. 1988. Comparative analysis of the fine structure of four isolates of trypanosomatids of the genus. *Phytomonas. J. Protozool.* **35**:365–370.
7. Attias, M., and Souza, W. 1986. Axenic cultivation and ultrastructural study of a *Phytomonas* sp. isolated from the milkweed plant. *Euphorbia hyssopifolia. J. Protozool.* **33**:84–87.
8. Ayala, S.C., Quintero, O.B., and Barreto, P. 1974. Tripanosomátideos de plantas laticiferas y sus insectos transmisores en Colombia y Costa Rica. *Rev. Biol. Trop.* **23**:5–15.
9. Barnett, H.C. 1946. The relation of the milkweed bug, *Oncopeltus fasciatus* (Dall.) to flagellosis in laticiferous plants. Thesis, University of Minnesota.
10. Bensaude, M. 1925. Flagellates in plants: a review of foreign literature. *Phytopathology* **15**:273–281.
11. Bezerra, J.L., and Figueiredo, J.M. 1982. Ocorrência de *Phytomonas staheli* McGhee et McGhee em coqueiro (*Cocos nucifera*) no estado da Bahia, Brasil. *Fitopatol. Bras.* **7**:139–143.
12. Blacklock, B. 1923. The aetiology of Kala-Azar and tropical sore. *Lancet* **205**:273–274.
13. Blakely, N. 1980. Divergence in seed resource use among neotropical milkweed bugs (Oncopeltus). *Oikos* **35**:8–15.
14. Bouet, G., and Roubaud, E. 1911. Sur la présence au Dahomey et le mode de transmission de *Leptomonas davidi* Lafont, flagellé parasite des Euphorbiacées. *C.R. Soc. Biol.* **70**:55–57.
15. Camargo, E.P., Coelho, J.A., and Moraes, G. 1978. *Trypanosoma* spp., *Leishmania* spp. and *Leptomonas* spp.: Enzymes of ornithine-arginine metabolism. *Exp. Parasitol.* **46**:141–144.
16. Camargo, E.P., Kastelein, P., and Roitman, I. 1990. Trypanosomatid parasites of plants (*Phytomonas*). *Parasitol. Today* **6**:22–25.
17. Camargo, E.P., Sbravate, C., Teixeira, M.M.G., Uliana, S., Soares, M., Affonso, H., and Floeter-Winter, L. 1992. Ribosomal DNA restriction analysis and synthetic oligonucleotide probing in the identification of genera of lower trypanosomatids. *J. Parasitol.* **78**:40–48.
18. Camargo, E.P., Silva, S., Roitman, I., Souza, W., Jankevicius, J.V., and Dollet, M. 1987. Enzymes of ornithine-arginine metabolism in trypanosomatids of the genus. *Phytomonas. J. Protozool.* **34**:439–441.
19. Carter, R.M. 1911. Non-ulcerating oriental sore: The cultural characteristics of the parasite as compared with a new similar parasite in *Erythesina fullo* (Thunb.), a pentatomid bug. *Ann. Trop. Med. Parasitol.* **5**:15–34.
20. Carter, W. 1973. *Insects in Relation to Plant Diseases*, 2nd ed. John Wiley, New York.
21. Chaplin, S.I. 1973. Reproductive isolation between two sympatric species of *Oncopeltus* (Hemiptera:Lygaeidae) in the tropics. *Ann. Entomol. Soc. Am.* **66**:997–1000.

22. Chenon, R.D. 1984. Recherches sur le genre *Lincus* Stal, Hemiptera Pentatomidae Discocephalinae, et son rôle éventuel dans la transmission de la Marchitez du palmier à huile et du Hart-Rot du cocotier. *Oleagineux* **39**:1–6.
23. Conchon, I., Campaner, M., Sbravate, C., and Camargo, E.P. 1989. Trypanosomatids, other than *Phytomonas* spp., isolated and cultured from fruit. *J. Protozool.* **36**:412–414.
24. De Leeuw, G.T.N., and Kastelein, P. 1987. Observation of flagellate protozoa (*Phytomonas* sp.) in *Cocos nucifera*, *Cecropia palmata* and *Euphorbia hirta*. *J. Phytopathol.* **120**:25–30.
25. Dollet, M. 1982. Les maladies des palmiers et cocotiers à protozoaires flagellés intraphloémiques en Amerique Latine (*Phytomonas* sp., Trypanosomatidae). *Oleagineux* **37**:9–12.
26. Dollet, M. 1984. Plant diseases caused by flagellate protozoa (*Phytomonas*). *Ann. Rev. Phytopathol.* **22**:115–132.
27. Dollet, M., Cambrony, D., and Gargani, D. 1982. Culture axénique in vitro de *Phytomonas* sp. (Trypanosomatidae) d'euphorbe transmis par *Stenocephalus agilis* Scop. (Coreide). *C.R. Acad. Sci. Paris Ser. III* **295**:547–550.
28. Dollet, M., Gargani, D., Cambrony, D., and Dzido, J.L. 1983. Observations of flagellate Protozoa (*Phytomonas* sp.) in the xylem of *Euphorbia lasiocarpa* in Ecuador. *Can. J. Bot.* **61**:237–240.
29. Dollet, M., Giannotti, J., and Ollagnier, M. 1977. Observation de protozoaires flagellés dans les tubes criblés de palmiers à huile malades. *C.R. Acad. Sci. Paris Ser. D* **284**:643–645.
30. Dolling, W.R. 1984. Pentatomid bugs (Hemiptera) that transmit a flagellate disease of cultivated palms in South America. *Bull. Entomol. Res.* **74**: 473–476.
31. Donovan, C. 1909. Kala-azar in Madras, especialy with regard to its connexion with the dog and the bug (*Conorrhinus*). *Lancet* **Nov. 20**: 1195–1196.
32. Duffey, S.S., and Scudder, G.G.E. 1972. Cardiac glycosides in North American Asclepiadaceae, a basis for unpalatability in brightly colored Hemiptera and Coleoptera. *J. Insect Physiol.* **18**:63–78.
33. Dussourd, D.E., and Eisner, T. 1987. Vein cutting behavior:insect counterploy to the latex defense of plants. *Science* **237**:898–901.
34. Fantham, H.B. 1925. Some parasitic Protozoa found in South Africa—VIII. *S. Afric. J. Sci.* **22**:346–354.
35. Fantham, H.B. 1926. Some flagellates in plants. *Trans. R. Soc. Trop. Med. Hyg.* **19**:347–348.
36. Figueiredo, E.N., Yoshida, N., Roitman, I., and Camargo, E.P. 1978. Enzymes of the ornithine-arginine metabolism in the genus. *Crithidia. J. Protozool.* **25**:546–549.
37. Fiorini, J.E., Silva, F.P.M., Brasil, R.P., Roitman, I., Angluster, J., and Souza, W. 1986. Detection of trypanosomatids in *Solanum gilo*, and *Solanum lycopersicum* in Alfenas M.G. Brazil. *Mem. Inst. O. Cruz* **81** (Suppl.):33.
38. Franchini, G. 1922. Sur un flagellé nouveau du latex de deux apocynées. *Bull. Soc. Pathol. Exot.* **15**:109–113.

39. Franchini, G. 1922. Sur un flagellé de Lygaeidae (*Crithidia oxycareni* n.sp.). *Bull. Soc. Pathol. Exot.* **15**:113–116.
40. Franchini, G. 1922. Sur un trypanosome du latex de deux espèces d'euphorbes. *Bull. Soc. Pathol. Exot.* **15**:18–23.
41. Franchini, G. 1922. Flagellés et amibes d'une urticacée éxotique. *Ficus parietalis. Bull. Soc. Pathol. Exot.* **15**:399–405.
42. Franchini, G. 1922. Nouvelles recherches sur les trypanosomes des euphorbes et sur leur culture. *Bull. Soc. Pathol. Exot.* **15**:299–303.
43. Franchini, G. 1922. Au sujet de la culture de l'*Herpetomonas* du *Pyrrochoris apterus. Bull. Soc. Pathol. Exot.* **15**:163–165.
44. Franchini, G. 1922. Flagellose du chou et des punaises du chou. *Bull. Soc. Pathol. Exot.* **22**:164–165.
45. Franchini, G. 1923. Sur les protozoaires des plantes. *Ann. Inst. Pasteur* **37**:879–885.
46. Franchini, G. 1923. Nouvelles recherches sur les protozoaires des plantes à latex. *Bull. Soc. Pathol. Exot.* **16**:655–659.
47. Franchini, G. 1923. Sur un flagellé d'une Asclepiadacée (*Araujia angustifolia*). *Bull. Soc. Pathol. Exot.* **16**:652–655.
48. Franchini, G. 1923. Sur un flagellé particulier d'une urticacée (*Ficus benjamina*). *Bull. Soc. Pathol. Exot.* **16**:650–652.
49. Franchini, G. 1931. Su di flagellato speciale des lattice del *Ficus hochstetteri* (Mig.) A. Rich. *Arch. Ital. Sci. Med. Col.* **12**:637–643.
50. França, C. 1914. La flagellose des Euphorbes. *Arch. Protistenk.* **34**:108–132.
51. França, C. 1919. L'insect transmetteur de *Leptomonas davidi. Bull. Soc. Pathol. Exot.* **12**:513–514.
52. França, C. 1920. La flagellose des euphorbes II. *Ann. Inst. Pasteur* **34**:432–465.
53. França, C. 1921. Sur deux phytoflagellés. (*L. elmassiani* et *L.bordasi* sp.n.). *Ann. Soc. Belg. Med. Trop.* **1**:245–254.
54. Freymuller, E., Milder, R., Jankevicius, J.V., Jankevicius, S.I., and Camargo, E.P. 1990. Ultrastructural studies on the trypanosomatid *Phytomonas serpens* in the salivary glands of a phytophagous hemipteran. *J. Protozool.* **37**:225–229.
55. Galli-Valerio, B. 1914. Notes de parasitologie et de technique parasitologique. *Z. Bakter. Parasitenk. Infektion.* **75**:46–53.
56. Galli-Valerio, B. 1920. Le cycle evolutif probable de *l'Herpetomonas pyhrrhocoris* Zotta et Galli-Valerio (Note préalable). *Schweiz. Med. Wochenschrift* **21**:401–402.
57. Gibbs, A.J. 1950. *Crithidia familiaris* n.sp. in *Cenaeus carnifex* Fabr. (Hemiptera). *Parasitology* **40**:322–327.
58. Gibbs, A.J. 1951. *Crithidia sandoni* sp.nov., in *Holopterna alata* (Hemiptera). *J. Parasitol.* **37**:587–593.
59. Gibbs, A.J. 1951. *Leptomonas capsularis* n.sp. and other flagellates parasitic in *Cletus ochraceus* (Hemiptera). *Parasitology* **41**:128–133.
60. Gibbs, A.J. 1957. *Leptomonas serpens* n.sp., parasitic in the digestive tract and salivary glands of *Nezara viridula* (Pentatomidae) and in the sap of *Solanum lycopersicum* (tomato) and other plants. *J. Parasitol.* **47**:297–303.
61. Glasgow, H. 1914. The gastric caeca and the caecal bacteria of Heteroptera. *Biol. Bull., Marine Biol. Lab. Woods Hole* **26**:101–155.

62. Hanson, W.L., and McGhee, R.B. 1961. The biology and morphology of *Crithidia acanthocephali* n.sp., *Leptomonas leptoglossi* n.sp. and *Blastocrithidia euchisti* n.sp. *J. Protozool.* **8**:200–204.
63. Hanson, W.L., McGhee, R.B., and Blake, J.D. 1966. Experimental infection of various latex plants of the family Asclepiadaceae with *Phytomonas elmassiani. J. Protozool.* **13**:324–327.
64. Harvey, R.B., and Lee, S.B. 1943. Flagellates of laticiferous plants. *Plant Physiol.* **18**:633–655.
65. Hewitt, R. 1940. Phytomonad flagellates from mexican milkweeds. *J. Parasitol.* **26**:160.
66. Hindle, E., and Lewis, M.A. 1912. Note on *"Crithidia" cleti* n.sp., parasitic in the alimentary canal of *Cletus varius*, Dall. *Parasitology* **5**:109–113.
67. Holmes, F.O. 1924. Herpetomonad flagellates in the lates of milkweed in Maryland. *Phytopathology* **14**:146–151.
68. Holmes, F.O. 1925. The relationship of Herpetomonas elmassiani (Migone) to its plant and insect hosts. *Biol. Bull., Marine Biol. Lab. Woods Hole* **49**:323–337.
69. Holmes, F.O. 1931. *Herpetomonas bancrofti* n.sp. from the latex of a ficus in queensland. *Contrib. Boyce Thompson Inst. Plant Res.* **3**:375–384.
70. Holmes, F.O. 1931. Protozoa of latex plants. In R. Hegner and J. Andrews, eds., *Problems and Methods of Research in Protozoology*. MacMillan, New York, pp. 257–275.
71. Jankevicius, J.V., Jankevicius, S.I., Campaner, M., Conehon, I., Maeda, L.A., Teixeira, M.M.G., Freymuller, E., and Camargo, E.P. 1989. Life cycle and culturing of *Phytomonas serpens* (Gibbs), a Trypanosomatid parasite of tomatoes. *J. Protozool.* **36**:265–271.
72. Jankevicius, S.I., Jankevicius, J.V., Menezes, M.C.M.D., Lima, H., and Menezes, J.R. 1987. Presence of Protozoa of the genus *Phytomonas* in leguminous crops. *Fitopatol. Bras.* **12**:152.
73. Jankevicius, S.I., Jankevicius, J.V., Menezes, M.C.M.D., Torrezan, H.C., Menezes, J.R., and Rezende, M.I. 1987. *Phytomonas* sp. found in leguminous crops. *Mem. Inst. Oswaldo Cruz* **82** (Suppl.):35.
74. Kastelein, P. 1985. Transmission of *Phytomonas* sp. (Trypanosomatidae) by the bug *Edessa loxdali* (Pentatomidae). *Surin. Landb.* **33**:62–64.
75. Kastelein, P., and Camargo, E.P. 1990. Trypanosomatid protozoa in fruit of Solanaceae in Southeastern Brazil. *Mem. Inst. Oswaldo Cruz* **85**:413–417.
76. Kastelein, P., and Parsadi, M. 1984. Observations on cultures of the Protozoa *Phytomonas* sp. (Trypanosomatidae) associated with the laticiferous *Allamanda cathartica* L. (Apocynaceae). *Surin. Landb.* **32**:85–89.
77. Kastelein, P., and Parsadi, M. 1988. Axenic cultivation of *Phytomonas* (Trypanosomatidae) associated with laticiferous plants in Suriname. *J. Protozool.* **35**:533–536.
78. Lafont, A. 1909. Sur la présence d'un Leptomonas, parasite de la classe des flagellés dans le latex de l'*Euphorbia pilulifera. C.R. Soc. Biol.* **66**:1011–1013.
79. Lafont, A. 1910. Sur la présence d'un *Leptomonas*, parasite de la classe des flagellés, dans le latex de trois euphorbiacées. *Ann. Inst. Pasteur* **24**:205–219.

80. Lafont, A. 1911. Sur la transmission du *Leptomonas, davidi* des euphorbes par un hémiptère, *Nysius euphorbiae*. *C.R. Soc. Biol.* **70**:58–59.
81. Leger, A. 1911. Présence de *Leptomonas*, *davidi* Lafont dans l'*Euphorbia pilulifera* du Haut-Senegal et Niger. *Bull. Soc. Pathol. Exot.* **4**:626–627.
82. Léger, L., and Duboscq, O. 1910. *Selenococcidium intermedium* Lég. et Dub. et la systematique des sporozoaires. *Arch. Zool. Exp. Gen.* **40**:187–238.
83. Lipa, J.J. 1958. Pierwotniaki zyjace w roslinach (Protozoan living in plants). *Wszechswiat* **1**:13–15.
84. Lipa, J.J. 1966. *Blastocrithidia raabei* sp.n., a flagellate parasite of *Mesocerus marginatus* L. (Hemiptera: Coreidae). *Acta Protozool.* **4**:19–23.
85. Lipa, J.J. 1968. Some observations on flagellate parasites of hemipterans *Corimelaena*, *Euchistus*, *Gerris*, *Leptocoris* and *Oncopeltus* in the United States. *Acta Protozool.* **6**:58–69.
86. Lipa, J.J., and Steinhaus, E.A. 1962. Further report on identifications of Protozoa pathogenic for insects. *Acta Parasitol. Polonica* **10**:165–175.
87. Louise, C., Dollet, M., and Mariau, D. 1986. Recherches sur le hartrot du cocotier, maladie à *Phytomonas* (Trypanosomatidae) et sur son vécteur *Lincus* sp. (Pentatomidae) en Guyane. *Oleagineux* **41**:437–449.
88. Lutz, A., Araujo, H.C.S., and Fonseca Filho, O. 1918. Viagem científica no rio Paraná e a Assunción com volta por Buenos Aires, Montevideo e Rio Grande. *Mem. Inst. Oswaldo Cruz* **10**:104–173.
89. McCoy, R.E., and Martinez-Lopez, G. 1982. *Phytomonas staheli* associated with coconut and oil palm diseases in Colombia. *Plant Dis.* **66**:675–677.
90. McCulloch, I. 1915. An outline of the morphology and life history of *Crithidia leptocoridis*, sp.nov. *Univ. Calif. Public. Zool.* **16**:1–22.
91. McCulloch, I. 1917. *Crithidia euryophthalmi*, sp.nov., from the hemipteran bug, *Euryophthalmus convivus* Stål. *Univ. Calif. Public. Zool.* **18**:75–88.
92. McGhee, R.B. 1959. The infection of avian embryos with Crithidia species and *Leishmania tarentolae*. *J. Infect. Dis.* **105**:18–25.
93. McGhee, R.B., and Cosgrove, W.B. 1980. Biology and physiology of the lower Trypanosomatidae. *Microbiol. Rev.* **44**:140–173.
94. McGhee, R.B., and Hanson, W.L. 1962. Growth and reproduction of *Leptomonas oncopelti* in the milkweed bug, *Oncopeltus fasciatus*. *J. Protozool.* **9**:488–493.
95. McGhee, R.B., and Hanson, W.L. 1964. Comparison of the life cycle of *Leptomonas oncopelti* and *Phytomonas elmassiani*. *J. Protozool.* **11**:555–562.
96. McGhee, R.B., Hanson, W.L., and Schmittner, S. 1969. Isolation, cloning and determination of biologic characteristics of five new species of *Crithidia*. *J. Protozool.* **16**:514–520.
97. McGhee, R.B., and McGhee, A.H. 1971. The relation of migration of *Oncopeltus fasciatus* to distribution of *Phytomonas elmassiani* in the eastern United States. *J. Protozool.* **18**:344–352.
98. McGhee, R.B., and McGhee, A.H. 1979. Biology and structure of *Phytomonas staheli* sp.n., a trypanosomatid located in sieve tubes of coconut and oil palms. *J. Protozool.* **26**:348–351.
99. McGhee, R.B., and Postell, F.J. 1982. Transmission of the trypanosomatid flagellate *Phytomonas davidi*, a symbiont of the Euphorbiaceae, by the

hemipteran bug *Pachybrachius bilobata scutellatus*. *J. Protozool.* **29**:445–448.

100. Menara, A., Dollet, M., Gargani, D., and Louise, C. 1988. Culture *in vitro* sur cellules d'invertebrés, des *Phytomonas* sp. (Trypanosomatidae) associés au hartrot, maladie du cocotier. *C.R. Acad. Sci. Paris* **307**:597–602.
101. Migone, L.E. 1916. Un nouveau flagellé des plants: *Leptomonas elmassiani*. *Bull. Soc. Pathol. Exot.* **9**:356–359.
102. Miles, P.W. 1959. The salivary secretions of a plant sucking bug, *Oncopeltus fasciatus* (Dall.) (Hemiptera: Lygaeidae). I. The types of secretion and their role during feeding. *J. Insect Physiol.* **3**:243–255.
103. Miles, P.W. 1960. The salivary secretions of a plant sucking bug, *Oncopeltus fasciatus* (Dall.) (Heteroptera: Lygaeidae). III. Origins in the salivary gland. *J. Insect Physiol.* **4**:271–282.
104. Morisita, T. 1954. Notes of the haemoflagellate (Trypanosomidae; *Crithidia*) from the bug (*Erythesina fullo*). *Jpn. J. Med. Sci. Biol.* **7**:135–137.
105. Noc, F., and Stevenel, L. 1911. Présence à Martinique de *Leptomonas davidi* Lafont. *Bull. Soc. Pathol. Exot.* **4**:461–464.
106. Noguchi, H., and Tilden, E.B. 1926. Comparative studies of Herpetomonads and Leishmanias. *J. Exp. Med.* **44**:307–325.
107. Packchanian, A. 1957. The isolation and cultivation of hemoflagellates in pure culture from six species of insects. *Texas Rep. Biol. Med.* **15**:399–410.
108. Parthasarathy, M.V., and Slobbe, W.G. 1978. Hartrot or Fatal Wilt of palms. I. Coconuts (*Cocos nucifera*). *Principes* **22**:1–14.
109. Parthasarathy, M.V., Slobbe, W.G., and Soudant, C. 1976. Trypanosomatid flagellate in the phloem of diseased coconut palms. *Science* **192**:1346–1348.
110. Patton, W.S. 1908. *Herpetomonas lygaeae*. *Arch. Protistenk.* **13**:1–18.
111. Perthuis, B., Chenon, R.D., and Merland, E. 1985. Mise en évidence du vecteur de la marchitez sorpresiva du palmier à huile, la punaise *Lincus lethifer* Dolling (Hemiptera, Pentatomidae, Discocephalinae). *Oleagineux* **40**:473–475.
112. Podlipaev, S.A. 1986. *Phytomonas elmassiani* (Matigophora: Trypanosomamonadina) from the plant *Cynanchum sibiricum* (Asclepiadaceae) in Central Asia and Kazakhstan. In Russian. *Proc. Zool. Inst. Acad. Sci. USSR* **144**:61–65.
113. Podlipaev, S.A. 1988. *Blastocrithidia raabei rostrata* subsp.n. (Mastigophora, Trypanosomamonadida) parasite of a bug *Coreus marginatus*. In Russian. *Zool. J. Acad. Nauk USSR* **67**:1407–1411.
114. Podlipaev, S.A. 1990. Catalogue of world fauna of Trypanosomatidae (Protozoa). In Russian. *Proc. Zool. Inst. Acad. Sci. USSR* **217**:1–177.
115. Podlipaev, S.A., and Frolov, A.O. 1987. Description and laboratory cultivation of *Blastocrithidia miridarum* sp.n. (Mastigophora, Trypanosomatidae). In Russian. *Parasitologiya* **21**:545–552.
116. Podlipaev, S.A., Frolov, A.O., and Kolesnikov, A.A. 1990. *Proteomonas inconstans* n.gen., n.sp. (Kinetoplastida, Trypanosomatidae) a parasite of the bug *Calocoris sexgutatus* (Hemiptera: Miridae). In Russian. *Parasitologiya* **24**:339–346.
117. Poisson, R. 1930. Sur un *Herpetomonas* parasite en Normandie de *Spilostethus (Lygaeus) saxatilis* (Scop.) (Hemiptères Lygeoideae). A propos des phytoflagelloses. *C.R. Soc. Biol.* **103**:1057–1061.

118. Poisson, R. 1930. *Herpetomonas tortum* n.sp., parasite intestinal des *Camptopus lateralis* (Germ.) (Hémiptères Coreidae, Alydaria) des environs de Banyuls. Rôle possible de cet insect comme agent transmetteur de phytoflagellose. *C.R. Soc. Biol.* **103**:1061–1064.
119. Prasad, D.V.R.K., and Kalavati, C. 1987. Morphology and life-cycle of two new species of *Leptomonas*, *L. indica* sp.n. and *L. bakeri* sp.n. from lygaeid insects. *Acta Protozool.* **26**:315–324.
120. Reichenow, E. 1940. Ostafrikanische Beobachtungen an Trypanosomiden. *Arch. Protistenk.* **94**:267–287.
121. Rezende, M.L.V., Borges, R.E.L., Bezerra, J.L., and Oliveira, D.P. 1986. Transmissão da murcha de *Phytomonas* a coqueiros e dendezeiros por *Lincus lobuliger* (Hemiptera, Pentatomidae): resultados preliminares. *Rev. Theobroma* **16**:149–154.
122. Robertson, M. 1912. Notes on some flagellate infections found in certain hemiptera in Uganda. *Proc. Rog. Soc. London B* **85**:234–240.
123. Rodhain, J. 1926. Phytoflagellé du latex d'une Asclepiadacée congolaise. *Ann. Soc. Belg. Med. Trop.* **6**:271–274.
124. Rodhain, J., and Bequaert, J. 1911. Présence de *Leptomonas* dans le latex d'une *Euphorbe* congolaise. *Bull. Soc. Pathol. Exot.* **4**:198–200.
125. Rodhain, J., Pons, C., Branden, F., and Bequaert, J. 1913. Leptomonas d'asilides et trypanosomides intestinaux de réduves et d'hémiptères phytophages au Katanga. *Rev. Zool. Afric.* **2**:291–301.
126. Ruiz, A. 1958. Contribuiciōn al estudio del genero *Phytomonas* Donovan en Costa Rica. I. *Phytomonas tortuosa* n.sp. *Rev. Biol. Trop.* **6**:87–93.
127. Ruiz, A. 1958. Contribuiciōn al estudio del genero *Phytomonas* Donovan en Costa Rica. II. *Phytomonas elmassiani* (Migone, 1916) Wenyon, 1926. *Rev. Biol. Trop.* **6**:263–272.
128. Ruiz, A. 1961. Contribuiciōn al estudio del genero *Phytomonas* Donovan en Costa Rica. III. *Phytomonas davidi* (Lafont, 1909) Donovan, 1909. *Rev. Biol. Trop.* **9**:227–231.
129. Sbravate, C., Campaner, M., Camargo, L.E.A., Conchon, I., Teixeira, M.M.G., and Camargo, E.P. 1989. Culture and generic identification of trypanosomatids of phytophagous hemiptera in Brazil. *J. Protozool.* **36**:543–547.
130. Scudder, G.G.E., and Duffey, S.S. 1972. Cardiac glycosides in the Lygaeinae. *Can. J. Zool.* **50**:35–52.
131. Shortt, H.E. 1923. Record of kala-azar research work carried on at the King Edward VII Memorial Pasteur Institute, Shillong, during 1922. *Indian J. Med. Res.* **10**:1150–1168.
132. Slansky, F.J., and Panizzi, A.R. 1987. In F.J. Slansky and J.G. Rodriguez, eds., *Nutritional Ecology of Seed-Sucking Insects*. Wiley & Sons, New York, pp. 283–320.
133. Slobbe, W.G. 1977. Phloem inhabiting *Phytomonas* in diseased coffee, coconut palms and African oil palms. *Surin. Landb.* **25**:4–13.
134. Slobbe, W.G., Parthasarathy, M.V., and Hesen, J.A.J. 1978. Hartrot or fatal wilt of palms. II. Oil palms (*Elaeis guineensis*) and other palms. *Principes* **22**:15–25.
135. Stahel, G. 1931. Zur Kenntnis der Siebröhrenkrankheit (Phloemnekrose) des Kaffeebaumes in Surinam. II. *Phytopathol. Z.* **4**:539–548.

136. Strong, R.P. 1924. Investigations upon flagellate infections. *Am. J. Trop. Med.* **4**:345–386.
137. Sweet, M.H. 1960. The seed bugs: a contribution to the feeding habits of the Lygaeidae. *Ann. Entomol. Soc. Am.* **53**:317–321.
138. Sweet, M.H. 1964. The biology and ecology of the Rhypanochrominae of New England (Heteroptera: Lygaeidae) Part I. *Entomol. Am.* **43**:1–124.
139. Sweet, M.H. 1964. The biology and ecology of the Rhypanochrominae of New England (Heteroptera: Lygaeidae) Part II. *Entomol. Am.* **44**:1–201.
140. Teixeira, M.M.G., and Camargo, E.P. 1989. Monoclonal antibodies for the identification of trypanosomatids of the genus *Phytomonas*. *J. Protozool.* **36**:262–264.
141. Teodoro, G. 1927. Osservazioni sul *Pyrrhocoris apterus* L., con particolare riguardo alla flagellosi. *Redia* **16**:45–50.
142. Tieszen, K.L., Molyneux, D.H., and Abdel-Hafez, S.K. 1986. Host-parasite relationships of *Blastocrithidia familiaris* in *Lygaeus pandurus* Scop. (Hemiptera: Lygaeidae). *Parasitology* **92**:1–12.
143. Tuzet, O., and Laporte, M. 1965. *Blastocrithidia vagoi*, n.sp. parasite de l'hémiptère hetéroptère *Eurydema ventralis* Kol. *Arch. Zool. Exp. Gen.* **105**:77–81.
144. Uribe, C. 1926. *Crithidia ortheae* n.sp. from reduvids of the genus *Orthea*. *J. Parasitol.* **12**:199–205.
145. Urueta, E.J. 1987. Investigaciónes sobre la Marchitez Sorpressiva de la palma africana en los Llanos Orientales de Colombia. *Rev. Siall* **4**:4–16.
146. Vainstein, M.H., and Roitman, I. 1986. Cultivation of *Phytomonas français* associated with poor development of root system in cassava. *J. Protozool.* **33**:511–513.
147. Vaidya, S.K., and Ranade, D.R. 1984. Occurrence of *Leptomonas pyrrhocoris* in the hemolymph of *Odontopus nigricornis* (Insecta, Heteroptera, Pyrrhocoridae). *Cur. Sci.* (Bangalore) **53**:606–607.
148. Vermeulen, H. 1963. A wilt of *Coffea liberica* in Surinam and its association with a flagellate, *Phytomonas leptovasorum* Stahel. *J. Protozool.* **10**:216–222.
149. Vermeulen, H. 1968. Investigations into the cause of the phloem necrosis disease of *Coffea liberica* in Surinam, South America. *Neth. J. Plant Pathol.* **74**:202–218.
150. Vickerman, K. 1962. Observations in the life cycle of *Phytomonas elmassiani* (Migone) in East Africa. *J. Protozool.* **9**:26–33.
151. Villain, G. 1925. Note sur le kala-azar du Honan (Chine centrale). *Bull. Soc. Pathol. Exot.* **18**:669–676.
152. Wallace, F.G. 1966. The trypanosomatids of insects and arachnids. *Exp. Parasitol.* **18**:124–193.
153. Wallace, F.G. 1977. *Leptomonas seymouri* sp.n. from the cotton stainer *Dysdercus suturellus*. *J. Protozool.* **24**:483–484.
154. Wallace, F.G. 1979. Biology of the Kinetoplastida of arthropods. In W.H.R. Lumsden and D.A. Evans, eds., *Biology of the Kinetoplastida*, Academic Press, London, Vol. 1, pp. 213–240.
155. Wallace, F.G., Camargo, E.P., McGhee, R.B., and Roitman, I. 1983. Guidelines for the description of new species of lower trypanosomatids. *J. Protozool.* **30**:308–313.

156. Yoshida, N., Jankevicius, J.V., Roitman, I., and Camargo, E.P. 1978. Enzymes of the ornithine-arginine metabolism in the genus Herpetomonas. *J. Protozool.* **25**:550–555.
157. Zotta, G. 1912. Sur un flagellé du type *Herpetomonas* chez *Pyrrhocoris apterus*. *Ann. Sci. Univ. Jassy* **7**:210–223.
158. Zotta, G. 1923. *Leptomonas familiaris* n.sp., parasite du tube digestif de *Lygaeus familiaris*, en Roumanie. *C.R. Soc. Biol.* **88**:285–287.
159. Zotta, G. 1924. La leptomonadiose spontanée chez *Cynanchum acutum*, Asclepiadée autochtone in Roumanie. *C.R. Soc. Biol.* **90**:141–143.

13
Molecular Biology of Plant Virus–Vector Interactions

Roger Hull

Introduction

The movement of a virus from one plant host to another presents several special problems. Most plants are sessile and, thus, unless hosts are growing close enough to each other to have direct contact, virus transmission requires some agent. Most of the plant surfaces are covered with a cuticle and the cells are enclosed in walls made primarily of cellulose. Therefore, the transmitting agent has to be able to penetrate these barries to enable infection. Plant viruses have developed a wide range of systems that overcome these problems and that involve many types of transmitting agents or vectors. One of the major features of these systems is the specificity between the virus and its vector, a theme that will be reiterated throughout this chapter. The specificity is not only at the level of vector type, family, genus, or species but can often be even at the level of biotype. As a general rule viruses that are transmitted by one type of vector (see below) are not transmitted by any of the others. This specificity indicates that there has been coevolution of the virus and its vector that involves the development of specific interactions that control the ability of the vector to transmit the virus. Transmission is also affected by the interactions between the vector and the host. I will not deal with this aspect in this chapter; for a discussion of the interactions between aphids and plants the reader is referred to Harris (39, 40).

The recent advances in the understanding of the molecular biology of at least one member of most plant virus groups has started to give us insight into the details of the interactions between viruses and their vectors. This is revealing the mechanisms by which viruses have close associations with their vectors, mechanisms that could be an "Achilles heel" forming a target that could be used for nonconventional protection of plants against viruses.

Roger Hull, John Innes Institute, Colney Lane, Norwich, NR4 7UH, United Kingdom.

In this chapter I will introduce the various types of virus vectors and describe characteristics of virus–vector interactions. I then discuss the virus genes that are involved in these interactions and what is known about the molecular details of these interactions.

Types of Virus Vectors

The overall picture of plant virus transmission is reviewed by Matthews (58) who distinguished two major forms, direct passage in higher plant material and transmission by organisms other than higher plants. The former includes transmission by mechanical means, by vegetative propagation and seed, and by pollen. Of these, pollen can be considered to be a vector as it can transmit the virus from an infected to a healthy plant. All the systems categorized as "organisms other than higher plants" are also vectors. The major groups of these other organisms are listed in Table 13.1.

Virus–Vector Interactions

The characteristics of the basic interactions between viruses and their vectors have been studied in most detail with viruses transmitted by insects, most particularly those transmitted by aphids. The basic concepts of virus–vector interactions were introduced by Watson and Roberts (92) and since then have been modified and refined many times, often causing controversy. Watson and Roberts (92) coined the terms "persistent" and "nonpersistent" to describe the length of time the vector was able to transmit the virus after acquisition. Sylvester (78) also recognized a "semipersistent" form of transmission that differed in detail from the "nonpersistent." A greater understanding of the regions of the vector showed that the "persistent" viruses passed through the insect's gut wall and hemocoel to accumulate in the salivary glands and thus became termed "circulative"; "nonpersistent" viruses did not pass through the insect's interior but associated with the anterior portion of the feeding canal and became termed "noncirculative" or "stylet-borne" (38, 53) or even "foregut-borne" (58). It was also recognised that there were two types of circulative viruses, those that replicated in the vector termed "propagative" and the "nonpropagative," which did not replicate. Because I am considering vectors in general and not just insects I will use the general terms *externally borne* or noncirculative for those viruses that do not pass through the vector's interior, and *internally borne* or circulative for those that pass through the vector; for the externally borne viruses I recognize two subcategories, *nonpersistent* and *semipersistent*, for the internally borne viruses there are the *nonpropagative* and the

TABLE 13.1. Types of vector.

Direct passage in higher plant material	
Pollen	
Transmission by organisms other than higher plants	
Invertebrates	
Nematoda	
Dorylaimidae	
Trichodoridae	
Arthropoda	
Insecta (approx. number of vector species)[a]	
Orthoptera (27)	*Diptera* (2)
Dermaptera (1)	*Thysanoptera* (4)
Coleoptera (38)	*Hemiptera* (4)
Lepidoptera (4)	*Homoptera* (>300)
	Aphididae
	Cicadellidae
	Delphacidae
	Membracidae
Arachnida	
Eriophyidae	
Fungi	
Plasmodiophoromycetes	*Polymyxa* spp.
Chytridomycetes	*Olpidium* spp.

[a] From Matthews (58).

propagative. The characteristics of these various virus–vector relationships are given in Table 13.2.

As noted above, these categories of virus–vector interactions were developed for aphid-transmitted viruses. With increasing knowledge of virus–vector interactions they have also been shown to apply to many other types of vector.

TABLE 13.2. Virus–vector relationships.

Externally borne	Internally borne
Acquired after short feed (seconds)	Acquired after longer feed (minutes to hours)
No latent period	Often latent period
Virus retained for short period	Virus retained for long period (often life)
Virus lost on moult	Virus not lost on moult
Often many vector species	Often few vector species
Vector specificity	Vector specificity
Virus spread in host finding flight	Virus spread by insects breeding on host
Virus–vector interactions in foregut (exoskeleton association)	Virus passes through gut wall into hemocoel and accumulates in salivary glands
	Propagative, virus multiplies in vector
	Nonpropagative, virus does not multiply in vector

TABLE 13.3. Plant virus groups and vectors.

Virus group	Vector type[a]	Virus–vector interaction[b]
Alfalfa mosaic virus	Aphid	EB-NP
Anthriscus yellows virus	Aphid	EB-SP, HC?
Badnavirus	Mealybug	EB-NP
Baymovirus	Fungi [Plasmodiophorales]	
Bromovirus	(Beetles)	
Capillovirus	Unknown	
Carlavirus	Aphid (whitefly)	EB-NP
Carmovirus	Unknown	
Caulimovirus	Aphid	EB-SP, HC-S
Closterovirus	Aphid, (whitefly)	EB-SP
Comovirus	Beetle	
Cryptovirus	None	
Cucumovirus	Aphid	EB-NP
Dianthovirus	Unknown	
Fabavirus	Aphid	EB-NP
Fijivirus	Planthoppers	IB-P
Furovirus	Fungi [Plasmodiophorales]	IB, HC-I
Geminivirus		
Subgroup A	Leafhopper	IB-NP
Subgroup B	Whitefly	IB-NP
Subgroup C	Leafhopper	IB-NP
Hordeivirus	Unknown	
Ilarvirus	Pollen	
Luteovirus	Aphid	IB-NP, HC-I
Machlovirus	Leafhopper	EB-SP, HC?
Marafivirus	Leafhopper	IB-P
Necrovirus	Fungi [Chytridales]	EB, IB
Nepovirus	Nematode	EB-NP
Pea enation mosaic virus	Aphid	IB-NP, HC-I
Phytoreovirus	Leafhopper, planthopper	IB-P
Plant Rhabdovirus	Leafhopper, planthopper, aphid	IB-P
Potexvirus	Mechanical	
Potyvirus	Aphid	WB-NP, HC-S
Sobemovirus	Beetle	
Tenuivirus	Planthopper	IB-P
Tobamovirus	Mechanical	
Tobravirus	Nematode	
Tombusvirus	Unknown	
Tospovirus	Thrips	IB-P
Tymovirus	Beetle	

[a] Names in parentheses are minor vectors; those in brackets indicate vector family.
[b] EB, externally borne; NP, nonpersistent; SP, semipersistent; IB, internally borne; NP, nonpropagative; P, propagative; HC, helper component; I, integral; S, separate.

Specific Vectors

As pointed out in the Introduction, in most cases each virus has its own specific type of vector. The type of vector is usually consistent among members of a virus group. In fact, one of the aims of the classification of viruses into groups is to predict the vector. The major vector types of the plant virus groups are listed in Table 13.3. Below are given some comments on the virus–vector interactions for the various types of vector.

Pollen

There are two forms of association of viruses with pollen (reviewed by 57). In the first, exemplified by prunus necrotic ringspot ilarvirus, the virus particles are borne on the exterior exine surface of the pollen grains and are thought to infect the new host when the pollen tube is penetrating the stigma. The particles of other viruses, e.g., tobacco ring spot nepovirus, are found within the cytoplasm of pollen grains and, thus, are internally borne. Nothing is known of the molecular mechanisms that determine the pollen transmissibility of a virus.

Fungi

Evidence for both major forms of virus–vector interactions are found in the fungal transmission of viruses (reviewed by 3). Tobacco necrosis and cucumber necrosis viruses are absorbed in a specific manner onto the zoospore surface of *Olpidium* sp. (85) whereas tobacco stunt virus is found in the cytoplasm of fungal spores (44, 55).

The furoviruses are transmitted by Plasmodiophora fungi, mainly *Polymyxa* spp. It is generally accepted that beet necrotic yellow vein and soil-borne wheat mosaic furoviruses are carried internally within the fungal resting spores (1, 73, 80) and most probably within the zoospores (81).

Nematodes

Nepoviruses are retained in their nematode vectors for weeks or even months but are lost on moulting and thus can be classified as externally borne. Little is known about the association of tobraviruses and their Trichodoridae vectors.

Arthropoda

ARACHNIDA

Mites

The mite vectors of plant viruses belong to the Eryophyid family, with both nymphs and adults being able to transmit (see 57). The most studied mite-transmitted virus is wheat streak mosaic virus transmitted by *Eriophyes tulipae*, which, it is suggested, has a circulative relationship. However, the possibility of transmission by regurgitation or introduction of defecated virus through feeding punctures cannot be ruled out.

INSECTA

Aphids

Aphids are the most important group of insect vectors of plant viruses with about 250 of the approximately 370 viruses with invertebrate vectors being transmitted by them (58). Of these the majority are transmitted in the externally borne manner, but there are several important examples of internally borne aphid-transmitted viruses. For a recent review of aphid transmission see Harris (40).

Leafhoppers, Planthoppers, Treehoppers

Hoppers are vectors of viruses from seven groups. In the majority of cases the viruses form an internally borne relationship with their vector, often propagative. In one case (machloviruses) the relationship is externally borne, semipersistent. For a recent review of hopper transmission of viruses the reader is referred to Nault and Ammar (67).

Whiteflies

Whiteflies are becoming of increasing importance as virus vectors, especially in tropical plants. Only two genera of whitefly, *Bemisia* and *Trialeurodes*, are known to transmit viruses to plants. The transmission characteristics of whitefly-transmitted viruses indicate that they have an internally borne nonpropagative relationship with their vector.

Thrips

Only three viruses are transmitted by thrips. Tomato spotted wilt tospovirus (TSWV) forms an internally borne propagative relationship with its vector, *Frankliniella occidentalis* (93). One feature of TSWV transmission is that the virus can be acquired only by the larval stage. Tobacco ringspot nepovirus (TRSV) and tobacco streak ilarvirus are also reported to be transmitted by thrips (50, 61), but little is known about their relationships with the vector. The transmission of TRSV is unusual in that it is also transmitted by nematodes.

Beetles

Viruses from four groups of plant viruses that have stable particles occurring in high concentrations are transmitted by Chrysomelid beetles. The vectors acquire the viruses very rapidly, which would indicate an externally borne relationship. However, the vector efficiency of the beetle increases with longer feeding times as does the retention time (26). Furthermore, virus particles are found in the hemolymph soon after commencement of feeding and insects can transmit virus injected into the hemocoel. RNase activity in the regurgitant fluid is a key factor in determining whether or not a virus will have a beetle vector (29, 30). Thus, the transmission of viruses by beetles seems to be a complex process and there is no clear picture of the virus–vector interaction.

Other Insects

Several viruses, including some economically important members of the badnavirus group, are transmitted by mealybugs. The transmission of cocoa swollen shoot badnavirus suggest that it forms an externally borne interaction with its vector.

Two viruses are transmitted by psyllids. Beet leafcurl rhabdovirus multiplies in its vector having an internally borne propagative relationship (71). The little information available on the transmission of beet savoy virus by *Piesma cinereum* indicates an internally borne nonpropagative interaction (57).

Some sobemoviruses are transmitted by mirid bugs in a manner similar to that described above for beetle transmission of these viruses (31).

Sites of Interactions in Vectors

Externally Borne Viruses

The nonpersistent characteristics of the externally borne interactions, especially transmission not surviving a moult, indicate that viruses transmitted in this manner interact with the exoskeleton of arthropods or nematodes or with external membranes or surfaces of fungal or pollen vectors. Nothing is known about the latter, but there are some observations on the location of virus particles with the former.

Electron microscopic studies have shown the apparent adsorption of various nepoviruses in defined regions of their nematode vectors. Particles of raspberry ringspot, tomato black ring, and artichoke Italian latent viruses have been found associated with the stylet lumen or guiding sheath of *Longidorus* spp. (82, 84), whereas those of arabis mosaic (AMV), grapevine fanleaf, tobacco ringspot, and strawberry latent ringspot viruses were located in the cuticular lining of the buccal capsule or the oesophagus of *Xiphinema* spp. (19, 60, 83). One of the features of these locations is that they are sharply delimited, implying that there are specific chemical groupings with which the virus particles associate (41). Particles of AMV appeared to be associated with carbohydrate-like material in *Xiphenema diversicaudatum* (76).

There have been numerous attempts to locate potyviruses in the mouth parts and anterior region of the feeding canal of their aphid vectors (reviewed by 41). Ideas on the mechanism of transmission have varied from contamination of the distal portion of the maxillary stylets (16, 17) to association with the inner surface of the mandibles (54) or regurgitation of ingested sap from the sucking pump (38). Berger and Pirone (11) showed that ^{125}I-labeled potyvirus particles could be detected specifically associated with the maxillary stylets and with portions of the alimentary canal anterior to the gut. One of the problems with these types of experiments is that they show where the virus accumulates but do not give any indications as to whether that virus is transmissible. Furthermore, the minimum amount of virus required for transmission is in the femtogram range (70), which makes detection by currently available techniques difficult, if not impossible.

Particles thought to be those of anthriscus yellows virus, which also requires a helper component, were found in the lining of the ventral wall of a specific region of the pharynx of the aphid vector, *Cavariella aegopodii* (66). The particles were surrounded by densely staining material that was embedded in a matrix of lightly staining material.

Internally Borne Viruses

Internally borne viruses have to cross one or more barrier to enter their vector and further barrier(s) to move from the vector to the host. Nothing is known about these barriers in pollen or fungi. In arthropod vectors there are two major layers of barriers, the gut wall and the salivary glands. Consideration of these barriers in relation to propagative viruses is given in the chapter by Ammar (4) in this volume.

The most detailed study of interactions of a nonpropagative virus with these barriers has been with various luteoviruses (reviewed by 34). The passage of propagative viruses across the gut wall involves virus replication. No such replication takes place with nonpropagative viruses and, thus, two possible routes have to be considered, across the cell or between cells. In an electron microscopic study, Gildow (33) showed that the particles of the RPV strain of barley yellow dwarf luteovirus associated only with the cell membranes of the hindgut of the aphid vector *Rhopalosiphon padi*. It was suggested that the particles entered the hindgut cells by endocytosis into coated pits and coated vesicles and accumulated in tubular vesicles and lysosomes. Particles were then released into the hemocoel by fusion of the tubular vesicles with the basal plasmalemma.

In a similar study (32) on the interaction between beet western yellows and potato leaf roll luteoviruses and salivary glands of their aphid vector *Myzus persicae*, virus particles were seen in the basal lamina and plasmalemma invaginations of accessory salivary cells. Particles were also frequently seen in tubular vesicles in the cytoplasm near salivary canals and in coated pits connected to the canal membrane.

From these studies the route that luteoviruses take across the two barrier tissues in their aphid vector would appear to be by incorporation into coated vesicles. Thus, the main sites of interaction that virus particles have are with the plasma-membrane on the gut side of gut wall cells and with the plasma membrane on the hemocoel side of the salivary gland accessory cells. Although the picture of the sites of interactions is clear for luteoviruses, it would be dangerous to extrapolate this information to other internally borne nonpropagative viruses.

Virus Genes Involved in Vector Interactions

In all cases of vector transmission the virus particles are the transmitted entities. Thus, the specific interactions with the vector must be determined by the surface molecules of the virus particles. Most plant viruses have simple particles comprising the viral genome, either RNA or DNA, encapsidated in one to three species of capsid protein in a helical arrangement in rod-shaped particles or with icosahedral symmetry in isometric and bacilliform particles. Some viruses have a more complex structures including some with the nucleoprotein being surrounded by a lipoprotein membrane.

Thus, for most viruses it is the surface of the capsid protein that interacts with the vector. This interaction can be either direct, or indirect via another virus-encoded protein termed the helper component or transmission factor. Aspects of this subject have been reviewed by Pirone (60).

Other viral genes can affect transmission indirectly. The efficiency of transmission can be influenced by virus concentration, which is controlled by virus replication. An example of this effect is shown by Atreya et al. (7). Similarly, acquisition of a virus can be affected by the localization of that virus in the host in relation to the site(s) of interaction of the vector with the host. Virus genes, such as those involved in cell-to-cell movement, can influence the localization of a virus in a given host.

Capsid Proteins

The capsid protein is known to be directly involved in the noncirculative aphid transmission of cucumoviruses. Chen and Francki (21) showed that tobacco mosaic virus RNA encapsidated in the capsid protein of tomato aspermy cucumovirus was aphid transmissible. Gera et al. (28) made heterologous reassembled particles between the genomes and capsid proteins of a highly aphid-transmitted (HAT) and a poorly aphid-transmitted (PAT) strain of cucumber mosaic virus (CMV). They showed that the efficiency of aphid transmission segregated with the source of capsid protein. There is no indication of differences in size of the capsid proteins of strains of CMV, which differ in aphid transmissibility and which could be attributed to readthrough into a helper component (see below). Recently, it has been suggested from a comparison of the sequences of proteins from aphid transmissible and nontransmissible CMV strains that amino acid 168 is involved in aphid transmission (M. Shintaku and P. Palukaitis, quoted in 69).

It is likely that the aphid transmissibility of alfalfa mosaic virus is also controlled directly by the viral capsid protein.

While there is no indication of the involvement of a helper component in the nematode transmission of nepoviuses, various observations point to the viral capsid protein controlling vector specificity (reviewed by 41).

There is evidence for the direct involvement of capsid proteins in the internally borne geminiviruses. Briddon et al. (18) exchanged the capsid protein gene of the whitefly-transmitted African cassava mosaic geminivirus (ACMV) for that of the leafhopper-transmitted beet curly top geminivirus (BCTV). Leafhoppers did not acquire the chimeric virus from infected plants but, when the virus was injected into the hemocoel, the leafhoppers could transmit it. This demonstrates that the BCTV capsid protein determined the passage from the hemocoel to the salivary glands. There were three suggestions for the inability of the leafhoppers to acquire the chimeric virus on feeding. First, it was in *Nicotiana benthamiana*, which is a poor host for the leafhopper *Circulifera tenellus*; second, it is possible that the ACMV/BCTV chimera was located differently in the plant to BCTV alone and that the feeding behavior of the leafhopper did not enable it to acquire it; third, it may be that another factor was required for the chimera to pass the gut wall barrier.

Although most of the evidence points to the involvement of the integral helper component (see below) being involved in the transmission of luteoviruses, there is one observation that suggests that the capsid protein may play a role in the interactions (74). These authors showed that a transmission-neutralizing monoclonal antibody mapped to an antigenic domain in the capsid protein of barley yellow dwarf virus.

The capsid protein is also involved in transmission mediated by separate helper components (see below), in a manner presumed to be by molecular interactions between the helper component and the virus particle. There is considerable information available on the region of the capsid protein of potyviruses that is presumed to be involved in this interaction. Harrison and Robinson (42) identified an amino acid triplet, DAG (aspartic acid–alanine–glycine), near the N-terminus of aphid-transmissible strains of potyviruses that was DAS in an aphid nontransmissible strain of tobacco etch virus. Since then other natural mutations of the DAG triplet have been associated with loss of aphid transmissibility, e.g., plum pox virus DAL (56), tobacco vein mottling virus (TVMV) DAE (8), papaya ringspot and zucchini yellow mosaic viruses (ZYMV) DTG (27, 72). Using infectious cDNA clones the region containing the DAG motif of an aphid transmissible isolate of TVMV was substituted with that of the mutant DAE, resulting in loss of aphid transmissibility (8). Amino acid substitutions in the glycine of the triplet abolished transmissibility, whereas various substitutions of the first two amino acids of the triplet resulted in a reduction in the efficiency of aphid transmission as did substitution of the lysine residue adjacent to the glycine (6). Aphid transmissibility was restored to a nontransmissible isolate of ZYMV by mutating the DTG to DAG (27).

Potato aucuba mosaic potexvirus (PAMV) is transmitted by aphids from plants coinfected with PVY though not when it is on its own (51). The nucleotide sequence of the coat protein gene of PAMV showed that it contained a DAG motif in its N-terminal region. A recombinant isolate in which the N-terminal region of potato virus X (PVX) coat protein gene was replaced by that of PAMV was aphid transmissible in the presence of PVY, whereas the wild-type PVX was not (10).

These data point to the N-terminal region of potyviruses, and especially the DAG sequence, being important in the interaction with the helper component but do not yet throw light on the molecular nature of the interaction.

Helper Proteins

Helper components or transmission factors have been recognised or suggested for members of 9 virus groups (Table 13.3). There are two forms of helper component, those that are encoded separately from the coat protein and those that are formed by readthrough of a weak stop codon on the coat protein and thus are integral with that protein. The separate helper components are thought to interact with both the viral coat protein and the specific site in the vector. The integral helper components need only interact with the vector.

Separate Helper Proteins

The two groups of viruses for which the genome organization is known and that are externally borne on their aphid vector, the caulimoviruses and the potyviruses, have separate helper components. This protein is known as the helper component in the potyvirus system and as the acquisition factor, helper component, or aphid transmission factor in caulimoviruses (for recent review see 69).

Caulimoviruses

Analysis of deletion mutants and the formation of recombinants between aphid transmissible and nontransmissible isolates of cauliflower mosaic virus (CaMV) gave a strong indication that the helper component is encoded by gene II (Fig. 13.1) (5, 35, 94). The recent complementation of aphid transmissibility by the gene II product expressed in insect cells has given direct proof that this is the helper component (12). CaMV gene II encodes a protein (P18) of M_r 18,000 and there is no evidence for posttranslational modifications such as glycosylation or phosphorylation. Computer-based analyses of the amino acid sequence of this protein from CaMV (63) and two other caulimoviruses indicate that it has two domains, an N-terminal domain that is mainly β-sheet and turn, and a C-terminal domain that is primarily α-helix (Fig. 13.2); the two domains are separated by a predicted random region. CaMV P18 accumulates as electron-lucent inclusion bodies in infected cells (25), which have a paracrystalline structure (14).

There are two types of nontransmissible isolates of CaMV, CM4-184 with a deletion of most of gene II (45), and CM1841 and Campbell, which have a substitution at amino acid 94 (95). P18 produced by the latter accumulates to only very low levels in infected plants and does not form inclusion bodies (25, 37). It has been recently shown that insect cell-expressed CM1841 P18 has biological activity and, thus, it is likely that the amino acid substitution results in a loss of stability of the protein (13).

Evidence for the association of CaMV P18 with virus particles comes from the fractionation studies of Espinoza et al. (24). This, and the recent work of Blanc et al. (14), indicates that extraction with 200 m*M* Tris-HCl, pH 7.6, 100 m*M* EGTA, 50 m*M* $MgCl_2$, disrupts the electron-lucent P18 inclusion bodies and releases virus particles with P18 associated.

Potyviruses

The helper components of several potyviruses have been characterized extensively (see 69 for review). They are proteins with subunit M_r in the range 53,000–58,000 processed from near the N-terminus of the viral polyprotein (Fig. 13.1). The protein has two functions, the N-terminal portion being the helper component and the C-terminal part having protease activity, thus being termed HC-Pro (20). Comparison of the sequences of HC-Pros from several potyviruses showed that the N-terminal domain is variable and the C-terminal domain is highly conserved (87). The termini of the two domains are rich in cysteine

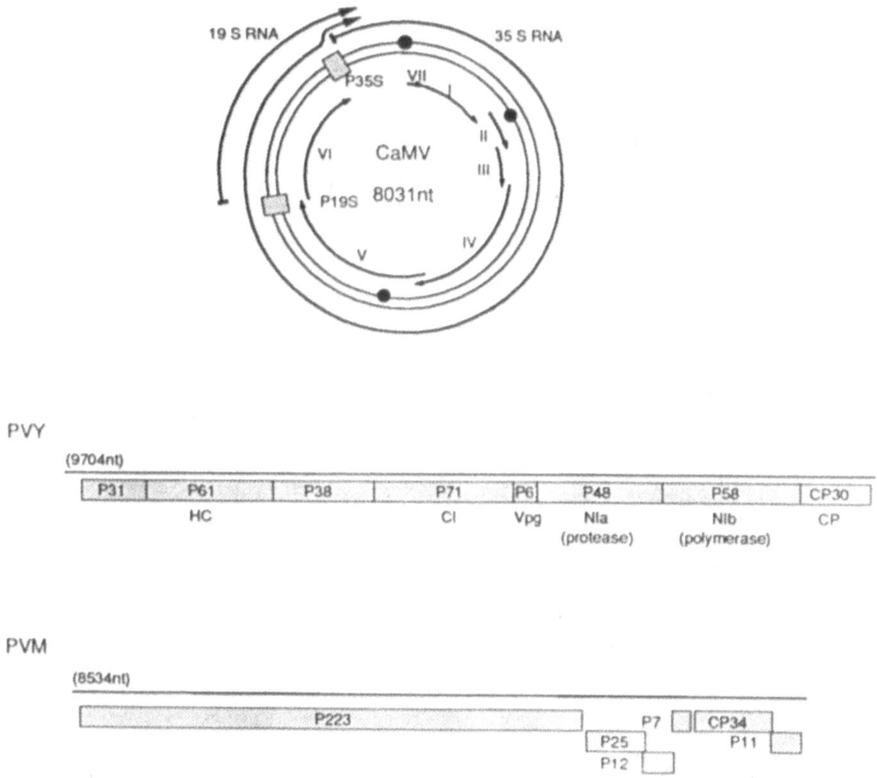

FIGURE 13.1. Genome organizations of viruses with, or suspected of having, separate helper components. At the top is the circular double-stranded DNA genome of cauliflower mosaic virus. The inner rings of arcs are the open reading frames I–VII with II being the helper component. The double circle is the DNA genome with replication priming sites represented by dots and transcription start sites by boxes. The outer arrowed circles and arcs are the RNA transcripts. The middle diagram is the genome organization of potyviruses exemplified by potato virus Y. The RNA genome (single line) is translated as a polyprotein which is cleaved into the indicated products. The helper component is P61. The bottom diagram shows the genome organization of the carlavirus PVM.

residues, that in the N-terminal domain being suggested to form a zinc-finger motif (75). The helper components of tobacco vein mottle virus and potato virus Y (PVY) have been purified and shown to be biologically active in the dimer form (86).

Certain isolates of PVY and ZYMV are nontransmissible due to a defect in their helper component. The sequence of the helper component of the nontransmissible strain of PVY (PVC) differed from transmissible PVY and four other potyviruses in two amino acids conserved among the transmissible viruses (87). Mutation of one of these amino acids at

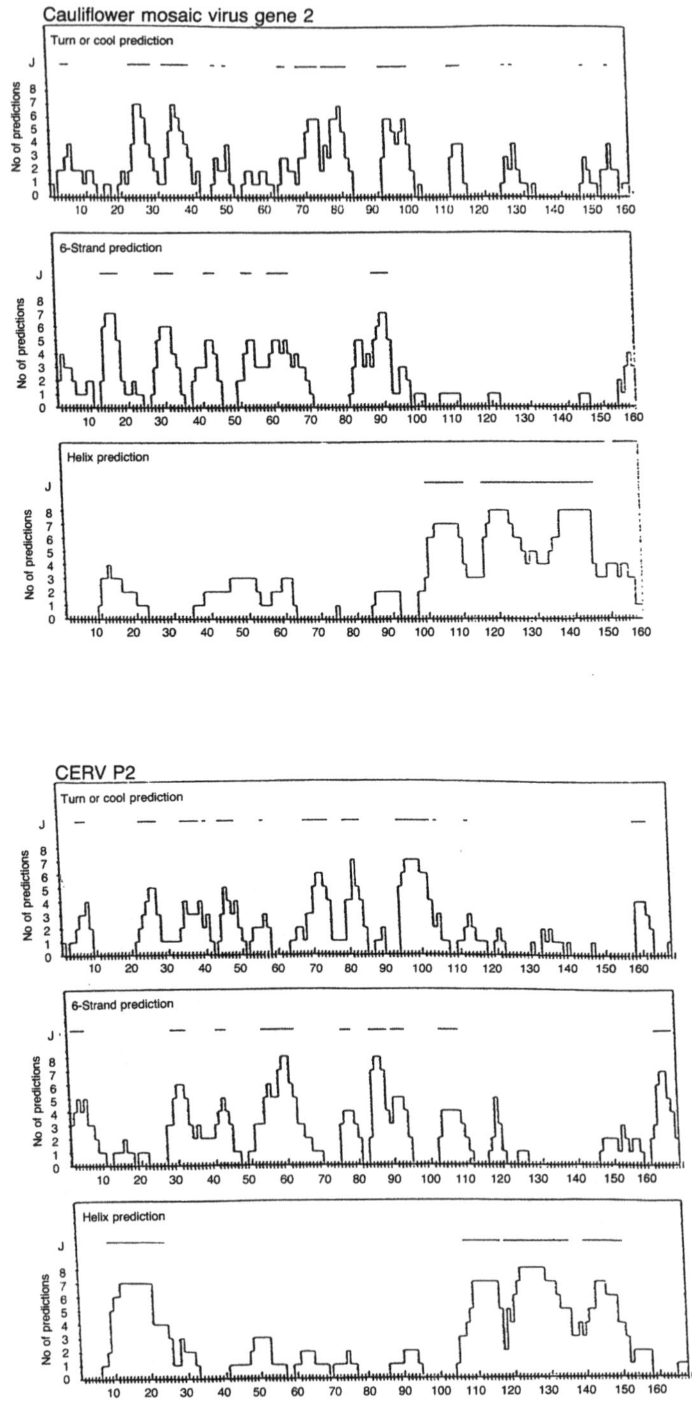

FIGURE 13.2. *Caption on p. 375.*

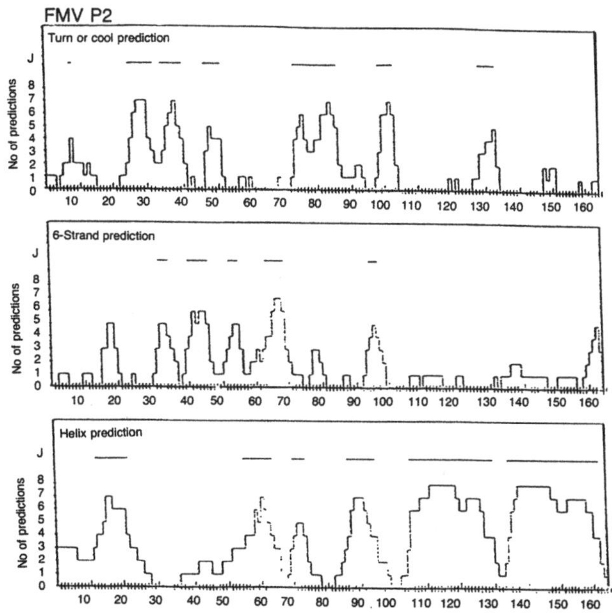

FIGURE 13.2. Secondary structures of the helper components from three caulimoviruses, cauliflower mosaic, carnation etched ring, and figwort mosaic viruses predicted using the program PREDICT (23). For each protein three structural features, turn or coil, β-strand, and α-helix are predicted using eight algorithms and the number of predictions of each feature for each residue position is plotted. The line above each plot shows a strong prediction. Note the β-strand, turn prediction in the N-termini, and the α-helix prediction in the C-termini.

position 307 of TVMV helper component from lysine (K) to glutamic acid (E) abolished transmissibility (7). Since this amino acid position is in the N-terminal cysteine-rich region it is possible that this mutation affected the ability of the helper component to form active dimers. A similar substitution (K→E) was found in a comparison of the sequences of a nontransmissible and a transmissible strain of ZYMV; it was suggested that the nontransmissible strain was defective in its interaction with the aphid rather than the virion (36).

Other Viruses with Helper Components

Helper components are implicated in the transmission of several other viruses, but the genes involved have not yet been identified.

Anthriscus yellows virus (AYV) is transmitted by aphids in the externally borne, semipersistent manner and assists the transmission of parsnip yellow fleck virus (for review see 41). Data from transmission experiments indicate the involvement of a separate helper component. As noted above, particles suggested to be those of AYV were found embedded in densely staining material in vector insects; virus-like particles are found in inclusion bodies with similar staining characteristics in infected plants (64).

Rice tungro spherical virus (RTSV) resembles AYV in that it is externally borne with a semipersistent relationship, but with a leafhopper, not an aphid, vector and it assists the transmission of another virus, rice tungro bacilliform virus. There is evidence that RTSV and another semipersistent leafhopper-borne virus, maize chlorotic dwarf virus, have helper components (43, 49). The recently published sequence of RTSV (77) does not reveal any obvious gene that could be the helper component but shows that there are several putative genes with no function yet attributed.

There are suggestions that carlaviruses and closteroviruses might involve helper components in their transmission (65), but there is no real experimental evidence. The carlavirus genome organization has been determined (see Fig. 13.1) and there is one putative gene, P11, to which no function has yet been ascribed.

Mechanisms of Separate Helper Components

The mechanism(s) by which separate helper components function is still open to speculation (69). The most widely accepted hypothesis is that the helper component has two domains, one interacting with the virus particle and one with the specific site in the vector. The mutagenesis data would support this, as would the predicted two domain structure of caulimovirus helper components. One feature of the externally borne transmission mechanism that requires further consideration is that of the release of virus particles that have been acquired at the specific retention site in the vector. It is interesting to note that the binding site on potyviral capsid protein is at the N-terminus, which opens up the suggestion that a protease might be involved in the release from the retained complex (42).

Integral Helper Components

Several of the internally borne viruses have helper components that are effectively extensions of the capsid protein. They are produced from readthrough of a weak termination codon at the C-terminus of the capsid protein, the readthrough taking place in about 10% of the translations.

Recently the genomes of several members of the luteovirus group have been sequenced (52, 59, 62, 88, 89, 91) and, although they fall into two groups, they have similar genome organizations (see Fig. 13.3). The capsid protein gene is separated from the adjacent gene by an amber termination codon and there is evidence that this stop codon is read through to give a larger protein (see 9, 62). In the readthrough protein

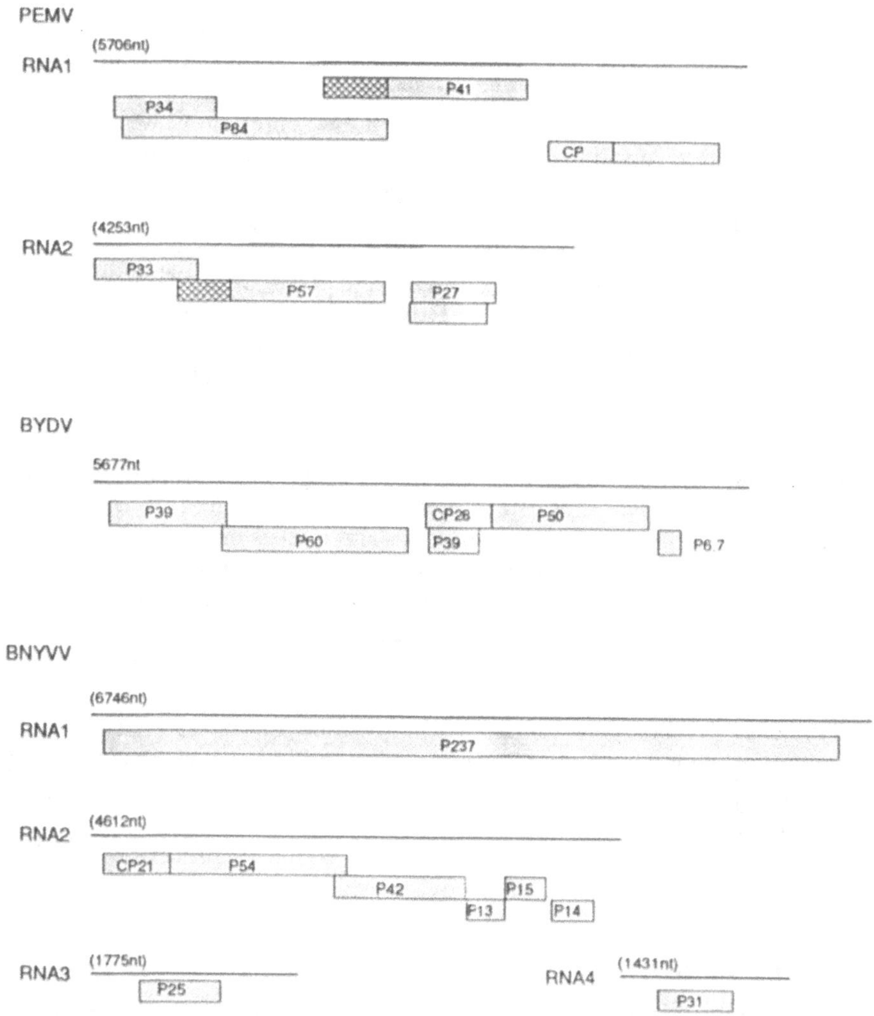

FIGURE 13.3. Genome organizations of viruses that have integral helper components. The top part shows the two RNAs of pea enation mosaic virus. Note the capsid protein and readthrough protein at the 3′ end of RNA 1. The middle part shows barley yellow dwarf luteovirus. The capsid protein and readthrough protein are toward the 3′ end. The bottom part shows the four RNAs of beet necrotic yellow vein furovirus. The capsid protein and readthrough protein are at the 5′ end of RNA 2.

the stop codon is followed by a region rich in proline residues. It is thought that this proline-rich region could form an interdomain linker, which would enable the read through region to fold separately from the capsid protein (62). Comparison of the sequences of several isolates of barley yellow dwarf virus that differ is vector specificity showed that there was more variation in the readthrough portion than in the capsid protein (90).

Pea enation mosaic virus (PEMV) is a two-component virus that is aphid transmitted in the internally borne manner and is also mechanically transmissible. Sequence of the larger RNA showed that it has a genome organization similar to that of luteoviruses (22) (Fig. 13.3). Analyses of coat proteins of PEMV particles showed several species (2, 47) that, together with the sequence data, suggested a situation similar to that described above for luteoviruses. However, some recent observations have thrown doubt on this interpretation (Salgueiro and Hull, unpublished observations). It appears that in infected plants, the readthrough molecule is not attached covalently to the capsid protein; in virus preparations covalent attachment is found, but this may be due to cross-linking during virus extraction. A reappraisal of the data on luteoviruses indicates that the situation is not as clear as previously thought. Whatever the actual structure of luteovirus and PEMV particles there is reasonable evidence to suggest that the readthrough portion is involved in receptor recognition in the vector. The possible involvement of the capsid protein of luteoviruses noted above raises the possibility that one protein may operate at the gut wall and the other at the salivary glands.

The coat protein gene of BNYVV encoded on RNA-2 has a weak termination codon, which allows translational readthrough (15) (Fig. 13.3). From a study of mutants of BNYVV, Tamada and Kusume (81) suggested that the readthrough protein was essential for fungal transmission of this virus. It would appear to be a transmission helper protein, analogous to those of luteoviruses.

Other Proteins

The internally borne propagative viruses are presumed to have surface proteins that interact with receptor sites of cells at various locations within their vectors. These will include the gut wall and salivary gland barriers discussed above and also the plasma membranes of other cells types that these viruses enter and multiply within.

It can be seen from Table 13.3 that there are six groups of viruses that multiply in their vectors. Viruses in two of these groups, the plant rhabdoviruses and the tospoviruses, have membrane-enveloped particles with glycoprotein spikes extending from the membrane. By analogy with animal rhabdoviruses and Bunyaviruses it is most likely that these glycoproteins interact with cell surface receptors in the vector. Furthermore, removal of the glycoproteins decreases the infectivity of potato yellow dwarf rhabdovirus for insect cells but not for its host plant (46). The genes encoding the glycoproteins of members of both these

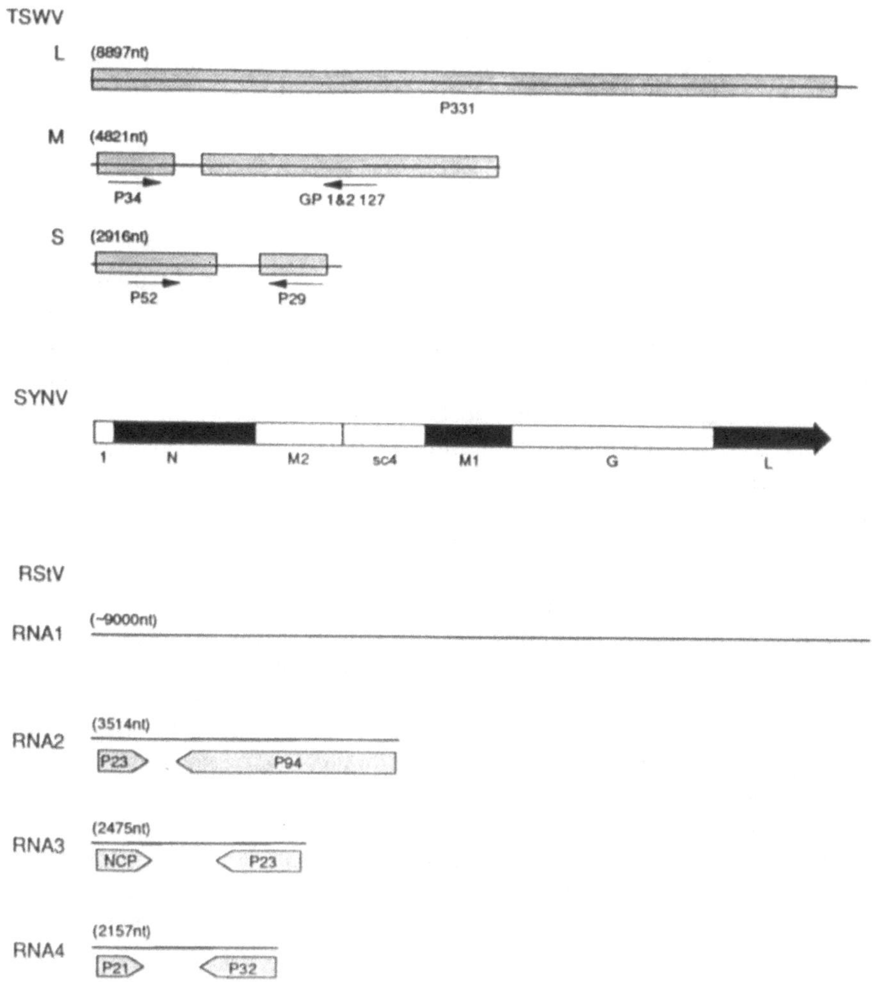

FIGURE 13.4. Genome organizations of viruses with other proteins possibly involved in internally borne transmission. Tomato spotted wilt virus (TSWV) has three RNA species. The ambisense M RNA encodes two glycoproteins at the 3′ end in the complementary sense. Sonchus yellow net rhabdovirus (SYNV) has a genome organization similar to those of animal rhabdoviruses and encodes surface glycoproteins (G). Rice stripe tenuivirus (RStV) encodes putative glycoproteins as P94 on RNA 2.

groups have been identified (Fig. 13.4). It is possible that the particles of tenuiviruses, a virus group that has several features in common with Bunyaviruses, are membrane-bound. The recently published sequence of RNA 2 of rice stripe tenuivirus (79) (Fig. 13.4) shows that the 94K predicted protein shows amino acid similarity to glycoproteins of some Bunyaviruses.

The plant reoviruses have genomes comprising 12 double-stranded RNA segments. Loss of vector transmissibility of wound tumor reovirus is associated with loss of two segments, S2 and S5, the gene products of which are located in the virus outer coat (reviewed in 68).

Discussion

Major advances are being made in the understanding of the interactions between plant viruses and their vectors and it is likely that these will continue. There are still various questions to be answered. These include the details of the interactions involved in externally borne transmission, especially that in insects and nematodes. What controls the differences between nonpersistent and semipersistent relationships? What are the mechanisms of retention and of release of externally borne viruses? What is the mode of function of the separate helper components? As noted above, the current idea is that they form a bridge between the virus particle and the retention site within the vector. However, it should be borne in mind that there could be other mechanisms such as potentiating the entry of the virus particles into host cells. Why has the separate helper component system been developed? Does it have an advantage over transmission dependent on direct interaction between the capsid and the vector? How did the helper component gene(s) evolve?

The detailed understanding of the mechanisms involved in virus–vector interactions will raise a range of questions on virus evolution. Answering these questions will depend not only on molecular biological information but also on knowledge of the evolutionary pressures presented by viral epidemiology and the need to obtain efficiency and reliability of transmission. Thus, factors such as vector behavior and vector–host interactions will also have to be taken into account.

The increase in knowledge of virus–vector interactions will open up the possibilities of new approaches to controlling virus diseases. One of the recent new concepts in plant virology has been that of nonconventional protection in which "genes" are introduced into the host that interfere with the functioning of the virus (for review see 48). The understanding of the molecular biology of virus–vector interactions will identify targets that can be interfered with, thus preventing virus transmission.

Acknowledgments. I thank Dr. D. Baulcombe for the preprint of his paper and Mr. M. Harvey for preparing the figures.

References

1. Abe, H., and Tamada, T. 1986. Association of beet necrotic yellow vein virus with isolates of *Polymyxa betae* Keskin. *Ann. Phytopathol. Soc. Jpn.* **252**:235–247.
2. Adam, G., Sander, E., and Shepherd, R.J. 1979. Structural differences between PEMV strains affecting transmissibility by *Acyrthosiphon pisum*. *Virology* **92**:1–14.
3. Adams, M.J. 1991. Transmission of plant viruses by fungi. *Ann. Appl. Biol.* **118**:479–492.
4. Ammar, E.D. 1994. Propagative transmission of plant and animal viruses by insects: Factors affecting vector specificity and competence. In K.F. Harris, ed., *Advances in Disease Vector Research*, Vol. 10. Springer-Verlag, New York.
5. Armour, S.L., Melcher, V., Pirone, T.P., Lyttle, D.G., and Essenberger, R.C. 1983. Helper component for aphid transmission encoded by region II of cauliflower mosaic virus DNA. *Virology* **129**:25–30.
6. Atreya, P.L., Atreya, C.D., and Pirone, T.P. 1991. Amino acid substitutions in the coat protein result in loss of insect transmissibility of a plant virus. *Proc. Natl. Acad. Sci. U.S.A.* **88**:7887–7891.
7. Atreya, C.D., Atreya, P.L., Thornbury, D.W., and Pirone, T.P. 1992. Site-directed mutations in the potyvirus HC-Pro gene affect helper component activity, virus accumulation, and symptom expression in infected tobacco plants. *Virology* **191**:106–111.
8. Atreya, C.D., Raccah, B., and Pirone, T.P. 1990. A point mutation in the coat protein abolishes aphid transmissibility of a potyvirus. *Virology* **178**:161–165.
9. Bahner, I., Lamb, J., Mayo, M.A., and Hay, R.T. 1990. Expression of the genome of potato leafroll virus: Readthrough of the coat protein termination codon *in vivo*. *J. Gen. Virol.* **71**:2251–2256.
10. Baulcombe, D.C., Lloyd, J., Manoussopoulos, I.N., Roberts, I.M., and Harrison, B.D. 1993. The signal for potyvirus-dependent aphid transmission of potato aucuba mosaic virus and the effect of its transfer to potato virus X. *J. Gen. Virol.* **74**:1245–1253.
11. Berger, P.H., and Pirone, T.P. 1986. The effect of helper component on the uptake and localization of potyviruses in *Myzus persicae*. *Virology* **153**:256–261.
12. Blanc, S., Cerutti, M., Usmany, M., Vlak, J.M., and Hull, R. 1993. Biological activity of cauliflower mosaic virus aphid transmission factor expressed in a heterologous system. *Virology* **192**:643–650.
13. Blanc, S., Cerutti, M., Chaabihi, H., Louis, C., Devauchelle, G., and Hull, R. 1993. Gene II product of an aphid non-transmissible isolate of cauliflower mosaic virus expressed in a baculovirus system posesses aphid transmission factor activity. *Virology* **192**:651–654.
14. Blanc, S., Schmidt, I., Kuhl, G., Esperandieu, P., Lebeurier, G., Hull, R., Cerutti, M., and Louis, C. 1993. Paracrystalline structure of cauliflower mosaic virus aphid transmission factor produced both in plants and in an heterologous system and relationship with a solubilized form. *Virology* (in press).

15. Bouzoubaa, S., Ziegler, V., Beck, D., Guilley, H., Richards, K., and Jonard, G. 1986. Nucleotide sequence of beet necrotic yellow vein virus RNA-2. *J. Gen. Virol.* **67**:1689–1700.
16. Bradley, R.H.E. 1966. Which of an aphid's stylets carry transmissible virus? *Virology* **29**:396–401.
17. Bradley, R.H.E., and Ganong, R.Y. 1955. Evidence that potato virus Y is carried near the tip of the stylets of the aphid vector *Myzus persicae* (Sulz). *Can. J. Microbiol* **1**:775–782.
18. Briddon, R.W., Pinner, M.S., Stanley, J., and Markham, P.G. 1990. Geminivirus coat protein gene replacement alters insect specificity. *Virology* **177**:85–94.
19. Brown, D.J.F., and Trudgill, D.L. 1983. Differential transmissibility of arabis mosaic and strains of strawberry latent ringspot viruses by populations of *Xiphenema diversicaudatum* (Nematoda: Dorylaimida) from Scotland, Italy and France. *Rev. de Nematol.* **6**:229–238.
20. Carrington, J.C., Cary, S.M., Parks, T.D., and Dougherty, W.G. 1989. A second proteinase encoded by a plant potyvirus genome. *EMBO J.* **8**:365–370.
21. Chen, B., and Francki, R.I.B. 1990. Cucumovirus transmission by the aphid *Myzus persicae* is determined solely by the viral coat protein. *J. Gen. Virol.* **71**:934–944.
22. Demler, S.A., and de Zoeten, G.A. 1991. The nucleotide sequence and luteovirus-like nature of RNA 1 of an aphid non-transmissible strain of pea enation mosaic virus. *J. Gen. Virol.* **72**:1819–1834.
23. Eliopoulos, E.E., Geddes, A.J., Brett, M., Pappin, D.J.C., and Findlay, J.B.C. 1982. A structural model for the chromophore binding domain of rhodopsin. *Int. J. Biol. Macromol.* **4**:263.
24. Espinoza, A.M., Markham, P.G., Maule, A.J., and Hull, R. 1988. *In vitro* biological activity associated with the aphid transmission factor of cauliflower mosaic virus. *J. Gen. Virol.* **69**:1819–1830.
25. Espinoza, A.M., Medina, V., Hull, R., and Markham, P.G. 1991. Cauliflower mosaic virus gene II product forms distinct inclusion bodies in infected plant cells. *Virology* **185**:337–344.
26. Fulton, J.P., Gergerich, R.C., and Scott, H.A. 1987. Beetle transmission of plant viruses. *Annu. Rev. Phytopathol.* **25**:111–123.
27. Gal-On, A., Antignus, Y., Rosner, A., and Raccah, B. 1992. A zucchini yellow mosaic virus coat protein gene mutation restores aphid transmissibility but has no effect on multiplication. *J. Gen. Virol.* **73**:2183–2187.
28. Gera, A., Lobenstein, G., and Raccah, B. 1979. Protein coats of two strains of cucumber mosaic virus affect transmission by *Aphis gossypii*. *Phytopathology* **69**:396–399.
29. Gergerich, R.C., Scott, H.A., and Fulton, J.P. 1986. Evidence that ribonuclease in beetle regurgitant determines the transmission of plant viruses. *J. Gen. Virol.* **67**:367–370.
30. Gergerich, R.C., and Scott, H.A. 1988. The enzymatic function of ribonuclease determines plant virus transmission by leaf-feeding beetles. *Phytopathology* **78**:270–272.

31. Gibb, K.S., and Randles, J.W. 1988. Studies on the transmission of velvet tobacco mottle virus by the mirid *Cyrtopeltis nicotianae*. *Ann. Appl. Biol.* **112**:427–437.
32. Gildow, F.E. 1982. Coated-vesicle transport of luteoviruses through salivary glands of *Myzus persicae*. *Phytopathology* **72**:1289–1296.
33. Gildow, F.E. 1985. Transcellular transport of barley yellow dwarf virus into the hemocoel of the aphid vector, *Rhopalosiphum padi*. *Phytopathology* **75**:292–297.
34. Gildow, F.E. 1987. Virus-membrane interactions involved in circulative transmission of luteoviruses by aphids. *Current Topics Vector Res.* **4**:93–120.
35. Givord, L., Xiong, C., Giband, M., Koenig, I., Hohn, T., Lebeurier, G., and Hirth, L. 1984. A second cauliflower mosaic virus gene product influences the structure of the viral inclusion body. *EMBO J.* **3**:1423–1427.
36. Grumet, R., Bada, R., and Hamar, S. 1992. Analysis of the zucchini yellow mosaic virus (ZYMV) potyviral helper component, possible identification of an aphid-interaction domain. *Phytopathology* **82**:1176.
37. Harker, C.L., Woolston, C.J., Markham, P.G., and Maule, A.J. 1987. Cauliflower mosaic virus aphid transmission factor protein is expressed in cells infected with some aphid non-transmissible isolates. *Virology* **160**:252–254.
38. Harris, K.F. 1977. An ingestion-egestion hypothesis of non-circulative virus transmission. In K.F. Harris and K. Maramorosch, eds., *Aphids as Virus Vectors*. Academic Press, New York, pp. 166–208.
39. Harris, K.F. 1983. Sternorrynchous vectors of plant viruses: virus-vector interactions and transmission mechanisms. *Adv. Virus Res.* **28**:113–140.
40. Harris, K.F. 1990. Aphid transmission of plant viruses. In C.L. Mandahar, ed., *Plant Viruses*, Vol. 2: *Phytopathology*. CRC, Boca Raton, FL, pp. 177–204.
41. Harrison, B.D., and Murant, A.F. 1984. Involvement of virus-coded proteins in transmission of plant viruses by vectors. In M.A. Mayo and K.A. Harrap, eds., *Vectors in Virus Biology*. Academic Press, London, pp. 1–36.
42. Harrison, B.D., and Robinson, D.J. 1988. Molecular variations in vector-borne plant viruses: epidemiological significance. *Philos. Trans. Roy. Soc. London Ser. B* **321**:447–462.
43. Hibino, H., and Cabauatan, P.Q. 1987. Infectivity neutralization of rice tungro-associated viruses acquired by leafhopper vectors. *Phytopathology* **77**:473–476.
44. Hiruki, C. 1987. Recovery and identification of tobacco stunt virus from air-dried resting spores of *Olpidium brassicae*. *Plant Pathol.* **36**:224–228.
45. Howarth, A.J., Gardner, R.C., Messing, J., and Shepherd, R.J. 1981. Nucleotide sequence of naturally occurring deletion mutant of cauliflower mosaic virus. *Virology* **112**:678–685.
46. Hsu, H.T., Nuss, D.L., and Adam, G. 1983. Utilization of insect tissue culture in the study of the molecular biology of plant viruses. *Current Topics Vector Res.* **1**:189–214.
47. Hull, R. 1977. Particle differences related to aphid-transmissibility of a plant virus. *J. Gen. Virol.* **34**:183–187.

48. Hull, R., and Davies, J.W. 1992. Approaches to non-conventional control of plant virus diseases. *Crit. Rev. Plant Sci.* **11**:17–33.
49. Hunt, R.E., Nault, L.R., and Gingery, R.E. 1988. Evidence for infectivity of maize chlorotic dwarf virus and for a helper component in its leafhopper transmission. *Phytopathology* **78**:499–504.
50. Kaiser, W.J., Wyatt, S.D., and Pesho, G.R. 1982. Natural hosts and vectors of tobacco streak virus in Eastern Washington. *Phytopathology* **72**:1508–1512.
51. Kassanis, B. 1961. The transmission of potato aucuba mosaic virus by aphids from plants also infected by potato viruses A or Y. *Virology* **13**:93–97.
52. Keese, P., Martin, R.R., Kawchuk, L., Waterhouse, P.M., and Gerlach, W.L. 1990. Nucleotide sequences of an Australian and a Canadian isolate of potato leafroll luteovirus and their relationships with two European isolates. *J. Gen. Virol.* **71**:719–724.
53. Kennedy, J.S., Day, M.F., and Eastop, V.F. 1962. *A Conspectus of Aphids as Vectors of Plant Viruses*. Commonwealth Inst. Entomology, London.
54. Lim, W.L., de Zoeten, G.A., and Hagedorn, D.J. 1977. Scanning electron-microscopic evidence for the attachment of a non-persistently transmitted virus to its vector's stylets. *Virology* **79**:121–128.
55. Lin, M.T., Campbell, R.N., Smith, P.R., and Temmink, J.H.M. 1970. Lettuce big vein virus transmission by single sporangium isolates of *Olpidium brassicae*. *Phytopathol.* **60**:1630–1634.
56. Maiss, E., Timpe, U., Brisske, A., Jelkmann, W., Casper, R., Himmler, G., Mattanovich, D., and Katinger, H.W.D. 1989. The complete nucleotide sequence of plum pox virus RNA. *J. Gen. Virol.* **70**:513–524.
57. Mandahar, C.L. 1990. Virus transmission. In C.L. Mandahar, ed., *Plant Viruses*, Vol. 2: *Phytopathology*. CRC, Boca Raton, FL, pp. 205–253.
58. Matthews, R.E.F. 1991. *Plant Virology*, 3rd ed. Academic Press, New York.
59. Mayo, M.A., Robinson, D.J., Jolly, C.A., and Hyman, L. 1989. Nucleotide sequence of potato leafroll luteovirus RNA. *J. Gen. Virol.* **70**:1037–1051.
60. McGuire, J.M., Kim, K.S., and Doutit, L.B. 1970. Tobacco ringspot virus in the nematode *Xiphenema americana*. *Virology* **42**:212–216.
61. Messieha, M. 1969. Transmission of tobacco ringspot virus by thrips. *Phytopathology* **59**:943–945.
62. Miller, W.A., Waterhouse, P.M., and Gerlach, W.L. 1988. Sequence and genome organization of barley yellow dwarf virus genomic RNA. *Nucl. Acids Res.* **16**:6097–6111.
63. Modjahedi, N., Volovitch, M., Mazzolini, L., and Yot, P. 1985. Comparison of the predicted secondary structure of aphid transmission factor for transmissible and non-transmissible cauliflower mosaic virus. *FEBS Lett.* **181**:223–228.
64. Murant, A.F., and Roberts, I.M. 1977. Virus-like particles in phloem tissue of chervil (*Anthriscus cerefolium*) infected with anthriscus yellows virus. *Ann. Appl. Biol.* **85**:403–406.
65. Murant, A.F., Raccah, B., and Pirone, T.P. 1988. Transmission by vectors. In R.G. Milne, ed., *The Plant Viruses*. Vol. 4: *The Filamentous Plant Viruses*. Plenum Press, New York, pp. 237–273.
66. Murant, A.F., Roberts, I.M., and Elnagar, S. 1976. Association of virus-like particles with the foregut of the aphid, *Cavariella aegopodii*, transmitting the

semi-persistent viruses anthriscus yellows and parsnip yellow fleck. *J. Gen. Virol.* **31**:47–57.

67. Nault, L.R., and Ammar, E.D. 1989. Leafhopper and planthopper transmission of plant viruses. *Annu. Rev. Entomol.* **34**:503–529.
68. Nuss, D.L. 1984. Molecular biology of wound tumor virus. *Adv. Virus Res.* **29**:57–93.
69. Pirone, T.P. 1991. Viral genes and gene products that determine insect transmissibility. *Sem. Virol.* **2**:81–87.
70. Pirone, T.P., and Thornbury, D.W. 1988. Quantity of virus required for aphid transmission of a potyvirus. *Phytopathology* **78**:104–107.
71. Proesler, G. 1980. Piesmids. In K.F. Harris and K. Maramorosch, eds., *Vectors of Plant Pathogens*. Academic Press, New York, pp. 97–113.
72. Quemada, H., L'Hostis, B., Gonsalves, D., Reardon, I.M., Hinrickson, R., Hiebert, E., Siew, L.C., and Slightom, J.L. 1990. The nucleotide sequences of the 3-terminal regions of papaya ringspot virus strains W and P. *J. Gen. Virol.* **69**:1497–1508.
73. Rao, A.S., and Brakke, M.K. 1969. Relation of soil-borne wheat mosaic virus and its fungal vector, *Polymyxa graminis*. *Phytopathology* **59**:581–587.
74. Rizzo, T.M., and Gray, S.M. 1991. Localization of capsid protein domains which may regulate aphid transmission of barley yellow dwarf virus. *Phytopathology* **81**:1174.
75. Robaglia, C., Durand-Tardif, M., Tronchet, M., Boudazin, G., Astier-Manifacier, S., and Casse-Delbart, F. 1989. Nucleotide sequence of potato virus Y (N strain) genomic RNA. *J. Gen. Virol.* **70**:935–947.
76. Robertson, W.M., and Henry, C.E. 1986. An association of carbohydrates with particles of arabis mosaic virus retained within *Xiphenema diversicaudatum*. *Ann. Appl. Biol.* **109**:299–305.
77. Shen, P., Kaniewski, M., Smith, C., and Beachy, R.N. 1993. Nucleotide sequence and genomic organization of rice tungro spherical virus. *Virology* **193**:621–630.
78. Sylvester, E.S. 1956. Beet yellows virus transmission by the green peach aphid. *J. Econ. Entomol.* **49**:789–800.
79. Takahashi, M., Toriyama, S., Hamamatsu, C., and Ishihama, A. 1993. Nucleotide sequence and possible ambisense coding strategy of rice stripe virus RNA segment 2. *J. Gen. Virol.* **74**:769–773.
80. Tamada, T. 1975. Beet necrotic yellow vein virus. *CMI/AAB Descriptions of Plant Viruses*, No. 144.
81. Tamada, T., and Kusume, T. 1991. Evidence that the 75K readthrough protein of beet necrotic yellow vein virus RNA-2 is essential for transmission by the fungus *Polymyxa betae*. *J. Gen. Virol.* **72**:1497–1504.
82. Taylor, C.E., and Robertson, W.M. 1969. The location of raspberry ringspot and tomato blackring viruses in the nematode vector, *Longidorus elongatus* (de Man). *Ann. Appl. Biol.* **64**:233–237.
83. Taylor, C.E., and Robertson, W.M. 1970. Sites of virus retention in the alimentary tract of the nematode vectors, *Xiphinema diversicaudatum* (Micol.) and *X. index* (Thorne and Allen). *Ann. Appl. Biol.* **66**:375–380.
84. Taylor, C.E., Robertson, W.M., and Roca, F. 1976. Specific association of artichoke Italian latent virus with the odontostyle of its vector, *Longidorus attenuatus*. *Nematol. Medit.* **4**:23–30.

85. Temmink, J.H.L., Campbell, R.N., and Smith, P.R. 1970. Specificity and site of *in vitro* acquisition of tobacco necrosis virus by zoospores of *Olpidium brassicae*. *J. Gen. Virol.* **9**:201–213.
86. Thornbury, D.W., Hellmann, G.M., Rhoades, R.E., and Pirone, T.P. 1985. Purification and characterization of potyvirus helper component. *Virology* **144**:260–267.
87. Thornbury, D.W., Patterson, C.A., Dessens, J.T., and Pirone, T.P. 1990. Comparative sequences of the helper component (HC) region of potato virus Y and the HC-defective strain potato virus C. *Virology* **178**:573–578.
88. van der Wilk, F., Huisman, M.J., Cornelissen, B.J.C., Huttinga, H., and Goldbach, R. 1989. Nucleotide sequence and organization of potato leafroll genomic RNA. *FEBS Lett.* **245**:51–56.
89. Veidt, I., Lot, P., Leisner, M., Scheidecker, D., Guilley, H., Richards, K., and Jonard, G. 1988. Nucleotide sequence of beet western yellows virus RNA. *Nucl. Acids Res.* **16**:9917–9933.
90. Vincent, J.R., Lister, R.M., and Larkins, B.A. 1991. Nucleotide sequence analysis and genomic organization of the NY-RPV isolate of barley yellow dwarf virus. *J. Gen. Virol.* **72**:2347–2355.
91. Vincent, J.R., Ueng, P.P., Lister, R.M., and Larkins, B.A. 1990. Nucleotide sequences of coat protein genes for three isolates of barley yellow dwarf virus and their relationships to the luteovirus coat protein sequences. *J. Gen. Virol.* **71**:2791–2799.
92. Watson, M.A., and Roberts, F.M. 1939. A comparative study of the transmission of *Hyocyamus* virus 3, potato virus Y and cucumber mosaic virus by the vector *Myzus persicae* (Sulz), *M. circumflexus* (Buckton and *Macrosiphum gei* (Koch). *Proc. R. Soc. London Ser. B* **127**:543–576.
93. Wijkamp, I., van Lent, J., Kormelink, R., Goldbach, R., and Peters, D. 1993. Multiplication of tomato spotted wilt virus in its insect vector *Frankliniella occidentalis*. *J. Gen. Virol.* **74**:341–349.
94. Woolston, C.J., Covey, S.N., Penswick, J.R., and Davies, J.W. 1983. Aphid transmission and a polypeptide are specified by defined region of the cauliflower mosaic virus genome. *Gene* **23**:15–23.
95. Woolston, C.J., Czaplewski, L.G., Markham, P.G., Goad, A.S., Hull, R., and Davies, J.W. 1987. Location and sequence of a region of cauliflower mosaic virus gene II responsible for aphid transmissibility. *Virology* **160**:246–251.

Index

F

N

O

Q

R

SPRINGER NATURE

GPSR Compliance

The European Union's (EU) General Product Safety Regulation (GPSR) is a set of rules that requires consumer products to be safe and our obligations to ensure this.

If you have any concerns about our products, you can contact us on ProductSafety@springernature.com

In case Publisher is established outside the EU, the EU authorized representative is:

Springer Nature Customer Service Center GmbH
Europaplatz 3
69115 Heidelberg, Germany

Kolibri 360 GmbH
Ferdinand-Jühlke-Straße 7
99095 Erfurt, Deutschland
produktsicherheit@kolibri360.de